Sylvia S. Mader

with contributions by
Patrick L. Galliart
North Iowa Area Community College

fourth edition

UNDERSTANDING

Human
anatomy & physiology

Boston   Burr Ridge, IL   Dubuque, IA   Madison, WI   New York   San Francisco   St. Louis
Bangkok   Bogotá   Caracas   Lisbon   London   Madrid
Mexico City   Milan   New Delhi   Seoul   Singapore   Sydney   Taipei   Toronto

McGraw-Hill Higher Education

A Division of The McGraw-Hill Companies

UNDERSTANDING HUMAN ANATOMY AND PHYSIOLOGY, FOURTH EDITION

Published by McGraw-Hill, an imprint of The McGraw-Hill Companies, Inc., 1221 Avenue of the Americas, New York, NY 10020. Copyright © 2001, 1997 by The McGraw-Hill Companies, Inc. All rights reserved. No part of this publication may be reproduced or distributed in any form or by any means, or stored in a database or retrieval system, without the prior written consent of The McGraw-Hill Companies, Inc., including, but not limited to, in any network or other electronic storage or transmission, or broadcast for distance learning.

Some ancillaries, including electronic and print components, may not be available to customers outside the United States.

♲ This book is printed on recycled, acid-free paper containing 10% postconsumer waste.

234567890 QWKQWK 098765432

ISBN 0-07-290756-8
ISBN 0-07-118081-8 (ISE)

Vice president and editor-in-chief: *Kevin T. Kane*
Publisher: *Michael D. Lange*
Sponsoring editor: *Kristine Tibbetts*
Developmental editor: *Patricia Hesse*
Marketing managers: *Michelle Watnick/Heather K. Wagner*
Senior project manager: *Marilyn M. Sulzer*
Production supervisor: *Sandy Ludovissy*
Designer: *K. Wayne Harms*
Cover designer: *James A. O'Neal*
Cover photo: *Laurie Peak/Tony Stone Images*
Senior photo research coordinator: *Lori Hancock*
Photo research: *Connie Mueller*
Supplement coordinator: *Tammy Juran*
Compositor: *GTS Graphics, Inc.*
Typeface: *10/12 Giovanni Book*
Printer: *Quebecor Printing Book Group/Dubuque, IA*

The credits section for this book begins on page 429 and is considered an extension of the copyright page.

Library of Congress Cataloging-in-Publication Data

Mader, Sylvia S.
   Understanding human anatomy and physiology / Sylvia Mader, Patrick Galliart. — 4th ed.
      p.   cm.
   Includes bibliographical references and index.
   ISBN 0-07-290756-8
   1. Human physiology.   2. Human anatomy.   I. Galliart, Patrick.   II. Title.

QP34.5 .M353   2001
612—dc21   99-086647
                                             CIP

INTERNATIONAL EDITION ISBN 0-07-118081-8
Copyright © 2001. Exclusive rights by The McGraw-Hill Companies, Inc., for manufacture and export. This book cannot be re-exported from the country to which it is sold by McGraw-Hill. The International Edition is not available in North America.

www.mhhe.com

Brief Contents

Contents

Part V

Reproduction and Development 337

Boxed Readings

Preface

Understanding Human Anatomy and Physiology is written for students who are taking a one-semester course in anatomy and physiology. It covers all the basic information necessary for a general understanding of the structure and function of the human body.

The writing style and depth of presentation are appropriate for students who have little background in science and who are just beginning to pursue a career in an allied health field. Each chapter presents the topic clearly, simply, and distinctly so that students will feel capable of mastering the chapter learning objectives.

Understanding Human Anatomy and Physiology excels in pedagogical features which are described in the Guided Tour.

The Fourth Edition Changes

New and vibrant illustrations are a part of an art program that will motivate students because of its appeal. The Visual Focus illustrations and the Working Together illustrations have also been redone to increase their vitality. The illustrations in certain chapters, such as the cell chapter and the skeletal system chapter, have icons which will help students relate the part to the whole.

All chapters have been revised in this edition to update where necessary and to increase student learning of difficult concepts. The end of chapter and the end of text material have also been expanded. There are more objective questions in most chapters and 50 additional Further Readings have been added to Appendix D.

This edition uses three different icons at the end of figure legends, or the titles of sections, to alert students to technology resources. The icons are:

- 𝒳 Indicates The Dynamic Human: the 3D Visual Guide to Anatomy and Physiology, a CD-ROM.
- ▣ Indicates the WCB Life Science Animations, a videotape series that covers key physiological processes.
- ▤ Indicates animations in the Essential Study Partner CD-ROM, an interactive student study tool.

The Correlation section of the Preface, p. xxii, correlates these resources to the text.

The New Technology

Many technology aids are available for use with *Understanding Human Anatomy and Physiology*

New to this edition, the Mader home page contains an Online Learning Center with instructor and student resources such as Web links, Quizzes, Study Guide, Case Studies, Clinical Applications and Matching Activities provides additional resources student will enjoy and appreciate. Each chapter in the text ends with Website Link, a section that gives the Mader home page internet address.

The Essential Study Partner, free with each text, is an interactive student study tool which contains more than 100 animations and more than 800 learning activities. This powerful CD-ROM contains a text guide correlated to the material presented in *Understanding Human Anatomy and Physiology*. A film icon placed in the text beside topics and concepts that are animated on this resource.

The Organization of the Text

This edition of *Understanding Human Anatomy and Physiology*. has a renewed emphasis on homeostasis. A significant portion of chapter 1 is devoted to explaining the concept of homeostasis and outlining the role the systems of the body play in to maintaining homeostasis. The Working Together illustrations that appear throughout the text describe how each organ system works with other systems to achieve homeostasis.

Part I: Human Organization

Chapter 1 explains the organization of the human body and the terms used to describe the location of body parts. It introduces the various organ systems and the concept of homeostasis, an equilibrium that is maintained by these systems.

Chapters 2 through 4 describe the chemistry of cell, cell structure and function, body tissues and membranes. Chapter 5 reviews the structure, functions, and disorders of the skin. This chapter has a Working Together illustration.

Part II: Support and Movement

The two chapters in this section concern the skeletal system and the muscular system, which support and protect the body and allow its parts to move. Both chapters have a Working Together illustration.

Chapter 6 considers the functions of the skeletal system before taking up the axial skeleton, the appendicular skeleton and the joints. Lists, tables, and oversize illustrations facilitate student learning. Chapter 7 considers the functions of the muscular system and the contraction of muscle fibers before reviewing the skeletal muscles of the body. The sliding filament theory is explained in an easy-to-understand manner.

Part III: Integration and Coordination

Separate chapters are devoted to the nervous system, the senses, and the endocrine system. The nervous and endocrine systems are vitally important to the coordination of body systems and therefore to homeostasis.

The first part of chapter 8 describes the structure and function of a neuron, a description of the central nervous system precedes that of the peripheral nervous system. In this chapter, illustrations coordinate closely with the discussion of brain structure and function. Chapter 9 is divided into general receptors (skin, visceral and proprioceptors); chemoreceptors (taste and smell); photoreceptors (those of the eye); mechanoreceptors (hearing and balance). The explanations of how we taste, smell, see and hear in this chapter are well presented. Chapter 10 considers the cellular mechanism of hormonal action before taking up the endocrine glands in turn. A table of the principal endocrine gland and their hormones is central to this chapter. Human hormonal disorders, such as diabetes mellitus, are emphasized.

Part IV: Maintenance of the Body

In this part chapter 11 reviews the composition of blood and functions of blood before taking up blood groups and typing. Chapter 12 first considers the anatomy of the heart before the vascular system and disorders of the circulatory system. Chapter 13 includes a description of the lymphatic system as well as a modern discussion of the defense mechanisms. In chapter 14, a description of the anatomy of the respiratory system precedes mechanisms of breathing and gas exchange. Chapter 15 describes the organs of the digestive system, mechanical and chemical digestion, and nutrition. Chapter 16 reviews the organs of the urinary system before explaining urine formation and the regulatory functions of the kidneys. Working Together illustrations appear before each chapter summary except for chapter 11. The functions of blood are included in the Working Together illustration for the circulatory system.

Part V: Reproduction and Development

This part includes chapters on the reproductive system, human development and genetics.

In chapter 17, the male reproductive system is discussed before the female reproductive system. There is also a discussion of birth control measure and infertility. This chapter has a Working Together illustration. Chapter 18 begins with a description of fertilization, the extraembryonic membranes and the functions of the placenta before the events of development and birth are outlined. Chapter 19 gives a simplified view of the human inheritance and biochemical genetics. It also includes a look at biotechnology, a technique that is now utilized to produce medications and carry out gene therapy.

About the Author

Dr. Sylvia Mader has successfully helped students learn the structure and function of the human body for more than 20 years. A brilliant and prolific writer, Dr. Mader was a respected and well-loved instructor before she began her writing career. Her descriptive writing style, carefully constructed pedagogy, and emphasis on concepts as well as terminology provides students with a firm grasp of anatomy and physiology. In her 20 year career with McGraw-Hill, she has written an impressive collection of textbooks including Human Biology, Sixth Edition, Inquiry into Life, Ninth Edition, and Biology, Seventh Edition, in addition to this text. Throughout the years, her goal remains the same—"to give students what they need to best understand and learn the basics."

GUIDED TOUR

Before you begin your study of anatomy and physiology, spend a little time looking over the next few pages. They provide a quick guide to the learning tools found throughout the text that have been designed to enhance your understanding of human anatomy and physiology.

CHAPTER OUTLINE

Each chapter begins with a chapter outline that lists the learning objectives appropriate to each major section of the chapter. Page referencing of the major sections makes it easier for students to coordinate the learning objectives with the text material.

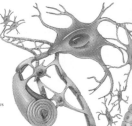

chapter **8**

The Nervous System

■ **Chapter Outline and Learning Objectives**

After you have studied this chapter, you should be able to:

Nervous System (p. 136)
■ Describe the three functions of the nervous system.
■ Describe the structure and function of the three types of neurons and four types of neuroglial cells.
■ Explain how a nerve impulse is conducted along a nerve and across a synapse.
■ Describe the structure of a nerve and the differences between the three different types of nerves.
■ Describe the structure of a reflex arc and the function of a reflex.

Central Nervous System (p. 144)
■ Describe the major parts of the brain and the lobes of the cerebral cortex. State functions for each structure.
■ Describe in detail the structure of the spinal cord, and state its functions.
   ...yers of meninges, and state
   ...ospinal fluid is formed and

...vous System (p. 152)
...rs of cranial nerves, and give a
...e and function of spinal nerves.
...the autonomic nervous system.
...the sympathetic and
...ssions in four ways, and give examples
...ects on specific organs.

...g (p. 158)
...siological changes occur in the nervous

Working Together (p. 158)
■ The nervous system works with other systems of t...
maintain homeostasis.

Visual Focus
Synapse Structure and Function (p. 141)

Medical Focus
The Left and Right Brain (p. 146)
EEG (p. 149)
Spinal Cord Injuries (p. 150)

MedAlert
Alzheimer Disease (p. 160)

Cells found in the central nervous system (brain a...
cord).

Part

ICONS

Following various figures icons direct you to media that further explain the concept.

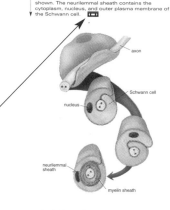

figure 8.3 Myelin and neurilemmal sheaths. The myelin sheath forms when Schwann cells wrap themselves around a nerve fiber in the manner shown. The neurilemmal sheath contains the cytoplasm, nucleus, and outer plasma membrane of the Schwann cell.

axon

Schwann cell

nucleus

neurilemmal sheath

myelin sheath

The dendrites and axons of neurons are sometimes called processes or neuron **fibers.** Outside the CNS, long fibers are covered by a white **myelin sheath** formed by **Schwann cells,** which wrap themselves tightly around these fibers. The portions of the Schwann cells that contain cytoplasm and nuclei form a **neurilemmal sheath** (fig. 8.3). The neurilemmal sheath promotes regeneration of a nerve fiber if it is injured. Gaps between Schwann cells form the nodes of Ranvier and are important in nerve cell conduction (see fig. 8.2).

Multiple sclerosis (MS) is a disease of the myelin sheath. Lesions develop that soon become hardened scleroses, or scars, that interfere with normal conduction of nerve impulses. The effects are widespread because of the number of fibers covered by a myelin sheath. The disease is chronic and tends to worsen with time.

figure 8.4 Neuroglial cells in the central nervous system.

neuron

capillary

astrocyte

oligodendroglial cell

axon

microglial cell

ependymal cell

Neuroglial Cells

Schwann cells are a type of neuroglial cell found outside the CNS. Microglial cells, astrocytes, and oligodendroglial cells and ependymal cells are among the neuroglia found inside the CNS (fig. 8.4). *Microglial cells* are small, phagocytic cells that migrate to the site of an injury, where they engulf microbes and clean away debris. *Astrocytes* have cytoplasmic processes that provide structural support by forming bridges between neurons and capillaries. They may have a nutritive function involving the transport of molecules between capillaries and neurons. *Oligodendroglial cells* produce the myelin sheath that envelops neurons in the CNS. *Ependymal cells* line cavities in the CNS and assist in the production and circulation of the cerebrospinal fluid that fills the cavities of the CNS.

Nervous tissue contains neurons and neuroglial cells. Each type of neuron has three parts (dendrite[s], cell body, and axon) but is specific as to function. Neuroglial cells support, protect, and nourish the neurons.

INTERNAL SUMMARY STATEMENTS

Summary statements are strategically placed throughout the chapter to immediately reinforce the concepts just discussed. These internal summary statements will aid your retention of the chapter's central concepts.

BOLDFACED TERMS

Basic Key Terms and Clinical Key Terms appear in boldface print as they are introduced in the text. Phonetic pronunciations follow the more challenging boldfaced terms. The terms are immediately defined in context. Key terms are listed with their pronunciations and page referenced at the end of the chapter. All boldfaced terms are defined in the glossary at the end of the book.

VISUAL FOCUS ILLUSTRATIONS

Visual Focus illustrations describe complex processes and use boxed statements to help you follow the events being depicted.

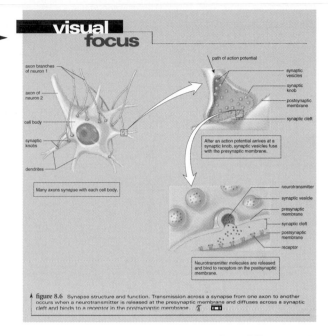

visual focus

axon branches of neuron 1

axon of neuron 2

cell body

synaptic knobs

dendrites

Many axons synapse with each cell body.

path of action potential

synaptic vesicles

synaptic knob

postsynaptic membrane

synaptic cleft

After an action potential arrives at a synaptic knob, synaptic vesicles fuse with the presynaptic membrane.

neurotransmitter

synaptic vesicle

presynaptic membrane

synaptic cleft

postsynaptic membrane

receptor

Neurotransmitter molecules are released and bind to receptors on the postsynaptic membrane.

figure 8.6 Synapse structure and function. Transmission across a synapse from one axon to another occurs when a neurotransmitter is released at the presynaptic membrane and diffuses across a synaptic cleft and binds to a receptor in the postsynaptic membrane.

Neurotransmitters and Neurological Disorders

Several neurological illnesses, such as **Parkinson disease** and **Huntington disease,** are due to an imbalance in neurotransmitters within the brain. Parkinson disease is characterized by a wide-eyed, unblinking expression, an involuntary tremor of the fingers and thumbs, muscular rigidity, and a shuffling gait. All of these symptoms are due to a deficiency of dopamine. Huntington disease is characterized by a progressive deterioration of the individual's nervous system, which eventually leads to constant thrashing and writing movements, and finally, to insanity and death. The problem is believed to be due to a malfunction of GABA, another neurotransmitter of the brain. As

working together   Nervous System

Integumentary System
Brain controls nerves that regulate size of cutaneous blood vessels, activate sweat glands and arrector pili muscles.

Skin protects nerves, helps regulate body temperature; skin receptors send sensory input to brain.

Skeletal System
Receptors send sensory input from bones and joints to brain.

Bones protect sense organs, brain, and spinal cord; store Ca^{2+} for nerve function.

Muscular System
Brain controls nerves that innervate muscles; receptors send sensory input from muscles to brain.

Muscle contraction moves eyes, permits speech, and creates facial expressions.

Endocrine System
Hypothalamus is part of endocrine system; nerves innervate certain glands of secretion.

Sex hormones affect development of brain.

Circulatory System
Brain controls nerves that regulate the heart and dilation of blood vessels.

Blood vessels deliver nutrients and oxygen to neurons, carry away wastes.

How the Nervous System works with other body systems

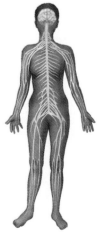

Lymphatic System/Immunity
Microglial cells engulf and destroy pathogens.

Lymphatic vessels pick up excess tissue fluid; immune system protects against infections of nerves.

Respiratory System
Respiratory centers in brain regulate breathing rate.

Lungs provide oxygen for neurons and rid the body of carbon dioxide produced by neurons.

Digestive System
Brain controls nerves that innervate smooth muscle and permit digestive tract movements.

Digestive tract provides nutrients for growth, maintenance, and repair of neurons and neuroglial cells.

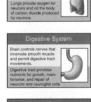

Urinary System
Brain controls nerves that innervate muscles that permit urination.

Kidneys maintain blood levels of Na^+, K^+, and Ca^{2+}, which are needed for nerve conduction.

Reproductive System
Brain controls onset of puberty; nerves are involved in erection of penis and clitoris, contraction of ducts that carry gametes, and contraction of uterus.

Sex hormones masculinize or feminize the brain, exert feedback control over the hypothalamus, and influence sexual behavior.

WORKING TOGETHER ILLUSTRATIONS

Working Together illustrations describe how each organ system works with the other systems to achieve homeostasis. These illustrations help you see the body as a working whole.

Sample page 148

The **pons** is a "bridge;" it contains bundles of axons traveling between the cerebellum and the rest of the CNS. In addition, the pons functions with the medulla to regulate breathing rate and has reflex centers concerned with head movements in response to visual and auditory stimuli.

The **midbrain** encloses the cerebral aqueduct. Aside from acting as a relay station for tracts passing between the cerebrum and the spinal cord or cerebellum, the midbrain has reflex centers for visual, auditory, and tactile responses.

Diencephalon

The hypothalamus and thalamus are in a portion of the brain known as the **diencephalon** (di"en-sef'ah-lon), where the **third ventricle** is located. The **hypothalamus** (hi"po-thal'ah-mus), which forms the floor of the third ventricle, maintains homeostasis, or the constancy of the internal environment, and contains centers for regulating hunger, sleep, thirst, body temperature, water balance, and blood pressure. The hypothalamus controls the pituitary gland and thereby serves as a link between the nervous and endocrine systems.

The **thalamus**, in the lateral walls of the third ventricle, is the last portion of the brain for sensory input before the cerebrum. It serves as a central relay station for sensory impulses traveling upward from other parts of the body and brain to the cerebrum. It receives all sensory impulses (except those associated with the sense of smell) and channels them to appropriate regions of the cortex for interpretation.

The brain stem and the diencephalon contain reflex centers that are involved in controlling internal organs, and tracts that channel impulses into the brain or between brain regions. The hypothalamus maintains homeostasis and controls the secretions of the pituitary gland.

Cerebellum

The **cerebellum** (ser"ĕ-bel'um), which lies below the posterior portion of the cerebrum, is separated from the brain stem by the fourth ventricle. The cerebellum has two parts called hemispheres that are joined by a constricted median portion. The surface of the cerebellum is gray matter, and the interior is largely white matter. The cerebellum functions in muscle coordination, integrating impulses received from higher centers to ensure that all of the skeletal muscles work together to produce smooth and graceful motion. The cerebellum is also responsible for maintaining normal muscle tone and transmitting impulses that maintain posture. It receives information about body position from the inner ear and then sends impulses to the muscles, whose contraction maintains or restores balance.

The cerebellum controls balance and complex muscular movements.

Cerebrum

The **cerebrum** (ser'ĕ-brum) is the largest and most superior part of the brain. It is the only area of the brain responsible for consciousness. The cerebrum is divided into halves known as the right and left **cerebral hemispheres**; each hemisphere contains a lateral ventricle.

The outer layer of the cerebrum, called the *cortex*, is gray and contains cell bodies and short fibers. The cortex has convolutions known as **gyri**, which are separated by shallow grooves called *sulci* and deep grooves called *fissures*. The cerebrum is almost divided by a deep, longitudinal fissure. At the base of this fissure lies the **corpus callosum**, a bridge of myelinated fibers that joins the two hemispheres. Left-brain and right-brain abilities are examined in the Medical Focus reading on this page.

Each cerebral hemisphere has four lobes: **frontal, parietal** (pah-ri'ĕ-tal), **temporal**, and **occipital** (ok-sip'ĭ-tal) (fig. 8.11), which are named for the bones that cover them. Each lobe has particular functions (table 8.3).

Medical Focus

The Left and Right Brain

Some years ago, Roger W. Sperry and Michael Gazzaniga severed the corpus callosum in some of their patients who suffered from epilepsy. The corpus callosum connects the left and right sides of the brain. From these procedures and further experimentation, they learned that the left brain and right brain have different abilities.

Sperry and Gazzaniga found that the left brain contains centers for speech and is responsible for language ability. Therefore, patients could report what was seen by the left half of each eye but not what was seen by the right half of each eye. In contrast, patients were unable to report verbally on left-hand activities because the left hand is controlled by the right half of the brain.

Sperry and Gazzaniga also determined that the right brain is far superior to the left brain in dealing with spatial relationships. For example, the left hand, not the right hand, is better able to recognize and remember objects by their shape. In addition to spatial relationships, the right brain also appears to be involved with musical and artistic activities and the expression of emotions.

148   Part III Integration and Coordination

THE *ESSENTIAL STUDY PARTNER* CD-ROM

The filmstrip icon next to various topics encourage you to use the interactive CD-ROM that accompanies your textbook. *The Essential Study Partner* is packed with hundreds of animations and learning activities. The animations will bring to life those concepts that are difficult to envision. The quizzing will help you grasp difficult topics.

BOXED READINGS

Most chapters have a MedAlert reading, which examines a particular medical condition in some detail. These end with critical thinking questions for you to answer. (Answers to these questions are included in Appendix C.)

Also, Medical Focus readings appear in each chapter. The clinical terms used in the boxed readings appear as Clinical Key Terms at the end of the chapter and are defined in the glossary at the end of the book.

Sample page 160

Alzheimer Disease

Alzheimer disease (AD) is a disorder characterized by a gradual loss of reason that begins with memory lapses and ends with an inability to perform any type of daily activity. Personality changes signal the onset of AD. A normal 50- to 60-year-old adult might forget the name of a friend not seen for years. People with AD, however, forget the name of a neighbor who visits daily. With time, they have trouble traveling and cannot perform simple errands. People afflicted with AD become confused and tend to repeat the same question. Signs of mental disturbances eventually appear, and patients gradually become bedridden and die of a complication, such as pneumonia.

A normal neuron (nerve cell), and a neuron damaged by Alzheimer disease (AD), are shown in figure 8A. The AD neuron has two abnormalities not seen in the normal neuron: (1) Bundles of fibrous protein, called neurofibrillary tangles, surround the nucleus in the cell, and (2) protein-rich accumulations, called amyloid plaques, envelop the axon branches. These abnormal neurons are especially seen in the portions of the brain that are involved in reason and memory (frontal lobe and limbic system). To see the abnormal brain neurons, brain tissue must be examined microscopically after the patient dies.

A chemical test can be used to check brain tissue for the presence of a protein called Alzheimer disease associated protein (ADAP), which is believed to be the protein contained in the neurofibrillary tangles. If ADAP is proven to be the protein involved in AD, individuals could be tested for this protein by obtaining a spinal tap of cerebrospinal fluid.

Over a life span of 100 years, the likelihood of developing AD is 16% for people with no family history of AD, and 24% for those having first-degree relatives with AD. This difference in susceptibility suggests that AD might have a genetic basis. Researchers have discovered that in some families whose members have a 50% chance of AD, a genetic defect exists on chromosome 21. This is of extreme interest because Down syndrome (p. 381) results from the inheritance of three copies of chromosome 21, and people with Down syndrome tend to develop AD. Further, the genetic defect affects the normal production of amyloid precursor protein (APP), which may be the cause of the amyloid plaques.

Acetylcholine is a chemical that stimulates neurons to carry nerve impulses, and it appears that this chemical may be in short supply in the brains of patients with AD. Drugs that enhance acetylcholine production are currently being tested in AD patients. Experimental drugs that prevent neuron degeneration are also being tested. For example, it is possible that nerve growth factor, a substance that is made by the body and that promotes the growth of neurons, will one day be available to AD patients.

Questions

1. Why are drugs that enhance acetylcholine production being tested in AD patients?
2. What evidence suggests that AD might have a genetic basis?
3. How does the AD neuron differ from a normal neuron?

160   Part III Integration and Coordination

NEW TERMS LIST

The Selected New Terms list at the end of each chapter is divided into two parts. The first part lists many of the Basic Key Terms that appear in boldface in the chapter, and the second part lists Clinical Key Terms that appear in boldface in the chapter. All terms are defined in the glossary at the end of the book.

F. Nerves. A nerve contains bundles of long fibers covered by fibrous, connective tissue layers. In the CNS, bundles of long fibers are found in tracts. White matter is composed of myelinated fibers, and gray matter is composed of cell bodies and unmyelinated fibers.

G. Reflexes and the reflex arc. Reflexes (automatic reactions to internal and external stimuli) depend on the reflex arc. Some reflexes are important for avoiding injury, and others are necessary for normal physiological functions.

II. Central Nervous System
A. Ventricles of the brain. The brain has four ventricles. The lateral ventricles are found in the left and right cerebral hemispheres. The third ventricle is found in the diencephalon. The fourth ventricle is found in the brain stem.
B. Brain stem. The brain stem contains the medulla oblongata, pons, and midbrain. The medulla oblongata contains vital centers for regulating heartbeat, breathing, and blood pressure. The pons assists the medulla oblongata in regulating the breathing rate. The midbrain contains tracts that conduct impulses to and from the higher parts of the brain.

C. Diencephalon. The hypothalamus helps control the functioning of most internal organs and controls the secretions of the pituitary gland. The thalamus receives sensory impulses from all parts of the body and channels them to the cerebrum.
D. Cerebellum. The cerebellum controls balance and complex muscular movements.
E. Cerebrum. Consciousness is under the control of the cerebrum, the most highly developed portion of the brain. It is responsible for higher mental processes, including the interpretation of sensory input and the initiation of voluntary muscular movements.
F. Limbic system. The limbic system includes portions of the cerebrum, the thalamus, and the hypothalamus. It is involved in learning and memory and in causing the emotions that guide behavior.
G. Spinal cord. The spinal cord is located in the vertebral column in cross sections composed of white matter and gray matter. White matter contains bundles of nerve fibers, called tracts, that conduct nerve impulses to and from the higher centers of the brain. Gray matter is mainly made up of short fibers and cell bodies. The spinal

cord is a center for reflex action and allows communication between the brain and the peripheral nerves leaving the spinal cord.
H. Meninges and cerebrospinal fluid. The CNS is protected by the meninges and the cerebrospinal fluid.

III. Peripheral Nervous System
A. Somatic nervous system. Cranial nerves take impulses to and/or from the brain. Spinal nerves take impulses to and from the spinal cord.
B. Autonomic nervous system. The ANS controls the functioning of internal organs without need of conscious control.
1. The divisions of the autonomic nervous system: (1) function automatically and usually subconsciously in an involuntary manner, (2) innervate all internal organs, and (3) utilize two motor neurons and one ganglion for each impulse.
2. The sympathetic division brings about the responses associated with the "fight-or-flight" response.
3. The parasympathetic division brings about the responses associated with normally restful activities.

Study Questions

1. What are the two main divisions of the nervous system? How are these divisions subdivided? (pp. 136–37)
2. What are the types of neurons and neuroglial cells? How are they similar, and how are they different? (pp. 137–39)
3. What does the term *nerve impulse* mean, and how is a nerve impulse brought about? (p. 140)
4. What is a neurotransmitter? Where is it stored, and how does it function? How is it destroyed? Name several well-known neurotransmitters. (pp. 140–41)
5. Describe the structure of a nerve, and state the location of nerves and tracts. (p. 142)

6. What is the path of a spinal reflex that involves three neurons? What is the function of reflexes? (pp. 142–43)
7. Where are the ventricles of the brain located? (p. 144)
8. Name the various parts of the brain, state where the parts are located, and give their functions. (pp. 144–48)
9. What does it mean to say that the cerebral cortex can be mapped? Discuss this in relation to the primary motor areas and the primary sensory areas. (pp. 146–49)
10. Describe the anatomy of the spinal cord. What are the functions of the gray and white matter in the spinal cord? (p. 150)

11. What are the three different meninges, and what is their function? (pp. 150–51)
12. What is cerebrospinal fluid? Where is it made, and how does it circulate? (p. 151)
13. What are the different cranial nerves, and what is the function of each? (pp. 152–53)
14. What are the structure and function of the spinal nerves? (p. 152)
15. What is the autonomic nervous system, and what are its two major divisions? Describe several similarities and differences between these divisions. (pp. 155–58)

Chapter 8 The Nervous System 163

Selected New Terms

Basic Key Terms

acetylcholine (ACh) (as"ĕ-til-ko'lēn), p. 140
acetylcholinesterase (AChE) (as"ĕ-til-ko'lin-es'ter-ās), p. 140
arachnoid membrane (ah-rak'noid mem'brān), p. 150
autonomic nervous system (aw"to-nom'ik ner'vus sis'tem), p. 152
axon (ak'son), p. 137
cell body (sel bod'e), p. 137
central nervous system (sen'tral ner'vus sis'tem), p. 136
cerebellum (ser"ĕ-bel'um), p. 146
cerebral hemisphere (ser'ĕ-bral hem'ĭ-sfēr), p. 146
cerebrospinal fluid (ser"e-bro-spi'nal floo'id), p. 151
cerebrum (ser'ĕ-brum), p. 146
cranial nerve (kra'ne-al nerv), p. 152
dendrite (den'drīt), p. 137
diencephalon (di"en-sef'ah-lon), p. 146
dorsal-root ganglion (dor'sal root gang'gle-on), p. 142
dura mater (du'rah ma'ter), p. 150
hypothalamus (hi"po-thal'ah-mus), p. 146
limbic system (lim'bik sis'tem), p. 149
medulla oblongata (mĕ-dul'ah ob"long-ga'tah), p. 144
meninges (mĕ-nin'jēz), p. 150
midbrain (mid'brān), p. 146
neurilemmal sheath (nu"rĭ-lem'al shēth), p. 139
neuron (nu'ron), p. 137
neurotransmitter (nu"ro-trans'mit-er), p. 140
parasympathetic division (par"ah-sim"pah-thet'ik dĭ-vizh'un), p. 155
peripheral nervous system (pĕ-rif'er-al ner'vus sis'tem), p. 136

pia mater (pi'ah ma'ter), p. 150
pons (ponz), p. 146
reflex (re'fleks), p. 142
Schwann cell (schwon sel), p. 139
sensory receptor (ri-sep'ter), p. 142
somatic nervous system (so-mat'ik ner'vus sis'tem), p. 152
spinal nerve (spi'nal nerv), p. 152
sympathetic division (sim"pah-thet'ik dĭ-vizh'un), p. 155
synapse (sin'aps), p. 140
ventricle (ven'trĭ-k'l), p. 144

Clinical Key Terms

Alzheimer disease (altz'hi-mer dĭ-zēz'), pp. 142, 160
ankle-jerk reflex (an'kl-jerk re'fleks), p. 143
cerebral palsy (ser'ĕ-bral pal'ze), p. 148
dermatome (der'mah-tōm), p. 152
electroencephalogram (e-lek"tro-in-sef'lah-gram), p. 149
epidural hematoma (ep"ĭ-du'ral he"mah-to'mah), p. 150
Huntington disease (hun'ting-tun dĭ-zēz'), p. 141
hydrocephalus (hi"dro-sĕ'fah-lus), p. 151
knee-jerk reflex (ne-jerk re'fleks), p. 143
multiple sclerosis (mul'tĭ-pul skler-o'sis), p. 139
paraplegia (par-ah-ple'je-ah), p. 150
Parkinson disease (par'kin-sun dĭ-zēz'), p. 141
quadriplegia (kwah-drah-ple'je-ah), p. 150
stroke (strōk), p. 149
subdural hematoma (sub"du'ral he"mah-to'mah), p. 150

Summary

I. Nervous System
A. Divisions of the nervous system. The nervous system is divided into the central nervous system (brain and spinal cord) and the peripheral nervous system (somatic and autonomic nervous systems). The CNS lies in the midline of the body, and the PNS is located peripherally to the CNS.

B. Functions of the nervous system. The nervous system permits sensory input, performs integration, and stimulates motor output.
C. Cells of nervous tissue. Nervous tissue contains neurons and neuroglial cells. Each type of neuron has three parts (dendrites, cell body, and axon) but is specific as to function. Neuroglial cells

support, protect, and nourish the neurons.
D. Nerve impulses. All neurons transmit the same type of nerve impulse: a change in polarity that flows along the membrane of a nerve fiber.
E. Synapse. Transmission of a nerve impulse across a synapse is dependent on the release of a neurotransmitter into a synaptic cleft.

162 Part III Integration and Coordination

CHAPTER QUESTIONS

Two types of questions—Study Questions and Objective Questions—appear at the close of each chapter. Answering the Study Questions results in a sequential review of chapter material. The Objective Questions allow you to quiz yourself with fill-in-the-blank and matching questions. The Objective Questions are answered in the *Instructor's Manual.*

CHAPTER SUMMARY

A summary at the end of each chapter offers a concise review of chapter material. You may read the Summary before beginning the chapter to preview the topics of importance, and you may also use it to refresh their memory after you have a firm grasp of the chapter's concepts.

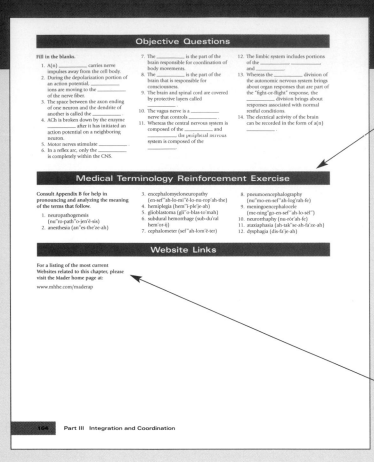

MEDICAL TERMINOLOGY REINFORCEMENT EXERCISE

An understanding of medical terminology is critical to students of anatomy and physiology. A medical terminology reinforcement exercise at the end of each chapter reinforces the principles covered in Appendix B, "Understanding Medical Terminology." Answers to these exercises are included in the *Instructor's Manual*.

WEBSITE LINK

The web address located at the end of the chapter is a reminder to you that additional study questions and links to anatomy- and physiology-related topics appear on the Mader home page.

APPENDICES

Appendix A: Reference Figures: The Human Organism.
The reference figures show the major organs of the human torso. The first plate illustrates the anterior surface and reveals the superficial muscles on one side. Each subsequent plate exposes deeper organs, including those of the thoracic, abdominal, and pelvic cavities. As you read the systems chapters of the text, you can refer to these plates to help visualize the locations of various organs.

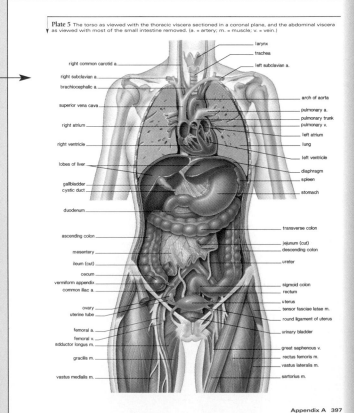

Understanding Medical Terminology

Learning Objectives

Upon completion of this section, you should be able to:

1. Discuss the importance of medical terminology and how it can be incorporated into the study of the human body.
2. Differentiate between a prefix, suffix, root word, and compound word.
3. Link word parts to form medical terms.
4. Differentiate between singular and plural endings of medical terms.
5. Practice pronunciation of medical words.
6. Dissect (cut apart) compound medical words into parts to analyze the meaning.
7. Recognize the more commonly used prefixes, suffixes, and root words used in medical terminology.

Introduction to Medical Terminology

As students of medical science, we are the inheritors of a vast fortune of knowledge. This fortune, amassed by giants of eighteenth- and nineteenth-century scholarship, was nurtured largely in the atmospheres of universities in which Latin and Greek were the languages of lecture and writing. Scientists then strove to define a universal language in which to communicate their findings. Latin and Greek, studied throughout Europe, became the languages of choice for scholars whose native tongue was English, German, French, Spanish, and so on, because they all read Latin and Greek. So, many seminal works in medicine were first penned in Latin, and their vocabularies remain to this day.

Anatomy and physiology were born in the eighteenth century in the midst of a glut of quacks, frauds, charlatans, myths, and superstitions. Honest scholars sought proofs to banish practices that should have been questioned by reason and proved wrong by experience. These scholars were among the first to connect disease with the failure of function or structure of body tissue; thus, the race to name and define all anatomical structures began.

Problems arose, inevitably, with the discovery of heretofore unknown tissue. Names were virtually created from parts or existing words by combining parts until they approximated an acceptable description. Medical terminology is simply a catalog of parts that allows us to take apart and reassemble the special language of medicine. The study of medical terminology is easier than it first seems.

Medical words have three basic parts: prefix, root word, and suffix. A prefix comes before a root word and alters the meaning. For example, the prefix *hyper-* means over or above. Hyper/kinetic means overactive, hyper/esthesia is overly sensitive, hyper/tension is high blood pressure, and hyper/trophy is overdevelopment.

A suffix is attached to the end of a root word and changes the meaning of the word. For example, the suffix *-itis* means inflammation. Inflammation can occur at almost any part of the body, so *-itis* can be added to root words to make hundreds of words. Dermat/itis is inflammation of the skin, rhin/itis is inflammation of the nose, gastr/itis is inflammation of the stomach, and so on.

A root word is the main part of the word. Once the root word is known for each part of the anatomy, the prefixes and suffixes can be used to analyze or build many medical words. The root word for heart is *cardi*. A few terms in which *cardi* appears are: cardi/algia means pain in the heart, cardio/omegaly means enlarged heart, brady/cardia means slow heart, and peri/cardio/centesis means puncture to aspirate fluid from around the heart.

Many medical words have, in addition to a prefix and/or a suffix, more than one word part. These are called compound words and can be analyzed by breaking them into parts. For example, hysterosalpingo-oophorectomy is made up of three root words and a suffix. *Hyster* is the root word for uterus, *salping* is the root word for tube, *oophor* is the root word for ovary, and *-ectomy* is the suffix for cut out. Now we know that hysterosalpingo-oophorectomy means the surgical excision of the uterus, tube, and ovary.

To facilitate pronunciation, word parts need to be linked together. The linkage for word parts is *o* and may be referred to as a combining form. For example, linking the root *cardi* with the suffix *-pathy* would produce a word that would be difficult to pronounce; therefore, an *o* is used to link the root word with the suffix. The complete word is written cardiopathy and pronounced kar″dēop'ah-the, and the combining form is cardi/o.

When a word is only a root or ends with a root, the word ending depends on whether the word is a noun or an

400

Appendix B: Understanding Medical Terminology
This appendix gives an overview of the basics of medical terminology and introduces you to the correct pronunciation of medical terms.

Appendix C: Answers to MedAlert Questions
This appendix gives answers to the questions that appear at the end of each MedAlert reading.

Answers to MedAlert Questions

Chapter 3

Cell Structure and Function

1. The longer we live, the more time there is to acquire "promoters" of cancer.
2. Smokers take carcinogens into the respiratory tract.
3. Certain foods are known to inhibit cancer, while others are known to promote cancer.

Chapter 6

The Skeletal System

1. It takes time for a joint to be "overworked."
2. Artificial hips do not have the flexibility of natural ones.

Chapter 7

The Muscular System

1. Exercise promotes regular bowel movements because it encourages movement of intestinal contents.
2. Exercise requires energy; therefore, it uses up body fat.
3. It would improve longevity because the heart does not work as hard.

Chapter 8

The Nervous System

1. Neurons that release this neurotransmitter are ʰ ˡᵉ ⁿ ̣.
2. AD runs in families.
3. The AD neuron has neurofibrillary tangles and amyloid plaques.

Chapter 9

The Senses

1. and 2. If the macula lutea degenerates, a person cannot see detail or color.
3. Vision in dim light is dependent on the rods, which are found outside the macula.

Chapter 12

The Circulatory System

1. Such a diet reduces blood cholesterol levels. High blood cholesterol levels cause plaque, particularly in coronary arteries.
2. During a myocardial infarction, a thromboembolism, or clot, lodges in a coronary artery that has already been narrowed by plaque. The portion of the heart deprived of blood dies, and surrounding tissue may be damaged.

Chapter 13

The Lymphatic System and Immunity

1. An HIV blood test is used to detect the presence of antibodies in the blood that are directed against HIV. A positive HIV test indicates prior exposure to the virus. A T4 cell count examines the number of T4 cells in the blood. AIDS is characterized by a T4 cell count below 200 per cubic millimeter.
2. Individuals with AIDS have a drastically weakened immune system and are unable to fend off infections that are normally nonfatal.

408   Appendix C

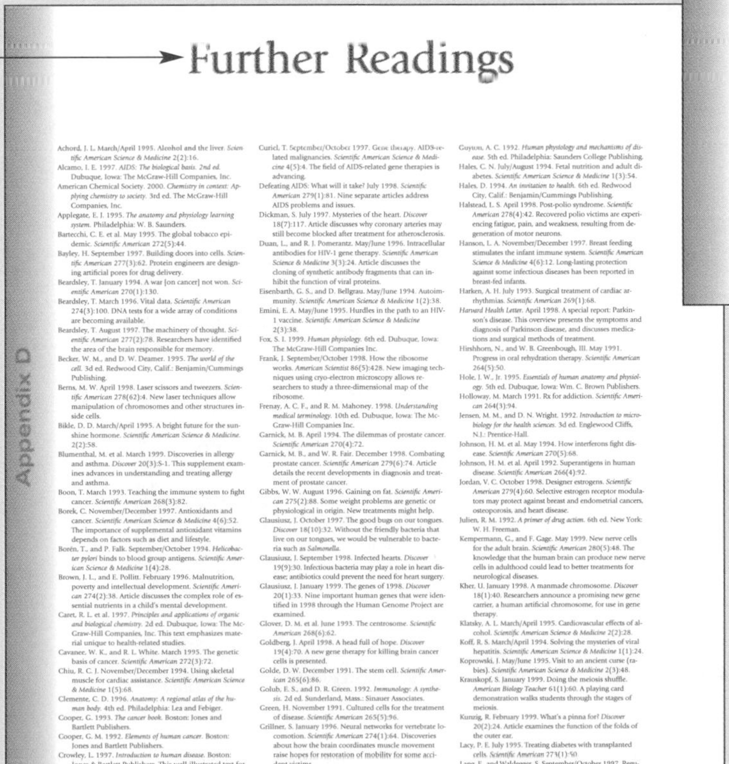

Appendix D: Further Readings
This appendix lists articles and books to give you more information about a particular topic or if you need references for a research paper.

Further Readings

Achord, J. L. March/April 1995. Alcohol and the liver. *Scientific American Science & Medicine* 2(2):16.

Alcamo, I. E. 1997. *AIDS: The biological basis.* 2nd ed. Dubuque, Iowa: The McGraw-Hill Companies, Inc.

American Chemical Society. 2000. *Chemistry in context: Applying chemistry to society.* 3rd ed. The McGraw-Hill Companies, Inc.

Applegate, E. J. 1995. *The anatomy and physiology learning system.* Philadelphia: W. B. Saunders.

Bartecchi, C. E. et al. May 1995. The global tobacco epidemic. *Scientific American* 272(5):44.

Bayley, H. September 1997. Building doors into cells. *Scientific American* 277(3):62. Protein engineers are designing artificial pores for drug delivery.

Beardsley, T. January 1994. A war [on cancer] not won. *Scientific American* 270(1):130.

Beardsley, T. March 1996. Vital data. *Scientific American* 274(3):100. DNA tests for a wide array of conditions are becoming available.

Beardsley, T. August 1997. The machinery of thought. *Scientific American* 277(2):78. Researchers have identified the area of the brain responsible for memory.

Becker, W. M., and D. W. Deamer. 1995. *The world of the cell.* 3d ed. Redwood City, Calif: Benjamin/Cummings Publishing.

Berns, M. W. April 1998. Laser scissors and tweezers. *Scientific American* 278(4):62. New laser techniques allow manipulation of chromosomes and other structures inside cells.

Bilde, D. D. March/April 1995. A bright future for the sunshine hormone. *Scientific American Science & Medicine.* 2(2):58.

Blumenthal, M. et al. March 1999. Discoveries in allergy and asthma. *Discover* 20(3):S-1. This supplement examines advances in understanding and treating allergy and asthma.

Born, T. March 1993. Teaching the immune system to fight cancer. *Scientific American* 268(3):82.

Borek, C. November/December 1997. Antioxidants and cancer. *Scientific American Science & Medicine* 4(6):52. The importance of supplemental antioxidant vitamins depends on factors such as diet and lifestyle.

Borén, T., and P. Falk. September/October 1994. *Helicobacter pylori* binds to blood group antigens. *Scientific American Science & Medicine* 1(4):28.

Brown, J. L., and E. Pollitt. February 1996. Malnutrition, poverty and intellectual development. *Scientific American* 274(2):38. Article discusses the complete role of essential nutrients in a child's mental development.

Caret, R. L. et al. 1997. *Principles and applications of organic and biological chemistry.* 2d ed. Dubuque, Iowa: The McGraw-Hill Companies, Inc. This text emphasizes material unique to health-related studies.

Cavanee, W. K., and R. L. White. March 1995. The genetic basis for cardiac assistance. *Scientific American* 272(3):72.

Chiu, R. C. J. November/December 1994. Using skeletal muscle for cardiac assistance. *Scientific American Science & Medicine* 1(5):68.

Clemente, C. D. 1996. *Anatomy: A regional atlas of the human body.* 4th ed. Philadelphia: Lea and Febiger.

Cooper, G. 1993. *The cancer book.* Boston: Jones and Bartlett Publishers.

Cooper, G. M. 1992. *Elements of human cancer.* Boston: Jones and Bartlett Publishers.

Crowley, L. 1997. *Introduction to human disease.* Boston: Jones & Bartlett Publishers. This well-illustrated text for study in the allied health fields describes diseases and their symptoms, diagnoses, and treatments.

Curiel, T. September/October 1997. Gene therapy. AIDS-related malignancies. *Scientific American Science & Medicine* 4(5):4. The field of AIDS-related gene therapies is advancing.

Defeating AIDS: What will it take? July 1998. *Scientific American* 279(1):81. Nine separate articles address AIDS problems and issues.

Dickman, S. July 1997. Mysteries of the heart. *Discover* 18(7):117. Article discusses why coronary arteries may still become blocked after treatment for atherosclerosis.

Duan, L., and R. J. Pomerantz. May/June 1996. Intracellular antibodies for HIV-1 gene therapy. *Scientific American Science & Medicine* 3(3):24. Article discusses the cloning of synthetic antibody fragments that can inhibit the function of viral proteins.

Eisenbarth, G. S., and D. Bellgrau. May/June 1994. Autoimmunity. *Scientific American Science & Medicine* 1(2):38.

Emini, E. A. May/June 1995. Hurdles in the path to an HIV-1 vaccine. *Scientific American Science & Medicine* 2(3):38.

Fox, S. I. 1999. *Human physiology.* 6th ed. Dubuque, Iowa: The McGraw-Hill Companies Inc.

Frank, I. September/October 1998. How the ribosome works. *American Scientist* 86(5):428. New imaging techniques using cryo-electron microscopy allows researchers to study a three-dimensional map of the ribosome.

Fresnay, A. C. F., and R. M. Mahoney. 1998. *Understanding medical terminology.* 10th ed. Dubuque, Iowa: The McGraw-Hill Companies Inc.

Garnick, M. B. April 1994. The dilemmas of prostate cancer. *Scientific American* 270(4):72.

Garnick, M. B., and W. R. Fair. December 1998. Combating prostate cancer. *Scientific American* 279(6):74. Article details the recent developments in diagnosis and treatment of prostate cancer.

Gibbs, W. V. August 1996. Gaining on fat. *Scientific American* 275(2):88. Some weight problems are genetic or physiological in origin. New treatments might help.

Glausiusz, J. October 1997. The good bugs on our tongues. *Discover* 18(10):32. Without the friendly bacteria that live on our tongues, we would be vulnerable to bacteria such as *Salmonella.*

Glausiusz, J. September 1998. Infected hearts. *Discover* 19(9):30. Infectious bacteria may play a role in heart disease; antibiotics could prevent the need for heart surgery.

Glausiusz, J. January 1999. The genes of 1998. *Discover* 20(1):33. Nine important human genes that were identified in 1998 through the Human Genome Project are examined.

Glover, D. M. et al. June 1993. The centrosome. *Scientific American* 268(6):62.

Goldberg, J. April 1998. A head full of hope. *Discover* 19(4):70. A new gene therapy for killing brain cancer cells is presented.

Golde, D. W. December 1991. The stem cell. *Scientific American* 265(6):86.

Golub, E. S., and D. R. Green. 1992. *Immunology: A synthesis.* 2d ed. Sunderland, Mass.: Sinauer Associates.

Green, H. November 1991. Cultured cells for the treatment of disease. *Scientific American* 265(5):96.

Grillner, S. January 1996. Neural networks for vertebrate locomotion. *Scientific American* 274(1):64. Discoveries about how the brain coordinates muscle movement raise hopes for restoration of mobility for some accident victims.

Gunstream, S. E. 2000. *Anatomy and physiology.* 2d ed. Dubuque, Iowa: Wm. C. Brown Publishers.

Guyton, A. C. 1992. *Human physiology and mechanisms of disease.* 5th ed. Philadelphia: Saunders College Publishing.

Hales, C. N. July/August 1994. Fetal nutrition and adult diabetes. *Scientific American Science & Medicine* 1(3):54.

Hales, D. 1994. *An invitation to health.* 6th ed. Redwood City, Calif.: Benjamin/Cummings Publishing.

Halstead, L. S. April 1998. Post-polio syndrome. *Scientific American* 278(4):42. Recovered polio victims are experiencing fatigue, pain, and weakness, resulting from degeneration of motor neurons.

Hanson, L. A. November/December 1997. Breast feeding stimulates the infant immune system. *Scientific American Science & Medicine* 4(6):12. Long-lasting protection against some infectious diseases has been reported in breast-fed infants.

Harken, A. H. July 1993. Surgical treatment of cardiac arrhythmias. *Scientific American* 269(1):68.

Harvard Health Letter. April 1998. A special report: Parkinson's disease. This overview presents the symptoms and diagnosis of Parkinson's disease, and discusses medications and surgical methods of treatment.

Hirshhorn, N., and W. B. Greenbough, III. May 1991. Progress in oral rehydration therapy. *Scientific American* 264(5):50.

Hole, J. W., Jr. 1995. *Essentials of human anatomy and physiology.* 5th ed. Dubuque, Iowa: Wm. C. Brown Publishers.

Holloway, M. March 1991. Rx for addiction. *Scientific American* 264(3):94.

Jensen, M. M., and D. N. Wright. 1992. *Introduction to microbiology for the health sciences.* 3d ed. Englewood Cliffs, N.J.: Prentice-Hall.

Johnson, H. M. et al. May 1994. How interferons fight disease. *Scientific American* 270(5):68.

Johnson, H. M. et al. April 1992. Superantigens in human disease. *Scientific American* 266(4):92.

Jordan, V. C. October 1998. Designer estrogens. *Scientific American* 279(4):60. Selective estrogen receptor modulators may protect against breast and endometrial cancers, osteoporosis, and heart disease.

Julien, R. M. 1992. *A primer of drug action.* 6th ed. New York: W. H. Freeman.

Kempermann, G., and F. Gage. May 1999. New nerve cells for the adult brain. *Scientific American* 280(5):48. The knowledge that the human brain can produce new nerve cells in adulthood could lead to better treatments for neurological diseases.

Khet, U. January 1998. A manmade chromosome. *Discover* 18(1):40. Researchers announce a promising new gene carrier, a human artificial chromosome, for use in gene therapy.

Klarsky, A. L. March/April 1995. Cardiovascular effects of alcohol. *Scientific American Science & Medicine* 2(2):28.

Koff, R. S. March/April 1994. Solving the mysteries of viral hepatitis. *Scientific American Science & Medicine* 1(1):24.

Koprowski, I. May/June 1995. Visit to an ancient culture (rabies). *Scientific American Science & Medicine* 2(3):48.

Krauskopf, S. January 1990. Doing the meiosis shuffle. *American Biology Teacher* 61(1):60. A playing card demonstration walks students through the stages of meiosis.

Kurnig, R. February 1999. What's a pinna for? *Discover* 20(2):24. Article examines the function of the folds of the outer ear.

Lacy, P. E. July 1995. Treating diabetes with transplanted cells. *Scientific American* 273(1):50.

Lang, F., and Waldegger, S. September/October 1997. Regulating cell volume. *American Scientist* 85(5):456. Changes in cell volume may threaten organ or tissue function.

410   Appendix D

INDEX

A thorough and complete index at the end of the book directs you to the page or pages on which various topics are discussed.

GLOSSARY

The end-of-book glossary defines all the bold-faced terms, including the Basic Key Terms and the Clinical Key Terms that appear at the end of each chapter. The glossary is page-referenced.

Essential Study Partner CD-ROM

A free study partner that engages, investigates, and reinforces what you are learning from your textbook. You'll find the **Essential Study Partner** for *Understanding Human Anatomy and Physiology* to be a complete, interactive student study tool packed with hundreds of animations and learning activities. From quizzes to interactive diagrams, you'll find that there has never been a better study partner to ensure the mastery of core concepts. Best of all, it's **FREE** with your new textbook purchase.

The topic menu contains an interactive list of the available topics. Clicking on any of the listings within this menu will open your selection and will show the specific concepts presented within this topic. Clicking any of the concepts will move you to your selection. You can use the UP and DOWN arrow keys to move through the topics.

The unit pop-up menu is accessible at anytime within the program. Clicking on the current unit will bring up a menu of other units available in the program.

To the right of the arrows is a row of icons that represent the number of screens in a concept. There are three different icons, each representing different functions that a screen in that section will serve. The screen that is currently displayed will highlight yellow and visited ones will be checked.

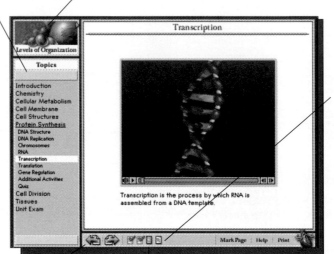

The film icon represents an animation screen.

Along the bottom of the screen you will find various navigational aids. At the left are arrows that allow you to page forward and backward through text screens or interactive exercise screens. You can also use the LEFT and RIGHT arrows on your keyboard to perform the same function.

The activity icon represents an interactive learning activity.

The page icon represents a page of informational text.

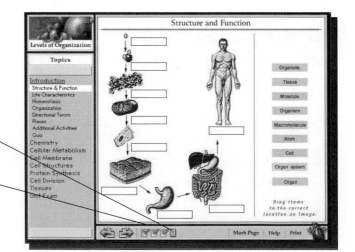

Contact your McGraw-Hill sales representative for more information or visit *www.mhhe.com*.

Visual Resource Library CD-ROMs

These CD-ROMs are electronic libraries of educational presentation resources that instructors can use to enhance their lectures. View, sort, search, and print catalog images, play chapter-specific slideshows using PowerPoint, or create customized presentations when you:

- Find and sort thumbnail image records by name, type, location, and user-defined keywords
- Search using keywords or terms
- View images at the same time with the Small Gallery View
- Select and view images at full size
- Display all the important file information for easy file identification
- Drag and place or Copy and Paste into virtually any graphics, desktop publishing, presentation, or multimedia application

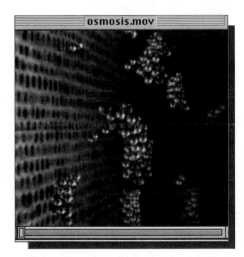

Life Science Animations Visual Resource Library CD-ROM

This instructor's tool, containing more than 125 animations of important biological concepts and processes—found in the *Essential Study Partner* and *Dynamic Human CD-ROMs*—is perfect to support your lecture. The animations contained in this library are not limited to subjects covered in the text, but include an expansion of general life science topics.

Inquiry and Human Biology Visual Resource Library CD-ROM

This helpful CD-ROM contains many photographs and illustrations from the text. You'll be able to create interesting multimedia presentations with the use of these images, and students will have the ability to easily access the same images in their texts to later review the content covered in class.

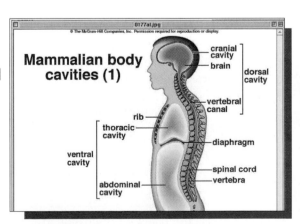

Contact your McGraw-Hill sales representative for more information or visit *www.mhhe.com*.

PageOut

Proven. Reliable. Class-tested.

Over 6,000 professors have chosen PageOut to create course websites. And for good reason: PageOut offers powerful features, yet is incredibly easy to use.

Now you can be the first to use an even better version of PageOut. Through class-testing and customer feedback, we have made key improvements to the grade book, as well as the quizzing and discussion areas. Best of all, PageOut is still free with every McGraw-Hill textbook. And students needn't bother with any special tokens or fees to access your PageOut website.

Customize the site to coincide with your lectures.

Complete the PageOut templates with your course information and you will have an interactive syllabus online. This feature lets you post content to coincide with your lectures. When students visit your PageOut website, your syllabus will direct them to components of McGraw-Hill web content germane to your text, or specific material of your own.

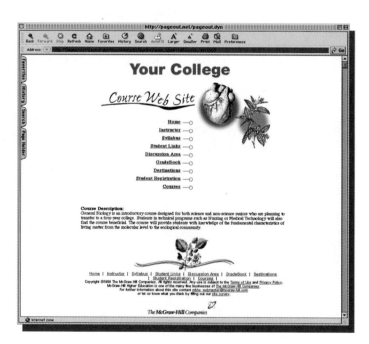

New Features based on customer feedback:

- Specific question selection for quizzes

- Ability to copy your course and share it with colleagues or as a foundation for a new semester

- Enhanced grade book with reporting features

- Ability to use the PageOut discussion area, or add your own third party discussion tool

- Password protected courses

Short on time? Let us do the work.

Send your course materials to our McGraw-Hill service team. They will call you by phone for a 30 minute consultation. A team member will then create your PageOut website and provide training to get you up and running. Contact your McGraw-Hill Publisher's Representative for details.

Contact your local McGraw-Hill sales representative for more information or visit *www.mhhe.com*.

The Online Learning Center

Your Password to Success

http://www.mhhe.com/biosci/ap/maderap/

This text-specific website allows students and instructors from all over the world to communicate. Instructors can create a more interactive course with the integration of this site, and students will find tools that help them improve their grades and learn that biology can be fun.

Student Resources

Study questions
Quizzing with immediate feedback
Links to chapter-related websites
Case studies
Interactive art labeling exercises
Critical thinking exercises

Instructor Resources

Instructor's Manual
Activities that can be assigned
   as coursework
Links to related websites to expand on
   particular topics
Classroom activities
Lecture outlines
Case studies

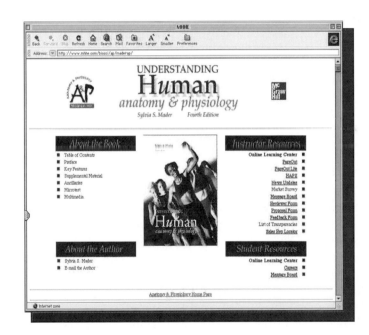

**Imagine the advantages of having so many
learning and teaching tools
all in one place—all at your fingertips—FREE.**

Contact your McGraw-Hill sales
representative for more information
or visit *www.mhhe.com.*

Supplemental Materials

Instructor's Manual—Test Item File

The *Instructor's Manual* was written by Patrick Galliart, North Iowa Area Community College, an experienced instructor of anatomy and physiology. This manual is designed to assist instructors as they plan and prepare for classes using *Understanding Human Anatomy and Physiology.* Each chapter includes an outline and chapter synopsis, concepts to emphasize, suggested student activities, and a listing of audiovisual materials. A good selection of objective test questions and several essay questions are provided for each chapter. (ISBN 0–07–290758–4)

Transparencies

One hundred full-color transparencies are available with *Understanding Human Anatomy and Physiology.* Chosen by the author, they were judged to be the illustrations that instructors would most like to have as adjuncts to their lectures. (ISBN 0–07–290761–4)

Microtest III

McGraw-Hill provides a computerized test generator for use with this text. It allows you to quickly create tests based on questions provided by McGraw-Hill and requires no programming experience. The questions are provided on diskette in a test item file. McGraw-Hill also provides support services, via mail or phone, to assist in the use of the test generator software, as well as in the creation and printing of tests. (Macintosh ISBN 0–07–290759–2) (Windows ISBN 0–07–290760–6)

Essential Study Partner

This CD-ROM is an interactive student study tool packed with hundreds of animations and learning activities. From quizzes to interactive diagrams, your students will find that there has never been a more exciting way to study anatomy and physiology. A self-quizzing feature allows students to check their knowledge of a topic before moving on to a new module. Additional unit exams give students the opportunity to review coverage for a more complete understanding. The ESP is packaged free with textbooks. (ISBN 0–07–237158–7)

Online Learning Center

Students and instructors gain access to a world of opportunities through this password-protected web site. Students will find quizzes, activities, links, and much more. Instructors will find all the enhancement tools needed for teaching online, or for incorporating technology in the traditional course. http://www.mhhe.com/biosci/ap/maderap/

Anatomy and Physiology Laboratory Textbook, Essentials Version (2d ed.)

By Stanley E. Gunstream, Harold J. Benson, Arthur Talaro, and Kathleen P. Talaro, all of Pasadena City College, is an excellent lab text. It presents the fundamentals of human anatomy and physiology in an easy-to-read manner that is appropriate for students in allied health programs. Designed especially for the one-semester course, it features a simple, concise writing style, thirty-seven self-directing exercises, full-color photomicrographs in the *Histology Atlas,* and numerous illustrations in each exercise. The lab text is accompanied by an *Instructor's Handbook* and slides. (ISBN 0–697–11311–6)

McGraw-Hill Life Science Animations Videotape Series

This is a series of six videotapes containing sixty-six animations that cover many of the key processes discussed in a physiology course:

a. *Chemistry, Cells and Energetics* ISBN 0–697–25068–7
b. *Cell Division, Heredity, Genetics, Reproduction, and Development* ISBN 0–697–25069–5
c. *Animal Biology #1* ISBN 0–697–25070–9
d. *Animal Biology #2* ISBN 0–697–25071–7
e. *Plant Biology, Evolution, and Ecology* ISBN 0–697–26600–1
f. *Physiological Concepts of Life Science* ISBN 0–697–21512–1

Life Science Animations 3D CD-ROM

More than 120 animations that illustrate key biological processes are available at your fingertips on this exciting CD-ROM. This CD contains all of the animations found on the *Essential Study Partner* and much more. The animations can be imported into presentation programs, such as PowerPoint. Imagine the benefit of showing the animations during lecture. (ISBN 0–07–234296–X)

Life Science Animations 3D Videotape

Featuring 42 animations of key biologic processes, this tape contains 3D animations and is fully narrated. Various figures throughout this text are correlated to video animations. (ISBN 0–07–290652–9)

The Dynamic Human CD-ROM

Version 2.0 NT Compatible illustrates the important relationships between anatomical structures and their functions in the human body. Realistic computer visualization and three-dimensional visualizations are the premier features of this learning tool. (ISBN 0–07–235476–3)

Explorations in Human Biology CD-ROM

Consists of sixteen interactive modules that stress human physiology. Students can actively investigate vital processes as they explore the modules, which are filled with color, sound, and movement.
(Macintosh ISBN 0–697–22964–5)
(Windows ISBN 0–697–22963–7)

Anatomy and Physiology Videodisc

A four-sided videodisc containing more than thirty animations of physiological processes, as well as line art and micrographs. A bar code directory is also available. (ISBN 0–697–27716–X)

McGraw-Hill Visual Resource Library

A CD-ROM containing many of the line art figures found in *Human Anatomy & Physiology* with an easy-to-use interface program enabling the user to quickly move among the images and create a multimedia presentation. (ISBN 0–07–234411–3)

Life Science Living Lexicon CD-ROM

Contains a comprehensive collection of life science terms, including definitions of their roots, prefixes, and suffixes as well as audio pronunciations and illustrations. The Lexicon is student-interactive, featuring quizzing and notetaking capabilities. (ISBN 0–697–37993–0 hybrid)

The Virtual Psysiology Lab CD-ROM

Contains 10 dry labs of the most common and important physiology experiments. (ISBN 0–697–37994–9 hybrid)

Web-Based Cat Dissection Review for Human Anatomy and Physiology

By John Waters. (ISBN 0–07–232157–1)

Interactive Histology CD-ROM

By Bruce Wingerd. (ISBN 0–07–237308–3)

Laboratory Atlas of Anatomy and Physiology, third edition

By Eder and colleagues is a full-color atlas containing histology, human skeletal anatomy, human muscular anatomy, dissections, and reference tables. (ISBN 0–697–39480–8)

Study Cards for Anatomy and Physiology

By Kent M. Van De Graaff et al., 3d edition, is a boxed set of (300) 3-by-5-inch cards. It serves as a well-organized and illustrated synopsis of the structure and function of the human body. The study cards offer a quick and effective way for students to review human anatomy and physiology. (ISBN 0–697–26447–5)

BodyWorks CD-ROM

The ultimate multimedia reference guide to human anatomy and physiology. The CD contains text, movies, 3-D modules, and animations of the human body. At the end of most chapters, a screened box appears, informing the reader which section on the CD applies to that chapter. To order *BodyWorks* from SoftKey International Inc., call 1–800–845–8692.

McGraw-Hill Anatomy and Physlology Video Series

a. *Introduction to the Human Cadaver and Prosection,* ISBN 0–697–11177–6
b. *Introduction to Cat Dissection and Cat Musculature,* ISBN 0–697–11630–1
c. *Blood Cell Counting, Identification, and Grouping,* ISBN 0–697–11629–8
d. *Internal Organs and the Circulatory System of the Cat,* ISBN 0–697–13922–0

Survey of Infectious and Parasitic Diseases

By Kent M. Van De Graaff is a black-and-white booklet that presents the essential information on one hundred of the most common and clinically significant diseases. A one-page presentation that includes pronunciation, derivation, definition, life cycle, description, signs and symptoms, laboratory diagnoses, and prevention/treatment is devoted to each disease. (ISBN 0–697–27535–3)

Coloring Guide to Anatomy and Physiology

By Robert and Judith Stone emphasizes learning through the process of color association. The coloring guide provides a thorough review of anatomical and physiological concepts. (ISBN 0–697–17109–4)

Atlas of the Skeletal Muscles, 3/e

By Robert and Judith Stone is a guide to the structure and function of human skeletal muscles. The illustrations help students locate muscles and understand their actions. (ISBN 0–07–290332–5)

Ancillaries available to qualified adopters.

McGraw-Hill Life Science Animations

A set of six videotapes contains over fifty animations of physiological processes integral to the study of human anatomy and physiology. These videotapes cover such topics as cell division, genetics, and reproduction. A videotape icon 📼 appears in appropriate figure legends to alert the reader to these animations. A list of the figures that relate to the animations follows.

Life Science Animations 3D Videotape

Features include 42 animations of key biologic processes. This tape contains 3D animations and is fully narrated. A videotape icon also appears in appropriate figure legends. A list of the figures that relate to the 3-D videotape follows.

Correlation of the McGraw-Hill Life Science Animations Videotape Series (LSA) and Life Science Animations (3-D) Videotape 📼

Chapter 2

2.3	LSA #1—Formation of an Ionic Bond
	3-D Concept 1—Atomic Structure and Covalent and Ion Bonding
2.11	3-D Concept 7—Enzyme Action
2.15	3-D Concept 13—Mitosis

Chapter 3

3.2	LSA #2—Journey into a Cell
3.3	LSA #17—Protein Synthesis
3.4	LSA #4—Cellular Secretion
	3-D Concept 3—Cellular Secretion
3A	3-D Concept 4—Diffusion
3.5	LSA #7—The Electron Transport Chain
	3-D Concept 9—Electron Transport Chain
3.9	LSA #3—Endocytosis
3.11	LSA #12—Mitosis
	3-D Concept 10—Mitosis

Chapter 7

7.2	3-D Concept 40—Muscle Contraction

Chapter 8

8.3	LSA #22—Formation of Myelin Sheath
8.5	3-D Concept 39—Action Potential
8.6	LSA #25—Reflex Arcs

Chapter 10

10.1	LSA #28—Peptide Hormone Action
	3-D Concept 41—Hormone Action

Chapter 12

12.5	LSA #37—Blood Circulation
12A	LSA #38—Production of Electrocardiogram
12.10	LSA #37—Blood Circulation
12.11	LSA #37—Blood Circulation
12.12	LSA #37—Blood Circulation

Chapter 13

13.7, 13.9, 13.10	3-D Concept 34—How T-Lymphocytes Work
13.7	LSA #41—B-Cell Immune Response
13.8	LSA #42—Structure and Function of Antibodies
13.9	LSA #44—Relationship of Helper T-cells & Killer T-cells

Chapter 14

14.10, 14.11, 14.12	3-D Concept 37—Gas Exchange

Chapter 15

15.3	LSA #33—Peristalsis
15.1, 15.4, 15.6	3-D Concept 36—Digestion Overview

Chapter 16

16.4	3-D Concept 38—Kidney Function

Chapter 18

18.1	LSA #21—Human Embryonic Development
18.2	LSA #21—Human Embryonic Development
18.4	LSA #21—Human Embryonic Development
18.5	LSA #21—Human Embryonic Development

The Dynamic Human CD-ROM, Version 2.0

Interactively illustrates the complex relationships between anatomical structures and their functions in the human body. This program covers each body system, demonstrating clinical concepts, histology, and physiology. *The Dynamic Human* icon 𝍏 appears in appropriate figure legends to alert the reader to corresponding information. A list correlating figures to specific sections of *The Dynamic Human* follows.

Correlation of *The Dynamic Human* 🏃

The Essential Study Partner CD-ROM

An interactive student study tool packed with hundreds of animations and learning activities. The ESP film icon ▯ placed in the text beside topics and concepts that are ani- mated on the CD alerts students to that helpful resource. The Essential Study Partner Animation Correlation Guide follows.

▯ Essential Study Partner CD-ROM Animations Correlation Guide

ESP Unit	ESP Topic	ESP Concept	Text Page Correlation
Levels of Organization	Introduction		
		Directional Terms	4
		Planes	5
	Chemistry	Atomic Structure	17
		Chemical Bonding	18
		Nucleic Acids	29
	Cellular Metabolism	Enzymes	25
	Cell Membrane	Diffusion	39
		Osmosis	41
		Active Transport	41
		Exocytosis & Endocytosis	42
	Protein Synthesis	DNA Structure	29
	Cell Division	Meiosis	44
Support and Movement	Receptors & Membranes	Chemical Signals	140
	Muscular System	Neuromuscular Junction	112
		Sliding Filament Theory	113
Integration and Coordination	Functional Organization of Nervous Tissue	Impulse Processing	142
	Central Nervous System	Cerebrum	146
	The Senses	Sense of Smell	169
		Sense of Taste	168
		Sense of Hearing	179
		Sense of Balance	183
		Sense of Sight	169
	Endocrine System	Endocrine System and Hormone Action	187
		Pituitary Gland	190
		Pancreas	197
Transport	Blood	Blood Groupings	217
	Cardiovascular System	Heart	223
		Heart Dynamics	229

Acknowledgments

Steven G. Bassett
Seton Hill College

Regina Betette
Midlands Technical College

David T. Corey
Midlands Technical College

Joyce A. Foster
State Fair Community College

Clarence E. Fouche
Virginia Intermont College

Carolyn McCracken
Northeast State Technical Community College

Lewis M. Milner
North Central Technical College

Jacqueline R. Shepperson
Winston-Salem State University

Sara Sybesma Tolsma
Northwestern College

Ricky K. Wong
Los Angeles Trade-Technical College

chapter 1

Organization of the Body

Chapter Outline and Learning Objectives

After you have studied this chapter, you should be able to:

Anatomy and Physiology (p. 2)
■ Define anatomy and physiology, and explain how they are related.

Organization of Body Parts (p. 2)
■ Describe each level of organization of the body with reference to an example.

Anatomical Terms (p. 3)
■ Use the terms that describe the relative positions of body parts and the planes, sections, and regions of the body.
■ List the cavities of the body, and state their locations.
■ Name the organs located in each of the body cavities.

Organ Systems (p. 9)
■ List the organ systems of the body, and state the major organs associated with each.
■ Describe in general the functions of each organ system.

Homeostasis (p. 12)
■ Describe homeostasis, and explain in general how homeostasis is maintained.
■ Define disease, and explain the difference between a local and a systemic disease.

Medical Focus
Organ Donation (p. 10)

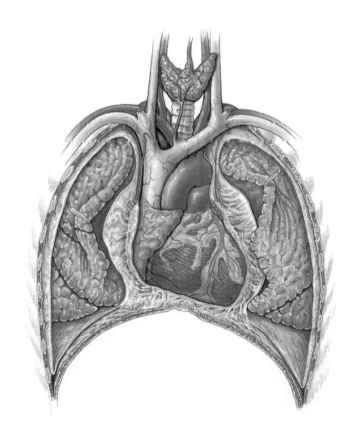

The human body is organized, with each part having a particular anatomical location. The heart and lungs are located in the thoracic (chest) cavity, as shown here.

Part I

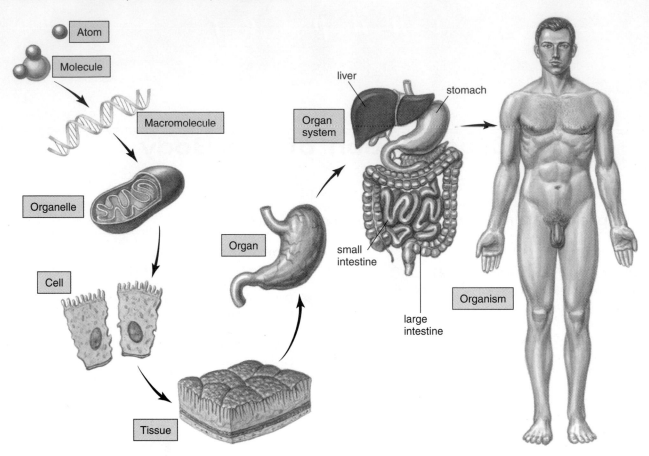

figure 1.1 Levels of organization of the human body. Each level is more complex than the previous level.

Atom

Molecule

Macromolecule

Organelle

Cell

Tissue

Organ

Organ system

liver

stomach

small intestine

large intestine

Organism

Anatomy and Physiology

Anatomy is the study of the structure of body parts. For example, the stomach is a J-shaped, pouchlike organ (fig. 1.1). The stomach wall has thick folds, which disappear as the stomach expands to increase its capacity. **Physiology** is the study of the function of body parts. For example, the stomach temporarily stores food, secretes digestive juices, and passes on partially digested food to the small intestine.

Anatomy and physiology are closely connected in that the structure of an organ suits its function. For example, the stomach's pouchlike shape and ability to expand are suitable to its function of storing food. In addition, the microscopic structure of the stomach wall is suitable to its secretion of digestive juices, as we shall see later in the text.

Anatomy is the study of the structure of body parts, and physiology is the study of the function of these parts. Structure is suited to the function of a part.

Organization of Body Parts

The structure of the body can be studied at different *levels of organization* (fig. 1.1). First, all substances, including body parts, are composed of chemicals made up of submicroscopic particles called **atoms.** The atoms most frequently found in the body are carbon, hydrogen, oxygen, and nitrogen. Atoms join together to form a **molecule,** which can join with other molecules to form **macromolecules.** The macromolecules in cells are called *biomolecules* in this text. For example, the molecules called amino acids join together to form a biomolecule called protein. Muscles contain a significant amount of protein; therefore, meat is a rich source of this basic nutrient.

Proteins and also fats contribute to the makeup of the **cell,** the basic unit of all living things. Cells are the smallest living portion of any organism, and it is at the cellular level that health and disease are best understood.

figure 1.2 Anatomical position and terms of location and position.

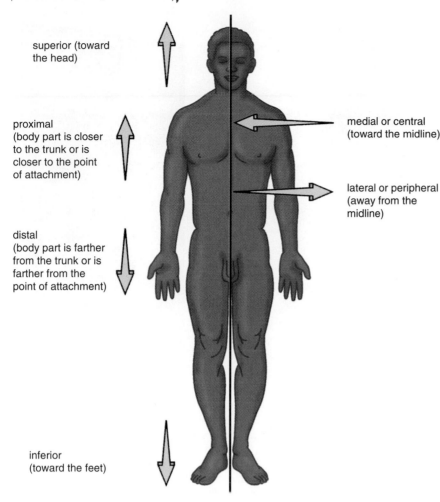

superior (toward the head)

proximal (body part is closer to the trunk or is closer to the point of attachment)

distal (body part is farther from the trunk or is farther from the point of attachment)

inferior (toward the feet)

medial or central (toward the midline)

lateral or peripheral (away from the midline)

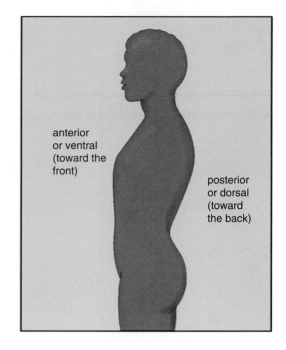

anterior or ventral (toward the front)

posterior or dorsal (toward the back)

Within cells are **organelles,** tiny membranous structures that perform cell functions. For example, the organelle called the nucleus is especially concerned with cell reproduction; another organelle called the mitochondrion supplies the cell with energy.

Cells are found in tissues, and tissues make up organs. A **tissue** is composed of similar types of cells and performs a specific function. An **organ** is composed of several types of tissues and performs a particular function within an **organ system.** For example, the stomach is an organ and is a part of the digestive system. It has a specific role in this system, in which the overall function is to supply the body with the nutrients needed for growth and repair. The other systems of the body (see fig. 1.8) also have specific functions.

All of the body systems together make up the **organism** —in this case, the human being. Human beings are complex animals, but this complexity can be broken down and studied at ever simpler levels. Each simpler level is organized and constructed in a particular way.

> The body has levels of organization that progress from atoms to molecules, macromolecules (biomolecules), cells, tissues, organs, organ systems, and finally, the organism.

Anatomical Terms

Certain terms are used to describe the location and position of body parts, imaginary planes and sections of the body, various regions, and body cavities. You should become familiar with these terms before your study of anatomy and physiology begins. Use of these terms assumes that the body is in the *anatomical position:* standing erect, with face forward, arms at the sides, and palms and toes directed forward, as illustrated in figure 1.2.

figure 1.3 Body planes. To observe internal parts, the body may be sectioned along these imaginary planes.

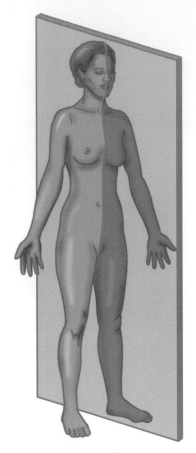
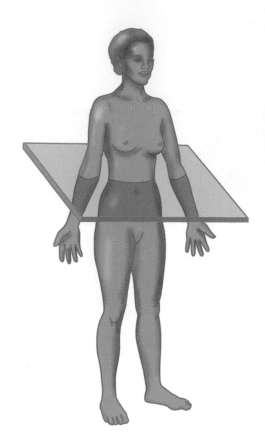
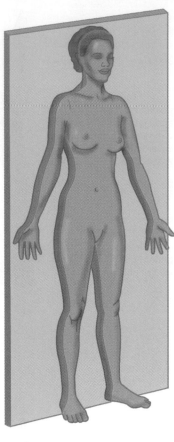

Sagittal (Vertical) Plane   **Transverse (Horizontal) Plane**   **Frontal (Coronal) Plane**

Relative Positions of Body Parts

The terms used in figure 1.2 describe the location of one body part in relation to another:

Superior means that a body part is located above another part, or toward the head. **Inferior** means that a body part is below another part, or toward the feet. The superior vena cava is in the chest, and the inferior vena cava is in the abdomen.

Anterior (ventral) means that a body part is located toward the front. **Posterior** (dorsal) means that a body part is located toward the back. The windpipe (trachea) is anterior to the esophagus, while the esophagus is posterior to the windpipe.

Medial means that a body part is nearer than another part to an imaginary midline of the body. **Lateral** means that a body part is farther to the side of the midline. The nose is medial to the eyes; however, the ears are lateral to the eyes.

Proximal means that a body part is closer to the point of attachment or closer to the trunk. **Distal** means that a body part is farther from the point of attachment or farther from the trunk. The upper arm is proximal to the elbow, and the lower arm is distal to the elbow.

Superficial (external) means that a body part is located near the surface. **Deep** (internal) means that the body part is located away from the surface. Superficial blood vessels are closer to the skin than those that lie deep in the abdominal cavity.

Central means that a body part is situated at the center of the body or an organ. **Peripheral** means that a body part is situated away from the center of the body or an organ. The central nervous system is located along the main axis of the body; the peripheral nervous system is outside the central nervous system.

The terms *superior/inferior, anterior/posterior, medial/lateral, proximal/distal, superficial/deep,* and *central/peripheral* describe the relative positions of body parts.

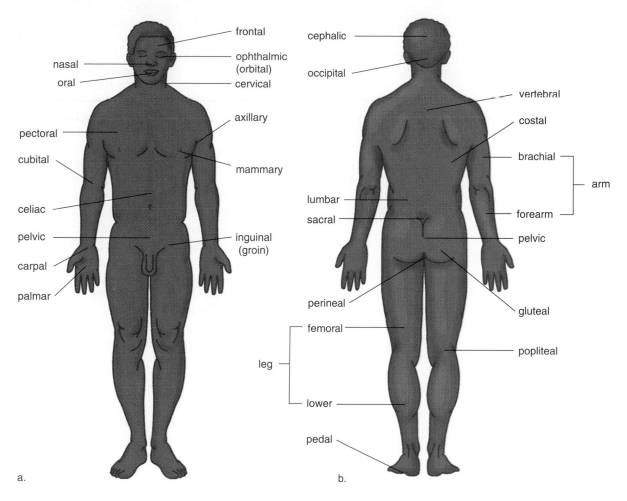

figure 1.4 Terms for body parts and areas. **a.** Anterior. **b.** Posterior.

Planes and Sections of the Body

To observe internal body parts, it is necessary to section (cut) the body. The body is customarily sectioned by these imaginary planes (fig. 1.3):

A **sagittal** (vertical) **plane** is a lengthwise cut that divides the body into right and left portions. A section that passes exactly through the midline is called a *midsagittal plane.*

A **transverse** (horizontal) **plane** is a cut that divides the body horizontally to give a cross section. A transverse cut divides the body into superior and inferior portions.

A **frontal** (coronal) **plane** is a cut that divides the body lengthwise into anterior and posterior portions.

The terms *longitudinal section* and *cross section* are often applied to body parts that have been removed and cut either lengthwise or straight across, respectively.

The body or its parts may be sectioned (cut) along certain imaginary planes. A sagittal (vertical) cut divides the body into right and left portions. A transverse (horizontal) cut is a cross section. A frontal (coronal) cut divides the body into anterior and posterior parts.

Regions of the Body

The human body can be divided into axial and appendicular portions. The **axial** portion includes the head, neck, and trunk. The trunk, or torso, contains the *thorax, abdomen,* and *pelvis.* The **appendicular** portion includes the limbs (arms and legs). Smaller regions of the body are named in the lists that follow and are illustrated in figure 1.4.

Head

   cephalic (head)

   cranial (skull)

   frontal (forehead)

Head, cont'd.
   occipital (back of head)
   oral (mouth)
   nasal (nose)
   ophthalmic (orbital, eyes)

Neck
   cervical (neck)

Thorax (chest)
   pectoral (chest)
   mammary (breast)
   axillary (armpit)
   vertebral (backbone)
   costal (ribs)

Abdomen celiac (abdomen)
   pelvic (lower portion of abdomen)
   gluteal (buttock)
   inguinal (groin)
   groin (depressed region of abdomen near thigh)
   lumbar (lower back)
   sacral (where vertebrae terminate)
   perineal (region between anus and external sex organs)

Limbs (arms and legs)
   brachial (upper arm)
   forearm (lower arm)
   carpal (wrist)
   cubital (elbow)
   palmar (palm)
   lower limb (leg)
   femoral (thigh)
   popliteal (back of knee)
   lower leg (distal to the knee)
   pedal (foot)

Other
   cutaneous (skin)

The body can be divided into axial and appendicular portions, each of which can be further subdivided into specific regions. For example, *brachial* refers to the upper arm, and *pedal* refers to the foot.

Cavities of the Body

The internal organs, called the **visceral** organs, are located within specific body cavities (fig. 1.5). The two main cavities are the **dorsal cavity** and the larger **ventral cavity**. The dorsal cavity can be subdivided into two parts: The **cranial cavity** within the skull contains the brain; and the **spinal cavity**, protected by the vertebrae, contains the spinal cord.

The ventral cavity is subdivided into the **thoracic** (thoras'ik) **cavity** and the **abdominopelvic cavity**. Membranes divide the thoracic cavity into the **pleural cavities**, containing the right and left lungs, and the **pericardial cavity**, containing the heart. The **mediastinum** (me"de-ah-sti'num) is the mass of tissues and organs between the pleural cavities. The thoracic cavity is separated from the abdominopelvic cavity by a horizontal muscle called the **diaphragm.**

The abdominopelvic cavity has two portions: the upper **abdominal cavity** and the lower **pelvic cavity**. The stomach, liver, spleen, gallbladder, and most of the small and large intestines are in the abdominal cavity. The pelvic cavity contains the rectum, the urinary bladder, the internal reproductive organs, and the rest of the large intestine. Males have an external extension of the abdominal wall, called the **scrotum,** where the testes are found.

The abdomen can be divided into four quadrants by running a transverse plane across the midsagittal plane at the point of the navel (fig. 1.6). Physicians commonly use these quadrants to identify the locations of patients' symptoms. The four quadrants are: (1) right upper (right superior) quadrant, (2) left upper (left superior) quadrant, (3) right lower (right inferior) quadrant, and (4) left lower (left inferior) quadrant.

The abdominopelvic cavity can also be divided into nine regions, as shown in figure 1.7. These regions are:

Umbilical, which is centrally located and surrounds the umbilicus (navel)

Lumbar, which are the regions to the right and left of the umbilical region (the term *lumbar* refers to the lower back, which occurs here)

Epigastric (*epi-*, above; *gastric*, stomach), which is the central region superior to the umbilical region

Hypochondriac (*hypo-*, under; *chondral*, cartilage), which are the regions to the right and left of the epigastric region and which are inferior to the cartilage of the rib cage

Hypogastric, which is the central region inferior to the umbilical region

Iliac, which are the regions to the right and left of the hypogastric region (the term *iliac* refers to the iliac [hip] bones that occur here)

figure 1.5 Mammalian body cavities. **a.** Lateral view. The dorsal cavity contains the cranial cavity and the spinal cavity. The brain is in the cranial cavity, and the spinal cord is in the spinal cavity. The ventral cavity is divided by the diaphragm into the thoracic cavity and the abdominopelvic cavity. The heart and lungs are in the thoracic cavity, and most other internal organs are in the abdominopelvic cavity. **b.** Frontal view of the thoracic cavity.

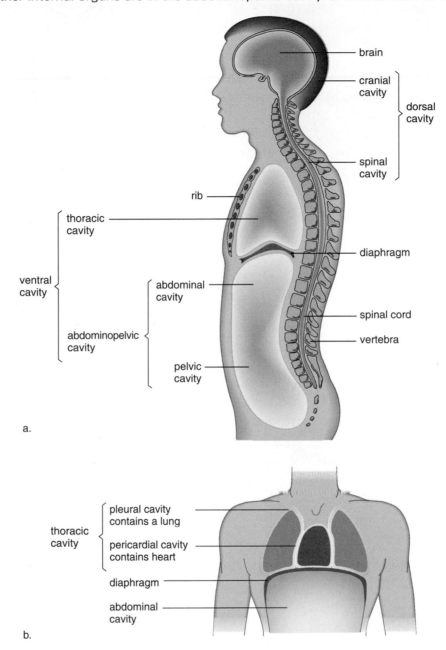

figure 1.6 Clinical subdivisions of the abdomen. These help physicians to identify the location of various symptoms. **a.** Photo showing approximate locations of quadrants. **b.** Quadrants superimposed over internal organs.

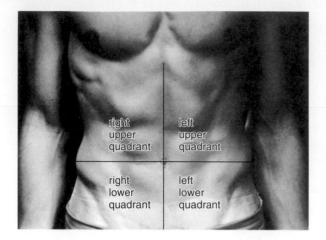

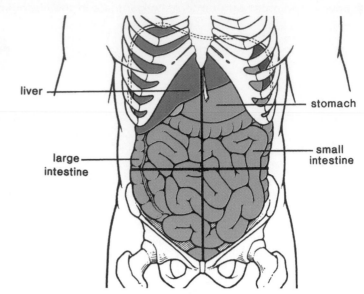

a.

b.

figure 1.7 Anatomical subdivisions of the abdominopelvic cavity. Anatomists divide the abdominopelvic cavity into these regions. **a.** Photo showing approximate locations of regions. **b.** Regions superimposed over internal organs.

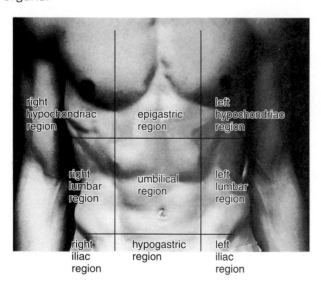

a.

b.

Organ Systems

The organs of the body work together in systems. Today, certain organs can be replaced by **organ transplantation,** during which a functioning organ is received from a donor, as discussed in the Medical Focus reading on page 10.

The reference figures in appendix A (see brown color bar) can serve as an aid to learning the 11 organ systems and their placement. Figure 1.8 introduces an illustration that will be used at the end of each of the organ system chapters. In this chapter, the illustration demonstrates the general functions of the body's organ systems. The corresponding illustrations in the organ system chapters will show how a particular organ system interacts with all the other systems. In this text, the organ systems of the body have been divided into the five categories that follow.

Integumentary System

The **integumentary system,** discussed in chapter 5, includes the skin and accessory organs, such as the hair, nails, sweat glands, and sebaceous glands. The skin protects underlying tissues, helps regulate body temperature, contains sense organs, and even synthesizes certain chemicals that affect the rest of the body.

> The integumentary system, which includes the skin, not only protects the body, but also has other functions.

Support and Movement

The **skeletal system** and the **muscular system** give the body support and are involved in the ability of the body and its parts to move.

The skeletal system, discussed in chapter 6, consists of bones of the skeleton and associated cartilage, as well as the ligaments that bind these structures together. The skeleton protects body parts. For example, the skull forms a protective encasement for the brain, as does the rib cage for the heart and lungs. Some bones produce blood cells, and all bones are a storage area for calcium and phosphorus salts. The skeleton, as a whole, serves as a place of attachment for the muscles.

Contraction of *skeletal muscles,* discussed in chapter 7, accounts for our ability to move voluntarily and to respond to outside stimuli. These muscles also maintain posture and are responsible for the production of body heat. *Cardiac muscle* and *smooth muscle* are called involuntary muscles because they contract automatically. Cardiac muscle makes up the heart, and smooth muscle is found within the walls of internal organs.

> The skeletal system contains the bones, and the muscular system contains the three types of muscles. The primary function of these systems is support and movement, but they have other functions as well.

Integration and Coordination

The **nervous system,** discussed in chapter 8, consists of the brain, spinal cord, and associated nerves. The nerves conduct nerve impulses from the sense organs to the brain and spinal cord. They also conduct nerve impulses from the brain and spinal cord to the muscles and glands.

The *sense organs,* discussed in chapter 9, provide us with information about the outside environment. This information is then processed by the brain and spinal cord, and the individual responds to environmental stimuli through the muscular system.

The **endocrine system,** discussed in chapter 10, consists of the hormonal glands that secrete chemicals that serve as messengers between body parts. Both the nervous and endocrine systems help maintain a relatively constant internal environment by coordinating and regulating the functions of the body's other systems. The nervous system acts quickly but has a short-lived effect; the endocrine system acts more slowly but has a more sustained effect on body parts. The endocrine system also helps maintain the proper functioning of male and female reproductive organs.

> The nervous system contains the brain, spinal cord, and nerves. Because the nervous system is in communication with both the sense organs and muscles, it allows us to respond to outside stimuli. The endocrine system contains the hormonal glands. The nervous and endocrine systems coordinate and regulate the activities of the body's other systems.

Maintenance of the Body

The internal environment of the body is the blood within the blood vessels and the tissue fluid that surrounds the cells. Five systems add substances to and/or remove substances from the blood: the circulatory, lymphatic, respiratory, digestive, and urinary systems.

The **circulatory system,** discussed in chapter 12, consists of the heart and the blood vessels that carry blood through the body. Blood transports nutrients and oxygen to the cells, and removes waste molecules to be excreted from the body. Blood also contains cells produced by the **lymphatic system,** discussed in chapter 13. The lymphatic system protects the body from disease.

The **respiratory system,** discussed in chapter 14, consists of the lungs and the tubes that take air to and from the lungs. The respiratory system brings oxygen into the lungs and takes carbon dioxide out of the lungs.

The **digestive system** (see fig. 1.1), discussed in chapter 15, consists of the mouth, esophagus, stomach, small intestine, and large intestine (colon), along with the accessory organs: teeth, tongue, salivary glands, liver, gall bladder, and pancreas. This system receives food and digests it into nutrient molecules, which can enter the cells of the body.

Medical Focus

Organic Donation

*t*ransplantation of the kidney, heart, liver, pancreas, lung, and other organs is now possible due to two major break-throughs. First, solutions have been developed that preserve donor organs for several hours. This made it possible for one young boy to undergo surgery for 16 hours, during which time he received five different organs. Second, rejection of trans-planted organs is now prevented by immunosuppressive drugs; therefore, organs can be donated by unrelated individu-als, living or dead. Living individuals can donate one kidney, a portion of their liver, and certainly bone marrow, which quickly regenerates.

After death, it is still possible to give the "gift of life" to someone else—over 25 organs and tissues from the same per-son can be used for transplants at that time. A liver transplant, for example, can save the life of a child born with biliary atre-sia, a congenital defect in which the bile ducts do not form. Dr. Thomas Starzl, a pioneer in this field, reports a 90% chance of complete rehabilitation among children who survive a liver transplant (fig. 1A). (He has also tried animal-to-human liver transplants, but so far, these have not been successful.) So many heart recipients are now alive and healthy that they have formed basketball and softball teams, demonstrating the nor-malcy of their lives after surgery.

One problem persists. Although the number of individu-als waiting for organs is greater than ever, only a small per-centage of people signify their willingness to donate organs at the time of their death. Organ and tissue donors must sign a donor card and carry it at all times (fig. 1B). In many states, the back of the driver's license acts as a donor card. Age is no draw-back, but the donor should have been in good health prior to death.

Organ and tissue donation does not interfere with funeral arrangements, and most religions do not object to the dona-tion. Family members should know ahead of time about the desire to become a donor because they will be asked to sign permission papers at the time of death. No money is received for the gift organs, which are removed by a team of surgeons from the nearest organ procurement center.

The United Network for Organ Sharing (UNOS), based in Richmond, Virginia, was established after the 1984 National Organ Transplant Act and has a computerized system for matching needy patients with available organs. The patients are ranked according to various medical criteria, and UNOS notifies the appropriate hospital of organ availability. Donor and recipient identities are confidential.

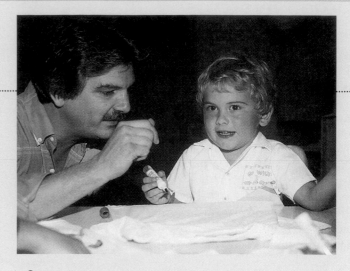

▲ **figure 1A** Transplant recipient. Children receiving a liver transplant to cure a birth defect have a good chance of survival and complete recovery.

ORGAN DONOR CARD

_____
Print or type name of donor

In the hope that I may help others, I hereby make this anatomical gift, if medically accept-able, to take effect upon my death. The words and marks below indicate my desires.

I give: (a) _____ any needed organs or parts

           (b) _____ only the following organs or parts

_____
Specify the organ(s) or part(s)

for the purposes of transplantation, therapy, medical research or education;

           (c) _____ my body for anatomical study if needed.

Limitations or special wishes, if any: _____

Signed by the donor and the following witnesses in the presence of each other:

_____   _____
Signature of Donor   Date of Birth of Donor

_____   _____
Date Signed   City & State

_____   _____
Witness   Witness

This is a legal document under the Uniform Anatomical Gift Act or similar laws. For further information consult your physician or

UNOS P.O. Box 13770 Richmond, Virginia 23225-8770

▲ **figure 1B** Organ donor card. These cards give the name of the donor and witnesses, one of whom should be a close family member.

Today over 27,000 Americans are waiting for a gift of life. Will they wait in vain?

Source: Data from Arkansas Regional Organ Recovery Agency (AURORA), Little, Rock, AR.

figure 1.8 Organ systems. The systems of the body work together to maintain homeostasis (the relative constancy of the body's internal environment).

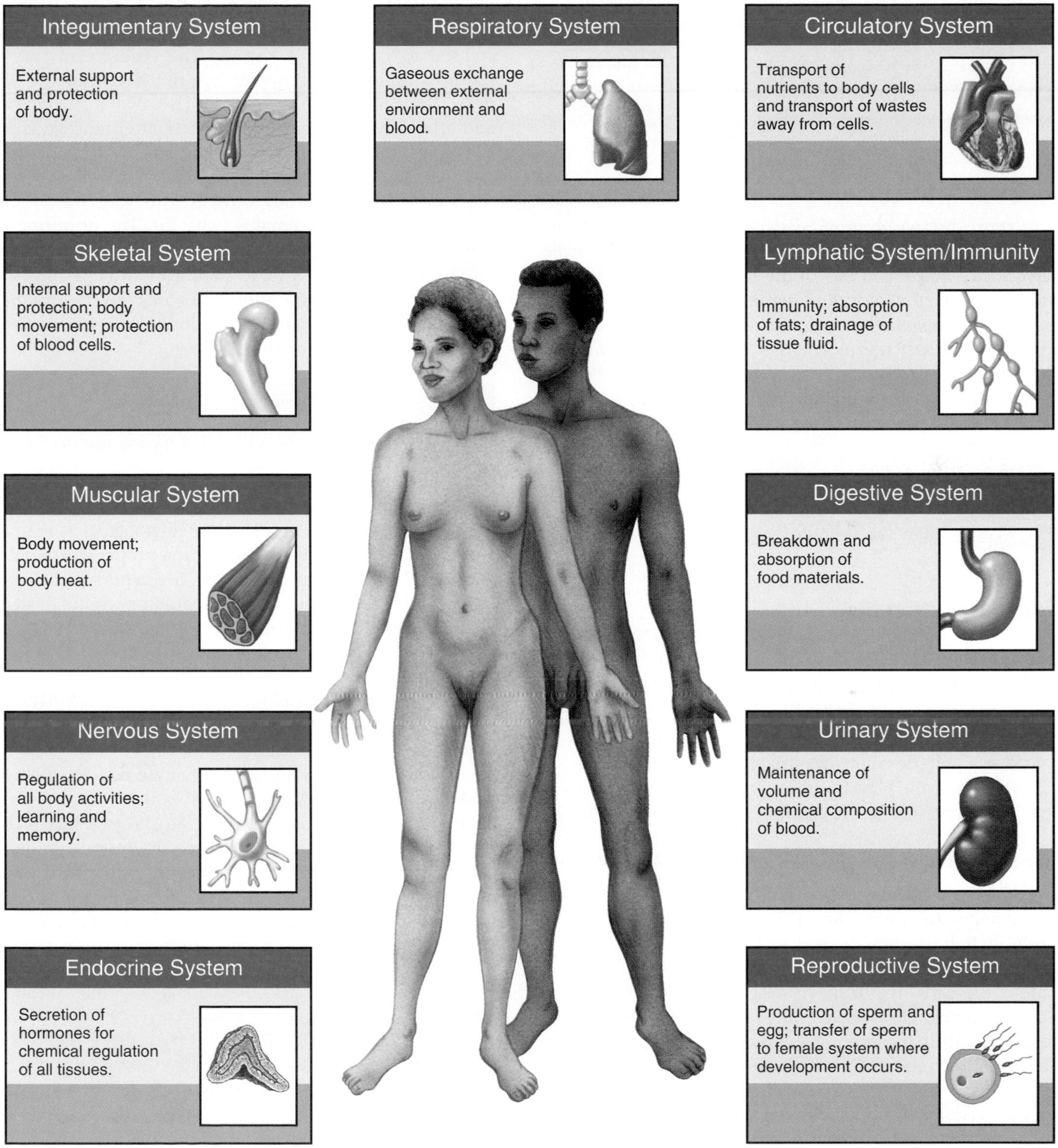

Integumentary System

External support and protection of body.

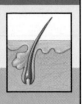

Skeletal System

Internal support and protection; body movement; protection of blood cells.

Muscular System

Body movement; production of body heat.

Nervous System

Regulation of all body activities; learning and memory.

Endocrine System

Secretion of hormones for chemical regulation of all tissues.

Respiratory System

Gaseous exchange between external environment and blood.

Circulatory System

Transport of nutrients to body cells and transport of wastes away from cells.

Lymphatic System/Immunity

Immunity; absorption of fats; drainage of tissue fluid.

Digestive System

Breakdown and absorption of food materials.

Urinary System

Maintenance of volume and chemical composition of blood.

Reproductive System

Production of sperm and egg; transfer of sperm to female system where development occurs.

The **urinary system,** discussed in chapter 16, contains the kidneys and the urinary bladder. This system rids the body of nitrogenous wastes and helps regulate the fluid level and chemical content of the blood.

> The circulatory system (heart and vessels), lymphatic system, respiratory system (lungs and conducting tubes), digestive system (mouth, esophagus, stomach, small and large intestines, and associated organs), and the urinary system (kidneys and bladder) all perform specific processing and transporting functions to maintain the normal conditions of the body.

Reproduction and Development

The **reproductive system,** discussed in chapter 17, involves different organs in the male and female. The *male reproductive system* consists of the testes, other glands, and various ducts that conduct semen to and through the penis. The *female reproductive system* consists of the ovaries, oviducts, uterus, vagina, and external genitalia. Both systems produce sex cells, but in addition, the female system receives the sex cells of the male and also nourishes and protects the fetus until the time of birth.

> The reproductive system in males (testes, other glands, ducts, and penis) and in females (ovaries, oviducts, uterus, vagina, and external genitalia) carries out those functions that give humans the ability to reproduce.

Homeostasis

Homeostasis (ho"me-o-sta′sis) means that the human body's internal environment remains relatively constant, regardless of the conditions in the external environment. For example:

1. Blood glucose concentration remains at about 0.1%.
2. The pH of the blood is always near 7.4 (see chapter 2).
3. Blood pressure in the brachial artery averages near 120/80.
4. Blood temperature averages around 37°C (98.6°F).

Although we are accustomed to using the word *environment* to mean the external environment of the body, it is important to realize that the body's internal environment, consisting of blood and the tissue fluid that bathes the cells, is ultimately responsible for our health and well-being. The body's ability to keep the internal environment within a certain range allows humans to live in a variety of habitats, such as arctic, desert, or tropical regions.

> Homeostasis is the relative constancy of the body's internal environment, which is composed of blood and the tissue fluid that bathes the cells.

Most systems of the body contribute to maintaining a constant internal environment. The digestive system takes in and digests food, providing nutrient molecules that enter the blood and replace the nutrients constantly being used up by body cells. The respiratory system adds oxygen to the blood and removes carbon dioxide. The amount of oxygen taken in and carbon dioxide given off can be increased to meet bodily needs. The chief regulators of blood composition, however, are the liver and the kidneys. They monitor the blood's chemical composition and alter it as required. Immediately after glucose enters the blood, the liver can remove the glucose for storage as glycogen. Later, glycogen can be broken down to replace the glucose used by body cells; in this way, the blood's glucose composition remains constant. The hormone insulin, secreted by the pancreas, regulates glycogen storage. The liver also removes toxic chemicals, such as ingested alcohol and drugs. These substances are converted to molecules that can be excreted by the kidneys, organs that are also under hormonal control.

> All of the body's organ systems contribute to homeostasis. Some, like the respiratory, digestive, and urinary systems, remove and/or add substances to blood.

Although homeostasis is, to a degree, controlled by hormones, the nervous system has ultimate control. The brain contains centers that regulate such factors as temperature and blood pressure. Maintaining proper temperature and blood pressure levels requires a receptor that detects unacceptable levels and signals a regulatory center. If a correction is required, the regulatory center then turns on an effector. The effector brings about a response that negates the original conditions that stimulated the receptor. In the absence of suitable stimulation, the receptor no longer signals the regulatory center (fig. 1.9). Therefore, this is called a **negative feedback** mechanism.

A useful analogy that helps illustrate how such a negative feedback mechanism works is the control of the heating system of a house by a thermostat (fig. 1.10). This negative feedback mechanism keeps the indoor temperature within a relatively narrow range. In a similar way, body temperature in humans is controlled by a portion of the brain that functions much like a thermostat, allowing slight fluctuations within narrow limits.

> The nervous and endocrine systems regulate the activities of other systems. Negative feedback is a self-regulatory mechanism by which systems and conditions of the body are controlled.

Disease represents a breakdown or upset in this normal self-regulation. When homeostasis fails, the body (or part of the body) no longer functions properly. The effects may be limited or widespread. A *local disease* is more or

figure 1.9 Negative feedback mechanism. A stimulus causes a sensory receptor to signal a regulatory center in the brain. The regulatory center signals an effector to respond, and the response cancels the stimulus.

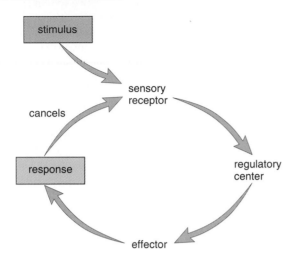

figure 1.10 Mechanical negative feedback mechanism. A thermostat signals a furnace to turn on or off, thereby maintaining a relatively stable room temperature.

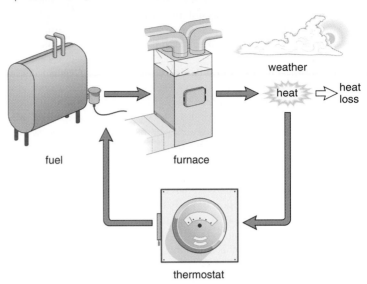

less restricted to a specific part of the body. On the other hand, a **systemic disease** affects the entire body or involves several organ systems. Diseases may also be classified on the basis of their severity and duration. **Acute diseases** occur suddenly and generally last a short time.

Chronic diseases tend to be less severe, develop slowly, and are long-term.

> Disease represents an upset in homeostasis.

Selected New Terms

Basic Key Terms

abdominal cavity (ab-dom′ĭ-nal kav′ĭ-te), p. 6
abdominopelvic cavity (ab-dom″ĭ-no-pel′vik kav′ĭ-te), p. 6
anatomy (ah-nat′o-me), p. 2
cranial cavity (kra′ne-al kav′ĭ-te), p. 6
distal (dis′tal), p. 4
frontal plane (frun′tal plān), p. 5
homeostasis (ho″me-o-sta′sis), p. 12
lateral (lat′er-al), p. 4
lumbar (lum′bar), p. 6
medial (me′de-al), p. 4
mediastinum (me″de-ah-sti′num), p. 6
negative feedback (neg′ah-tiv fēd′bak), p. 12
pelvic cavity (pel′vik kav′ĭ-te), p. 6

pericardial cavity (per″ĭ-kar′de-al kav′ĭ-te), p. 6
physiology (fiz″e-ol′o-je), p. 2
pleural cavity (ploo′ral kav′ĭ-te), p. 6
proximal (prok′sĭ-mal), p. 4
sagittal plane (saj′ĭ-tal plān), p. 5
spinal cavity (spi′nal kav′ĭ-te), p. 6
thoracic cavity (tho-ras′ik kav′ĭ-te), p. 6
transverse plane (trans-vers′ plān), p. 5
visceral (vis′er-al), p. 6

Clinical Key Terms

disease (dĭ-zez′), p. 12
organ transplantation (or′gan trans-plan-ta′shun), p. 9
systemic disease (sis-tem′ik dĭ-zēz), p. 13

Summary

I. **Anatomy and Physiology**
Anatomy is the study of the structure of body parts, and physiology is the study of the function of these parts. Structure is suited to the function of a part.

II. **Organization of Body Parts**
The body has levels of organization that progress from atoms to molecules, macromolecules (biomolecules), cells, tissues, organs, organ systems, and finally, the organism.

III. **Anatomical Terms**
Various terms are used to describe the location and position of body organs when the body is in the anatomical position (standing erect, with face forward, arms at the sides, and palms and toes directed forward).
   A. The terms *superior/inferior, anterior/posterior, medial/lateral, proximal/distal, superficial/deep,* and *central/peripheral* describe the relative positions of body parts.
   B. The body or its parts may be sectioned (cut) along certain imaginary planes. A sagittal (vertical) cut divides the body into right and left portions. A transverse (horizontal) cut is a cross section. A frontal (coronal) cut divides the body into anterior and posterior parts.
   C. The body can be divided into axial and appendicular portions, each of which can be further subdivided into specific regions. For example, *brachial* refers to the upper arm, and *pedal* refers to the foot.

D. The human body has two major cavities: the dorsal cavity and the ventral cavity. Each is subdivided into smaller cavities, within which specific visceral organs are located.

IV. **Organ Systems**
A number of systems function to maintain the normal conditions of the body. These systems have been categorized as follows:
   A. Integumentary system. This system, which includes the skin, not only protects the body, but also has other functions.
   B. Support and movement. The skeletal system contains the bones, and the muscular system contains the three types of muscles. The primary function of these systems is support and movement, but they have other functions as well.
   C. Integration and coordination. The nervous system contains the brain, spinal cord, and nerves. Because the nervous system is in communication with both the sense organs and muscles, it allows us to respond to outside stimuli. The endocrine system contains the hormonal glands. The nervous and endocrine systems coordinate and regulate the activities of the body's other systems.
   D. Maintenance of the body. The circulatory system (heart and vessels), lymphatic system (lymphatic vessels and nodes,

spleen, and thymus), respiratory system (lungs and conducting tubes), digestive system (mouth, esophagus, stomach, small and large intestines, and associated organs), and the urinary system (kidneys and bladder) all perform specific processing and transporting functions to maintain the normal conditions of the body.
   E. Reproduction and development. The reproductive system in males (testes, other glands, ducts, and penis) and in females (ovaries, oviducts, uterus, vagina, and external genitalia) carries out those functions that give humans the ability to reproduce.

V. **Homeostasis**
Homeostasis is the relative constancy of the body's internal environment, which is composed of blood and the tissue fluid that bathes the cells.
   A. All of the body's organ systems contribute to homeostasis. Some, like the respiratory, digestive, and urinary systems, remove and/or add substances to blood.
   B. The nervous and endocrine systems regulate the activities of other systems. Negative feedback is a self-regulatory mechanism by which systems and conditions of the body are controlled.
   C. Disease represents an upset in homeostasis.

Study Questions

1. Distinguish between the study of anatomy and the study of physiology. (p. 2)
2. Give an example that shows the relationship between the structure and function of body parts. (p. 2)
3. List the levels of organization within the human body in reference to a specific organ. (p. 2)
4. Distinguish between a midsagittal cut, a transverse cut, and a frontal cut. (pp. 4–5)
5. Distinguish between the axial and appendicular portions of the body. State at least two anatomical terms that pertain to the head, thorax, abdomen, and limbs. (p. 5)
6. Distinguish between the dorsal and ventral body cavities, and name two smaller cavities that occur within each. (p. 6)
7. Name the four quadrants of the abdominopelvic cavity. Use the following terms to divide the

abdominopelvic cavity into nine regions: *epigastric, umbilical, hypogastric, hypochondriac, lumbar,* and *iliac.* (p. 6)
8. Name the major organ systems, and describe the general functions of each. (pp. 9, 11–12)
9. List the major organs found within each organ system. (pp. 9, 11–12)
10. Define homeostasis, and explain its importance. (p. 12)

Objective Questions

I. Match the terms in the key to the relationships given in questions 1–5.
Key:
- a. superior
- b. inferior
- c. anterior
- d. posterior
- e. medial
- f. lateral
- g. proximal
- h. distal

1. the esophagus in relation to the stomach
2. the ears in relation to the nose
3. the shoulder in relation to the hand
4. the intestines in relation to the vertebrae
5. the rectum in relation to the mouth

II. Match the terms in the key to the body regions listed in questions 6–12.
Key:
- a. oral
- b. occipital
- c. gluteal
- d. cutaneous
- e. palmar
- f. cervical
- g. axillary

6. buttocks
7. palm
8. back of head
9. mouth
10. skin
11. armpit
12. neck

III. Match the terms in the key to the organs listed in questions 13–18.
Key:
- a. cranial cavity
- b. spinal cavity
- c. thoracic cavity
- d. abdominal cavity
- e. pelvic cavity

13. stomach
14. heart
15. urinary bladder
16. brain
17. liver
18. small intestine

IV. Match organ systems in the key to the organs listed in questions 19–25.
Key:
- a. digestive system
- b. urinary system
- c. respiratory system
- d. circulatory system
- e. reproductive system
- f. nervous system
- g. endocrine system

19. thyroid gland
20. lungs
21. heart
22. ovaries
23. brain
24. stomach
25. kidneys

V. Fill in the blanks.

26. A(n) _____ is composed of several types of tissues and performs a particular function.
27. The imaginary plane that passes through the midline of the body is called the _____ plane.
28. All the organ systems of the body together function to maintain _____ , a relative constancy of the internal environment.

Medical Terminology Reinforcement Exercise

Consult Appendix B for help in pronouncing, analyzing, and filling in the blanks to give a brief meaning to the terms that follow.

1. Suprapubic (soo″prah-pu′bik) means _____ the pubis.
2. Infraorbital (in″fra-hor′bi-tal) means _____ the eye orbit.
3. Gastrectomy (gas-trek′to-me) means excision of the _____ .
4. Celiotomy (se″le-ot′o-me) means incision (cut into) of the _____ .
5. Macrocephalus (mak″ro-sef′ah-lus) means large _____ .
6. Transthoracic (trans″tho-ras′ik) means across the _____ .
7. Bilateral (bi-lat′er-al) means two or both _____ .
8. Ophthalmoscope (of-thal′mo-skōp) is an instrument to view inside the _____ .
9. Dorsalgia (dor-sal′je-ah) means pain in the _____ .
10. Endocrinology (en″do-kri-nol′o-je) is the _____ of the endocrine system (secretions within).

Website Link

For a listing of the most current websites related to this chapter, please visit the Mader home page at:

www.mhhe.com/maderap

chapter 2

Chemistry of Life

Chapter Outline and Learning Objectives

After you have studied this chapter, you should be able to:

■ Explain why chemistry is pertinent to a study of the body.

Elements and Atoms (p. 17)
■ Name and describe the subatomic particles of an atom, and indicate which one accounts for the occurrence of isotopes.

Molecules and Compounds (p. 18)
■ Distinguish between ionic and covalent reactions, and between ionic and covalent bonds.
■ List and discuss the functions of ions in the body.

Some Important Inorganic Molecules (p. 20)
■ Describe the structure of water, and give examples of how it functions in the body.
■ Relate the term *electrolyte* to the presence of ions in body fluids and tissues.
■ Define the terms *acid* and *base*. Describe the pH scale, and explain the significance of buffers.

Some Important Organic Molecules (p. 23)
■ Compare and contrast the structures and functions of carbohydrates, lipids, proteins, and nucleic acids.
■ Explain what enzymes are, and describe their role in the body.
■ Describe how the structure and function of DNA and RNA differ.
■ Describe the structure of ATP, and explain how ATP functions in the body.

Medical Focus
Imaging the Body (p. 19)
Nutrition Labels (p. 26)

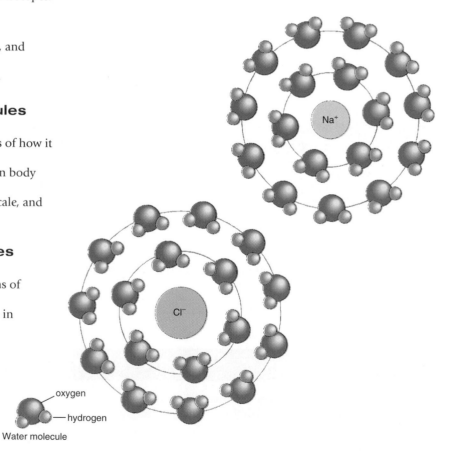

oxygen

hydrogen

Water molecule

Strange as it may seem, life is dependent on the chemistry of water, a small molecule that contains one atom of oxygen and two atoms of hydrogen. Water helps substances like salt (NaCl) dissolve.

figure 2.1 Periodic table of the elements (shortened). Each element has an atomic number, an atomic symbol, and an atomic weight. The elements in color make up most (about 98%) of the body weight of organisms.

I	II	III	IV	V	VI	VII	VIII
1 **H** Hydrogen 1						atomic number — 2 atomic symbol — **He** Helium atomic weight — 4	
3 **Li** Lithium 7	4 **Be** Beryllium 9	5 **B** Boron 11	6 **C** Carbon 12	7 **N** Nitrogen 14	8 **O** Oxygen 16	9 **F** Fluorine 19	10 **Ne** Neon 20
11 **Na** Sodium 23	12 **Mg** Magnesium 24	13 **Al** Aluminum 27	14 **Si** Silicon 28	15 **P** Phosphorus 31	16 **S** Sulfur 32	17 **Cl** Chlorine 35	18 **Ar** Argon 40
19 **K** Potassium 39	20 **Ca** Calcium 40						

A few minutes of reflection on the body's dietary needs will convince us that humans are indeed made of chemicals. For example, the body needs calcium to maintain bones, iron to prevent anemia, and adequate amino acid intake to build muscles.

Because the human body is composed only of chemicals, anatomy and physiology students must have a basic understanding of chemistry.

Elements and Atoms

All matter, living and nonliving, is composed of **elements.** While there are 92 naturally occurring elements, only six elements—carbon, hydrogen, nitrogen, oxygen, phosphorus, and sulfur—make up most (about 98%) of the body weight of humans (fig. 2.1). The acronym CHNOPS helps us remember these six elements. Scientists use symbols to identify elements. For example, the letter *C* stands for carbon, and the letter *N* stands for nitrogen.

Elements contain tiny particles called **atoms.** Only one type of atom is present in each element, and therefore, the same name is used for both. An atom is the smallest unit of matter to enter into chemical reactions. Even though extremely small, atoms contain even smaller subatomic particles called protons, neutrons, and electrons. Figure 2.2 shows a model of a carbon atom. A central nucleus contains the protons and neutrons, while the electrons are located outside the nucleus in shells. The shells represent energy levels. The innermost shell has the lowest energy level and can hold two electrons. The other shell of a carbon atom can hold eight electrons. An atom is most stable when the outermost shell has eight electrons.

An atom has an atomic number that is equal to its number of protons. Notice in table 2.1 that protons have a positive (+) electrical charge and that electrons have a negative (−) charge. When an atom is electrically neutral, the number of protons equals the number of electrons. The carbon atom shown in figure 2.2 has an atomic number of six; therefore, it has six protons. Since it is electrically neutral, it also has six electrons.

table 2.1

Subatomic Particles		
Name	**Charge**	**Weight**
Electron	One negative unit	Almost no weight
Proton	One positive unit	One atomic mass unit
Neutron	No charge	One atomic mass unit

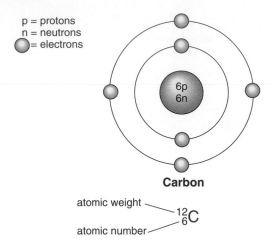

figure 2.2 Carbon atom. A carbon atom has six protons and six neutrons; therefore its atomic weight is 12 atomic mass units. The atomic number of carbon is six; therefore, it has six protons and six electrons.

p = protons
n = neutrons
= electrons

Carbon

atomic weight
atomic number
$^{12}_{6}C$

Isotopes: Atomic Weights Vary

Subatomic particles are so light that their weight is indicated by special units called atomic mass units (see table 2.1). The atomic weight, or mass, of each atom is noted in the periodic table beneath the atomic symbol (see figure 2.1). The atomic weight of an atom equals the number of protons plus the number of neutrons, since electrons have almost no weight, while protons and neutrons each have a weight of one atomic mass unit. The atomic weight of carbon is 12, and since carbon's atomic number (number of protons) is six, it is easy to calculate that carbon has six neutrons.

The atomic weights given in a periodic table represent the average weight for each kind of atom. Actually, atoms of the same type may differ in the number of neutrons and, therefore, in their weight. Atoms that have the same atomic number and differ only in the number of neutrons are called **isotopes** (i'so-tōps). Isotopes of carbon can be written in the following manner, where the subscript stands for the atomic number and the superscript stands for the atomic weight:

$$^{12}_{6}C \quad ^{13}_{6}C \quad ^{14}_{6}C*$$

*radioactive

Carbon 12 has six neutrons, carbon 13 has seven neutrons, and carbon 14 has eight neutrons.

Certain isotopes, called **radioactive isotopes,** are unstable—they break apart to form smaller atoms and release radiation in the process. This radiation can be detected with a special counter or scanner. Among those isotopes of carbon listed in the previous paragraph, only carbon-14 is radioactive. Radioactive isotopes are widely used in medical diagnostic procedures. For example, because the thyroid gland utilizes iodine, radioactive iodine can be administered to allow a physician to determine the thyroid's activity level. Other types of radioactive isotopes are used in order to form images of internal organs, as discussed in the Medical Focus reading on page 19.

All matter is composed of elements, each containing just one type of atom. Each atom has an atomic symbol, atomic number (number of protons), and atomic weight (number of protons and neutrons). The isotopes of some atoms are radioactive.

Molecules and Compounds

Most atoms do not have enough electrons to fill their outer energy shells. These atoms bond with one another. After bonding occurs, a **molecule** results, in which each atom has enough electrons to fill its outer energy shell. A molecule can contain atoms of the same kind, as when an oxygen atom joins with another oxygen atom to form oxygen gas. Or the atoms can be different, as when an oxygen atom joins with two hydrogen atoms to form water. When the atoms are different, a **compound** results.

Two types of bonds join atoms: **ionic** (i-on'nik) **bonds** and **covalent** (ko-va'lent) **bonds.** During an ionic reaction, atoms give up or take on electrons in order to achieve a stable outermost energy shell. Figure 2.3 depicts a reaction between a sodium (Na) and chlorine (Cl) atom in which chlorine takes an electron from sodium. **Ions** are particles that carry either a positive (+) or negative (−) charge. The sodium ion carries a plus charge because it now has one more proton than electrons, and the chloride ion carries a negative charge because it now has one less proton than electrons. The attraction between oppositely charged sodium and chloride ions forms an ionic bond. Ionic bonds typically are found in inorganic compounds.

We are quite familiar with the inorganic compound sodium chloride because it is table salt. Intake of too much table salt may be associated with the occurrence of **hypertension** (high blood pressure) in some persons. When sodium is at a higher than normal concentration in blood, the kidneys retain water, and this causes blood pressure to rise.

Besides sodium and chloride ions, many other ions are significant in the human body (table 2.2). Calcium (Ca^{2+}) is needed for strong bones, potassium (K^+) is required for muscle contraction, including the beating of the heart, and lack of sufficient iron (Fe^{2+}) in the blood is called **iron deficiency anemia.** Anemia is characterized by a tired feeling and is due to a low red blood cell count or to an insufficient amount of hemoglobin in red blood cells. Iron is a part of the hemoglobin molecule that transports oxygen.

In another type of reaction, called a *covalent reaction,* atoms form a molecule by sharing electrons. The bond

Imaging the Body

X rays, which are produced when high-speed electrons strike a heavy metal, have long been used to image body parts. Dense structures like bone absorb X rays well and show up as light areas, whereas soft tissues absorb X rays to a lesser extent and show up as dark areas on photographic film. During CAT (computerized axial tomography) scans, X rays are sent through the body at various angles, and a computer uses the X-ray information to form a series of cross sections (fig. 2A). CAT scanning has all but eliminated the need for exploratory surgery.

PET (positron emission tomography) is a variation on CAT scanning. Radioactively labeled substances are injected into the body; metabolically active tissues tend to take up these substances and then emit gamma rays. A computer uses the gamma-ray information to again generate cross-sectional images of the body, but this time, the image indicates metabolic activity, not structure. PET scanning is used to diagnose brain disorders, such as a brain tumor, Alzheimer disease, epilepsy, or whether a stroke has occurred.

During MRI (magnetic resonance imaging), the patient lies in a massive, hollow, cylindrical magnet and is exposed to short bursts of a powerful magnetic field. This causes the protons in the nuclei of hydrogen atoms to align. Then, when exposed to strong radio waves, the protons move out of alignment and produce signals. A computer changes these signals into an image (fig. 2B). Tissues with many hydrogen atoms (such as fat) show up as bright areas, while tissues with few hydrogen atoms (such as bone) appear black. This is the opposite of an X ray, which is why MRI is more useful than an X ray for imaging soft tissues. However, many people cannot undergo MRI, since the magnetic field can actually pull a metal object out of the body, such as a tooth filling, a prosthesis, or a pacemaker!

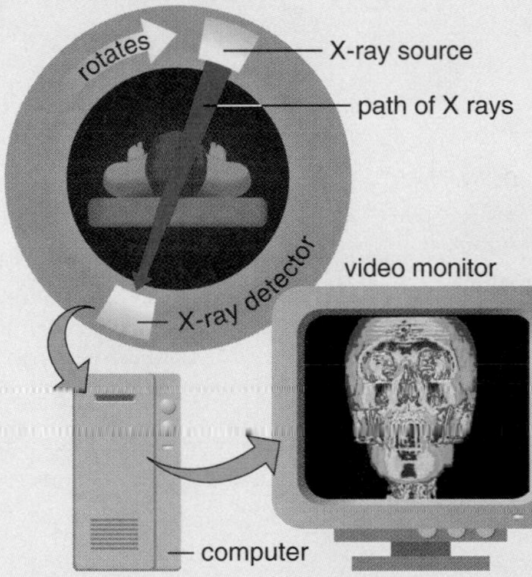

figure 2A CAT scan.

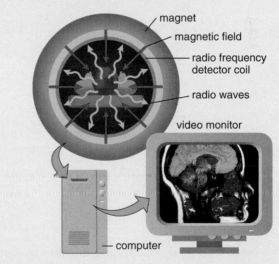

figure 2B MRI.

figure 2.3 Formation of the salt sodium chloride (NaCl). During this ionic reaction, an electron is transferred from the sodium atom to the chlorine atom. Both resulting ions carry a charge, as shown.

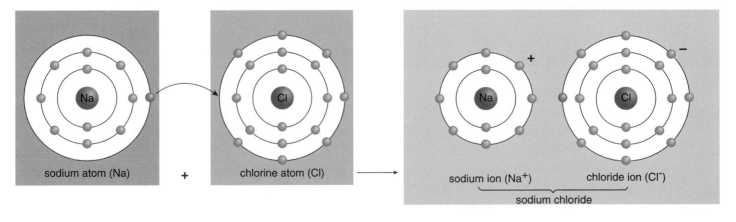

sodium atom (Na) + chlorine atom (Cl) → sodium ion (Na$^+$) chloride ion (Cl$^-$)

sodium chloride

table 2.2

Significant Ions in the Body

Name	Symbol	Special Significance
Sodium	Na^+	Found in body fluids; important in muscle contraction and nerve conduction
Chloride	Cl^-	Found in body fluids
Potassium	K^+	Found primarily inside cells; important in muscle contraction and nerve conduction
Phosphate	PO_4^{3-}	Found in bones, teeth, and the high-energy molecule ATP
Calcium	Ca^{2+}	Found in bones and teeth; important in muscle contraction
Iron	Fe^{2+}	Found in hemoglobin; combines with oxygen
Bicarbonate	HCO_3^-	
Hydrogen	H^+	Important in acid-base balance
Hydroxide	OH^-	
Ammonium	NH_4^+	

figure 2.4 Formation of water. Following a covalent reaction, oxygen is sharing electrons with two hydrogen atoms.

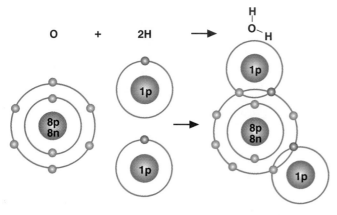

oxygen atom (O) + hydrogen atoms (2H) ⟶ water (H_2O)

figure 2.5 Hydrogen bonding between water molecules. The polarity of the water molecules allows hydrogen bonds (dotted lines) to form between the molecules.

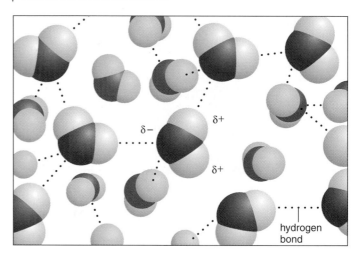

hydrogen bond

that forms when atoms share electrons is called a covalent bond. For example, when oxygen reacts with two hydrogen atoms, water (H_2O) is formed (fig. 2.4).

Atoms react with one another to form molecules. Following one type of reaction, positively and negatively charged ions form a molecule in which they are ionically bonded. Following another type of reaction, atoms sharing electrons form a molecule in which they are covalently bonded.

Some Important Inorganic Molecules

Inorganic molecules are characterized by the presence of a small number of atoms ionically bonded together (see fig. 2.3). In water, however, the atoms are covalently bonded.

Water

Water is the most abundant molecule in living organisms, making up 60–70% of total body weight. Water's physical and chemical properties make life, as we know it, possible.

Sometimes, covalently bonded atoms share electrons unevenly; that is, the electrons spend more time circling the nucleus of one atom than the other. In water, the electrons spend more time circulating the larger oxygen (O) atom than the smaller hydrogen (H) atoms. As a result, the hydrogen atoms have a partial positive charge, and the oxygen atom has a partial negative charge.

Because the water molecule has charged atoms, it is called a polar molecule. Polar molecules are similar to magnets in that they have positive and negative poles. Hydrogen bonding occurs between water molecules because they are polar (fig. 2.5). A **hydrogen bond** occurs whenever a covalently bonded hydrogen is attracted to a negatively charged atom some distance away. The hydrogen bond is represented by a dotted line in figure 2.5 because the bond is relatively weak and can be broken. Polar molecules like water do not form hydrogen bonds with and are unattracted to nonpolar molecules.

> In some covalent bonds, the electrons are shared unequally, and the result is a polar molecule. Hydrogen bonding can occur between polar molecules.

Characteristics of Water

Hydrogen bonds cause water molecules to be cohesive—to cling together. Without hydrogen bonding between molecules, water would easily boil and easily freeze, making life impossible. Instead, water is a liquid at body temperature. It absorbs a great deal of heat before evaporating and releases this heat as it cools down. This property helps to keep body temperature within normal limits and even accounts for the cooling effect of sweating. The cohesiveness of water also allows it to fill tubular vessels, which makes water an excellent transport medium for distributing substances and heat throughout the body.

Water, being a polar molecule, acts as a solvent and dissolves various chemical substances, particularly other polar molecules. This property of water greatly facilitates chemical reactions in cells.

> Because of hydrogen bonding, water heats up and cools down slowly, and this helps keep body temperature within normal limits. Water is a polar molecule and acts as a solvent; it dissolves various chemical substances and facilitates chemical reactions.

Dissociation

Polarity also causes water molecules to tend to **ionize,** or split up, in the following manner:

$$\text{H—O—H} \quad \rightleftharpoons \quad \text{H}^+ \quad + \quad \text{OH}^-$$

water   hydrogen   hydroxide
            ion   ion

The hydrogen ion (H^+) has lost an electron; the hydroxide ion (OH^-) has gained the electron. Because very few molecules actually dissociate, few H^+ and OH^- result.

Electrolytes

Salts, acids, and bases are compounds that dissociate; that is, they ionize in water. For example, when a salt such as sodium chloride (NaCl) is put into water, the negative ends of the water molecules are attracted to the sodium ions, and the positive ends of the water molecules are attracted to the chloride ions. This causes the

sodium chloride to break apart, or dissociate, into individual ions:

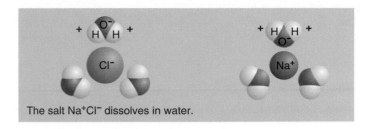

The salt Na⁺Cl⁻ dissolves in water.

Substances that break apart and release ions when put into water are called **electrolytes** (e-lek′tro-līts), since the ions can conduct an electrical current. The electrolyte balance in the blood and body tissues is important for good health because it affects the functioning of vital organs, such as the heart and the brain.

> Substances like salts, acids, and bases that dissociate (that is, release ions when in water) are called electrolytes. The electrolyte balance in the blood and body tissues is important for good health.

Acids and Bases

When water ionizes, it releases an equal number of hydrogen ions and hydroxide ions. Only a few water molecules at a time are dissociated, and the actual number of ions is very small (0.0000001 of the water molecules in a pure sample have dissociated).

Lemon juice, vinegar, tomatoes, and coffee are all familiar acids. **Acids** (as′idz) are molecules that dissociate in water, releasing hydrogen ions. For example, hydrochloric acid (HCl) dissociates in the following manner:

$$\text{HCl} \rightarrow \text{H}^+ + \text{Cl}^-$$

Because almost all hydrochloric acid molecules dissociate in water, hydrochloric acid is called a strong acid. If hydrochloric acid is added to a beaker of water, the number of hydrogen ions increases.

Milk of magnesia and ammonia are common bases. **Bases** are molecules that either take up hydrogen ions or release hydroxide ions. For example, sodium hydroxide (NaOH) dissociates in the following manner:

$$\text{NaOH} \rightarrow \text{Na}^+ + \text{OH}^-$$

Because almost all sodium hydroxide molecules dissociate in water, sodium hydroxide is called a strong base. If sodium hydroxide is added to a beaker of water, the number of hydroxide ions increases. These hydroxide ions can combine with hydrogen ions to form water. Thus, when a base is added to water, the number of hydrogen ions decreases.

figure 2.6 pH scale. The diagonal line indicates the proportionate number of hydrogen ions to hydroxide ions. Any pH value above 7 is basic, while any pH value below 7 is acidic.

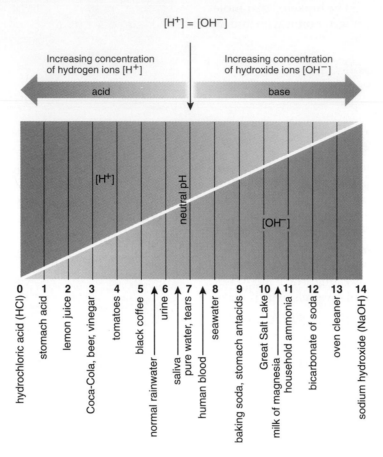

pH

The **pH** scale ranges from 0–14. Any pH value below 7 is acidic, with the degree of acidity increasing toward the lower numbers. Any pH value above 7 is basic (or alkaline), with the degree of basicity increasing toward the higher numbers. A pH of exactly 7 is neutral. Water has an equal number of H^+ and OH^-, and therefore, one ion of each is formed when water dissociates. The fraction of water molecules that dissociate is 10^{-7} (0.0000001), which is the source of the pH value for neutral solutions. The pH scale was devised to simplify discussion of the hydrogen ion concentration $[H^+]$ without using cumbersome numbers. For example:

1. $1 \times 10^{-6}[H^+] = $ pH 6   Each lower pH unit has 10
2. $1 \times 10^{-7}[H^+] = $ pH 7   times the amount of H^+ as
3. $1 \times 10^{-8}[H^+] = $ pH 8   the next higher unit.

Of the three values listed here, pH 6 has the greater number of H^+ and is acidic, pH 7 has an equal number of H^+ and OH^- and is neutral, and pH 8 has the lower number of H^+ and is basic. Figure 2.6 gives the complete pH scale with proper notations.

All living things need to maintain the hydrogen ion concentration $[H^+]$, or pH, at a constant level. For example, the pH of the blood is held constant at about 7.4, or a person becomes ill. If this pH value drops below 7.35, the person is said to have **acidosis** (as″ĭ-do′sis); if it rises above 7.45, the condition is called **alkalosis** (al″kah-lo′sis). Without medical intervention, a person usually cannot survive if the pH drops to 6.9 or rises to 7.8 for more than a few hours.

The presence of buffers helps keep the pH constant. A **buffer** is a chemical or a combination of chemicals that can take up excess H^+ or excess OH^-. When an acid is added to a buffered solution, the buffer takes up excess H^+, and when a base is added to a buffered solution, the buffer takes up excess OH^-. Therefore, the pH changes minimally whenever a solution is buffered.

Acids have a pH that is less than 7, and bases have a pH that is greater than 7. The presence of buffers helps keep the pH of body fluids constant at about neutral, or pH 7, because a buffer can absorb both hydrogen ions (H^+) and hydroxide ions (OH^-).

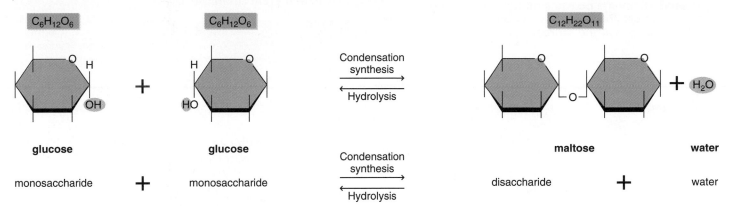

figure 2.7 Condensation synthesis and hydrolysis of maltose, a disaccharide. During condensation synthesis, a bond forms between the two glucose molecules, and the components of water are removed. During hydrolysis, the components of water are added, and the bond is broken.

Some Important Organic Molecules

The chemistry of carbon accounts for the formation of the very large number of organic molecules associated with living organisms. Carbon shares electrons with as many as four other atoms. Many times, carbon atoms share electrons with each other to form rings or chains of carbon atoms. These act as a skeleton for the *biomolecules*—carbohydrates, lipids (fats), proteins, and nucleic acids. Biomolecules are polymers or chains of unit molecules joined together.

Carbohydrates

Carbohydrates (kar″bo-hi′drats) supply short-term but quick energy for all organisms, including humans. Carbohydrate molecules are characterized by the presence of the atomic grouping CH_2O, in which the ratio of hydrogen atoms (H) to oxygen atoms (O) is approximately 2:1.

Monosaccharides and Disaccharides

If the number of carbon atoms in a molecule is low (from three to seven), then the carbohydrate is a simple sugar, or **monosaccharide** (mon″o-sak′ah-rīd). These molecules are often designated by the number of carbon atoms they contain; for example, **glucose,** with six carbon atoms, is called a hexose. Other common monosaccharides are fructose, found in fruits, and galactose, a constituent of milk. These three monosaccharides (glucose, fructose, and galactose) all occur as ring structures with the molecular formula $C_6H_{12}O_6$, but the exact shape of the ring differs, as does the arrangement of the hydrogen and the hydroxide groups attached to the ring.

A **disaccharide** (di-sak′ah-rīd) (*di-*, two; *saccharide,* sugar) contains two monosaccharides. Synthesis of a disaccharide is a **condensation** reaction because water is produced as the two monosaccharides join to form the disaccharide (fig. 2.7). Degradation of a disaccharide is a **hydrolysis** (hi-drol′ĭ-sis) reaction because water is used to split a bond.

Maltose is a disaccharide that forms from two glucose molecules. When glucose and fructose join, the disaccharide sucrose forms. Sucrose, which is ordinarily derived from sugarcane and sugar beets, is commonly known as table sugar. Eating a candy bar provides quick energy because sucrose provides glucose. Cells use glucose as their primary energy source.

Polysaccharides

A **polysaccharide** (pol″e-sak′ah-rīd) is a carbohydrate that contains a large number of monosaccharide molecules. Three polysaccharides are common in animals and plants: glycogen, starch, and cellulose. All of these are polymers of glucose, just as a necklace might be made up of only one type of bead. Even though all three polysaccharides contain only glucose, they are distinguishable from one another.

Glycogen (gli′ko-jen), a molecule having many side branches (fig. 2.8), is the storage form of glucose in humans. After a meal, the liver stores glucose as glycogen; between meals, the liver releases glucose so that the blood glucose level is always 0.1%. If the blood contains more glucose, it spills over into the urine, signaling that the condition **diabetes** (di″ah-be′tēz) exists.

The polymers *starch* and *cellulose* are found in plants. Plants store glucose as starch, a polymer similar in structure to glycogen except that it has few side branches. Starch is an important source of glucose energy in the diet because it can be hydrolyzed to glucose by digestive enzymes. In cellulose, often called fiber, the glucose units are joined by a slightly different type of linkage from that of glycogen and starch.

figure 2.8
Glycogen structure and function. Glycogen is a highly branched polymer of glucose molecules. The branching allows breakdown to proceed at several points simultaneously. The electron micrograph shows glycogen granules in liver cells. Glycogen is the storage form of glucose in humans.

Glucose is an immediate source of energy in cells. It is the unit molecule for glycogen, starch, and cellulose. Glycogen stores energy in the body, starch is a dietary source of energy, and cellulose is fiber in the diet.

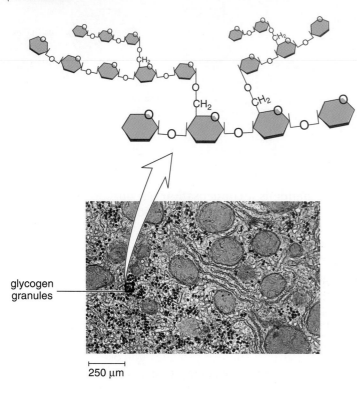

glycogen granules

250 µm

Lipids

Many **lipids** (lip'idz) are nonpolar, which means that, unlike water, their electrons are evenly distributed. This difference in structure makes lipids insoluble in water. For example, when water and oil are mixed together and set aside, they immediately begin to separate. This is true of fats, the most familiar lipids, such as lard, butter, and oil. In the body, fats act as long-term energy storage. Adipose tissue is composed of cells that contain many molecules of fat.

Fats (Triglycerides)

A fat contains two types of unit molecules: **glycerol** (glis'er-ol) and **fatty acids** (fat'e as'idz). Each fatty acid has a hydrocarbon chain (carbon atoms with hydrogens attached) that ends in an acid group (fig. 2.9). Fatty acids are either *saturated* or *unsaturated*. Saturated fatty acids have no double bonds between the carbon atoms. The carbon chain is saturated, so to speak, with all the hydrogens that can be held. Unsaturated fatty acids have double bonds in the carbon chain wherever there is only one hydrogen atom per carbon atom. Unsaturated fatty acids are most often found in vegetable oils and account for the liquid nature of these oils. Vegetable oils are hydrogenated (hydrogens are added) to make margarine. Polyunsaturated margarine contains a large number of unsaturated, or double, bonds.

figure 2.9
Fat formation from three fatty acids and glycerol. Fatty acids can be saturated (no double bonds) or unsaturated (have double bonds). When a fat molecule forms, three fatty acids combine with glycerol and three water molecules are produced.

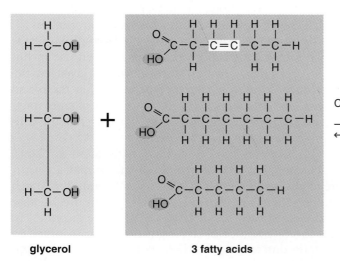

glycerol   3 fatty acids

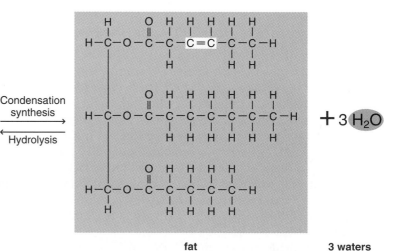

fat   3 waters

Glycerol is a compound with H — C — OH attached to three carbon atoms. When fat is formed by dehydration synthesis, the —OH groups react with the acid portions of three fatty acids so that three molecules of water are formed. The reverse of this reaction represents hydrolysis of the fat molecule into its separate components (fig. 2.9).

Phospholipids

Phospholipids, as their name implies, contain a phosphate group. Essentially, phospholipids are constructed like fats, except in place of one fatty acid there is a phosphate group, or a grouping that contains both phosphorus and nitrogen. The phosphate group carries a charge and becomes the so-called polar "head" of the molecule. The rest of the molecule becomes the nonpolar "tail." The plasma membrane (cell membrane) is a phospholipid double layer of molecules in which the polar heads are attracted to and face the water. The nonpolar tails are water repelling and face each other (see fig. 3.1).

Steroids

Steroids are lipids with a structure that differs entirely from that of fats. Steroid molecules have a backbone of four fused carbon rings, but each one differs primarily by the arrangement of the atoms in the rings and the type of functional groups attached to them. Cholesterol (fig. 2.10a) is the precursor of several other steroids, such as aldosterone, a hormone that helps to regulate the sodium level of blood, and the sex hormones, such as estrogen and testosterone (fig. 2.10b), which help to maintain male and female characteristics.

Evidence has been accumulating for years that a diet high in saturated fats and cholesterol leads to deposits of fatty material inside the lining of blood vessels. These deposits reduce blood flow and result in circulatory disorders. In **familial hypercholesterolemia** (fah-mǐ'le-al hi"per-ko-les-ter-ol-e'me-ah), the individual is unable to remove cholesterol from the bloodstream and suffers from heart attacks.

As discussed in the Medical Focus reading on page 26, packaged food is now required to have nutrition labels. These labels list the calories from fat per serving and the percentage of daily value from saturated fat and cholesterol.

> Lipids include nonpolar fats (long-term, energy-storage molecules that form from glycerol and three fatty acids) and the related phospholipids, which have a charged group. Steroids are lipids with an entirely different structure from that of fats.

Proteins
Functions

Proteins are huge biomolecules that sometimes have mainly a structural function. For example, in humans, keratin is a protein that makes up hair and nails, and collagen is a protein found in connective tissue, including cartilage, bone, and the fibrous connective tissue of ligaments and tendons. Muscles contain proteins that account for muscles' ability to contract.

Some proteins function as **enzymes** (en'zīmz), necessary contributors to the chemical workings of the cell and, therefore, of the body. Enzymes are organic catalysts that speed chemical reactions. As figure 2.11 shows, enzymes

figure 2.10 Steroid diversity. **a.** Cholesterol, like all steroid molecules, has four adjacent rings, but their effects on the body largely depend on the attached groups indicated in red. **b.** Testosterone is the male sex hormone.

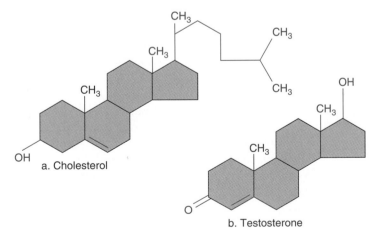

a. Cholesterol

b. Testosterone

figure 2.11 Specificity of enzymatic reactions. Certain reactants fit onto the surface of a particular enzyme, like a key fits into a lock. The reaction occurs, and when it is over, the product of the reaction leaves the enzyme's surface. The same enzyme can be used over and over again to bring about a particular reaction.

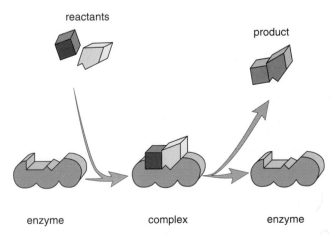

reactants

product

enzyme   complex   enzyme

Medical Focus

Nutrition Labels

*a*s of May 1994, packaged foods must have a nutrition label like the one depicted in figure 2C. The nutrition information given in the label shown here is based on the serving size (that is, 1 1/4 cup, 57 grams) of a cereal. One serving of the cereal provides 220 Calories, of which 20 are from fat. A Calorie* is a measurement of energy. The recommended amounts of nutrients listed at the bottom of the label are based on a typical diet of 2,000 Calories for women and 2,500 Calories for men.

Fats are the nutrient with the highest energy content: 9 Calories per gram compared to 4 Calories per gram for carbohydrates and proteins. The body stores fat under the skin and around the organs for later use. A 2,000-Calorie diet should contain no more than 65 grams (595 Calories) of fat. Dietary fat has been implicated in cancer of the colon, pancreas, ovary, prostate, and breast. Dietary consumption of saturated fat and cholesterol, in particular, should be controlled. Cholesterol and saturated fat contribute to the formation of deposits called plaque, which clog arteries and lead to cardiovascular disease, including high blood pressure.

For these reasons, knowing how a serving of the cereal will contribute to the maximum recommended daily amount of fat, saturated fat, and cholesterol is important. This information is found in the listing under *% Daily Value:* The total fat in one serving of the cereal provides 3% of the recommended amount of fat for the day. How much will a serving of the cereal contribute to the maximum recommended daily amount of saturated fat? of cholesterol?

Carbohydrates (sugars and polysaccharides) are the quickest, most readily available energy source for the body. Because carbohydrates are not usually associated with health problems, they should comprise the largest proportion of the diet. Complex carbohydrates in breads and cereals are preferable to the simple carbohydrates found in candy and ice cream because they are likely to contain dietary fiber (nondigestible plant material). Insoluble fiber has a laxative effect and seems to reduce the risk of colon cancer; soluble fiber combines with the cholesterol in food and prevents the cholesterol from entering the body proper. The nutrition label in figure 2C indicates that one serving of the cereal provides 15% of the recommended daily carbohydrates.

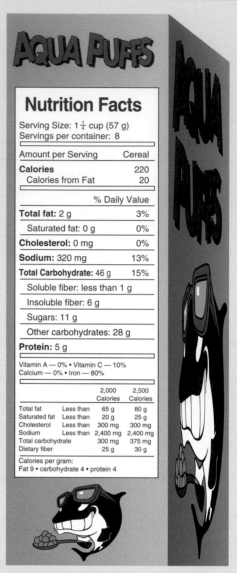

figure 2C Nutrition label on a cereal box.

The body does not store amino acids for the production of proteins, which are found particularly in muscles but also in all cells of the body. A woman should have about 44 grams of protein and a man about 56 grams of protein per day. Red meat is rich in protein, but meat is usually also high in saturated fat. Therefore, obtaining protein from plant origins (for example, whole-grain cereals, dark breads, legumes) to a greater extent than is customary in the United States is preferable. The cereal nutrition label in figure 2C shows that 5 grams of protein are obtained from each serving.

The amount of dietary sodium (as in table salt) in a food product is of concern because excessive sodium intake has been linked to high blood pressure in some people. Sodium intake should be no more than 2,400 milligrams per day. What percentage of this maximum amount does a serving of the cereal in figure 2C provide?

Vitamins are organic molecules required in small amounts in the diet for good health. Each vitamin has a recommended daily intake, and the nutrition label on food products tells what percentage of the recommended amount is provided by one serving. The nutrition label for the cereal in figure 2C indicates that, while the cereal provides no vitamin A or calcium, one serving does contain 10% of the suggested daily intake of vitamin C and 80% of the recommended daily iron intake.

*A calorie is the amount of heat required to raise the temperature of 1 gram of water 1°C. Food energy is measured in Calories (the capital C means 1,000 calories).

figure 2.12 Formation of a peptide. Two amino acids on the left-hand side of the equation that differ by their *R*-groups join. A peptide bond forms, and a water molecule is produced. During hydrolysis, water is added, and the peptide bond is broken.

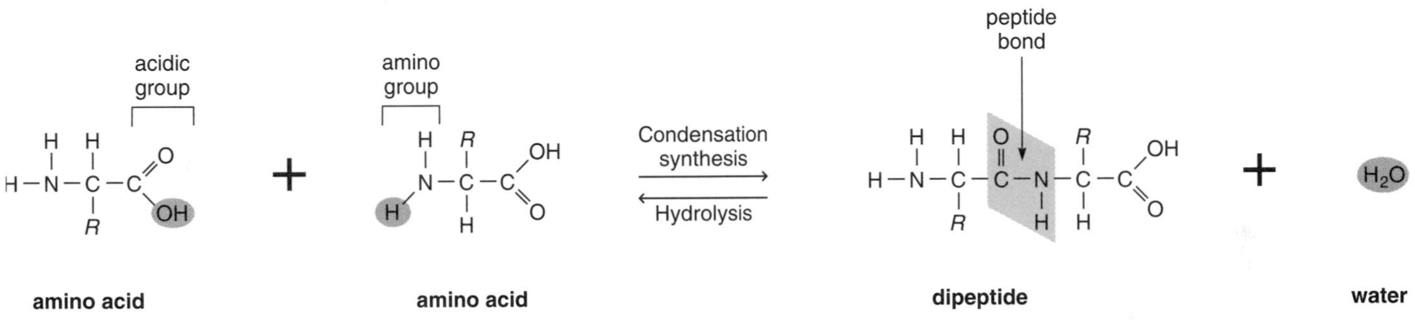

have shapes that fit the reactants (molecules reacting with one another). The reaction occurs when the reactants are close to one another on the enzyme surface. Enzymes also catalyze reactions that break down reactants into their molecular subunits. Enzymes work so quickly that a reaction that might normally take several hours or days without an enzyme takes only a fraction of a second when an enzyme is present.

Structure

The unit molecules found in proteins are called **amino acids** (ah-me'no as'idz). The name *amino acid* refers to the fact that the molecule has two functional groups: an *amino group* and an *acid group*.

Amino acids differ from one another by their *R groups*, the remainder of the molecule. In amino acids, *R* groups vary from a single hydrogen atom (—H) to a complicated ring. Because about 20 different common amino acids are found in the proteins of living things, there are also about 20 different types of *R* groups.

A **peptide bond** (pep'tīd bond) joins two amino acids together. As shown in figure 2.12, during protein **synthesis** (sin'the-sis) the acid group of one amino acid reacts with the amino group of another amino acid, and water is given off. A dipeptide contains only two amino acids, but a *polypeptide* can contain hundreds of amino acids. Polypeptides have three levels of structure (fig. 2.13*a–c*). The primary structure is the sequence of amino acids in that particular polypeptide. The secondary structure is often a helix, held in place by hydrogen bonding. The tertiary structure is the final three-dimensional shape of the polypeptide.

Some proteins have only one polypeptide chain, while others have more than one type of polypeptide chain, each with its own primary, secondary, and tertiary structures. These separate polypeptides are arranged to give some proteins a fourth level of structure, termed the quaternary structure (fig. 2.13*d*). Hemoglobin is a complex protein with a quaternary structure.

Proteins have both structural and enzymatic functions in the human body. Amino acids are the unit molecules for peptides and polypeptides.

figure 2.13 Levels of organization in protein structure. **a.** Primary structure of a protein is the order of the amino acids. **b.** Secondary structure is often a helix, in which hydrogen bonding occurs along the length of a polypeptide, as indicated by the dotted lines. **c.** In globular proteins, the tertiary structure is a twisting and turning, which takes place because of bonding between the *R* groups. **d.** The quaternary structure occurs when a protein contains two or more linked polypeptides.

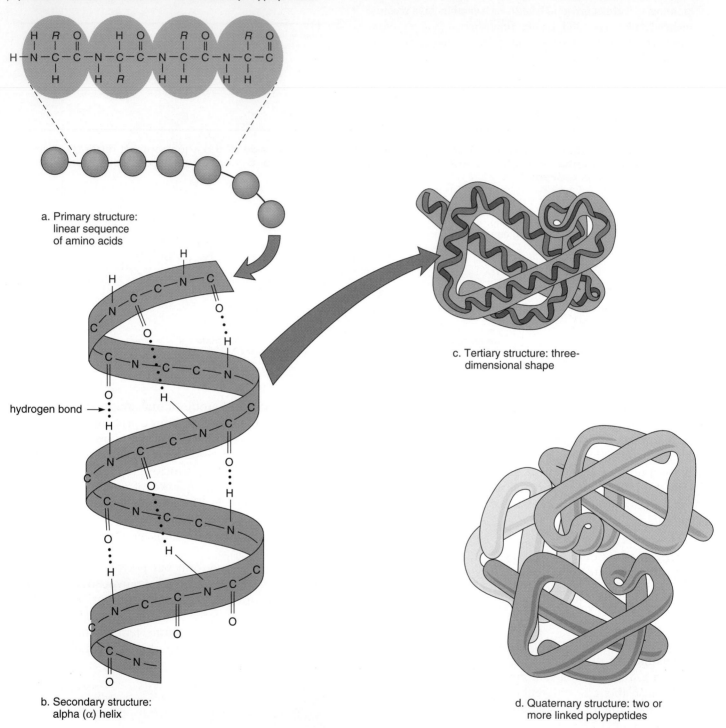

a. Primary structure:
linear sequence
of amino acids

hydrogen bond

b. Secondary structure:
alpha (α) helix

c. Tertiary structure: three-
dimensional shape

d. Quaternary structure: two or
more linked polypeptides

figure 2.14 Nucleotide and nucleic acid structure.
a. A nucleotide has three unit molecules: a phosphate group, a 5-carbon sugar, and a nitrogen base.
b. When nucleotides join, they form a nucleic acid strand. The molecule has a sugar-phosphate backbone, and the nitrogen bases project to the side.

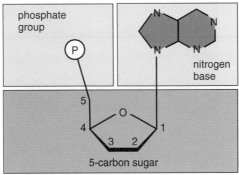

a. Nucleotide

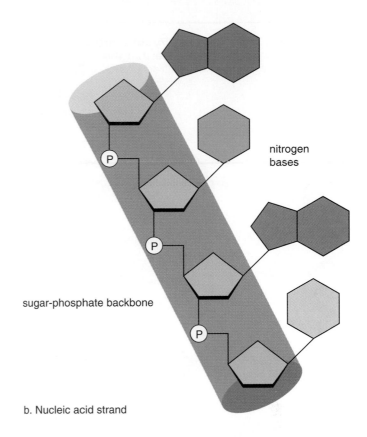

b. Nucleic acid strand

Nucleic Acids

Nucleic acids (nu-kla'ik as'idz) are huge biomolecules with very specific functions in cells. **Genes,** the hereditary factors that we receive from our parents and that control the characteristics of the cell and organism, are composed of a nucleic acid called **DNA (deoxyribonucleic acid)** (de-ok'sĭ-ri"bo-nu-kla"ik as'id). DNA is a molecule that stores coded information. Another important nucleic acid—**RNA (ribonucleic acid)**—works in conjunction with DNA to bring about protein synthesis in cells.

Both DNA and RNA are polymers of nucleotides joined together. Every nucleotide is a molecular complex of three types of unit molecules: a phosphate, a 5-carbon sugar, and a nitrogen base (fig. 2.14*a*). The sugar in DNA is deoxyribose, while that in RNA is ribose, which accounts for the difference in their names. The nitrogen bases in DNA are adenine (A), guanine (G), thymine (T), and cytosine (C). The bases in RNA are the same, except that uracil (U) is substituted for thymine (table 2.3).

When nucleotides join together, they form a linear molecule called a strand, composed of a sugar-phosphate backbone, with the nitrogen bases projecting to one side (see fig. 2.14*b*). RNA is single-stranded, but DNA is double-stranded. The two strands of DNA are twisted in the form of a double helix and are held together by hydrogen bonds between the bases (fig. 2.15). An unwound DNA helix resembles a ladder: The steps of the ladder are the hydrogen-bonded nitrogen bases.

The sequence of nitrogen bases in DNA serves as a code for directing the sequence of bases in RNA and then the sequence of amino acids in a protein. In other words,

table 2.3	**DNA Structure Compared to RNA Structure**	
	DNA	**RNA**
Sugar	Deoxyribose	Ribose
Bases	Adenine, guanine, thymine, cytosine	Adenine, guanine uracil, cytosine
Strands	Double-stranded	Single-stranded
Helix	Yes	No

the genes we inherit determine the types of proteins we can produce in our cells.

> Both DNA and RNA are polymers of nucleotides; only DNA is double-stranded. DNA makes up the genes, and along with RNA, controls protein synthesis.

ATP

ATP (adenosine triphosphate) (ah-den'o-sēn tri-fos'fāt) (fig. 2.16) is a very special type of nucleotide. It is composed of the base adenine and the sugar ribose (together called adenosine) and three phosphate groups. The ⁊ lines in the formula for ATP indicate high-energy phosphate bonds; when these bonds are broken, an unusually large amount of energy is released. Because of this property, ATP is the energy currency of cells: When cells "need" energy, they "spend" ATP.

figure 2.15 Overview of DNA structure. **a.** Double helix. **b.** Complementary base pairing between strands. **c.** Ladder configuration. Notice that the uprights are composed of phosphate and sugar molecules and that the rungs are complementary paired bases.

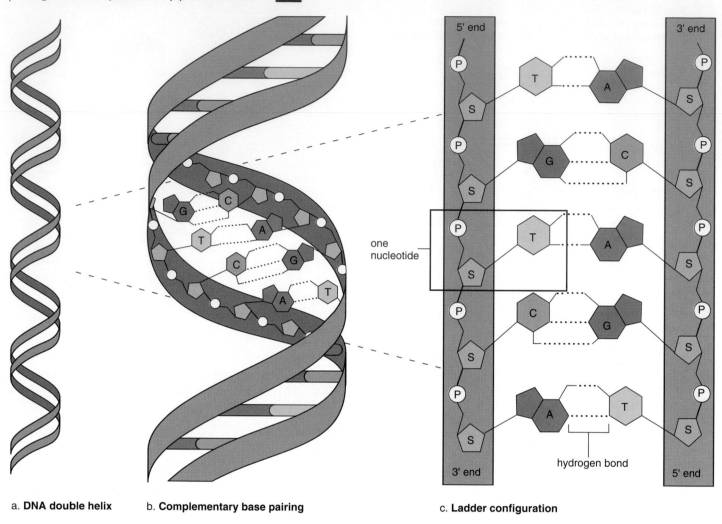

a. **DNA double helix**   b. **Complementary base pairing**   c. **Ladder configuration**

figure 2.16 ATP, the energy molecule in cells. ATP has two high-energy phosphate bonds (indicated by ⌐ lines). When cells require energy, the last phosphate bond is broken, and energy and a phosphate molecule are released.

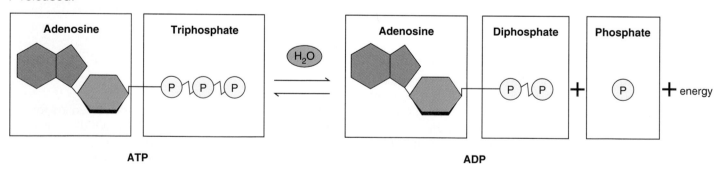

ATP is used in body cells for synthetic reactions, active transport, nerve impulse conduction, and muscle contraction. When energy is required for these processes, the end phosphate group is removed from ATP, breaking down the molecule to ADP (adenosine diphosphate) and Ⓟ (phosphate). The ATP molecule is rebuilt when ADP joins with Ⓟ. The energy needed to rebuild ATP comes from glucose metabolism, as we shall see in the next chapter.

> ATP is the energy "currency" of cells because it contains high-energy phosphate bonds.

Selected New Terms

Basic Key Terms

acid (as'id), p. 21
amino acid (ah-me'no as'id), p. 27
ATP (adenosine triphosphate) (ah-den'o-sēn tri-fos'fāt), p. 29
base (bās'), p. 21
buffer (buf'er), p. 22
carbohydrate (kar″bo-hi'drāt), p. 23
covalent bond (ko-va'lent bond), p. 18
disaccharide (di-sak'ah-rīd), p. 23
DNA (deoxyribonucleic acid) (de-ok'se-ri″-bo-nu-kla″ik as'id), p. 29
electrolyte (e-lek'tro-līt), p. 21
enzyme (en'zīm), p. 25
fatty acid (fat'e as'id), p. 24
glycerol (glis'er-ol), p. 24
glycogen (gli'ko-jen), p. 23
hydrogen bond (hi'dro-jen bond), p. 21
hydrolysis (hi-drol'ĭ-sis), p. 23
ion (i'on), p. 18
ionic bond (i-on'ik bond), p. 18

isotope (i'so-tōp), p. 18
lipid (lip'id), p. 24
monosaccharide (mon″o-sak'ah-rīd), p. 23
nucleic acid (nu-kla'ik as'id), p. 29
peptide bond (pep'tīd bond), p. 27
pH, p. 22
polysaccharide (pol″e-sak'ah-rīd), p. 23
protein (pro'tēn), p. 25
radioactive isotope (ra″de-o-ak'tiv i'so-tōp), p. 18
RNA (ribonucleic acid) (ri″bo-nu-kla'ik as'id), p. 29
synthesis (sin'the-sis), p. 27

Clinical Key Terms

acidosis (as″ĭ-do'sis), p. 22
alkalosis (al″kah-lo'sis), p. 22
diabetes (di″ah-be'tēz), p. 23
familial hypercholesterolemia (fah-mĭ'le-al hi″perko-les-ter-ol-e'me-ah), p. 25
hypertension (hi″per-ten'shun), p. 18
iron deficiency anemia (i'ern dĭ-fi'shun-se ah-ne'me-ah), p. 18

Summary

I. Elements and Atoms
All matter is composed of elements, each containing just one type of atom. Each atom has an atomic symbol, atomic number (number of protons), and atomic weight (number of protons and neutrons). The isotopes of some atoms are radioactive.

II. Molecules and Compounds
Atoms react with one another to form molecules. Following one type of reaction, positively and negatively charged ions form a molecule in which they are ionically bonded. Following another type of reaction, atoms sharing electrons form a molecule in which they are covalently bonded.

III. Some Important Inorganic Molecules
A. In some covalent bonds, the electrons are shared unequally, and the result is a polar molecule. Hydrogen bonding can occur between polar molecules.
B. Because of hydrogen bonding, water heats up and cools down slowly, and this helps keep body temperature within normal limits. Water is a polar molecule and acts as a solvent; it dissolves various chemical substances and facilitates chemical reactions.
C. Substances like salts, acids, and bases that dissociate (that is, release ions when in water) are called electrolytes. The electrolyte balance in the blood and body tissues is important for good health.
D. Acids have a pH that is less than 7, and bases have a pH that is greater than 7. The presence of buffers helps keep the pH of body fluids constant at about neutral, or pH 7, because a buffer can absorb both hydrogen ions (H^+) and hydroxide ions (OH^-).

IV. Some Important Organic Molecules
A. Glucose is an immediate source of energy in cells. It is the unit molecule for glycogen, starch, and cellulose. Glycogen stores energy in the body, starch is a dietary

source of energy, and cellulose is fiber in the diet.

B. Lipids include nonpolar fats (long-term, energy-storage molecules that form from glycerol and three fatty acids) and the related phospholipids, which have a charged group.

Steroids are lipids with an entirely different structure from that of fats.

C. Proteins have both structural and enzymatic functions in the human body. Amino acids are the unit molecules for peptides and polypeptides.

D. Both DNA and RNA are polymers of nucleotides; only DNA is double-stranded. DNA makes up the genes, and along with RNA, controls protein synthesis.

E. ATP is the energy "currency" of cells because it contains high-energy phosphate bonds.

Study Questions

1. Describe the composition of an atom, and explain the weight and charge of an atom's components. (p. 17)
2. Give an example of an ionic reaction, and define the term *ion.* (p. 18)
3. State the function of some important ions in the body. (p. 18, 20)
4. Give an example of a covalent reaction, and define the term *covalent bond.* (p. 18, 20)
5. What are electrolytes, and how are they important? (p. 21)
6. On the pH scale, which numbers indicate a basic solution? an acidic solution? (p. 22)
7. What are buffers, and why are they important to life? (p. 22)
8. Name a monosaccharide, disaccharide, and polysaccharide, and state appropriate functions. What is the most common unit molecule for these? (p. 23)
9. What type of unit molecules react to form a molecule of fat? Explain the difference between a saturated fatty acid and an unsaturated fatty acid. (p. 24)
10. Name several types of lipids, and state their functions. (pp. 24–25)
11. Describe the primary, secondary, and tertiary structures of proteins. What functions do proteins serve in the body? (pp. 27–28)
12. What are the two types of nucleic acids in cells, and what are their functions? What is the unit molecule of a nucleic acid? Name four differences between DNA and RNA. (p. 29)

Objective Questions

Fill in the blanks.

1. _____ are the smallest units of matter nondivisible by chemical means.
2. Isotopes differ by the number of _____ in the nucleus.
3. The two primary types of reactions and bonds are _____ and _____ .
4. A type of weak bond, called a _____ bond, exists between water molecules.

5. Acidic solutions contain more _____ ions than basic solutions, but they have a _____ pH.
6. Glycogen is a polymer of _____ , molecules that serve to give the body immediate _____ .
7. A fat hydrolyzes to give one _____ molecule and three _____ molecules.

8. The primary structure of a protein is the sequence of _____ ; the secondary structure is very often a _____ ; the tertiary structure is the final _____ of the protein.
9. _____ speed chemical reactions in cells.
10. Genes are composed of _____ , a nucleic acid made up of _____ joined together.

Medical Terminology Reinforcement Exercise

Consult Appendix B for help in pronouncing and analyzing the meaning of the terms that follow.

1. anisotonic (an-i″so-ton′ik)
2. dehydration (de″hi-dra′shun)
3. hypokalemia (hi″po-ka-le′me-ah)
4. hypovolemia (hi″po-vo-le′me-ah)
5. nonelectrolyte (non″e-lek′tro-līt)
6. hydrolysis (hi-drol′ĭ-sis)
7. lipometabolism (lip″o-me-tab′o-lizm)
8. hyperlipoproteinemia (hi″per-lip″ o-pro″te-in-e′me-ah)
9. hyperglycemia (hi″per-gli-se′me-ah)
10. hypoxemia (hi″pok-se′me-ah)

Website Link

For a listing of the most current Websites related to this chapter, please visit the Mader home page at:

www.mhhe.com/maderap

chapter 3

Cell Structure and Function

■ Chapter Outline and Learning Objectives

After you have studied this chapter, you should be able to:

The Plasma Membrane (p. 34)
■ Describe the structure and function of the plasma membrane.

The Nucleus (p. 35)
■ Describe the structure and function of the nucleus.
■ Describe the role of DNA in protein synthesis.

The Cytoplasm (p. 36)
■ Describe the roles of ribosomes and the three types of RNA in protein synthesis.
■ Describe the structures and roles of the endoplasmic reticulum and the Golgi apparatus in packaging and secretion.
■ Describe the structures of lysosomes and the role of these organelles in the breakdown of molecules.
■ Describe the structure of mitochondria and their role in producing ATP.
■ Describe the structures of centrioles, cilia, and flagella and their roles in cellular movement.

Plasma Membrane Transport (p. 39)
■ Describe how substances move across the plasma membrane, and distinguish between passive and active transport.

Cell Division (p. 42)
■ Give an overview of mitotic cell division, and explain the mechanism by which the chromosome number stays constant.
■ Contrast mitosis with meiosis in general terms.

Medical Focus
Hemodialysis (p. 40)

MedAlert
Cancer (p. 44)

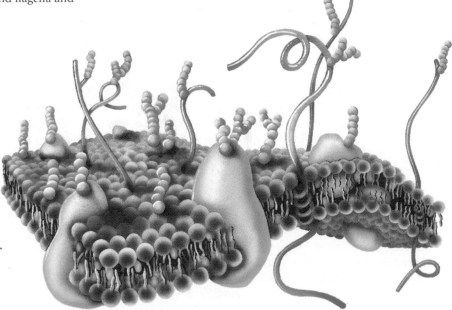

All cells are bounded by a plasma membrane that has a complex, complicated structure and that regulates the entrance of molecules into and out of the cell.

figure 3.1 Fluid-mosaic model of the plasma membrane. **a.** The membrane is composed of a phospholipid bilayer. The polar heads of the phospholipids are at the surfaces of the membrane; the nonpolar tails make up the membrane interior. **b.** Proteins are embedded in the membrane. Some of these function as receptors for chemical messengers, as conductors of molecules through the membrane, and as enzymes to speed metabolic reactions. Carbohydrate chains of glycolipids and glycoproteins are involved in cell-to-cell recognition.

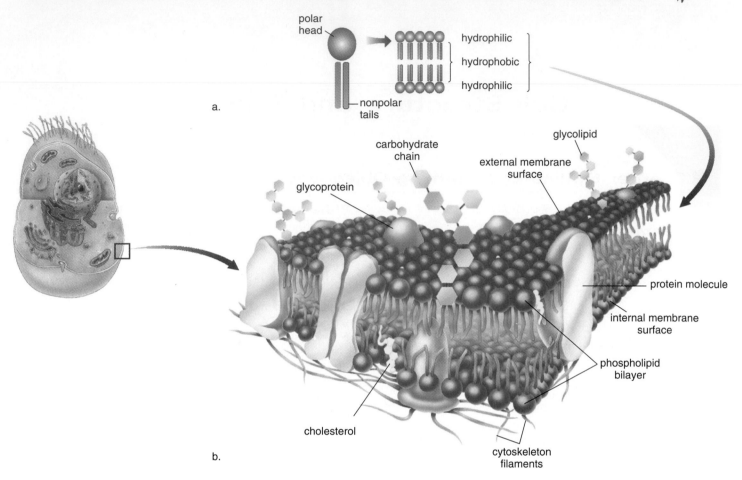

The cells of the body perform specific functions, and therefore, their structures vary greatly. Even so, because all cells have the same basic organization, we can begin the study of cell structure by examining a generalized cell. Knowledge of the generalized animal cell was obtained by using the light microscope and the electron microscope. The *light microscope*, which utilizes light to view the object, does not show much detail, but the *electron microscope*, which uses electrons to view the object, allows cell biologists to make out cell structure in great detail. The plasma membrane has also been examined using the electron microscope.

The Plasma Membrane

Our cells are surrounded by an outer **plasma membrane.** The plasma membrane is the boundary between the inside of the cell, termed the cytoplasm, and the outside of the cell. Plasma membrane integrity is necessary to the life of the cell.

The plasma membrane is a phospholipid bilayer with attached or embedded proteins. The phospholipid molecule has a polar head and nonpolar tails (fig. 3.1*a*). Because the polar heads are charged, they are *hydrophilic* (water-loving) and face outward, where they are likely to encounter a watery environment. The nonpolar tails are *hydrophobic* (water-fearing) and face inward, where there is no water. When phospholipids are placed in water, they naturally form a spherical bilayer because of the chemical properties of the heads and the tails.

At body temperature, the phospholipid bilayer is a liquid; it has the consistency of olive oil, and the proteins are able to change their positions by moving laterally. The fluid-mosaic model, a working description of membrane structure, suggests that the protein molecules have a changing pattern (form a mosaic) within the fluid phospholipid bilayer (fig. 3.1*b*). Our plasma membranes also contain a substantial number of *cholesterol* molecules. These molecules lend stability to the phospholipid bilayer and prevent a drastic decrease in fluidity at low temperatures.

figure 3.2 Cell structure. The generalized cell is based on electron micrographs.

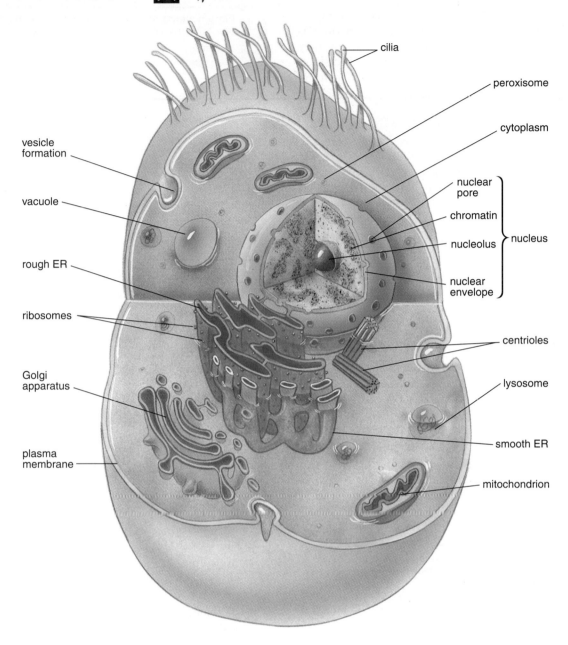

cilia

peroxisome

cytoplasm

vesicle formation

nuclear pore

chromatin

nucleolus

nucleus

nuclear envelope

vacuole

rough ER

ribosomes

centrioles

Golgi apparatus

lysosome

smooth ER

plasma membrane

mitochondrion

Short chains of sugars are attached to the outer surfaces of some protein and lipid molecules (called glycoproteins and glycolipids, respectively). These carbohydrate chains, specific to each cell, mark the cell as belonging to a particular individual and account for such characteristics as blood type or why a patient's system sometimes rejects an organ transplant. Some glycoproteins have a special configuration that allows them to act as a receptor for a chemical messenger like a hormone. Some plasma membrane proteins form channels through which certain substances can enter cells, while others are carriers involved in the passage of molecules through the membrane.

The Nucleus

The **nucleus** (nu'kle-us) is enclosed by a nuclear envelope that is continuous with the endoplasmic reticulum (ER), another part of the cell (fig. 3.2). Pores, or openings, in this nuclear envelope allow the passage of large molecules from the *nucleoplasm,* the fluid portion of the nucleus, to the cytoplasm.

The nucleus is the control center that oversees the cell's metabolic functioning and ultimately determines the cell's characteristics. Within the nucleus are masses of threads called **chromatin** (kro′mah-tin), so named because they take up stains and become colored. Chromatin is indistinct in the nondividing cell, but it condenses to rodlike structures called **chromosomes** (kro′mo-sōmz) just prior to cell division.

Chromosomes contain DNA, which makes up the genes. Recall from chapter 2 that DNA is double-stranded; each strand carries a particular sequence of nitrogen bases. These serve as a genetic code that is passed on to a type of RNA called **messenger RNA (mRNA)** for the purpose of directing protein synthesis in the cell. Another type of RNA, called **transfer RNA (tRNA),** is also made in the nucleus and functions in protein synthesis. Some of the synthesized proteins have a structural role, and some are enzymes involved in **metabolism,** which is all the chemical reactions that occur in the cell. In this way, DNA controls cell structure and function.

Occasionally, the sequence of bases in DNA changes, and this mistake, called a **mutation,** is copied by mRNA and may result in a faulty protein. The individual in which this occurs is said to have a **genetic disease** because the cells cannot function properly, and the result is a noticeable illness.

The **nucleolus** (nu-kle′o-lus) is a spherical body found in the nucleus (see fig. 3.2). Here, another type of RNA, called **ribosomal RNA (rRNA),** is formed and contributes to the manufacture of small granules called **ribosomes** (ri′bo-sōmz). After their formation in the nucleolus, ribosomes are transported from the nucleus to the cytoplasm, where they function in protein synthesis.

> The nucleus contains chromatin, which condenses into chromosomes just prior to cell division. Genes, composed of DNA, are on the chromosomes, and they code for the production of proteins in the cytoplasm. The nucleolus is involved in ribosome formation.

The Cytoplasm

The **cytoplasm** is the substance of the cell outside the nucleus. A number of **organelles,** small bodies with specific structures and functions, are located in the cytoplasm (see fig. 3.2 and table 3.1). Organelles help the cell carry out its many activities. For example, ribosomes carry out protein synthesis, the endomembrane system packages the proteins for secretion or for use inside the cell, and mitochondria convert the energy of glucose into the energy of ATP, the molecule that supplies metabolic energy. In addition to

table 3.1	Organelles (Simplified)	
	Name	**Function**
	Plasma membrane	Regulation of molecules passing into and out of a cell
	Nucleus	Control center of a cell
	Chromatin (chromosomes)	Structure that contains hereditary information
	Ribosome	Protein synthesis
	Endoplasmic reticulum	Transport by means of vesicles
	Rough	Protein synthesis
	Smooth	Lipid synthesis
	Golgi apparatus	Packaging and secretion
	Vacuole and vesicle	Containers for material
	Lysosome	Intracellular digestion
	Mitochondrion	Aerobic cellular respiration
	Centriole	Direction of microtubule organization; associated with cell division
	Cilium and flagellum	Particle movement along surface; movement of cell

these general duties, each cell also has specialized functions, which we will examine in later chapters.

Ribosomes and Protein Synthesis

In low-power electron micrographs, ribosomes appear to be small, dense granules, but at a higher resolution, two subunits can be seen (fig. 3.3). Each subunit contains proteins and ribosomal RNA (rRNA), one of the three types of RNA made in the nucleus, as was described earlier. Protein synthesis occurs at the ribosomes, and all three types of RNA play a role. Synthesis refers to the bonding of small organic molecules to make a larger one. In this case, amino acids are joined to make a polypeptide, a part of a protein.

Ribosomal RNA (rRNA) plays a structural role in protein synthesis in that ribosomes are structures that contain rRNA. Messenger RNA (mRNA) carries a message; it has a copy of the genetic code passed to it by DNA. As discussed in chapter 2, one polypeptide differs from another by the sequence of its amino acids. The sequence of nitrogen bases in mRNA specifies the order of the amino acids for the polypeptide being made. Transfer RNA (tRNA) transfers the amino acids to the ribosome; there is a separate tRNA for each of the different types of amino acids found in polypeptides.

> Ribosomes occur within the cytoplasm but are often attached to endoplasmic reticulum. Three types of RNA—ribosomal RNA (rRNA), messenger RNA (mRNA), and transfer RNA (tRNA)—play a role in protein synthesis at the ribosomes.

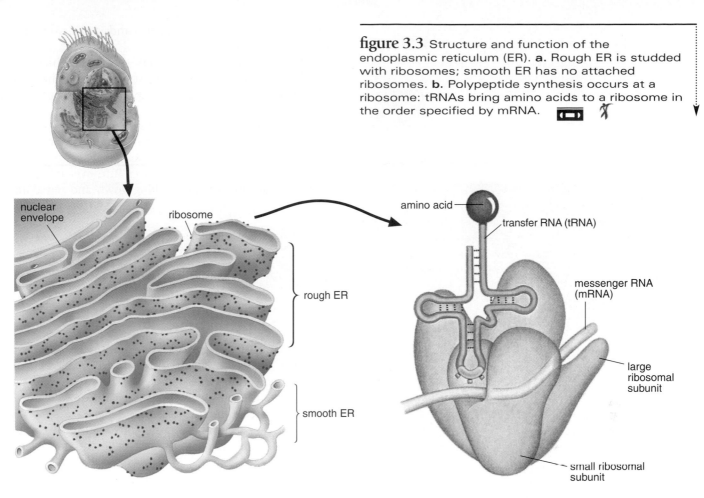

figure 3.3 Structure and function of the endoplasmic reticulum (ER). **a.** Rough ER is studded with ribosomes; smooth ER has no attached ribosomes. **b.** Polypeptide synthesis occurs at a ribosome: tRNAs bring amino acids to a ribosome in the order specified by mRNA.

a. Endoplasmic reticulum

b. Protein synthesis at a ribosome

Endomembrane System

The endoplasmic reticulum, the Golgi apparatus, lysosomes, and peroxisomes are structurally and functionally related membranous structures that comprise the endomembrane system. They work together to produce, transport, store, or secrete cellular products.

Endoplasmic Reticulum

The **endoplasmic reticulum** (en-do-plaz′mic rĕ-tik′u-lum) **(ER)** forms a membranous system of tubular canals that begins at the nuclear envelope and branches throughout the cytoplasm. Ribosomes are attached to portions of the endoplasmic reticulum. If ribosomes are present, the reticulum is called *rough ER*; if ribosomes are not present, it is called *smooth ER* (see fig. 3.3).

In certain cells, smooth ER contains enzymes that make lipids, such as steroid hormones. The administration of drugs causes an increase in the amount of smooth ER in the liver where smooth ER contains enzymes that detoxify drugs.

The ribosomes present on rough ER function in protein synthesis. Proteins that are exported (secreted) from the cell are stored temporarily in the channels of the reticulum. Small portions of the ER then break away to form **vesicles** (small, membranous sacs) that migrate to the Golgi apparatus, where the product is received, modified, and repackaged before being secreted.

Peroxisomes

Peroxisomes (pĕ-roks′ĭ-sōmz) are membranous sacs that contain enzymes. Hydrogen peroxide, a normal product of metabolism, is harmful if allowed to accumulate. Peroxisomal enzymes break down hydrogen peroxide to water and molecular oxygen. In the liver and kidney, peroxisomes detoxify harmful substances, such as formaldehyde and alcohol. Peroxisomes are believed to bud directly from the smooth ER.

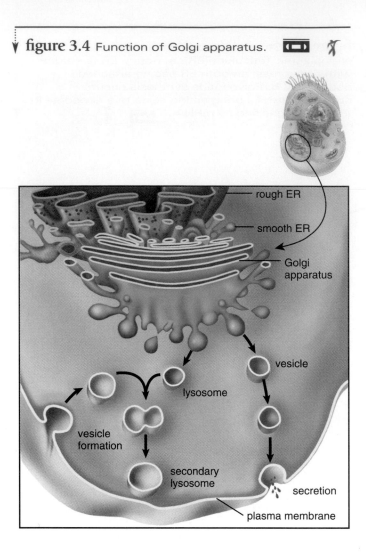

figure 3.4 Function of Golgi apparatus.

rough ER

smooth ER

Golgi apparatus

vesicle

lysosome

vesicle formation

secondary lysosome

secretion

plasma membrane

Golgi Apparatus

The **Golgi apparatus** (gol'je ap"ah-rǎ'tus) (fig. 3.4) is named for Camillo Golgi, the Italian scientist who first discovered its presence in cells. It is composed of a stack of six or more saccules that look like flattened vacuoles (large, membranous sacs). At the edges of the saccules are rounded vacuoles and vesicles.

The Golgi apparatus is especially well developed in cells that secrete (export) a product—for example, in the pancreatic cells that make digestive enzymes or the bronchial cells that produce mucus. When the Golgi apparatus packages a product for export, the product is enclosed within a vesicle that moves toward the plasma membrane, where the vesicle discharges its contents.

Lysosomes

A **lysosome** (li'so-sōm) is a special type of vesicle formed by the Golgi apparatus. All lysosomes carry out intracellular digestion and contain *digestive enzymes*. Following formation, the lysosome may fuse with an incoming vesicle that contains a substance to be digested. The products of

digestion enter the cytoplasm, and only nondigested residue is retained.

Tay Sachs (tā săks) is a genetic disease in which a newborn has a faulty lysosomal digestive enzyme. The cells fill with nonfunctioning lysosomes, and the death of the child follows.

Lysosomes carry out *autodigestion* when they dispose of worn-out or damaged cell components, such as mitochondria. This is an essential part of the normal process of cytoplasmic maintenance and turnover. Turnover refers to the cell's constant process of breaking down and remaking its parts.

> Proteins made for export by rough endoplasmic reticulum are packaged by the Golgi apparatus, which also makes lysosomes. Lysosomes contain digestive enzymes and sometimes function in autodigestion.

Mitochondria

A **mitochondrion** (mi"to-kon'dre-on), a rather complex organelle with an outer membrane and convoluted inner membrane, produces ATP molecules. As discussed previously, every cell needs a supply of ATP molecules to carry out general cell functions. Some cells have specialized functions. For example, muscle cells use ATP for muscle contraction, and nerve cells need ATP to conduct nerve impulses.

Mitochondria are often referred to as the powerhouses of the cell because, just as powerhouses burn fuel to produce electricity, mitochondria burn glucose products to produce ATP molecules. The word *burn* is used advisedly because mitochondria use up oxygen and give off carbon dioxide and water. Several systems of the body function to make this possible (fig. 3.5). The digestive system digests food, and as a result, glucose enters the blood vessels and is taken to the cell. Within the cell, glucose is first broken down to a molecule called pyruvate, and then pyruvate enters mitochondria. In the meantime, oxygen that has entered the lungs of the respiratory system is also transported to the cell, where it enters mitochondria. Following ATP formation, water and carbon dioxide exit the mitochondria and the cell. The lungs expel the carbon dioxide.

Since gas exchange is involved, mitochondria are said to carry on **aerobic cellular respiration.** One way to indicate the chemical transformation associated with aerobic cellular respiration is:

$$\text{carbohydrate} + \text{oxygen}$$
$$\downarrow$$
$$\text{carbon dioxide} + \text{water} + \text{ATP energy}$$

> Mitochondria are the sites of aerobic cellular respiration, a process that uses nutrients and oxygen to produce ATP, the type of chemical energy needed by cells.

figure 3.5 Relationship of glucose breakdown to the body proper. Glucose and oxygen are delivered to the cells by the bloodstream. Carbon dioxide and water are removed by the bloodstream. ATP remains in the cytoplasm as a source of energy for the cell to do work.

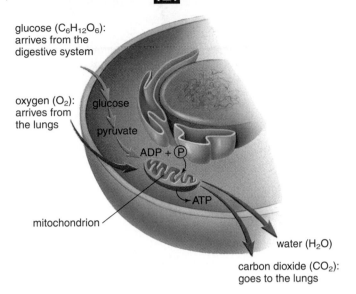

glucose ($C_6H_{12}O_6$):
arrives from the
digestive system

oxygen (O_2):
arrives from
the lungs

glucose

pyruvate

ADP + P

ATP

mitochondrion

water (H_2O)

carbon dioxide (CO_2):
goes to the lungs

Centrioles and Related Organelles

Centrioles

Centrioles (sen'tre-ōlz) are short cylinders that contain fine tubules called microtubules. Usually, two centrioles lie at right angles to one another near the nucleus (see fig. 3.2). Before a cell divides, the centrioles duplicate, and the members of each pair are also at right angles to one another.

Centrioles are believed to give rise to basal bodies that direct the formation of cilia and flagella. Centrioles may also be involved in the movement of material throughout the cells and in the organization of the spindle during cell division (p. 43).

Cilia and Flagella

Cilia (sil'e-ah) and **flagella** (flah-jel'ah) are plasma membrane extensions that contain microtubules. They can move either in an undulating fashion, like a whip, or stiffly, like an oar. Cells that have these organelles are capable of producing movement. For example, sperm cells, carrying genetic material to the egg, move by means of flagella (fig. 3.6). The cells that line the upper respiratory tract are ciliated. The cilia sweep debris trapped within mucus back up the throat, which helps keep the lungs clean.

Centrioles lie near the nucleus and may be involved in the production of the spindle during cell division and in the formation of cilia and flagella.

figure 3.6 Sperm. Sperm cells have long, whiplike flagella to move about.

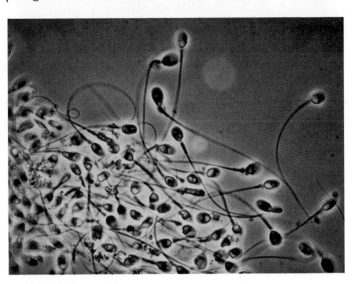

Plasma Membrane Transport

The plasma membrane allows only certain molecules to enter and exit the cytoplasm freely; therefore, the plasma membrane is said to be *selectively permeable*. Transport of molecules across the membrane can be active or passive. *Active transport* requires the use of ATP energy; *passive transport* does not. Passive transport mechanisms include diffusion, osmosis, and filtration.

Substances enter and exit cells by active transport or by passive transport (diffusion, osmosis, filtration).

Passive Transport

Diffusion

Diffusion is the passive movement of molecules from an area of higher concentration to an area of lower concentration. For example, if a perfume bottle is opened in one corner of a room, the perfume's scent will soon be apparent throughout the room because the perfume molecules have diffused from an area of high concentration (the corner) to areas of lower concentration. Another example of diffusion is putting a tablet of dye into water. The water eventually takes on the color of the dye as the tablet dissolves.

In the body, oxygen enters the blood from the alveoli (air sacs) of the lungs by diffusion. During kidney dialysis, waste molecules diffuse across a membrane from the area of higher concentration (the blood) to the area of lower concentration (the dialysate) (see the Medical Focus reading on page 40). Cells do not expend any energy when substances can simply diffuse across the plasma membrane.

Hemodialysis

hemodialysis (he"mo-di-al'i-sis) is a way to remove nitrogenous wastes and to regulate the pH of the blood when the kidneys are unable to perform these functions due to disease or injury. Dialysis occurs when small molecules pass through a membrane that will not allow the passage of large molecules. This same process occurs when a patient is hooked up to a kidney machine (fig. 3A). After the administration of an anticoagulant, blood from an artery moves through a tube to a cellulose acetate membrane that is continually washed by a so-

lution called the dialysate. Later, the blood is returned to the patient by way of a vein.

During hemodialysis, large substances like proteins and red blood cells cannot pass through the dialyzing membrane, but small molecules like urea, glucose, and salts can diffuse across. The composition of the dialysate is maintained at normal levels so that, if the blood lacks glucose and salts, they will pass from the dialysate to the blood. On the other hand, nitrogenous wastes, toxic chemicals, and drugs pass from the blood to the dialysate. In the course of a 6-hour hemodialysis, 50–250 grams of urea can be removed from a patient, an amount that greatly exceeds the urea clearance rate of normal kidneys. Most patients undergo treatments only about three times a week.

Drawbacks to hemodialysis include the need to use an anticoagulant, the possible damage to red blood cells, the risk of infection, the limitation on fluid and protein intake, and the time required.

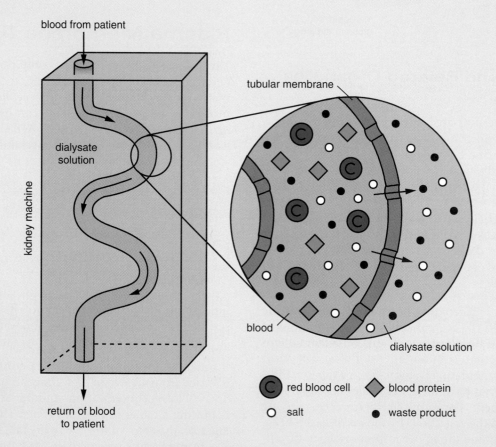

figure 3A Hemodialysis. During hemodialysis, red blood cells and blood proteins cannot pass through the tubular membrane. Salts and waste products pass through by diffusion to the dialysate solution whose composition regulates blood concentration.

figure 3.7 Effect of isotonic, hypotonic, and hypertonic solutions on a red blood cell. In an isotonic solution, the appearance of a red blood cell remains the same. In a hypotonic solution, the cell swells to bursting. In a hypertonic solution, the cell shrinks.

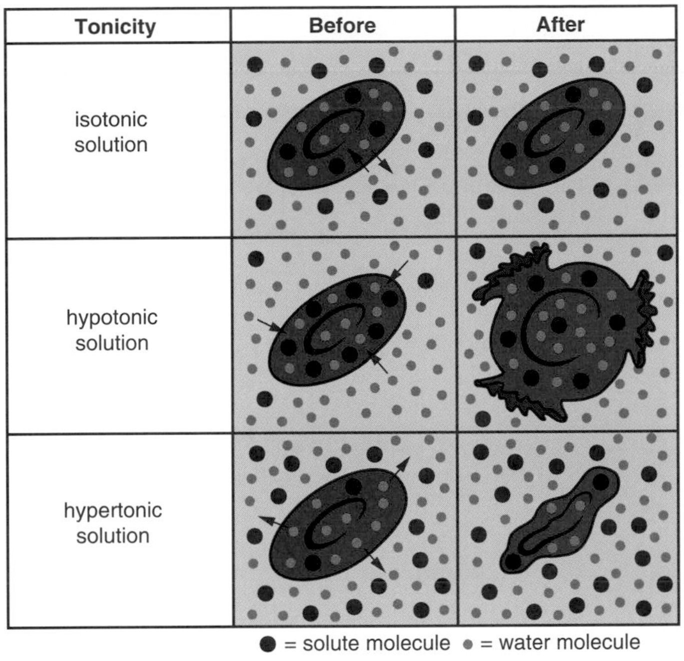

Tonicity	Before	After
isotonic solution		
hypotonic solution		
hypertonic solution		

● = solute molecule • = water molecule

figure 3.8 Active transport. An expenditure of energy is required to allow a carrier protein to actively transport molecules across the plasma membrane from an area of lower concentration to an area of higher concentration.

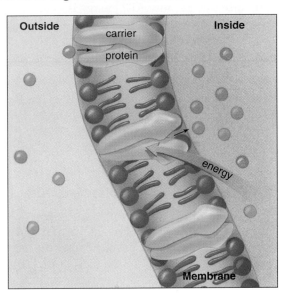

Some substances can simply diffuse across a plasma membrane. The diffusion of water is called osmosis. In an isotonic solution, cells neither gain nor lose water. In a hypotonic solution, cells swell. In a hypertonic solution, cells shrink.

Some molecules that cannot cross the phospholipid bilayer of the plasma membrane diffuse through plasma membrane protein channels. Passive transport of this sort is a form of facilitated diffusion.

Osmosis

Osmosis (oz-mo′sis) is the diffusion of water across a plasma membrane. It occurs whenever the concentrations of water on either side of a selectively permeable membrane are unequal. Normally, body fluids are **isotonic** (i″so-ton′ik) to cells (fig. 3.7); that is, the concentrations of substances (solutes) and water (solvent) on either side of the plasma membrane are equal, and cells, therefore, maintain their usual size and shape. For this reason, most intravenous solutions are also isotonic to cells.

If red blood cells are placed in a **hypotonic** (hi″po-ton′ik) solution, which has a higher concentration of water (lower concentration of solute) than do the cells, water will enter the cells, and they will swell to bursting. Bursting of red blood cells is called **hemolysis.** On the other hand, if red blood cells are placed in a **hypertonic** (hi″per-ton′ik) solution, which has a lower concentration of water (higher concentration of solute) than do the cells, water will leave the cells, and they will shrink. The shrinking of red blood cells is called **crenation.**

Filtration

Because capillary walls are only one cell thick, small molecules (water, small solutes) tend to passively diffuse across these walls, from areas of higher concentration to those of lower concentration. However, blood pressure aids matters by pushing water and dissolved solutes out of the capillary. This process is called **filtration.**

Filtration is easily observed in the laboratory when a solution is poured past filter paper into a flask. Large substances stay behind, but small molecules and water pass through.

Filtration of water and substances in the region of capillaries is largely responsible for the formation of tissue fluid, the fluid that surrounds the cells. Filtration is also at work in the kidneys when water and small molecules move from the blood to the inside of the kidney tubules.

Active Transport

In **active transport,** substances accumulate either inside or outside the cell in the region of *higher* concentration. For example, iodine collects in the cells of the thyroid gland; sugar is completely absorbed from the gut by the cells lining the digestive tract; and sodium is sometimes almost completely withdrawn from urine by cells lining the kidney tubules.

Carrier proteins and an expenditure of energy are both needed to transport substances from an area of lower concentration to an area of *higher* concentration (fig. 3.8). A **carrier** is a plasma membrane protein that specializes in combining with and transporting substances across the plasma membrane. Because ATP energy is needed to cause a carrier to combine with the substance to be transported, cells primarily involved in active transport, such as kidney cells, have a large number of mitochondria near the membrane where active transport is occurring.

> During active transport, which requires plasma membrane carriers and ATP energy, substances move against a concentration gradient and accumulate in the area of higher concentration.

Endocytosis and Exocytosis

At times, substances are taken into cells by vesicle formation. This is called **endocytosis** (en"do-si-to'sis) (fig. 3.9). When the material taken in is quite large, the process is called **phagocytosis** (fag"o-si-to'sis) (cell eating). Phagocytosis is common to amoeboid-type cells, such as macrophages. These white blood cells are called the body's scavengers because they engulf worn-out red blood cells and other types of debris. When cells take in material that is small enough to be dissolved or suspended in water, the process is called **pinocytosis** (pi"no-si-to'sis) (cell drinking).

Vesicles within the cytoplasm of the cell can fuse with the plasma membrane and release their contents to the outside of the cell. This is called **exocytosis** (ex"o-si-to'sis). Some cells of the nervous system release substances involved in the transfer of nerve impulses between adjacent cells via exocytosis.

Cell Division

The two types of cell division are termed mitosis and meiosis. Mitosis occurs during growth and repair, and meiosis occurs during gametogenesis, the production of gametes—that is, the sperm and eggs.

Mitosis

During ordinary cell division, called **mitosis,** a mother cell divides, producing two daughter cells. In humans, the mother cell has 46 chromosomes, and the two daughter cells that result also have 46 chromosomes.

Figure 3.10 shows a mother cell that, for simplicity's sake, has only two pairs of chromosomes. These chromosomes are at first single, containing a single DNA helix. Before cell division takes place, **DNA replication** occurs as the

figure 3.10 Mitosis overview. Following duplication, each chromosome in the mother cell contains two portions. During mitotic division, these portions separate, becoming daughter chromosomes so that the daughter cells have the same number and kinds of chromosomes as the mother cell.

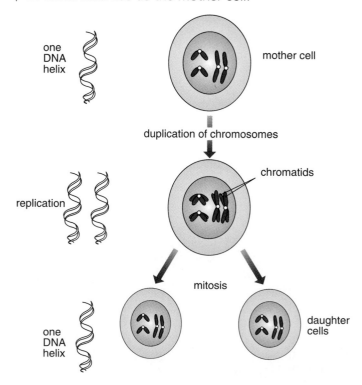

figure 3.9 Transport by vesicles. During endocytosis (either phagocytosis or pinocytosis), the plasma membrane forms a vesicle around the substance to be taken in. During exocytosis, a substance within a vesicle is deposited outside the cell when the vesicle fuses with the plasma membrane.

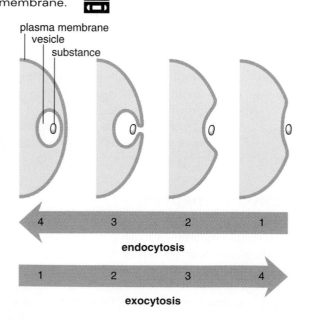

figure 3.11 Interphase and mitosis. Mitosis has four stages: prophase, metaphase, anaphase, and telophase. Interphase, which occurs between cell divisions, is not part of mitosis. Notice that the centrioles duplicate during interphase so that there are two pairs at the start of mitosis (compare early interphase to late interphase). Centrioles are believed to help organize the spindle, which is involved in chromosome movement. (The blue chromosomes were inherited from one parent, and the red chromosomes were inherited from the other parent.)

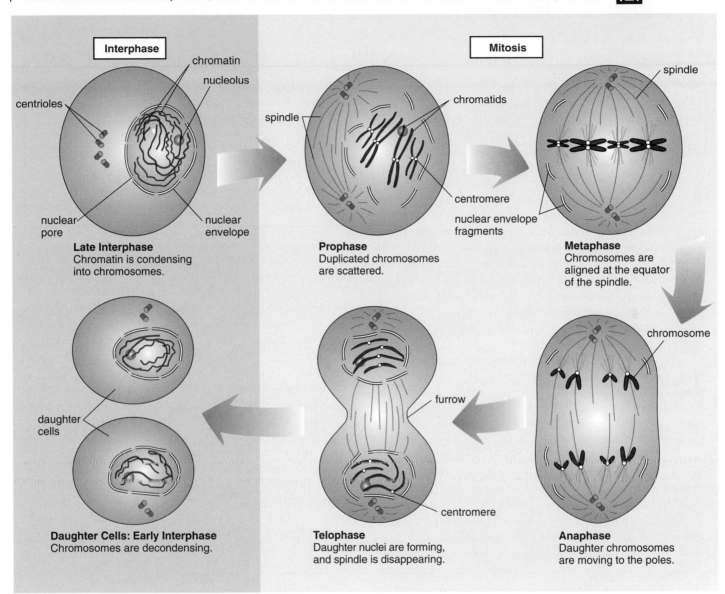

chromosomes duplicate. A duplicated chromosome contains two identical parts (called chromatids). During mitosis, these portions separate, becoming daughter chromosomes. A daughter chromosome contains one DNA helix again.

Mitosis (fig. 3.11) actually requires several stages, during which the nuclear envelope disappears and a **spindle** apparatus with spindle fibers forms. The chromosomes are attached to the spindle fibers by structures called centromeres. Once the chromosomes have moved to the center of the mother cell, the centromeres split, and the daughter chromosomes move toward the poles. Daughter nuclei form, and the cytoplasm divides by furrowing.

Mitosis is the type of cell division required for growth and tissue repair. The process of mitosis assures that each

cell in the body has the same number and kinds of chromosomes and, therefore, the same genes (since the genes are on the chromosomes).

Ordinarily, a cell divides only about 50 times; with maturity, a cell stops dividing. However, cancer cells never mature, and instead, continue to divide indefinitely as discussed in the MedAlert reading on page 44.

Mitosis ensures that each cell in the body is genetically identical. At the time of division, a chromosome has two portions. When these separate, each daughter cell receives the same number and kinds of chromosomes as the mother cell.

MedAlert

Cancer

Cancer cells are abnormal for two reasons: First, cancer cells exhibit uncontrolled and disorganized growth. In the body, a cancer cell divides to form a growth, or **tumor,** that invades and destroys neighboring tissue. This is in contrast to **benign tumors,** which stay in one place. To support their growth, cancer cells release a growth factor that causes neighboring blood vessels to branch into the cancer tissue. This phenomenon has been termed **vascularization,** and some modes of cancer treatment are aimed at preventing vascularization.

The second abnormal characteristic of cancer cells is that they detach from the tumor and spread to other sites in the body. Cancer cells invade the blood vessels or the lymphatic vessels and start new tumors elsewhere in the body. This process is called **metastasis** (mĕ-tas′tah-sis). If a tumor is found before metastasis has occurred, the chances of a cure are greatly increased. This is the rationale for early detection of cancer.

One theory says that the development of cancer is a two-step process involving (1) initiation and (2) promotion. **Carcinogens** (car-sĭ′nah-jenz), agents that cause gene mutations (changes), are initiators. These include viruses, radiation, and chemicals such as pesticides. Cigarette smoke plays a significant role in the development of lung cancer because it contains chemical carcinogens. A cancer promoter is any influence that causes a cell to start growing in an uncontrolled manner. A promoter might cause a second mutation or provide the environment for cells to form a tumor. For example, some evidence suggests that a diet rich in saturated fats and cholesterol is a cancer promoter. Considerable time may elapse between initiation and promotion, and this is one rea-

table 3A — Danger Signals for Cancer

C hange in bowel or bladder habits
A sore that does not heal
U nusual bleeding or discharge
T hickening or lump in breast or elsewhere
I ndigestion or difficulty in swallowing
O bvious change in wart or mole
N agging cough or hoarseness

son why cancer is seen more often in older rather than younger individuals.

Individuals should be aware of the seven danger signals for cancer (table 3A) and inform their doctors when they notice any one of these. Cancer can be detected by physical examination, assisted by various means of viewing the internal organs. Also, specific blood tests can detect tumors that secrete a particular chemical in the blood. For example, prostate specific antigen (PSA) appears to be a substance that increases in the blood according to the size of a prostate tumor.

Tumors can be surgically removed, but there is always the danger that they have already metastasized and are **malignant** (mah-lig′nant). When a growth is malignant, surgery is often preceded or followed by radiation therapy. Radiation destroys the more rapidly dividing cancer cells but causes less damage to the more slowly dividing normal cells. The use of radioactive protons is preferred over X ray because proton beams can be aimed directly at the tumor, like an automatic rifle hitting the bull's-eye of a target.

Meiosis

Meiosis is a special type of cell division that occurs only during the production of eggs and sperm. During meiosis, which takes two rounds of cell division, the chromosome number is reduced from 46 chromosomes to 23 chromosomes.

In females, meiosis occurs during **oogenesis** (o″o-jen′ĕ-sis) (egg production) in the ovaries. In males, meiosis occurs during **spermatogenesis** (sper″mah-to-jen′ĕ-sis) (sperm production) in the testes. Following sex-

ual intercourse, a sperm fertilizes an egg, and a new individual begins development. Because a sperm carries 23 chromosomes and an egg carries 23 chromosomes, the new individual has 46 chromosomes, which is the normal number for human beings. In this way, both parents contribute 23 chromosomes to the new individual.

> Meiosis is a special type of cell division that reduces the chromosome number. Meiosis occurs during oogenesis and spermatogenesis.

Chemotherapy is the use of drugs to kill cancer cells. Sometimes, cancer cells become resistant to chemotherapy (even when several drugs are used in combination). The plasma membrane in resistant cells contains a carrier that pumps toxic chemicals out of the cell. Researchers are testing drugs known to poison the pump in an effort to restore sensitivity to chemotherapy.

Immunotherapy and gene therapy are new, experimental ways of treating cancer. Immunotherapy is the use of an immune system component to treat a disease. For example, cancer patients are sometimes given interleukins, chemicals released by leukocytes, a type of white blood cell. Gene therapy is the substitution of "good genes" for defective or missing genes in order to treat a disease. Two types of genes are now implicated in causing cancer. An **oncogene** is a gene that causes a cell to begin to divide and grow abnormally. **Tumor-suppressor genes** are genes that ordinarily prevent a cell from dividing and growing abnormally. The hope is that, one day, oncogenes can be turned off and tumor-suppressor genes can be turned on by substituting a defective or missing gene in the cells of a person with cancer.

The evidence is clear that the risk of certain types of cancer can be reduced by adopting certain behaviors. In general, the avoidance of excessive radiation and carcinogenic chemicals is helpful. For example, avoiding excessive sunlight reduces the risk of skin cancer, and abstaining from smoking cigarettes and cigars reduces the risk of lung cancer, as well as other types of cancer.

Exercise and healthy dietary considerations are also believed to be important for cancer prevention. Recommendations include:

1. Lowering the total fat intake
2. Eating more high-fiber foods
3. Increasing consumption of foods that are rich in vitamins A and C
4. Reducing consumption of salt-cured and smoked foods
5. Including vegetables of the cabbage family in the diet
6. Consuming moderate amounts of alcohol

Questions

Why would you expect:

1. the incidence of cancer to increase with age?
2. smoking cigarettes to cause lung cancer?
3. diet to influence cancer development?

Selected New Terms

Basic Key Terms

active transport (ak′tiv trans′port), p. 41
centriole (sen′tre-ōl), p. 39
chromatin (kro′mah-tin), p. 36
chromosome (kro′mo-sōm), p. 36
cilia (sil′e-ah), p. 39
cytoplasm (si′to-plazm), p. 36
diffusion (dĭ-fu′zhun), p. 39
endocytosis (en″do-si-to′sis), p. 42
endoplasmic reticulum (en-do-plaz′mic rĕ-tik′u-lum), p. 37
exocytosis (ex″o-si-to′sis), p. 42
filtration (fil-tra′shun), p. 41
flagella (flah-jel′ah), p. 39
Golgi apparatus (gol′je ap″ah-ră′tus), p. 38
hypertonic (hi″per-ton′ik), p. 41
hypotonic (hi″po-ton′ik), p. 41
isotonic (i″so-ton′ik), p. 41
lysosome (li′so-sōm), p. 38
metabolism (mĕ-tab′o-lizm), p. 36
mitochondrion (mi″to-kon′dre-on), p. 38
mitosis (mi-to′sis), p. 42
mutation (mu-ta′shun), p. 36

nucleolus (nu-kle′o-lus), p. 36
nucleus (nu′kle-us), p. 35
oogenesis (o″o-jen′ĕ-sis), p. 44
organelle (or″gah-nel′), p. 36
osmosis (oz-mo′sis), p. 41
peroxisome (pĕ-roks′ĭ-sōm), p. 37
phagocytosis (fag″o-si-to′sis), p. 42
pinocytosis (pi″no-si-to′sis), p. 42
plasma membrane (plaz′mah mem′brăn), p. 34
ribosome (ri′bo-sōm), p. 36
spermatogenesis (sper″mah-to-jen′ĕ-sis), p. 44
spindle (spin′dl), p. 43

Clinical Key Terms

carcinogen (car-sĭ′nah-jen), p. 44
chemotherapy (ke″mo-ther′ah-pe), p. 45
genetic disease (jĕ-net′ik dĭ-zēz′), p. 36
hemodialysis (he″mo-di-al′ĭ-sis), p. 40
malignant (mah-lig′nant), p. 44
metastasis (mĕ-tas′tah-sis), p. 44
Tay Sachs (tā săks), p. 38
tumor (too′mor), p. 44

Summary

Cells differ in shape and function, but even so, a generalized cell can be described.

I. The Plasma Membrane

The plasma membrane, the outer boundary of the cell, consists of a double layer of phospholipid (fat) molecules in which protein molecules form a mosaic pattern.

II. The Nucleus

The nucleus contains chromatin, which condenses into chromosomes just prior to cell division. Genes, composed of DNA, are on the chromosomes, and they code for the production of proteins in the cytoplasm. The nucleolus is involved in ribosome formation.

III. The Cytoplasm

A. Ribosomes occur within the cytoplasm but are often attached to endoplasmic reticulum. Three types of RNA—ribosomal RNA (rRNA), messenger RNA (mRNA), and transfer RNA (tRNA)—play a role in protein synthesis at the ribosomes.

B. The endoplasmic reticulum, peroxisomes, Golgi apparatus, and lysosomes comprise the endomembrane system. Proteins made for export by rough endoplasmic reticulum are packaged by the Golgi apparatus, which also makes lysosomes. Lysosomes contain digestive enzymes and sometimes function in autodigestion.

C. Mitochondria are the sites of aerobic cellular respiration, a process that uses nutrients and oxygen to provide ATP, the type of chemical energy needed by cells.

D. Centrioles lie near the nucleus and may be involved in the production of the spindle during cell division and in the formation of cilia and flagella.

IV. Plasma Membrane Transport

Substances enter and exit cells by active transport or by passive transport (diffusion, osmosis, filtration).

A. Some substances can simply diffuse across a plasma membrane. The diffusion of water is called osmosis. In an isotonic solution, cells neither gain nor lose water. In a hypotonic solution, cells swell. In a hypertonic solution, cells shrink.

B. During filtration, diffusion of small molecules out of a blood vessel is aided by blood pressure.

C. During active transport, which requires plasma membrane carriers and ATP energy, substances move against a concentration gradient and accumulate in the area of higher concentration.

D. Endocytosis involves the uptake of substances by a cell through vesicle formation. Phagocytosis and pinocytosis are two forms of endocytosis. Exocytosis involves the release of substances from a cell as vesicles within the cell cytoplasm fuse with the plasma membrane.

V. Cell Division

A. During mitosis, each newly formed cell receives a copy of each kind of chromosome. Later, the cytoplasm divides by furrowing. Mitosis ensures that each cell in the body is genetically identical. At the time of division, a chromosome has two portions. When these separate, each daughter cell receives the same number and kinds of chromosomes as the mother cell.

B. Meiosis is a special type of cell division that reduces the chromosome number. Meiosis occurs during the production of eggs (oogenesis) and sperm (spermatogenesis).

Study Questions

1. Describe the fluid-mosaic model of membrane structure. (p. 34)
2. Describe the nucleus and its contents, and include the terms *DNA* and *RNA* in your description. (pp. 35–36)
3. What roles do DNA and the three types of RNA play in protein synthesis? (p. 36)
4. Describe the structure and function of endoplasmic reticulum (ER). Include the terms *smooth ER, rough ER,* and *ribosomes* in your description. (p. 37)
5. Describe the structure and function of the Golgi apparatus. Mention vesicles and lysosomes in your description. (p. 38)
6. Describe the structure and function of mitochondria. Mention the energy molecule ATP in your description. (p. 38)
7. Describe the structure and function of centrioles. Mention the mitotic spindle in your description. (p. 39)
8. Contrast passive transport (diffusion, osmosis, filtration) with active transport of molecules across the plasma membrane. (pp. 39–42)
9. Define osmosis, and discuss the effects of placing red blood cells in isotonic, hypotonic, and hypertonic solutions. (p. 41)
10. What are endocytosis and exocytosis? (p. 42)
11. Describe mitosis, and discuss the function of mitosis in humans. (pp. 42–43)
12. What role does meiosis play in human reproduction? (p. 44)

Objective Questions

I. **Match the organelles in the key to the functions given in questions 1–5.**

Key:
- a. mitochondria
- b. nucleus
- c. Golgi apparatus
- d. rough ER
- e. centrioles

1. packaging and secretion
2. cell division
3. powerhouses of the cell
4. protein synthesis
5. control center for cell

II. **Fill in the blanks.**

6. The fluid-mosaic model of membrane structure says that _____ molecules drift about within a double layer of _____ molecules.
7. Rough ER has _____ , but smooth ER does not.
8. Basal bodies that organize the microtubules within cilia and flagella are believed to be derived from _____ .
9. Water will enter a cell when it is placed in a _____ solution.
10. Active transport requires a protein _____ and _____ for energy.
11. Materials taken into a cell by _____ is also called cell eating.
12. At the conclusion of mitosis, each newly formed cell in humans contains _____ chromosomes.
13. At the conclusion of meiosis, each newly formed cell in humans contains _____ chromosomes.
14. The _____ , which is the substance outside the nucleus of a cell, contains bodies called _____ , each with a specific structure and function.

Medical Terminology Reinforcement Exercise

Consult Appendix B for help in pronouncing and analyzing the meaning of the terms that follow.

1. phagocytosis (fag"o-si-to′sis)
2. hemolysis (he"mol′ĭ-sis)
3. isotonic (i"so-ton′ik)
4. cytology (si-tol′o-je)
5. cytometer (si-tom′ĕ-ter)
6. nucleoplasm (nu′kle-o-plazm")
7. lysosome (li′so-sōm)
8. pancytopenia (pan"si-to-pe′ne-ah)
9. cytogenic (si-to-jen′ik)
10. erythrocyte (e-rith′ro-sīt)

Website Link

For a listing of the most current Websites related to this chapter, please visit the Mader home page at:

www.mhhe.com/maderap

chapter 4

Body Tissues and Membranes

Chapter Outline and Learning Objectives

After you have studied this chapter, you should be able to:

Body Tissues (p. 49)

■ Describe the general characteristics and functions of epithelial tissues.

■ Name the major types of epithelial tissues, and relate each one to a particular organ.

■ Describe the general characteristics and functions of connective tissues.

■ Name the major types of connective tissues, and relate each one to a particular organ.

■ Describe the general characteristics and functions of muscular tissues.

■ Name the major types of muscular tissues, and relate each one to a particular organ.

■ Describe the general characteristics and functions of nervous tissue.

Body Membranes (p. 59)

■ Name the different types of membranes, and relate each one to a particular location in the body.

■ State the names and locations of the serous membranes found in the ventral cavity.

Medical Focus

Classification of Cancers (p. 57)

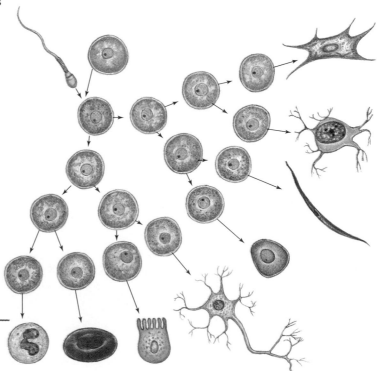

The fertilized egg divides and gives rise to all the different types of cells in the body.

figure 4.1 Simple squamous epithelium. The cells are thin and flat.

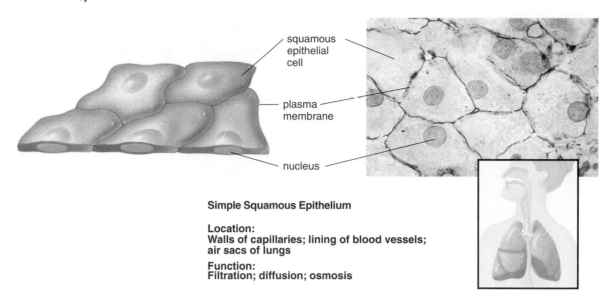

squamous
epithelial
cell

plasma
membrane

nucleus

Simple Squamous Epithelium

Location:
Walls of capillaries; lining of blood vessels;
air sacs of lungs

Function:
Filtration; diffusion; osmosis

Body Tissues

A **tissue** is composed of similarly specialized cells that perform a common function in the body. The tissues of the human body can be categorized into four major types: *epithelial tissue*, which covers body surfaces and lines body cavities; *connective tissue*, which binds and supports body parts; *muscular tissue*, which is specialized for contraction; and *nervous tissue*, which responds to stimuli and transmits impulses from one body part to another.

Epithelial Tissue

Epithelial (ep″ĭ-the′le-al) **tissue,** also called epithelium, forms a continuous layer, or sheet, over the entire body surface and most of the body's inner cavities. On the external surface, it protects the body from drying out, injury, and bacterial invasion. On internal surfaces, epithelial tissue may be specialized for other functions, in addition to protection. For example, in the respiratory tract, epithelial tissue sweeps up impurities by means of cilia, while along the digestive tract, it secretes mucus, which protects the organs of the digestive tract from the digestive enzymes. Epithelial tissue also efficiently absorbs molecules from kidney tubules because of fine, cellular extensions called microvilli.

The three main types of epithelial tissue are squamous, cuboidal, and columnar. **Squamous epithelium** (fig. 4.1) is composed of flat cells and lines the lungs and blood vessels. **Cuboidal epithelium** (fig. 4.2) has cube-shaped cells and lines the kidney tubules. **Columnar epithelium** (fig. 4.3) has pillar- or column-shaped cells, with nuclei usually located near the bottom, and is found lining the digestive tract.

An epithelium may have microvilli (tiny extensions from the cells) or cilia, depending on its particular function. For example, the oviducts are lined by ciliated columnar cells that propel the egg toward the uterus, or womb.

An epithelium may also be simple or stratified. Simple means that the cells occur in a single layer. **Stratified** means that the cells exist as layers piled one over the other. The nose, mouth, esophagus, anal canal, and vagina are all lined by stratified squamous epithelium. The outer layer of skin is also stratified squamous epithelium, but the cells are reinforced by keratin, a protein that provides strength. **Pseudostratified** epithelium appears to be layered; however, true layers do not exist because each cell touches the baseline. *Pseudostratified ciliated columnar epithelium* (fig. 4.4) lines the air passages of the respiratory system, including the nasal cavities and the trachea and its branches. Mucus-secreting goblet cells are scattered among the ciliated epithelial cells. A surface covering of mucus traps

figure 4.2 Simple cuboidal epithelium. The cells are cube-shaped.

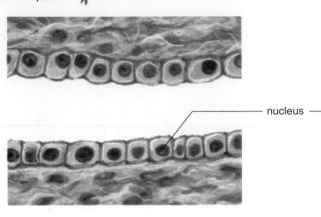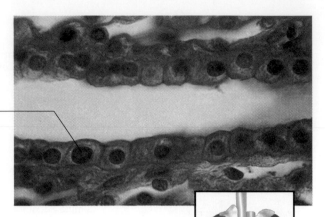

nucleus

Simple Cuboidal Epithelium

Location:
Surface of ovaries; linings of ducts of
glands; linings of kidney tubules

Function:
Secretion; absorption

figure 4.3 Simple columnar epithelium. The cells are more tall than wide.

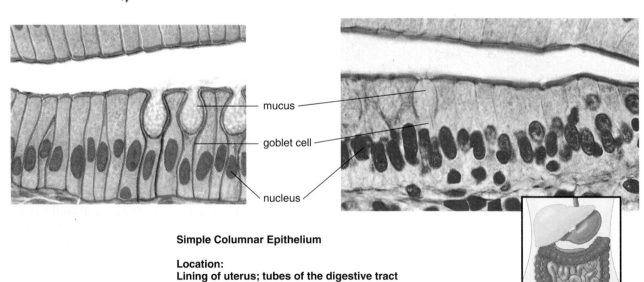

mucus

goblet cell

nucleus

Simple Columnar Epithelium

Location:
Lining of uterus; tubes of the digestive tract

Function:
Protection; secretion; absorption

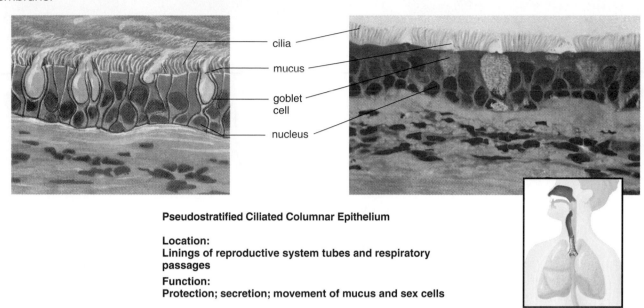

figure 4.4 Pseudostratified ciliated columnar epithelium. The cells have cilia and appear to be stratified, but each actually touches the basement membrane.

cilia

mucus

goblet cell

nucleus

Pseudostratified Ciliated Columnar Epithelium

Location:
Linings of reproductive system tubes and respiratory passages

Function:
Protection; secretion; movement of mucus and sex cells

foreign particles, and upward ciliary motion carries the mucus to the back of the throat, where it may be either swallowed or expectorated.

An epithelium like pseudostratified columnar epithelium that secretes a product is described as glandular. A gland can be composed of a single epithelial cell, as in the case of the mucus-secreting goblet cells found within the columnar epithelium lining the digestive tract (see fig. 4.3), or it can have many cells. Glands that secrete their products into ducts are called **exocrine glands** (for example, salivary glands and sweat glands), and those that secrete directly into the bloodstream are called **endocrine glands** (for example, the pituitary gland and the thyroid gland).

> Epithelial tissue is classified according to cell shape, which may be squamous, cuboidal, or columnar. The cells may be stratified and/or ciliated, and the tissue may be glandular.

Connective Tissue

Connective tissue binds structures together, provides support and protection, fills spaces, produces blood cells, and stores fat. The body uses this stored fat for energy, insulation, and organ protection. As a rule, connective tissue cells are widely separated by a noncellular **matrix** that varies in consistency from solid to semifluid to fluid. Whereas the functional and physical properties of epithelial tissues are derived from the characteristics of cells, connective tissue properties are largely derived from the characteristics of the matrix.

The matrix may have fibers of three types. White fibers contain collagen, a substance that gives the fibers flexibility and strength. Yellow fibers contain elastin, which while not as strong as collagen, is more elastic. Reticular fibers are very thin, highly branched, collagenous fibers that form delicate supporting networks.

Loose Connective Tissue

Loose (aerolar) **connective tissue** binds structures together (fig. 4.5). The cells of this tissue are mainly fibroblasts—large, star-shaped cells that produce extracellular fibers. In loose connective tissue, the fibroblasts are located some distance from one another and are separated by a jellylike matrix that contains many white and yellow fibers. The white fibers occur in bundles and are strong and flexible. The yellow fibers form highly elastic networks that return to their original length after stretching. Loose connective tissue commonly lies beneath an epithelium. In certain instances, the epithelium and its underlying connective tissue form a body membrane (p. 59).

Adipose tissue (fig. 4.6) is a type of loose connective tissue in which the fibroblasts enlarge and store fat and there is limited matrix. The fibroblasts of *reticular connective tissue* are called reticular cells, and the matrix contains only reticular fibers. This tissue, also called lymphoid tissue, is found

figure 4.5 Loose (areolar) connective tissue. This tissue has a loose network of fibers.

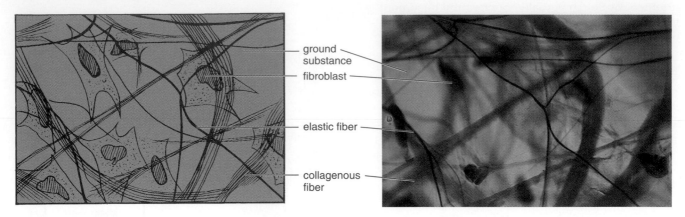

ground substance

fibroblast

elastic fiber

collagenous fiber

Loose (Areolar) Connective Tissue

Location:
Between muscles; beneath the skin; beneath most epithelial layers

Function:
Binds organs together

figure 4.6 Adipose tissue. The cells are filled with fat droplets.

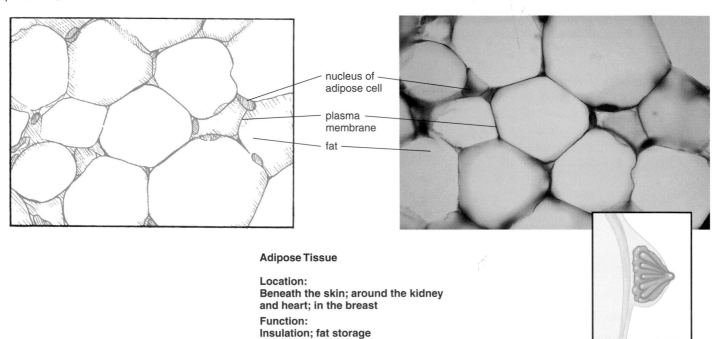

nucleus of adipose cell

plasma membrane

fat

Adipose Tissue

Location:
Beneath the skin; around the kidney and heart; in the breast

Function:
Insulation; fat storage

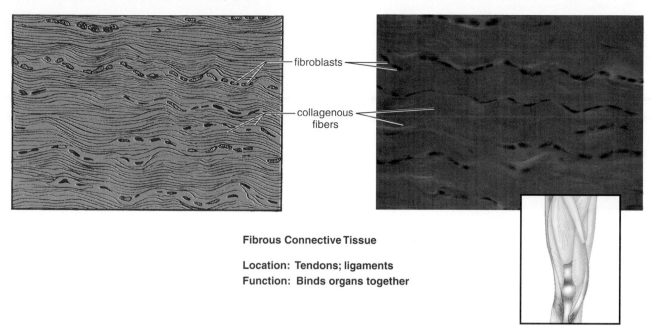

figure 4.7 Fibrous connective tissue. Parallel bundles of collagenous fibers are closely packed.

fibroblasts

collagenous fibers

Fibrous Connective Tissue

Location: Tendons; ligaments
Function: Binds organs together

in lymph nodes, the spleen, thymus, and red bone marrow. These organs are a part of the immune system because they store and/or produce white blood cells, particularly lymphocytes. All types of blood cells are produced in red bone marrow.

Fibrous Connective Tissue

Fibrous connective tissue has a matrix produced by fibroblasts that contain closely packed bundles of white collagenous fibers (fig. 4.7). This type of tissue has more specific functions than does loose connective tissue. For example, fibrous connective tissue is found in **tendons,** which connect muscles to bones, and **ligaments,** which connect bones to other bones at joints. Tendons and ligaments take a long time to heal following an injury because their blood supply is relatively poor.

> Loose and fibrous connective tissues, which bind body parts together, differ according to the type and abundance of fibers in the matrix.

Cartilage

In **cartilage** (kar'ti-lij), the cells (chondrocytes), which lie in small chambers called **lacunae** (lah-ku'ne), are separated by a matrix that is solid yet flexible. Unfortunately,

because this tissue lacks a direct blood supply, it heals very slowly. The three types of cartilage are classified according to the type of fiber in the matrix.

Hyaline cartilage (fig. 4.8) is the most common type of cartilage. The matrix, which contains only very fine collagenous fibers, has a glassy, white, opaque appearance. This type of cartilage is found in the nose, at the ends of the long bones and ribs, and in the supporting rings of the windpipe. The fetal skeleton is also made of this type of cartilage, although the cartilage is later replaced by bone.

Elastic cartilage has a matrix containing many elastic fibers, in addition to collagenous fibers. For this reason, elastic cartilage is more flexible than hyaline cartilage. Elastic cartilage is found, for example, in the framework of the outer ear.

Fibrocartilage has a matrix containing strong collagenous fibers. This type of cartilage absorbs shock and reduces friction between joints. Fibrocartilage is found in structures that withstand tension and pressure, such as the pads between the vertebrae in the backbone and the wedges found in the knee joint.

Bone

Bone (fig. 4.9) is the most rigid of the connective tissues. It has an extremely hard matrix of mineral salts, primarily

figure 4.8 Hyaline cartilage. The matrix is solid but flexible.

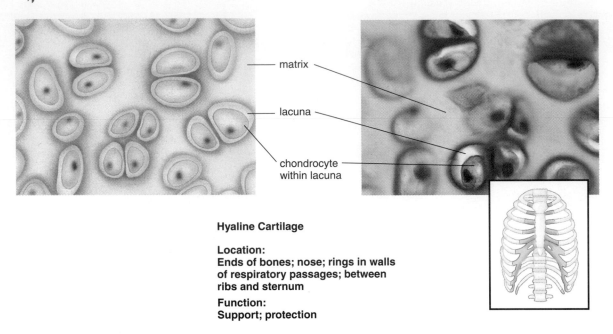

matrix

lacuna

chondrocyte within lacuna

Hyaline Cartilage

Location:
Ends of bones; nose; rings in walls of respiratory passages; between ribs and sternum

Function:
Support; protection

figure 4.9 Compact bone. Cells are arranged in a cylindrical manner about a central canal.

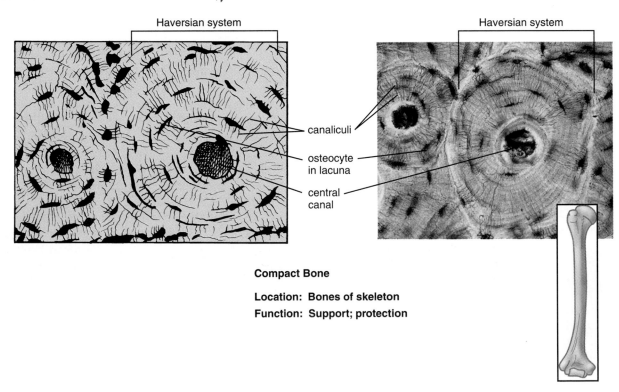

Haversian system

Haversian system

canaliculi

osteocyte in lacuna

central canal

Compact Bone

Location: Bones of skeleton
Function: Support; protection

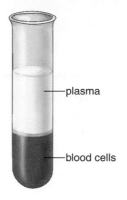

figure 4.10 Blood. Blood has two components: plasma and formed elements. Plasma is the liquid portion of the blood; red blood cells, white blood cells, and platelets are called the formed elements.

Blood sample

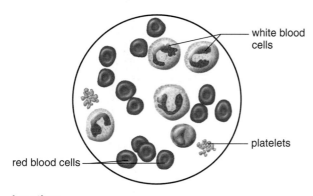

white blood cells

platelets

red blood cells

Location:
In the blood vessels

Function:
Supplies cells with nutrients and oxygen and takes away their wastes; fights infection

calcium salts, deposited around protein fibers. The minerals give bone rigidity, and the protein fibers provide elasticity and strength, much as steel rods do in reinforced concrete.

The outer portion of a long bone contains compact bone. In **compact bone,** bone cells (called osteocytes) are located in lacunae that are arranged in a concentric cylinder called a Haversian system (osteon). Haversian systems form around tiny tubes called central canals, which contain nerve fibers and blood vessels. The blood vessels bring the nutrients that allow bone to renew itself. The nutrients can reach all of the cells because minute canals (canaliculi) containing thin processes of the osteocytes connect the osteocytes with one another and with the central canals.

The ends of a long bone contain spongy bone, which has an entirely different structure. **Spongy bone** contains numerous bony bars and plates separated by irregular

spaces. Although lighter than compact bone, spongy bone is still designed for strength. Like braces used for support in buildings, the solid portions of spongy bone follow lines of stress. Blood cells are formed within red marrow found in spongy bone at the ends of certain long bones.

> Cartilage and bone are support tissues. Cartilage is more flexible than bone because the matrix is rich in protein, rather than the mineral salts found in bone.

Blood

Blood (fig. 4.10) is a connective tissue in which the cells are separated by a liquid matrix called **plasma.** Collectively, the blood cells are called formed elements. Blood cells are of two types: **red blood cells (erythrocytes),** which carry oxygen, and **white blood cells (leukocytes),** which aid in fighting infection. Also present are platelets, which are important to the initiation of blood clotting. Platelets are not complete cells; rather, they are fragments of giant cells found in the bone marrow.

In red bone marrow, cells called stem cells continually divide to produce new cells that mature into the different types of blood cells. The rate of cell division is high, as discussed in the Medical Focus reading on page 57.

Blood is unlike other types of connective tissue in that the intercellular matrix (that is, plasma) is not made by the cells of the tissue. Plasma is a mixture of different types of molecules that enter blood at various organs.

> Blood is a connective tissue in which the matrix is plasma.

Muscular Tissue

Muscular tissue is composed of muscle fibers that contain *actin* and *myosin* filaments, whose interaction accounts for the muscle contraction. Muscle contraction, in turn, accounts for movement. Three types of muscular tissue are found in the body: *skeletal, smooth,* and *cardiac muscle.*

Skeletal muscle (fig. 4.11) is attached to the bones of the skeleton and functions to move body parts. Skeletal muscle fibers are cylindrical and run the length of a muscle. They are multinucleated, with the nuclei appearing just inside the plasma membrane. The fibers also have characteristic light and dark bands perpendicular to the length of the cell. These bands give the muscle a striated appearance. Skeletal muscle is under conscious or voluntary control and contracts faster than any other muscle type.

Smooth muscle (fig. 4.12) is so named because it lacks striations. The spindle-shaped cells that make up smooth muscle are not under voluntary control and are said to be *involuntary.* Smooth muscle, which is found in the viscera (intestine, stomach, and so on) and in blood vessels, contracts more slowly than skeletal muscle, but can remain

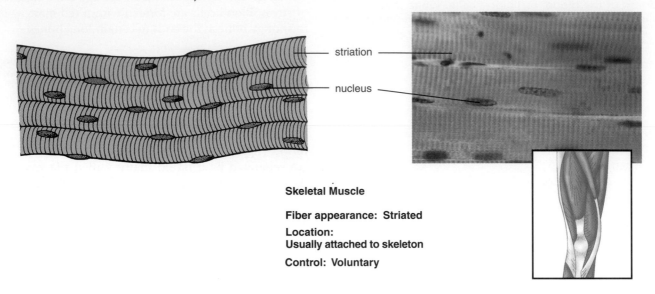

figure 4.11 Skeletal muscle. The cells are long, cylindrical, and multinucleated.

striation

nucleus

Skeletal Muscle

Fiber appearance: Striated

Location:
Usually attached to skeleton

Control: Voluntary

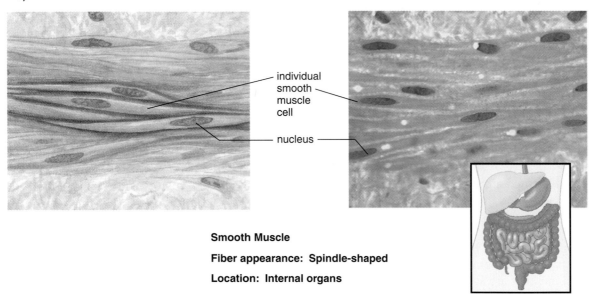

figure 4.12 Smooth muscle. The cells are spindle-shaped.

individual smooth muscle cell

nucleus

Smooth Muscle

Fiber appearance: Spindle-shaped

Location: Internal organs

figure 4.13 Cardiac muscle. The cells are cylindrical but branched.

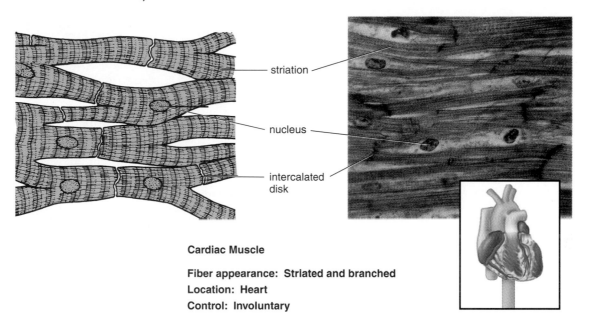

striation

nucleus

intercalated disk

Cardiac Muscle

Fiber appearance: Striated and branched
Location: Heart
Control: Involuntary

contracted for a longer time. The cells tend to form layers in which the thick middle portion of one cell is opposite the thin ends of adjacent cells. Consequently, the nuclei form an irregular pattern in the tissue.

Cardiac muscle (fig. 4.13), which is found only in the heart, is responsible for the heartbeat. Cardiac muscle seems to combine features of both smooth and skeletal muscle. Cardiac muscle has striations like those of skeletal muscle, but the contraction of the heart is involuntary for the most part. Cardiac muscle fibers also differ from skeletal muscle fibers in that they are branched and seemingly fused, one with the other, so that the heart appears to be composed of one large, interconnecting mass of muscle cells. Actually, however, cardiac muscle fibers are separate and individual, but they are bound, end-to-end, at *intercalated disks,* areas of folded plasma membrane between the cells.

> Muscular tissue contains actin and myosin filaments. These form a striated pattern in skeletal and cardiac muscle, but not in smooth muscle. Cardiac and smooth muscle are under involuntary control. Skeletal muscle is under voluntary control.

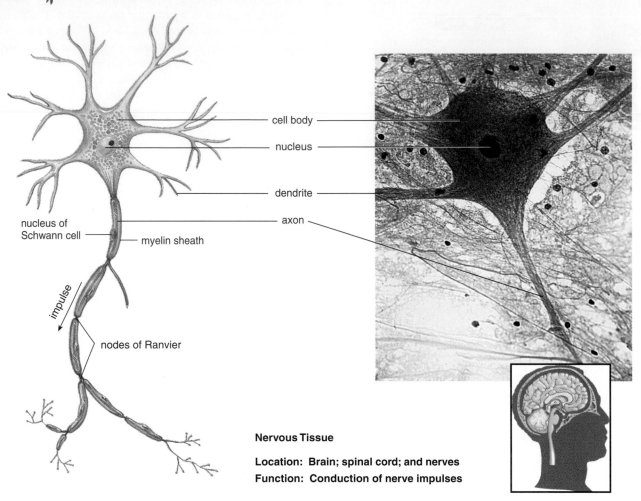

figure 4.14 Nervous tissue. Neurons are surrounded by neuroglial cells. Only neurons conduct nerve impulses.

cell body

nucleus

dendrite

axon

nucleus of
Schwann cell

myelin sheath

impulse

nodes of Ranvier

Nervous Tissue

Location: Brain; spinal cord; and nerves

Function: Conduction of nerve impulses

Nervous Tissue

Nervous tissue, found in the brain and spinal cord, contains specialized cells called neurons that conduct nerve impulses. A **neuron** (nu′ron) (fig. 4.14) has three parts: (1) a *dendrite* conducts signals to the cell body; (2) the *cell body* contains the nucleus and most of the cytoplasm of the neuron; and (3) the *axon* generally conducts nerve impulses away from the cell body.

Long axons are called *fibers*. Outside the brain and spinal cord, fibers are bound together by connective tissue to form **nerves.** Nerves conduct impulses from sense organs to the spinal cord and brain, where the phenomenon called sensation occurs. They also conduct nerve impulses away from the spinal cord and brain to the muscles, causing the muscles to contract.

In addition to neurons, nervous tissue contains **neuroglial** (nu-rog′le-al) **cells.** These cells maintain the tissue by supporting and protecting the neurons. **Schwann cells** are neuroglial cells that encircle all long nerve fibers that are outside of the brain or spinal cord. Each Schwann cell encircles only a small section (1 mm) of a nerve fiber. The gaps between Schwann cells are called the **nodes of Ranvier.** A nerve impulse moves from node to node. Collectively, the Schwann cells give nerve fibers a protective layer of fatty insulation called a **myelin sheath.** Because the myelin sheath is white, all nerve fibers appear to be white.

Nervous tissue contains conducting cells called neurons.
Neurons have processes called axons and dendrites.
Outside the brain and spinal cord, these long axons (fibers)
are found in nerves.

figure 4.15 The peritoneal cavity. **a.** The peritoneum lines the abdominal cavity (called parietal peritoneum) and covers its organs (called visceral peritoneum). **b.** A more complete diagram of the peritoneum within the abdominopelvic cavity.

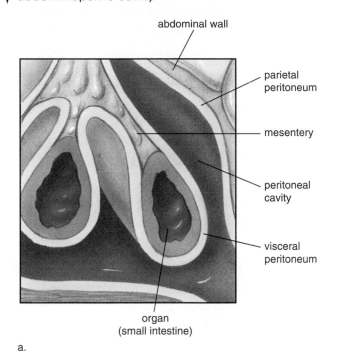

abdominal wall

parietal peritoneum

mesentery

peritoneal cavity

visceral peritoneum

organ (small intestine)

a.

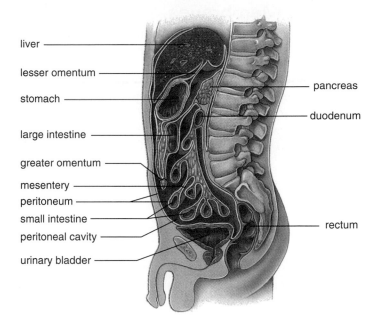

liver

lesser omentum

stomach

large intestine

greater omentum

mesentery

peritoneum

small intestine

peritoneal cavity

urinary bladder

pancreas

duodenum

rectum

b.

Body Membranes

Membranes line the internal spaces of organs and tubes that open to the outside, and they also line the body cavities discussed in chapter 1.

Mucous Membranes

Mucous membranes line the interior walls of the organs and tubes that open to the outside of the body, such as those of the digestive, respiratory, urinary, and reproductive systems. These membranes consist of an epithelium overlying a layer of connective tissue. The epithelium contains goblet cells that secrete mucus.

The mucus secreted by mucous membranes ordinarily protects interior walls from invasion by bacteria and viruses; hence, more mucus is secreted when a person has a cold and has to blow his or her nose. In addition, mucus usually protects the walls of the stomach and small intestine from digestive juices, but this protection breaks down when a person develops an *ulcer.*

Serous Membranes

Serous membranes line cavities, including the thoracic and abdominopelvic cavities, and cover internal organs like the heart. The term **parietal** (pah-ri′ĕ-tal) refers to the wall of the body cavity, while the term **visceral** (vis′er-al) pertains to

the internal organs. Therefore, parietal membranes line the interior of the thoracic and abdominopelvic cavities, and visceral membranes cover the organs.

Serous membranes consist of a layer of simple squamous epithelium overlying a layer of connective tissue. They secrete a watery fluid that keeps the membranes lubricated. Serous membranes support the internal organs and tend to compartmentalize the large thoracic and abdominopelvic cavities. This helps to hinder the spread of any infection.

In the thorax, the **pleural membranes** are serous membranes that line the thoracic cavity and then double back to cover the lungs. The parietal pleura lines the thoracic wall, while the visceral pleura adheres to the surface of the lungs. A well-known infection of these membranes is called *pleurisy.* A serous membrane is part of the **pericardium,** a covering for the heart.

In the abdomen, the interior wall and organs are lined by a serous membrane called the **peritoneum** (per″ĭ-to-ne′um). The parietal peritoneum lines the abdominopelvic cavity, and the visceral peritoneum covers the organs. The peritoneum comes together to form a double-layered **mesentery** (mes′en-ter″e) that supports the visceral organs. The greater omentum is a double-layered peritoneum that covers the intestines, and the lesser omentum is a double-layered peritoneum that runs between the stomach and the liver. Figure 4.15*a* is a diagram of the peritoneum, and figure 4.15*b* shows the visceral organs of the abdominopelvic cavity and supporting serous membranes.

Peritonitis is an infection of the peritoneum. Peritonitis is likely if an inflamed appendix bursts before it is removed.

Synovial Membranes

Synovial (sĭ-no′ve-al) **membranes** line freely movable joint cavities and are composed of connective tissues. They secrete synovial fluid into the joint cavity; this fluid lubricates the ends of the bones so that they can move freely. In *rheumatoid arthritis*, the synovial membrane becomes inflamed and grows thicker. Fibrous tissue then invades the joint and may eventually become bony so that the bones of the joint are no longer capable of moving.

Meninges

The **meninges** (mĕ-nin′jez) are membranes found within the dorsal cavity (p. 7). They are composed only of connective tissue and serve as a protective covering for the brain and spinal cord. *Meningitis* is a life-threatening infection of the meninges.

Cutaneous Membrane

The **cutaneous** (ku-ta′ne-us) **membrane,** or skin, forms the outer covering of the body. It consists of a thin outer layer of stratified squamous epithelium attached to a thicker underlying layer of connective tissue. The skin is discussed in detail in chapter 5.

> Various types of membranes line the body's cavities and organs and tubes that open to the outside.

Selected New Terms

Basic Key Terms

cartilage (kar′ti-lij), p. 53
connective tissue (kŏ-nek′tiv tish′u), p. 51
cutaneous membrane (ku-ta′ne-us mem′brān), p. 60
epithelial tissue (epi″ĭ-the′le-al tish′u), p. 49
lacuna (lah-ku′na), p. 53
matrix (ma′triks), p. 51
meninges (mĕ-nin′jēz), p. 60
mesentery (mes′en-ter″e), p. 59
mucous membrane (mu′kus mem′brān), p. 59
muscular tissue (mus′ku-lar tish′u), p. 55
myelin sheath (mi′ĕ-lin shēth), p. 58
nervous tissue (ner′vus tish′u), p. 58
neuroglial cell (nu-rog′le-al sel), p. 58
neuron (nu′ron), p. 58
parietal (pah-ri′ĕ-tal), p. 59

peritoneum (per″i-to-ne′um), p. 59
pseudostratified (su″do-strat′ĭ-fĭd), p. 49
serous membrane (ser′us mem′brān), p. 59
stratified (strat′ĭ-fĭd), p. 49
synovial membrane (sĭ-no′ve-al mem′brān), p. 60
visceral (vis′er-al), p. 59

Clinical Key Terms

biopsy (bi′op-se), p. 57
carcinoma (kar-sĭ-no′mah), p. 57
diagnosis (di-ig-no′sis), p. 57
leukemia (lu-ke′me-ah), p. 57
lymphoma (lim-fo′mah), p. 57
Pap smear (pap smēr), p. 57
pathologist (pah-thol′ah-jist), p. 57
sarcoma (sar-ko′mah), p. 57

Summary

I. Body Tissues
Body tissues are categorized into four types.
   A. Epithelial tissue. This tissue is classified according to cell shape, which may be squamous, cuboidal, or columnar. The cells may be stratified and/or ciliated, and the tissue may be glandular.
   B. Connective tissue.
      1. Loose and fibrous connective tissues, which bind body parts

together, differ according to the type and abundance of fibers in the matrix.
      2. Cartilage and bone are support tissues. Cartilage is more flexible than bone because the matrix is rich in protein, rather than the mineral salts found in bone.
      3. Blood is a connective tissue in which the matrix is plasma.

   C. Muscular tissue. This tissue contains actin and myosin microfilaments. These form a striated pattern in skeletal and cardiac muscle, but not in smooth muscle. Cardiac and smooth muscle are under involuntary control. Skeletal muscle is under voluntary control.
   D. Nervous tissue. This tissue contains conducting cells called neurons. Neurons have processes

called axons and dendrites. Outside the brain and spinal cord, these long axons (fibers) are found in nerves.

II. Body Membranes

Mucous membranes line the interior of organs and tubes that open to the outside. Serous membranes line the thoracic and abdominopelvic cavities, and cover the organs within these cavities. Synovial membranes line certain joint cavities. The skin forms a cutaneous membrane. Meninges are membranes that cover the brain and spinal cord.

Study Questions

1. What is a tissue? (p. 49)
2. Name the four major types of tissues. (p. 49)
3. What are the functions of epithelial tissue? Name the different kinds of epithelial tissue, and give a location for each. (pp. 49, 51)
4. What are the functions of connective tissue? Name the different kinds of connective tissue, and give a location for each. (pp. 51–55)

5. Contrast the structure of cartilage with that of bone, using the words *lacunae* and *central canal* in your description. (pp. 53–55)
6. Describe the composition of blood, and give a function for each type of blood cell. (p. 55)
7. What are the functions of muscular tissue? Name the different kinds of muscular tissue, and give a location for each. (pp. 55–57)

8. What types of cells does nervous tissue contain? Which organs in the body are made up of nervous tissue? (p. 58)
9. Name the different types of body membranes, and associate each type with a particular location in the body. (pp. 59–60)
10. Name the different types of serous membranes, and associate each type with a particular organ or organs. (p. 59)

Objective Questions

I. Fill in the blanks.
1. Most organs contain several different types of

   _____ .
2. Pseudostratified ciliated columnar epithelium contains cells that appear to be _____ , have projections called _____ , and are _____ in shape.
3. Connective tissue cells are widely separated by a _____ that usually contains

   _____ .

4. Both cartilage and blood are classified as _____ tissue.
5. A mucous membrane contains _____ tissue overlying _____ tissue.

II. Match the organs in the key to the epithelial tissues listed in questions 6–9.
Key:
   a. kidney tubules
   b. small intestine
   c. air sacs of lungs
   d. trachea (windpipe)

6. squamous
7. cuboidal
8. columnar
9. pseudostratified ciliated columnar

III. Match the muscle tissues in the key to the descriptions given in questions 10–12.
Key:
   a. skeletal muscle
   b. smooth muscle
   c. cardiac muscle
10. striated and branched, involuntary
11. striated and voluntary
12. visceral and involuntary

Medical Terminology Reinforcement Exercise

Consult Appendix B for help in pronouncing and analyzing the meaning of the terms that follow.

1. epithelioma (ep″ĭ the″le-o′mah)
2. fibrodysplasia (fi″bro-dis-pla′se-ah)
3. meningoencephalopathy (mĕ-ning″go-en-sef″al-lop′ah-the)
4. mesenteric (mes′en-ter-ik)
5. pericardiocentesis (per″i-kar″de-o-sen-te′sis)
6. peritonitis (per″ĭ-to-ni′tis)
7. intrapleural (in″tra-ploor′al)
8. neurofibromatosis (nu″ro-fi″bro″mah-to′sis)
9. submucosa (sub″mu-ko′sah)
10. polyarthritis (pol″e-ar-thri′tis)

Website Link

For a listing of the most current Websites related to this chapter, please visit the Mader home page at:

www.mhhe.com/maderap

chapter 5

The Integumentary System

▪ Chapter Outline and Learning Objectives

After you have studied this chapter, you should be able to:

Structure of the Skin (p. 63)
- Describe the regions of the skin and the subcutaneous layer.
- Name two main epidermal layers, and describe their structure and function.

Accessory Structures of the Skin (p. 65)
- Describe the structure and growth of hair and nails.
- Name three glands of the skin, and describe their structure and function.

Functions of the Skin (p. 67)
- List and discuss four functions of the skin.

Disorders of the Skin (p. 68)
- Name the three types of skin cancer, and state their cause.
- Name and describe four types of burns with regard to depth.
- Describe how the "rule of nines" may be used to estimate the extent of a burn.
- Describe the steps by which a skin wound heals.

Effects of Aging (p. 73)
- Anatomical and physiological changes occur in the integumentary system as we age

Working Together (p. 75)
- The integumentary system works with other systems of the body to maintain homoestasis (p. 74)

Visual Focus
Temperature Control (p. 69)

Medical Focus
Link between UV Radiation and Skin Cancer (p. 70)
Epidermis from the Laboratory (p. 71)

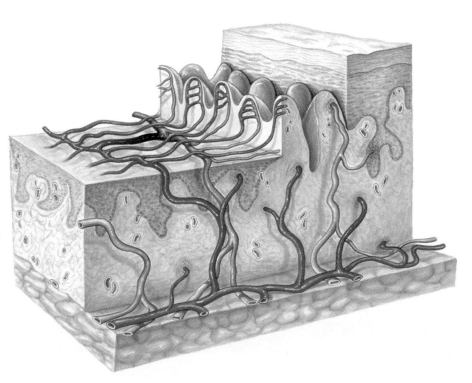

The skin is richly supplied with blood vessels.

figure 5.1 Skin anatomy. Skin is composed of two regions: the epidermis and the dermis. A subcutaneous layer is located beneath the skin.

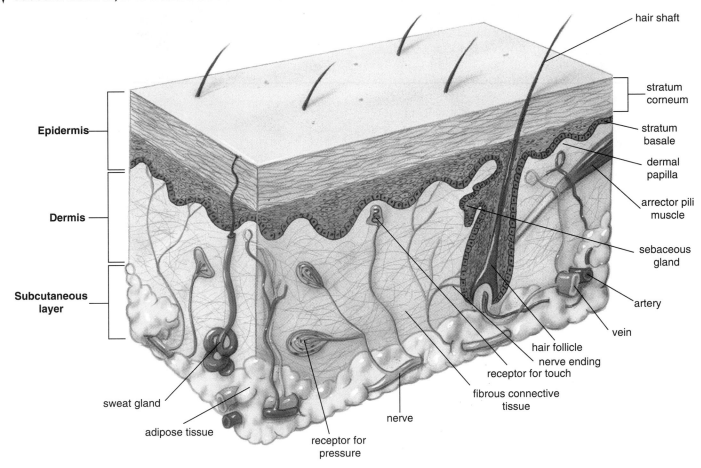

Epidermis

Dermis

Subcutaneous layer

hair shaft

stratum corneum

stratum basale

dermal papilla

arrector pili muscle

sebaceous gland

artery

vein

hair follicle

nerve ending

receptor for touch

fibrous connective tissue

nerve

receptor for pressure

adipose tissue

sweat gland

Structure of the Skin

The skin covers the entire exterior of the human body. In an adult, the skin has a surface area of about 1.8 square meters (20.83 square feet). Usually, the skin is only loosely attached to underlying muscle tissue, but where there are no muscles, the skin attaches directly to bone. For example, there are *flexion creases* where the skin attaches directly to the joints of the fingers.

The skin is sometimes called the **cutaneous membrane** or the **integument,** and since the skin has several accessory organs, it is also possible to speak of the **integumentary system.** The skin (fig. 5.1) has two regions: the epidermis and the dermis. A subcutaneous layer is found between the skin and any underlying structures, such as muscle.

> The skin has two regions: the epidermis and the dermis. A subcutaneous layer lies below the skin.

Epidermis

The **epidermis** (ep″ĭ-der′mis) is the outer and thinner region of the skin. It is made up of *stratified squamous epithelium* divided into several layers; the deepest layer is the stratum basale, and the most superficial layer is the stratum corneum.

Stratum Basale

The basal cells of the **stratum basale** lie just superior to the dermis and are constantly dividing and producing new cells that rise to the surface of the epidermis in two to four weeks. As the cells push away from the dermis, they move progressively farther away from the blood vessels in the dermis. Since these cells are not being supplied with nutrients and oxygen (the epidermis itself lacks blood vessels), they eventually die and are sloughed off.

Langerhans' cells are specialized cells produced in red bone marrow and found in the lower epidermis. These cells phagocytize microbes and then travel to lymphoid organs, where they stimulate the immune system to react.

Melanocytes (mel'ah-no-sīts) are another type of specialized cell located in the lower epidermis. Melanocytes produce **melanin** (mel'ah-nin), the pigment primarily responsible for skin color. Since the number of melanocytes is about the same in all individuals, variation in skin color is due to the amount of melanin produced and its distribution. When Caucasians sunbathe, melanocytes produce more melanin in an attempt to protect the skin from the damaging effects of the ultraviolet (UV) radiation in sunlight. The melanin is passed to other epidermal cells, and the result is tanning, or in some people, the formation of patches of melanin called freckles. A hereditary trait characterized by the lack of ability to produce melanin is known as **albinism** (al'bĭ-nizm). Individuals with this disorder lack pigment not only in the skin, but also in the hair and eyes. Another pigment, called carotene, is present in epidermal cells and in the dermis and gives the skin of people of Asiatic origin its yellowish coloration. The pinkish color of fair-skinned people is due to the pigment hemoglobin in the red blood cells in the capillaries of the dermis.

Stratum Corneum

As cells are pushed toward the surface of the skin, they become flat and hard, forming the tough, uppermost layer of the epidermis, the **stratum corneum.** Hardening is caused by keratinization, the cellular production of a fibrous, waterproof protein called **keratin** (ker'ah-tin). Over much of the body, keratinization is minimal, but the palms of the hands and the soles of the feet normally have a particularly thick outer layer of dead, keratinized cells.

The waterproof nature of keratin protects the body from water loss and water gain. The stratum corneum allows us to live in a desert or a tropical rain forest without damaging our inner cells.

The stratum corneum also serves as a mechanical barrier against microbe invasion. This function is assisted by the secretions of *sebaceous glands* (discussed in a later section of this chapter), which weaken or kill bacteria on the skin.

> The epidermis, the outer region of skin, is made up of stratified squamous epithelium. New cells continually produced in the stratum basale of the epidermis push outward and become the keratinized cells of the stratum corneum, which are sloughed off.

Dermis

The **dermis** is deeper and thicker than the epidermis and contains both loose and fibrous connective tissue. The upper layer of the dermis has fingerlike projections called dermal papillae. Dermal papillae project into and anchor the epidermis. In the overlying epidermis, they cause ridges, resulting in spiral and concentric patterns. The function of the epidermal ridges is to increase friction and thus provide a better gripping surface. Because they are unique to each person, fingerprints and footprints can be used for identification purposes.

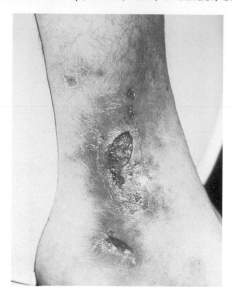

figure 5.2 A bedsore. The most frequent sites for bedsores are in the skin overlying a bony projection, such as on the hip, ankle, heel, shoulder, or elbow.

The dermis contains collagenous and elastic fibers. The *collagenous* fibers are flexible but offer great resistance to overstretching; they prevent the skin from being torn. The elastic fibers stretch to allow movement of underlying muscles and joints, but they maintain normal skin tension. The dermis also contains blood vessels that nourish the skin. Blood rushes into these vessels when a person blushes and is reduced in them when a person turns cyanotic or blue. Sometimes, blood flow to a particular area is restricted in bedridden patients, and consequently, they develop **decubitus** (dik-u'bit-es) **ulcers** (bedsores) (fig. 5.2). These can be prevented by changing the patient's position frequently and by massaging the skin to stimulate blood flow.

There are also numerous sensory nerve fibers in the dermis that take nerve impulses to and from the accessory structures of the skin, which are discussed later in this chapter.

> The dermis, which is composed of fibrous connective tissue, lies beneath the epidermis. It contains collagenous and elastic fibers, blood vessels, and nerve fibers.

Subcutaneous Layer

The **subcutaneous** (sub"ku-ta'ne-us) **layer,** or hypodermis, lies below the dermis. **Subcutaneous injections** are performed with **hypodermic needles.**

The subcutaneous layer is composed of loose connective tissue, including adipose (fat) tissue. **Fat** is an energy storage form that can be called upon when necessary. Adipose tissue also helps insulate the body from either gaining heat from the outside or losing heat from the inside.

figure 5.3 Hair follicle and hair shaft. **a.** A hair grows from the base of a hair follicle when epidermal cells undergo cell division and older cells move outward and become keratinized. **b.** A hair shaft penetrating the outer squamous cells of the epidermis.

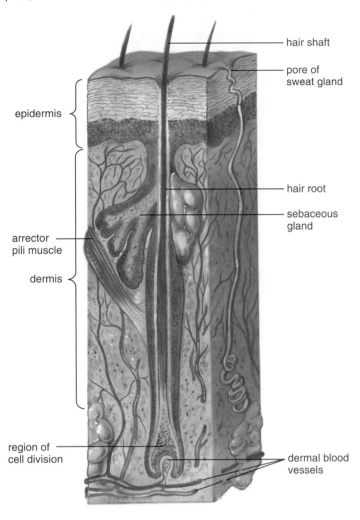

a. Hair follicle

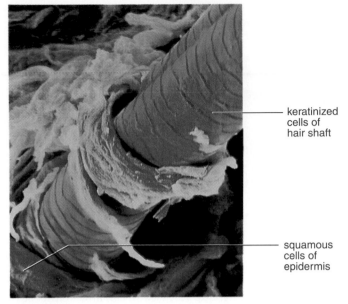

b. Hair shaft

A well-developed subcutaneous layer gives the body a rounded appearance and provides protective padding against external assaults. Excessive development of the subcutaneous layer results in obesity.

> The subcutaneous layer is made up of loose connective tissue and adipose tissue, which insulates the body from heat and cold.

Accessory Structures of the Skin

Hair, nails, and glands are structures of epidermal origin, even though some parts of hair and glands are largely found in the dermis.

Hair and Nails

Hair is found on all body parts except the palms, soles, lips, nipples, and portions of the external reproductive organs. Most of this hair is fine and downy, but the hair on the head includes stronger types as well. After puberty, when sex hormones are made in quantity, there is noticeable hair in the axillary and pelvic regions of both sexes. In the male, a beard develops, and other parts of the body may also become quite hairy. At the time of menopause, when female sex hormones are no longer produced, some women develop **hirsutism** (her'sah-tizm), a condition of excessive body and facial hair that results from the effect of male sex hormones. Hormonal injections and electrolysis to kill roots are possible treatments.

Hairs project from complex structures called **hair follicles.** These hair follicles contain numerous epidermal cells but are located in the dermis of the skin (fig. 5.3). Certain hair follicle cells continually divide, producing new cells that form a hair. At first, the cells are nourished by dermal blood vessels, but as the hair grows up and out of the follicle, they are pushed farther away from this source of nutrients, become keratinized, and die. The portion of a hair within the follicle is called the root, and the portion that extends beyond the skin is called the shaft.

The life of any particular hair is usually three to four months for an eyelash and three to four years for a scalp hair; then it is shed and regrows. In males, baldness occurs when the hair on the head fails to regrow. **Alopecia** (ah-lah-pe'she-ah), the term used to refer to hair loss, can have many causes. Male pattern baldness, or *androgenic alopecia,*

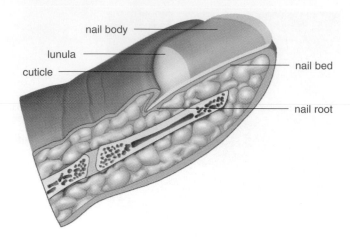

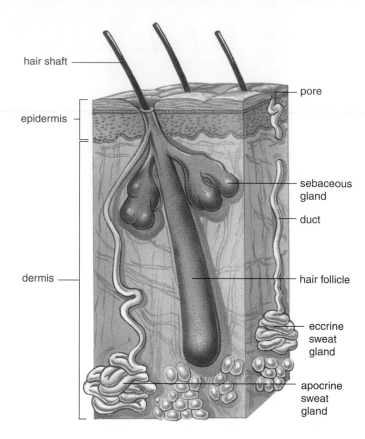

is an inherited condition. *Alopecia areata* is characterized by the sudden onset of patchy hair loss. It is most common among children and young adults, and can affect either sex.

Each hair has one or more oil, or sebaceous, glands, whose ducts empty into the follicle. A smooth muscle, the **arrector pili** (ah-rek'tor pil'i), attaches to the follicle in such a way that contraction of the muscle causes the hair to stand on end. If a person has had a scare or is cold, goose bumps develop, due to contraction of these muscles.

Nails (fig. 5.4) grow from special epithelial cells at the base of the nail in the region called the *nail root*. These cells become keratinized as they grow out over the nail bed. The visible portion of the nail is called the *nail body*. The cuticle is a fold of skin that hides the nail root. Ordinarily, nails grow only about 1 millimeter a week.

The pink color of nails is due to the vascularized dermal tissue beneath the nail. The whitish color of the half-moon–shaped base, or **lunula** (lu'nu-lah), results from the thicker germinal layer in this area.

> Both hair and nails are produced by the division of epidermal cells and consist of keratinized cells.

Glands

The *glands* in the skin are groups of cells specialized to produce and secrete a substance into ducts.

Sweat Glands

Sweat glands, or sudoriferous glands, are present in all regions of the skin. There can be as many as 90 glands per square centimeter on the leg, 400 glands per square centimeter on the palms and soles, and an even greater number on the fingertips. A sweat gland is a tubule that is coiled, particularly at its opening.

Some sweat glands—*apocrine glands* (fig. 5.5)—open into hair follicles in the anal region, groin, and armpits. These glands begin to develop at puberty, and a component of their secretion may act as a sex attractant. These glands become active when a person is under stress.

Other sweat glands—*eccrine glands*—open onto the surface of the skin. They become active when a person is hot and help lower body temperature. The sweat (perspiration) produced by these glands is mostly water, but it also contains salts and some urea, a waste substance. Therefore, sweat is a form of excretion. Ears contain modified sweat glands, called ceruminous glands, which produce cerumen, or earwax.

> Sweat glands are numerous and present in all regions of the skin. Sweating helps the body lower body temperature.

Sebaceous Glands

Most **sebaceous** (sĕ-ba'shus) **glands** are associated with a hair follicle. These glands secrete an oily substance called **sebum** (se'bum) that flows into the follicle and then out onto the skin surface. This secretion lubricates the hair and skin, and helps waterproof them.

Particularly on the nose and cheeks, the sebaceous glands may fail to discharge sebum and the secretions collect, forming whiteheads or blackheads. If pus-inducing bacteria are also present, a boil or pimple may result.

Acne vulgaris, the most common form of acne, is an inflammation of the sebaceous glands that most often occurs during adolescence. Hormonal changes during puberty cause the sebaceous glands to become more active at this time.

> Sebaceous glands are associated with a hair follicle and secrete sebum, which lubricates the hair and skin.

Mammary Glands

The **mammary glands** are modified sweat glands located within the breasts. A female breast contains 15 to 25 lobes, which are divided into lobules (see fig. 18.11). Each lobule contains many alveoli where milk is secreted and enters a milk duct, which leads to the nipple. Cells within the alveoli produce milk only after childbirth.

> Mammary glands are modified sweat glands that produce milk after childbirth.

Functions of the Skin

Protection

The skin forms a protective covering over the entire body, safeguarding underlying parts from physical traumas. The skin's outer dead cells also help prevent bacterial invasions. The oily secretions from sebaceous glands are acidic, which retards the growth of bacteria. Finally, since outer skin cells are dead and keratinized, the skin is waterproof, thereby preventing fluid (water) loss. The skin's waterproofing also prevents water from entering the body when the skin is immersed.

> Skin protects the body from physical trauma, bacterial invasion, and fluid gain or loss.

Synthesis of Vitamin D

When skin cells are exposed to sunlight, the ultraviolet (UV) rays assist them in producing vitamin D. The cells contain a precursor molecule that is converted to vitamin D in the body after UV exposure. Since only a small amount of UV radiation is needed to change the precursor molecule to vitamin D, the skin should not be exposed unnecessarily. Vitamin D leaves the skin and enters the liver and kidneys, where it is converted to a hormone called calcitriol. Calcitriol circulates throughout the body, regulating calcium and phosphorus metabolism where appropriate. Calcium and phosphorus are very important to the proper

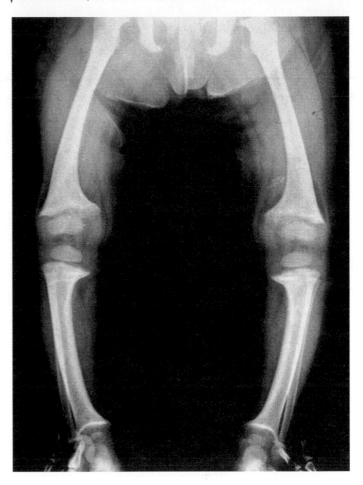

figure 5.6 X ray of child with rickets. Rickets develops from an improper diet and also from a lack of ultraviolet (UV) light (sunlight). Under these conditions, vitamin D does not form in the skin.

development and maintenance of the bones. Most milk today is fortified with vitamin D, which helps prevent the occurrence of **rickets** (fig. 5.6), or a defective mineralization of the skeleton.

> The skin contains a precursor molecule that is converted to vitamin D following exposure to UV radiation. A hormone derived from vitamin D helps regulate calcium and phosphorus metabolism, which is so important in bone development.

Sensory Reception and Communication

Specialized nerve endings in the dermis are sensory receptors for touch, pressure, pain, hot, and cold (see fig. 5.1). The fingertips contain the greatest number of touch receptors, which allows the fingers to be used for delicate tasks. The sensory receptors also account for the use of the skin as a means of communication between people.

Regulation of Body Temperature

As discussed in chapter 3, the energy content of nutrient molecules like glucose is converted to ATP energy within mitochondria (p. 38). When this conversion occurs, heat is released. Only about 40% of the energy available in a glucose molecule becomes ATP energy; the rest escapes as heat. Also, when ATP is broken down, as occurs when muscles contract, heat is released.

One of the best examples of homeostasis is regulation of normal body temperature (36.2–37.7°C [97–100°F]). The hypothalamus, the portion of the brain concerned with homeostasis, is involved in regulating body temperature. As figure 5.7 shows, the skin also regulates body temperature in two ways: (1) the blood vessels in the dermis can constrict to receive less blood or can dilate to receive more blood; and (2) the sweat glands can remain inactive or can secrete sweat as needed.

If body temperature starts to rise, the blood vessels dilate so that more blood is brought to the surface of the skin for cooling, and the sweat glands become active. Sweat absorbs body heat, and this heat is carred away as sweat evaporates. If the weather is humid, evaporation is hindered, but cooling can be assisted by a cool breeze.

If the outer temperature is cool, the sweat glands remain inactive, and the blood vessels constrict so that less blood is brought to the skin's surface. Whenever the body's temperature falls below normal, the muscles start to contract, causing shivering, which produces heat. The arrector pili muscles attached to hair follicles are also involved in this reaction, and this is why goose bumps occur when a person is cold. If the outside temperature is extremely cold and blood flow to the skin is severely restricted for an extended period, a portion of the skin will die, resulting in *frostbite*.

> The skin is involved in regulation of body temperature. Surface blood vessels dilate and the sweat glands are active when the body is too hot. Surface blood vessels constrict and the sweat glands are inactive when the body is cold.

Hyperthermia

Hyperthermia, a body temperature above normal, indicates that the body's regulatory mechanisms have been overcome. If *heat exhaustion* sets in, the individual becomes tired, complains of a headache, and may experience vomiting. Blood pressure may be low, and there may have been a loss of salts, due to profuse sweating. Bed rest and increased fluid and salt intake are helpful.

Heat stroke is characterized by an elevated temperature, up to 43°C (110°F). Dizziness, confusion, and delusions may occur. It is important to cool the body off immediately by immersing the person in cool water. Medical care is needed to restore the body's proper fluid and salt balance.

Fever is a special case of hyperthermia that can be brought on by an illness, such as an infection. Bacteria release toxic substances, some of which are called *pyrogens* (*pyro*, meaning fire, heat) because they cause fever. During the first stages of an infection, the body's thermostat in the hypothalamus is reset at a higher level. The person feels chilled, even though the body's temperature is increasing. Once the temperature has risen to the new setting, the body will maintain this temperature (that is, the fever will continue) until the infection has been brought to an end. At that time, the fever will break, as the skin becomes flushed and sweating occurs, and the patient is said to have passed the *crisis*.

Hypothermia

Hypothermia, a body temperature below normal, also indicates that the body's regulatory mechanisms have been overcome. At first hypothermia is characterized by uncontrollable shivering, incoherent speech, and a lack of coordination (body temperature 90–95°F). Then the pulse rate slows and hallucinations occur as unconsciousness develops (body temperature 80–90°F). Breathing becomes shallow, and there is diminished shivering; rigidity of the body develops. This degree of hypothermia is associated with a 50% mortality.

> Hyperthermia and hypothermia are two conditions that can result when the body's temperature regulatory mechanism is overcome. With hyperthermia, the body temperature rises above normal, and with hypothermia, the body temperature falls below normal.

Disorders of the Skin

The skin is subject to many disorders, some of which are more annoying than they are life threatening. For example, **athlete's foot** is caused by a fungal infection that usually involves the skin of the toes and soles. **Impetigo** (im″pĕ-ti′go) is a highly contagious disease occurring most often in young children. It is caused by a bacterial infection that results in pustules that crust over. **Eczema** (ek′zĕ-mah) and **psoriasis** (so-ri′ah-sis) are due not to an infection, but to overactive cell division, which results in areas of scaling and itching. Eczema is caused by sensitivity to various chemicals (soaps or detergents), to certain fabrics, or even to heat or dryness. Psoriasis is a chronic condition, possibly hereditary, in which the skin develops pink or reddish patches covered by silvery scales. Its cause is unknown, but outbreaks may be triggered by stress or infections. **Dandruff** is a skin disorder not caused by a dry scalp, as is commonly thought, but by an accelerated rate of keratinization in certain areas of the scalp, producing flaking and itching. **Urticaria** (ur″tĭ-ka′re-ah), or hives, is an allergic reaction characterized by the appearance of reddish, elevated patches and often by itching.

Skin Cancer

Skin cancer is categorized as either melanoma or nonmelanoma. **Melanoma** (fig. 5.8*a*), the malignant type of skin cancer, starts in the melanocytes and has the appearance of an unusual **mole.** Unlike a normal mole, which is

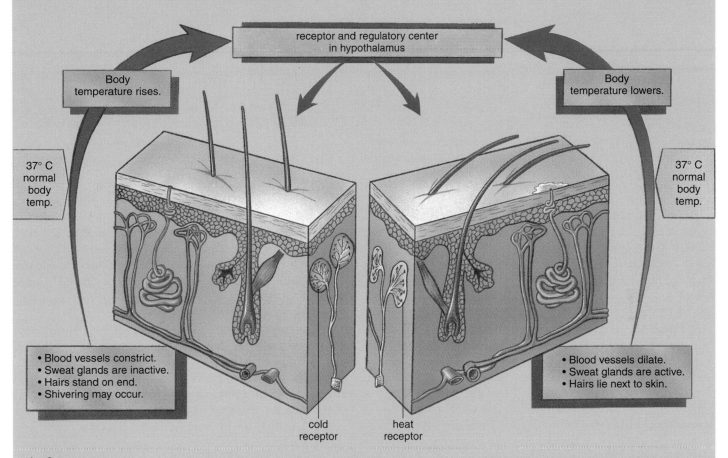

receptor and regulatory center
in hypothalamus

Body
temperature rises.

Body
temperature lowers.

37° C
normal
body
temp.

37° C
normal
body
temp.

- Blood vessels constrict.
- Sweat glands are inactive.
- Hairs stand on end.
- Shivering may occur.

- Blood vessels dilate.
- Sweat glands are active.
- Hairs lie next to skin.

cold
receptor

heat
receptor

figure 5.7 Temperature control. When body temperature rises, the regulatory center in the hypothalamus directs the blood vessels to dilate and the sweat glands to be active, thereby lowering the body temperature. When body temperature lowers, the regulatory center directs the blood vessels to constrict, may initiate shivering, and ceases to activate the sweat glands, thereby raising the body temperature.

figure 5.8 Skin cancer. In each of the three types shown, the skin clearly has an abnormal appearance.

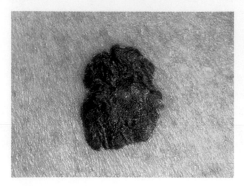

Melanoma

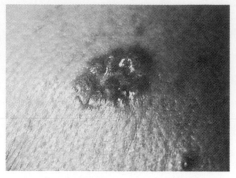

Basal cell carcinoma

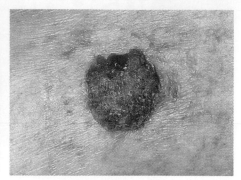

Squamous cell carcinoma

dark, circular, and confined, a melanoma mole looks like a spilled ink spot, and a single melanoma mole may display a variety of shades. A melanoma mole can also itch, hurt, or feel numb. The skin around the mole turns gray, white, or red. Melanoma is most common in fair-skinned persons, particularly if they have suffered occasional severe burns as children. Melanoma risk increases with the number of moles a person has. Most moles appear before the age of 14, and their appearance is linked to sun exposure. Melanoma rates have risen since the turn of the century, but the incidence has doubled in the last decade. Now, about 32,000 cases of melanoma are diagnosed each year, and one in five diagnosed persons dies within five years.

Nonmelanoma cancers called basal cell carcinoma and squamous cell carcinoma are much less likely to metastasize or spread than melanoma cancer. **Basal cell carcinoma**

(fig. 5.8b), the most common type of skin cancer, begins when ultraviolet (UV) radiation causes epidermal basal cells to form a tumor, while at the same time suppressing the immune system's ability to detect the tumor. The signs of a tumor are varied. They include an open sore that will not heal; a recurring reddish patch; a smooth, circular growth with a raised edge; a shiny bump; or a pale mark. About 95% of patients are easily cured, but reoccurrence is common.

Squamous cell carcinoma (fig. 5.8c) begins in the epidermis proper. While five times less common than basal cell carcinoma, it is more likely to spread to nearby organs, and the death rate is about 1% of cases. The signs of squamous cell carcinoma are the same as that for basal cell carcinoma, except that it may also show itself as a wart that bleeds and scabs.

Link Between UV Radiation and Skin Cancer

*t*he incidence of skin cancer is related to ultraviolet (UV) exposure. The sun gives off two types of UV rays: UV-A rays and UV-B rays. UV-A rays penetrate the skin deeply, affect connective tissue, and cause the skin to sag and wrinkle. UV-A rays are also believed to increase the effects of UV-B rays, which are the cancer-causing rays. UV-B rays are more prevalent at midday.

Even if you live in Alaska, you should take the following steps to protect yourself from the sun:

- Use a broad-spectrum sunscreen that protects you from both UV-A and UV-B radiation, with an SPF (sun protection factor) of at least 15. (This means that if you usually burn, for example, after a 20-minute exposure, it will take 15 times longer, or 5 hours, before you will burn.)

- Wear protective clothing. Choose fabrics with a tight weave, and wear a wide-brimmed hat. A baseball cap does not protect the rims of the ears.

- Stay out of the sun altogether between the hours of 10 A.M. and 3 P.M. This will reduce your annual exposure by as much as 60%.

- Wear sunglasses that have been treated to absorb both UV-A and UV-B radiation. Otherwise, sunglasses can expose your eyes to more damage than usual because pupils dilate in the shade.

- Avoid tanning machines. Although most tanning devices use only high levels of UV-A radiation, the deep layers of the skin become more vulnerable to UV-B radiation upon later exposure to the sun.

Raised growths on the skin, such as moles and **warts**, usually are not cancerous. Moles are due to an overgrowth of melanocytes, and warts are due to a viral infection.

As discussed in the Medical Focus reading on page 70, the development of skin cancers may be associated with exposure to the sun's rays.

> Skin cancer is associated with ultraviolet radiation and occurs in three forms. Malignant melanoma is the most dangerous form of skin cancer. Squamous cell carcinoma and basal cell carcinoma are less severe forms and can usually be removed surgically.

Burns

The epidermal injury known as a burn is usually caused by heat but can also be caused by radioactive, chemical, or electrical agents. Two factors affect burn severity: the depth of the burn and the extent of the burned area (fig. 5.9).

One way to classify burns is according to the depth of the burned area. In *first-degree burns*, only the epidermis is affected. There is redness and pain, but no blisters or swelling. A classic example of a first-degree burn is a moderate sunburn. The pain subsides within 48–72 hours, and the injury heals without further complications or scarring. The damaged skin peels off in about a week.

A *second-degree burn* extends through the entire epidermis and part of the dermis. There is not only redness and pain, but also blisters in the region of the damaged tissue. The deeper the burn, the more prevalent the blisters, which enlarge during the hours after the injury. Unless they become infected, most second-degree burns heal without complications and with little scarring in 10–14 days. If the burn extends deep into the dermis, it heals more slowly over a period of 30–105 days. The healing epidermis is extremely fragile, and scarring is common. First- and second-degree burns are sometimes referred to as partial-thickness burns.

Third-degree, or full-thickness, *burns* destroy the entire thickness of the skin. The surface of the wound is leathery and may be brown, tan, black, white, or red. There is no pain because the pain receptors have been destroyed, as have blood vessels, sweat glands, sebaceous glands, and hair follicles.

Epidermis from the Laboratory

t issue culture is the growth of cells in laboratory glassware. The epidermis is very suitable for regeneration in tissue culture because it largely consists of one type of cell—epidermal cells. In 1974, Howard Green of Harvard Medical School discovered the conditions that allow epidermal cells to grow in culture. In 1979, Nicholas O'Connor and John Mulliken treated badly burned adults at the Peter Bent Brigham Hospital (now called Brigham and Women's Hospital) in Boston with cultured grafts. Skin biopsies from these patients were brought to Green's laboratory, and their epidermal cells were cultured in small petri dishes. (A patient's own cells must be used because skin grafts donated by another person will be rejected by the patient's immune system.) In each dish, the cultured cells grew to become an epithelium, which was detached from the glass and attached to a gauze backing (fig. 5A). The preparations were brought to the hospital and applied to patients, and about half of the grafts took and generated epidermis. Patients under the care of John P.

Remensnyder at the Shriners Burn Institute of Boston have had as much as half of their epidermis regenerated from culture.

The amount of unburned epidermis needed for cultivation is small. A piece of epidermis 3 centimeters square can be expanded more than 5,000 times within three to four weeks, yielding nearly enough epithelium to cover the body surface of an adult human.

In addition to being used to treat third-degree burns, laboratory epidermis has been prepared for the removal of scars and skin ulcers and for regenerating linings of the mouth and urinary tract.

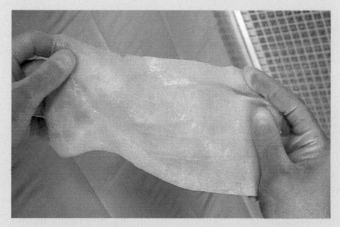

figure 5A Laboratory skin. Large sheet of skin grown in the laboratory from cells of a severely burned patient.

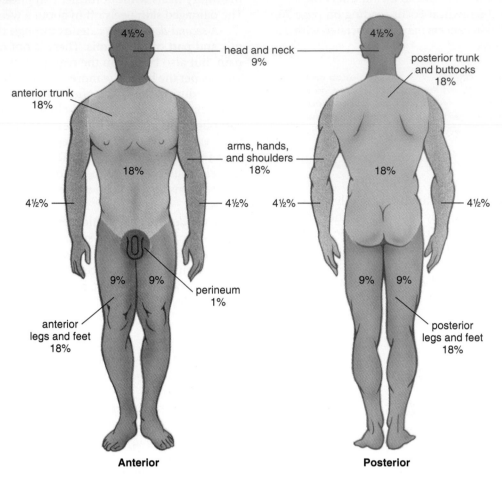

figure 5.9 The "rule of nines" for estimating the extent of burns.

head and neck
9%

4½%

posterior trunk
and buttocks
18%

anterior trunk
18%

arms, hands,
and shoulders
18%

18%

18%

4½%

4½%

4½%

4½%

perineum
1%

9% 9%

9% 9%

anterior
legs and feet
18%

posterior
legs and feet
18%

Anterior

Posterior

Fourth-degree burns involve tissues down to the bone. Obviously, the chances of a person surviving fourth-degree burns are not good unless a very limited area of the body is affected.

The major concerns with severe burns are fluid loss, heat loss, and bacterial infection. Fluid loss is counteracted by intravenous administration of a balanced salt solution. Heat loss is minimized by placing the burn patient in a warm environment. Bacterial infection is treated by isolation and the application of an antibacterial dressing.

As soon as possible, the damaged tissue is removed, and skin grafting is begun. The skin needed for grafting is usually taken from other parts of the patient's body. This is called *autografting,* as opposed to heterografting, in which the graft is received from another person. Autografting is preferred because rejection rates are very low. However, if the burned area is quite extensive, it may be difficult to acquire enough skin for autografting. As described in the Medical Focus reading on page 71, researchers are experimenting with cultured skin that is grown in the laboratory from only a few cells taken from the patient.

As noted earlier, the severity and treatment of burns depends not only on their depth but also on the amount of area involved. A useful technique for estimating the extent of a burn, called the "rule of nines," is often employed (fig. 5.9). In this method, the total body surface is divided into regions as follows: the head and neck, 9% of the total body surface; each upper limb, 9%; each lower limb, 18%; the front and back portions of the trunk, 18% each; and the perineum, which includes the anal and urogenital region, 1%.

> The severity of a burn depends on its depth and extent. First-degree burns affect only the epidermis. Second-degree burns affect the entire epidermis and a portion of the dermis. Third-degree burns affect the entire epidermis and dermis. The "rule of nines" provides a means of estimating the extent of a burn injury.

Wound Healing

Wound healing demonstrates the regenerative powers of epidermal tissue. The trauma may be extensive enough to open the blood vessels in the dermis. If so, the blood clot that forms is gradually converted to a scab. In the meantime, the basal layer of the epidermis begins to produce new cells at a faster than usual rate. Eventually, the wound

figure 5.10 The process of wound healing. **a.** A deep wound ruptures blood vessels, and blood flows out and fills the wound. **b.** After a blood clot forms, a protective scab develops. Fibroblasts and white blood cells migrate to the wound site. **c.** New epidermis forms, and fibroblasts promote tissue regeneration. **d.** Freshly healed skin.

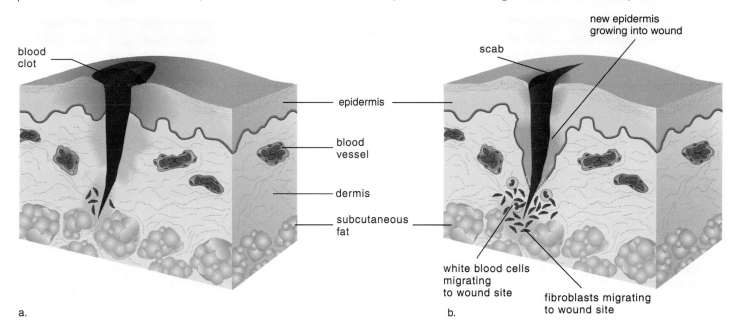

a.

b.

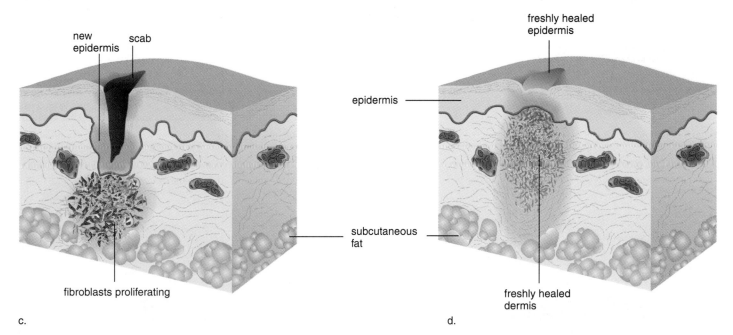

c.

d.

fills in (fig. 5.10), but if the wound is deep, scarring is likely. A scar is tissue composed of many collagenous fibers arranged so as to provide maximum strength. A scar does not contain the accessory organs of the skin and is usually devoid of feeling.

> The skin has regenerative powers and can grow back on its own if a wound is not too extensive.

Effects of Aging

As aging occurs, the epidermis maintains its thickness, but there is a decreased turnover of cells. The dermis becomes thinner, the dermal papillae flatten, and the epidermis is held less tightly to the dermis so that the skin is looser. Adipose tissue in the subcutaneous layer of the face and hands also decreases, which means that older people are more likely to feel cold.

Skeletal System

Skin protects bones; helps provide vitamin D for Ca^{2+} absorption.

Bones provide support for skin.

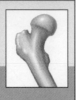

How the Integumentary System works with other body systems

Lymphatic System/Immunity

Skin serves as a barrier to pathogen invasion; Langerhans' cells phago-cytize pathogens; protects lymphatic vessels.

Lymphatic vessels pick up excess tissue fluid; immune system protects against skin infections.

Muscular System

Skin protects muscles; rids the body of heat produced by muscle contraction.

Muscle contraction provides heat to warm skin.

Respiratory System

Skin helps protect respiratory organs and helps regulate body temperature.

Gas exchange in lungs provides oxygen to skin and rids body of carbon dioxide from skin.

Nervous System

Skin protects nerves, helps regulate body temperature; skin receptors send sensory input to brain.

Brain controls nerves that regulate size of cutaneous blood vessels, activate sweat glands and arrector pili muscles.

Digestive System

Skin helps to protect digestive organs; helps provide vitamin D for Ca^{2+} absorption.

Digestive tract provides nutrients needed by skin.

Endocrine System

Skin helps protect endocrine glands.

Androgens activate sebaceous glands and help regulate hair growth.

Urinary System

Skin helps regulate water loss; sweat glands carry on some excretion.

Kidneys compensate for water loss due to sweating; activate vitamin D precursor made by skin.

Circulatory System

Skin prevents water loss; helps regulate body temperature; protects blood vessels.

Blood vessels deliver nutrients and oxygen to skin, carry away wastes; blood clots if skin is broken.

Reproductive System

Skin receptors respond to touch; modified sweat glands produce milk; skin stretches to accommodate growing fetus.

Androgens activate oil glands; sex hormones stimulate fat deposition, affect hair distribution in males and females.

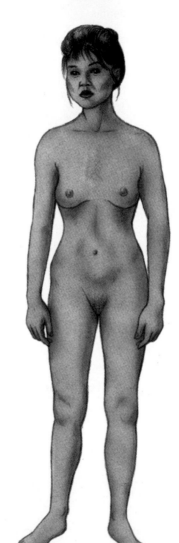

The fibers within the dermis change with age. The collagenous fibers become coarser, thicker, and farther apart; therefore, there is less collagen than before. Elastic fibers in the upper layer of the dermis are lost, and those in the lower dermis become thicker, less elastic, and disorganized. The skin wrinkles because (1) the epidermis is loose, (2) the fibers are fewer and those remaining are disorganized, and (3) the subcutaneous layer has less padding.

With aging, homeostatic adjustment to heat is limited because there is less vasculature (fewer blood vessels) and fewer sweat glands. The number of hair follicles decreases so that the hair on the scalp and extremities thins out. A reduced number of sebaceous glands results in the skin tending to crack.

As a person ages, the number of melanocytes decreases. This causes the hair to turn gray and the skin to become paler. In contrast, some of the remaining pigment cells are larger, and pigmented blotches appear in the skin.

Many of the changes that occur in the skin as a person ages appear to be due to sun damage. Ultraviolet radiation causes rough skin, mottled pigmentation, fine lines and wrinkles, deep furrows, numerous benign skin growths, and the various types of skin cancer previously discussed.

Working Together

The accompanying illustration shows how the integumentary system works with other organ systems of the body to maintain homeostasis. All the organ systems of the body are interrelated.

Selected New Terms

Basic Key Terms

arrector pili (ah-rek'tor pil'i), p. 66
cutaneous membrane (ku-ta'ne-us mem'brān), p. 63
dermis (der'mis), p. 64
epidermis (ep"ĭ-der'mis), p. 63
hair follicle (hār fol'ĭ-kl), p. 65
integument (in-teg'u-ment), p. 63
integumentary system (in-teg"u-men'tar-e sis'tem), p. 63
keratin (ker'ah-tin), p. 64
lunula (lu'nu-lah), p. 66
melanin (mel'ah-nin), p. 64
melanocyte (mel'ah-no-sīt), p. 64
sebaceous gland (sĕ-ba'shus gland), p. 66
sebum (se'bum), p. 66
subcutaneous layer (sub"ku-ta'ne-us la'er), p. 64
sweat gland (swet gland), p. 66

Clinical Key Terms

acne vulgaris (ak'ne vul-gar'is), p. 67
albinism (al'bĭ-nizm), p. 64

alopecia (ah-lah-pe'she-ah), p. 65
athlete's foot (ath'lēts fŭt), p. 68
basal cell carcinoma (bās'al sel kar-sĭ-no'mah), p. 70
dandruff (dan'druf), p. 68
decubitus ulcers (dik-u'bit-es ul'ser), p. 64
eczema (ek'zĕ-mah), p. 68
hirsutism (her'sah-tizm), p. 65
hyperthermia (hi"per-ther'me-ah), p. 68
hypodermic needle (hi-po-der'mik ne'dul), p. 64
hypothermia (hi"po-ther'me-ah), p. 68
impetigo (im"pĕ-ti'go), p. 68
melanoma (mel-ah-no'mah), p. 68
mole (mōl), p. 68
psoriasis (so-ri'ah-sis), p. 68
rickets (rik'ets), p. 67
squamous cell carcinoma (skwa'mus sel kar-sĭ-no'mah), p. 70
subcutaneous injection (sub"ku-ta'ne-us in-jek'shun), p. 64
urticaria (ur"tĭ-kar'e-ah), p. 68
warts (worts), p. 71

Summary

I. **Structure of the Skin**
The skin has two regions: the epidermis and the dermis. A subcutaneous layer lies below the skin.
 A. The epidermis, the outer region of the skin, is made up of stratified squamous epithelium. New cells continually produced in the stratum basale of the epidermis push outward and become the keratinized cells of the stratum corneum, which are sloughed off.
 B. The dermis, which is composed of fibrous connective tissue, lies beneath the epidermis. It contains collagenous and elastic fibers, blood vessels, and nerve fibers.
 C. The subcutaneous layer is made up of loose connective tissue and adipose tissue, which insulates the body from heat and cold.

II. **Accessory Structures of the Skin**
Accessory structures of the skin include hair, nails, and glands.

A. Both hair and nails are produced by the division of epidermal cells and consist of keratinized cells.
B. Sweat glands are numerous and present in all regions of the skin. Sweating helps the body lower body temperature.
C. Sebaceous glands are associated with a hair follicle and secrete sebum, which lubricates the hair and skin.
D. Mammary glands are modified sweat glands that produce milk after childbirth.

III. Functions of the Skin
A. Protection. Skin protects the body from physical trauma, bacterial invasion, and fluid gain or loss.
B. Synthesis of vitamin D. The skin contains a precursor molecule that is converted to vitamin D following exposure to UV radiation. A hormone derived from vitamin D helps regulate calcium and phosphorus metabolism involved in bone development.
C. Sensory reception and communication. The skin contains sensory receptors for touch, pressure, pain, hot, and cold, which help people to be aware of their surroundings.
D. Regulation of body temperature. Surface blood vessels dilate and the sweat glands are active when the body is too hot. Surface blood vessels constrict and the sweat glands are inactive when the body is cold.
E. Hyperthermia and hypothermia are two conditions that can result when the body's temperature regulatory mechanism is overcome. With hyperthermia, the body temperature rises above normal, and with hypothermia, the body temperature falls below normal.

IV. Disorders of the Skin
A. Skin cancer. Skin cancer, which is associated with ultraviolet radiation, occurs in three forms. Melanoma is the most dangerous form of skin cancer. Squamous cell carcinoma and basal cell carcinoma are less severe forms and can usually be removed surgically.
B. Burns. The severity of a burn depends on its depth and extent. First-degree burns affect only the epidermis. Second-degree burns affect the entire epidermis and a portion of the dermis. Third-degree burns affect the entire epidermis and dermis. The "rule of nines" provides a means of estimating the extent of a burn injury.
C. Wound healing. The skin has regenerative powers and can grow back on its own if a wound is not too extensive.

Study Questions

1. In general, describe the two regions of the skin. (pp. 63–64)
2. Describe the process by which epidermal tissue continually renews itself. (p. 63)
3. What function does the dermis have in relation to the epidermis? (p. 64)
4. What primary role does adipose tissue play in the subcutaneous layer? (p. 64)
5. Describe in general the structure of a hair follicle and a nail. How do hair follicles and nails grow? (pp. 65–66)
6. Describe the structure and function of sweat glands and sebaceous glands. (pp. 66–67)
7. Describe the structure of a mammary gland. (p. 67)
8. List and describe four functions of the skin. (pp. 67–68)
9. Name the three types of skin cancer, and explain why sunlight causes skin cancer. (pp. 68, 70)
10. Explain how to determine the severity of a burn. Describe the proper treatment for burns. (pp. 71–72)
11. Describe how a wound heals and a scar forms. (pp. 72–73)

Objective Questions

I. Match the terms in the key to the items in questions 1–5.
Key:
a. epidermis
b. dermis
c. subcutaneous layer
1. blood vessels and nerve fibers
2. fat cells
3. basal cells
4. location of sweat glands
5. many collagenous and elastic fibers

II. Fill in the blanks.
6. Sebaceous glands are associated with _____ in the dermis, and they secrete an oily substance called _____ .
7. Sweat glands are involved in body _____ regulation.
8. Skin protects against _____ trauma, _____ invasion, and _____ gain or loss.
9. Skin cells produce vitamin _____ , which is needed for strong bones.
10. The severity of a burn is determined by _____ and _____ .
11. The type of skin cancer with the highest death rate is _____ , while the most common form is _____ .

Consult Appendix B for help in pronouncing and analyzing the meaning of the terms that follow.

1. epidermomycosis (ep″ĭ-der″mo-mi-ko′sis)
2. melanogenesis (mel″ah-no-jen′ĕ-sis)
3. acrodermatosis (ak″ro-der″mah-to′sis)
4. pilonidal cyst (pi″lo-ni′dal sist)
5. mammoplasty (mam′o-plas″te)
6. antipyretic (an″ti-pi-ret′ik)
7. dermatome (der′mah-tōm)
8. hypodermic (hy″po-der′mik)
9. trichophagia (trik″o-fā′je-ah)
10. onychocryptosis (on″ĭ-ko-krip-to′sis)
11. hyperhydrosis (hi″per-hi-dro′sis)
12. rhytidectomy (rit″ĭ-dek′to-me)

Website Link

For a listing of the most current Websites related to this chapter, please visit the Mader home page at:

www.mhhe.com/maderap

chapter 6

The Skeletal System

■ Chapter Outline and Learning Objectives

After you have studied this chapter, you should be able to:

Skeleton: Overview (p. 79)
- Name at least five functions of the skeleton.
- Explain a classification of bones based on their shapes.
- Describe the anatomy of a long bone.
- Describe the growth and development of bones.
- Name and describe eight types of fractures, and state the four steps in fracture repair.

Axial Skeleton (p. 82)
- Distinguish between the axial and appendicular skeletons.
- Name the bones of the skull, and state the important features of each bone.
- Describe the structure and function of the hyoid bone.
- Name the bones of the vertebral column and the thoracic cage. Be able to label diagrams of them.
- Describe a typical vertebra, the atlas and axis, and the sacrum and coccyx.
- Name the three types of ribs and the three parts of the sternum.

Appendicular Skeleton (p. 91)
- Name the bones of the pectoral girdle and the pelvic girdle. Be able to label diagrams of them.
- Name the bones of the upper limb (arm) and the lower limb (leg). Be able to label diagrams that include surface features.
- Cite at least five differences between the female and male pelvises.

Joints (Articulations) (p. 100)
- Explain how joints are classified, and give examples of each type of joint.
- List the types of movements that occur at synovial joints.

Effects of Aging (p. 105)
- Anatomical and physiological changes occur in the skeletal system as we age.

Working Together (p. 105)
- The skeletal system works with other systems of the body to maintain homeostasis.

MedAlert
Arthritic Joints (p. 103)

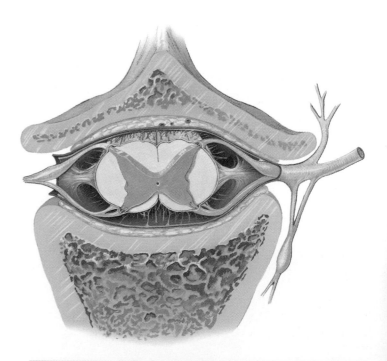

Vertebra and spinal cord in cross section. The vertebrae form the vertebral column, which protects the spinal cord from injury.

Part II

figure 6.1 Classification of bones. **a.** Long bones are longer than they are wide. **b.** Short bones are cube-shaped; their lengths and widths are about equal. **c.** Flat bones are platelike and have broad surfaces. **d.** Irregular bones have varied shapes with many places for connections with other bones. **e.** Round bones are circular.

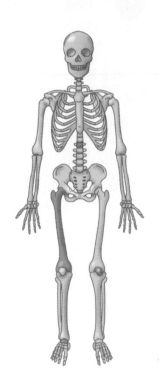

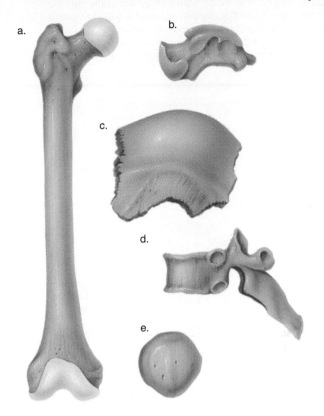

Skeleton: Overview

The skeletal system consists of the bones (206 in adults) and joints, along with the cartilage and ligaments that occur at the joints. The skeleton is divided into the axial skeleton (skull, hyoid bone, vertebral column, and thoracic cage) and the appendicular skeleton (girdles and limbs).

Functions

The skeleton has the following functions:

1. The skeleton—notably, the large, heavy bones of the legs—supports the body and its organs against the pull of gravity.
2. The skeleton protects soft body parts. For example, the skull forms a protective encasement for the brain.
3. Certain bones, such as those of the skull, ribs, and sternum (breastbone), produce blood cells in both adults and children.
4. Bones are storage areas for mineral salts, notably calcium salts.
5. Bones, especially those of the legs and arms, provide sites for muscle attachment and permit flexible body movement.

Anatomy of a Long Bone

Although the bones of the skeletal system vary considerably in shape as well as size (fig. 6.1), a long bone, such as one in the arm or leg, can be used to illustrate certain principles of bone anatomy (fig. 6.2). The bone is enclosed in a tough, fibrous, connective tissue covering called the **periosteum** (per″e-os′te-um), which is continuous with the ligaments and tendons that anchor bones. The periosteum contains blood vessels that enter the bone and service its cells. At both ends of a long bone is an expanded portion called an **epiphysis** (ĕ-pif′ĭ-sis); the portion between the epiphyses is called the **diaphysis** (di-af′ĭ-sis).

When bone is split open, as in figure 6.2, the section shows that the diaphysis is not solid but has a **medullary** (med′u-lār″e) **cavity** containing yellow marrow. Yellow marrow contains large amounts of fat. The medullary cavity is bounded at the sides by compact bone. The epiphyses contain spongy bone. Beyond the spongy bone, there is a thin shell of compact bone and, finally, a layer of cartilage called the **articular cartilage**. Articular cartilage is so named because here bone articulates (meets) another bone. **Articulation** is the joining together of bones at a

figure 6.2 Anatomy of a long bone. **a.** A long bone is encased by the periosteum except at the epiphyses, which are covered by articular cartilage. Spongy bone of the epiphyses may contain red bone marrow. The diaphysis contains yellow bone marrow and is bordered by compact bone. **b.** Detailed anatomy of spongy bone and compact bone are shown in the enlargement.

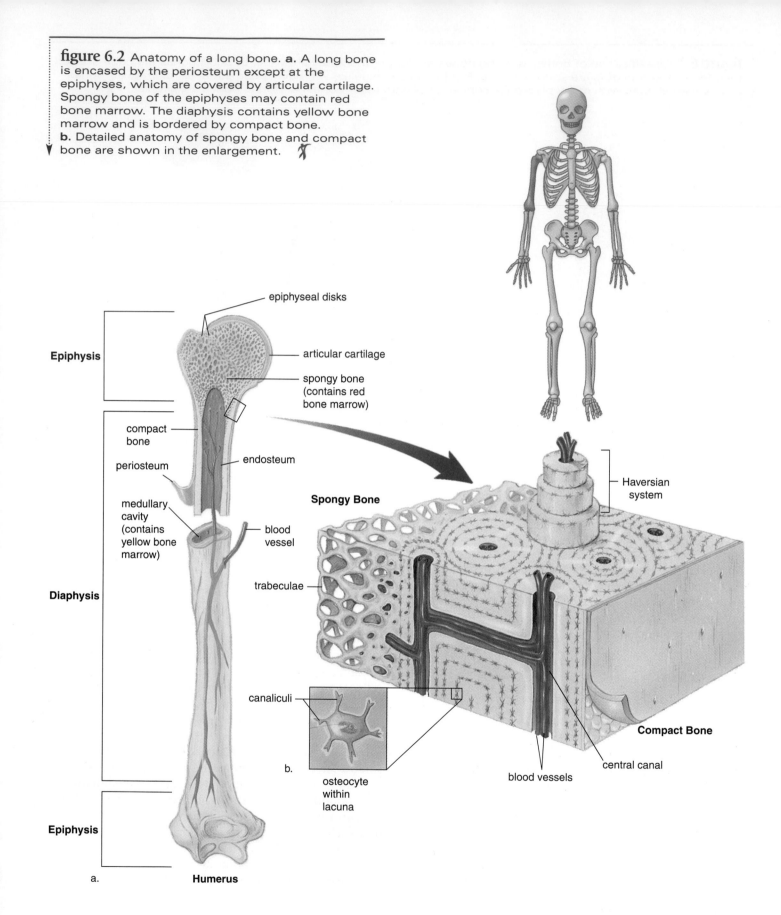

Epiphysis

epiphyseal disks

articular cartilage

spongy bone (contains red bone marrow)

compact bone

periosteum

endosteum

medullary cavity (contains yellow bone marrow)

blood vessel

Diaphysis

Spongy Bone

trabeculae

Haversian system

canaliculi

Compact Bone

b.

osteocyte within lacuna

central canal

blood vessels

Epiphysis

a.   **Humerus**

figure 6.3 Major stages in the development of an endochondral bone.

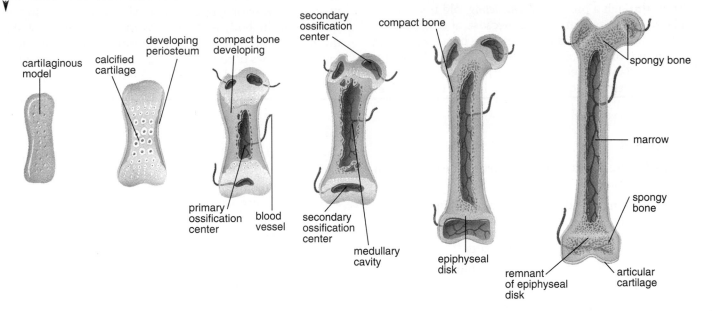

joint. The medullary cavity and the spaces of spongy bone are lined with *endosteum,* a thin layer of squamous epithelium.

Compact bone, or dense bone, as discussed previously (p. 55), contains **osteocytes** (os'te-o-sīts) (bone cells) in tiny chambers called *lacunae.* The osteocytes are arranged in a cylinder of concentric layers called *lamellae,* and the osteocytes and lamellae surrounding a single central canal comprise a *Haversian system.* Blood vessels and nerves from the periosteum enter the central canal. Osteocytes are connected to the central canal and to each other by passageways called *canaliculi.* The lacunae are separated by a matrix that contains collagenous protein fibers and mineral deposits, primarily of calcium and phosphorus salts.

Spongy bone, or cancellous bone, contains numerous bony bars and plates, called *trabeculae.* Although lighter than compact bone, spongy bone is still designed for strength. Like braces used for support in buildings, the trabeculae of spongy bone follow lines of stress.

In infants, **red bone marrow,** a specialized tissue that produces blood cells, is found in the cavities of most bones. In adults, red blood cell formation, called **hematopoiesis** (hem"ah-to-poi-e'sis), occurs in the spongy bone of the skull, ribs, sternum (breastbone), and vertebrae, and in the ends of the long bones.

A long bone has a shaft (diaphysis) and two ends (epiphyses). The diaphysis contains a medullary cavity with yellow marrow, and the epiphyses contain spongy bone with red marrow.

Growth and Development

Most of the skeleton is cartilaginous during prenatal development. The cartilage structures are shaped like the future bones and therefore provide "models" of the bones. Cartilaginous models are converted to bone when mineral salts are deposited in the matrix—first, by certain of the cartilage cells, and later, by bone-forming cells called **osteoblasts** (os'te-o-blasts). Conversion of cartilaginous models to bones is called **endochondral ossification.** Another type of ossification, called *intramembranous ossification,* occurs without a previous cartilaginous model. Facial bones and certain cranial bones form in this way.

Figure 6.3 shows the process of endochondral ossification in a long bone. At first, a primary ossification center is located in the middle of the bone. Later, a medullary cavity is surrounded by compact bone, and secondary ossification centers form in the epiphyses. A cartilaginous disk, called the **epiphyseal** (ep"ĭ-fiz'e-al) **disk,** remains in the epiphyses. Bone length is dependent on how long the cartilage cells within the disks continue to divide. Eventually, though, the disks disappear, and the bone stops growing as the individual attains adult height.

In the adult, bone is continually being broken down and built up again. After bone-absorbing cells, called **osteoclasts** (os'te-o-klasts"), break down bone, they remove worn cells and deposit calcium in the blood. Apparently, after a period of about three weeks, the osteoclasts disappear. Then, destruction caused by the osteoclasts is repaired by osteoblasts. As they form new bone, the osteoblasts take calcium from the blood. Eventually, some of these cells get caught in the matrix they secrete and are converted to osteocytes. Osteocytes are mature osteoblasts.

Thus, through a process of *remodeling*, old bone tissue is replaced with new bone tissue. Because of continual remodeling, the thicknesses of bones can change according to the amount of physical use or due to a change in certain hormonal balances (see chapter 10). Strange as it may seem, adults seem to require more calcium in the diet than do children in order to promote the work of osteoblasts.

Fractures

A **fracture** is a break in a bone. Fractures can be classified in the following manner:

Simple: Broken bone does not pierce the skin.

Compound: Broken ends of bone pierce the skin.

Complete: Bone is broken into two parts.

Partial: Bone is broken longitudinally but not separated into two parts.

Greenstick: A break on the outer arc of the bone is incomplete.

Impacted: Broken ends of bone are wedged into each other.

Comminuted: Bone breaks into several fragments.

Spiral: Due to twisting of bone, the break is ragged.

Repair of a Fracture

Repair of a fracture entails these steps:

1. *Hematoma.* Blood escapes from ruptured blood vessels and forms a hematoma (mass of clotted blood) in the space between the broken bones. The area is inflamed and swollen.
2. *Fibrocartilage callus.* Tissue repair begins, and fibrocartilage now fills the space between the ends of the broken bone. Collagenous fibers tie the ends of the bones together.
3. *Bony callus.* Osteoblasts produce trabeculae of spongy bone and convert the fibrocartilage callus to a bony callus that joins the broken bones together.
4. *Remodeling.* Osteoblasts build new compact bone at the periphery, and osteoclasts reabsorb the spongy bone and build a new medullary cavity.

Bone is a living tissue that is always being remodeled. Fractures are of various types, but repair requires four steps: (1) hematoma, (2) fibrocartilage callus, (3) bony callus, and (4) remodeling.

Axial Skeleton

The skeleton is divided into the axial skeleton and the appendicular skeleton (fig. 6.4). The **axial skeleton** lies in the midline of the body and contains the bones of the skull, the hyoid bone, the vertebral column, and the thoracic cage. The bones of the skeleton are not smooth; they have protuberances called processes and indentations called depressions, and various types of openings.

Skull

The skull is formed by the cranium and the facial bones. These bones contain **sinuses** (fig. 6.5), air spaces lined by mucous membranes, that reduce the weight of the skull and give the voice a resonant sound. The paranasal sinuses empty into the nose and are named for their locations; other sinuses include the maxillary, frontal, sphenoidal, and ethmoidal sinuses. The two mastoid sinuses drain into the middle ear. **Mastoiditis** (mas"toi-di'tis), a condition that can lead to deafness, is an inflammation of these sinuses.

Cranium

The cranium protects the brain and is composed of eight bones. These bones are separated from each other by immovable joints called **sutures**. Newborns have membranous regions called **fontanels,** where more than two bones meet. The largest of these is the anterior fontanel, which is located where the two parietal bones meet the two parts of the frontal bone. The fontanels permit the skull to compress during birth as the head passes through the birth canal. The anterior fontanel (often called the "soft spot") usually closes by the age of two years. The cranial bones are one frontal bone, two parietal bones, one occipital bone, two temporal bones, one sphenoid bone, and one ethmoid bone (figs. 6.6 and 6.7).

Frontal Bone One frontal bone forms the forehead, a portion of the nose, and the superior portions of the orbits (bony sockets of the eyes).

Parietal Bones Two parietal bones are just dorsal to the frontal bone. They form the roof of the cranium and also help form its sides.

Occipital Bone One occipital bone forms the most dorsal part of the skull and the base of the cranium. The spinal cord joins the brain by passing through a large opening in the occipital bone called the foramen magnum. The occipital condyles (table 6.1) are rounded processes on either side of the foramen magnum that articulate with the first vertebra of the spinal column.

figure 6.4 The skeleton. The skeleton of a human adult contains bones that belong to the axial skeleton (red labels) and those that belong to the appendicular skeleton (black labels).

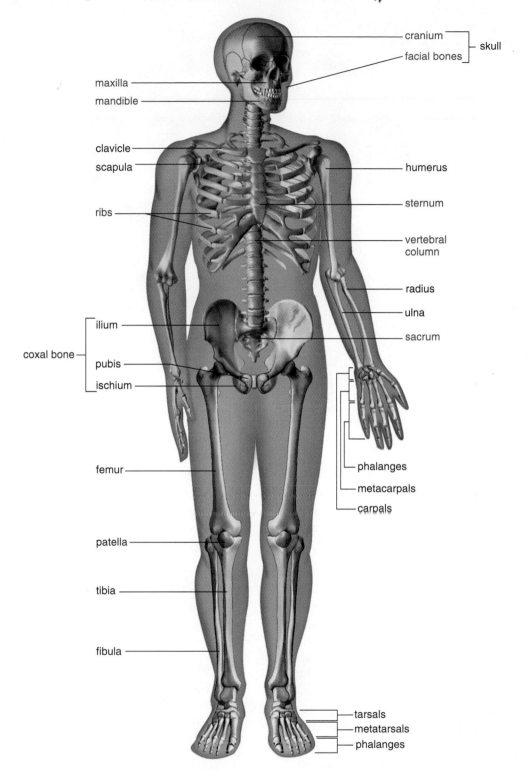

Temporal Bones Two temporal bones are just inferior to the parietal bones on the sides of the cranium. They also help form the base of the cranium (see figs. 6.6*b* and 6.7*a*). Each temporal bone has the following:

External auditory meatus, a canal that leads to the middle ear;

Mandibular fossa, which articulates with the mandible;

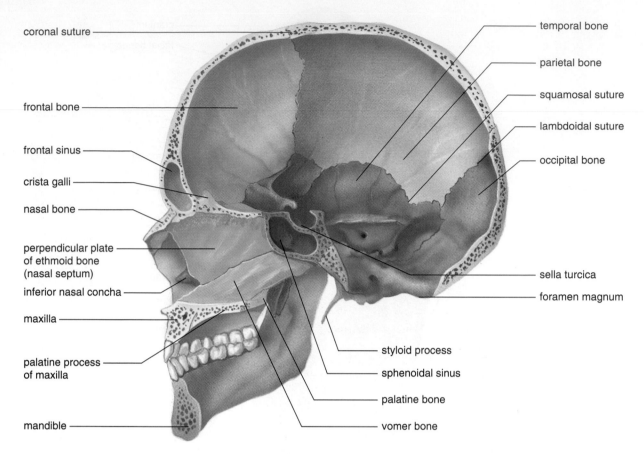

Mastoid process, which provides a place of attachment for certain neck muscles;

Styloid process, which provides a place of attachment for muscles associated with the tongue and larynx;

Zygomatic process, which projects anteriorly and helps form the cheekbone.

Sphenoid Bone   One sphenoid bone helps form the sides and base of the cranium, and the floors and sides of the orbits. Within the cranial cavity, the sphenoid bone has a saddle-shaped midportion called the **sella turcica** (see fig. 6.7*b*), which houses the pituitary gland in a depression.

Ethmoid Bone   One ethmoid bone forms part of the roof of the nasal cavity (see figs. 6.6 and 6.7*b*). The ethmoid bone contains the following:

Crista galli (cock's comb), a triangular process that serves as an attachment for membranes that enclose the brain;

Cribriform plate with tiny holes that serve as passageways for nerve fibers from the olfactory receptors;

Perpendicular plate (see fig. 6.5), which projects downward to form the nasal septum;

Superior and middle nasal conchae, which project toward the perpendicular plate. These projections support mucous membranes that line the nasal cavity.

> The skull is formed by the cranium and the facial bones. The cranium includes the frontal bone, two parietal bones, one occipital bone, two temporal bones, one sphenoid bone, and one ethmoid bone.

Facial Bones

Maxillae   The two maxillae form the upper jaw, and each has an alveolar process in which the teeth are located. Other processes, called the palatine processes, form the anterior portion of the *hard palate,* the roof of the mouth. The maxillae also contribute to the floors of the orbits and to the sides and floor of the nasal cavity.

Palatine Bones   The two palatine bones make up the posterior portion of the hard palate and the floor of the nasal cavity. A cleft palate results when the palatine bones have failed to fuse.

table 6.1

Surface Features of Bone

Processes

Term	Definition	Example
Articulating Surfaces		
Condyle (kon´dīl)	A large, rounded, articulating knob	The mandibular condyle of the mandible (fig 6.6*b*)
Head	A prominent, rounded, articulating proximal end of a bone	The head of the femur (fig. 6.19)
Projections for Muscle Attachment		
Crest	A narrow, ridgelike projection	The iliac crest of the coxal bone (fig. 6.18*b*)
Spine	A sharp, slender process	The spine of the scapula (fig. 6.13)
Trochanter (tro-kan´ter)	A massive process found only on the femur	The greater trochanter and lesser trochanter of the femur (fig. 6.19)
Tubercle (tu´ber-kl)	A small, rounded process	The greater tubercle of the humerus (fig. 6.14)
Tuberosity (tu˝bĕ-ros´ĭ-te)	A large, roughened process	The radial tuberosity of the radius (fig. 6.15)

Depressions and Openings

Term	Definition	Example
Foramen (fo-ra´men) (pl. *foramina*)	A rounded opening through a bone	The foramen magnum of the occipital bone (fig. 6.7*a*)
Fossa (fos´ah)	A flattened or shallow surface	The mandibular fossa of the temporal bone (fig. 6.7*a*)
Meatus (me-a´tus)	A tubelike passageway through a bone	The external auditory meatus of the temporal bone (fig 6.6*b*)
Sinus (si´nus)	A cavity or hollow space in a bone	The frontal sinus of the frontal bone (fig. 6.5)

Source: Data from Kent M. Van De Graaff and Stuart Ira Fox, *Concepts of Human Anatomy and Physiology*, 5th ed., 1999, p. 187.

Zygomatic Bones The two zygomatic bones form the sides of the orbits and the cheekbones. Each bone has a temporal process that joins the zygomatic process of a temporal bone.

Lacrimal Bones The two small, thin lacrimal bones are located on the medial walls of the orbits. A small groove between the orbit and the nasal cavity serves as a pathway for a tube that carries tears from the eyes to the nose.

Nasal Bones The two nasal bones are small, rectangular bones that form the bridge of the nose. The ventral portion of the nose is cartilage.

Vomer Bone The vomer bone joins with the perpendicular plate of the ethmoid bone to form the nasal septum (see fig. 6.5).

Inferior Nasal Conchae The two inferior nasal conchae are thin, curved bones that project into the nasal cavities and are attached to their lateral walls. Like the other con-chae mentioned previously, they support the mucous membranes that line the nasal cavity.

Mandible The mandible, or lower jaw, is the only movable portion of the skull. Its horseshoe-shaped body forms the chin. It also contains two upright projections called rami. Each ramus has a mandibular condyle that articulates with a temporal bone and a coronoid process, which serves as a place of attachment for the muscles used for chewing. The lower teeth are located on the alveolar arch of the mandible.

Hyoid Bone

The U-shaped hyoid bone is located superior to the larynx (voice box) in the neck. It is the only bone in the body that does not articulate with another bone. Instead, it is suspended from the styloid processes of the temporal bones by the stylohyoid muscles and ligaments. It anchors the tongue and serves as the site for the attachment of several muscles associated with swallowing.

figure 6.6 Skull anatomy. **a.** Anterior view.
b. Lateral view.

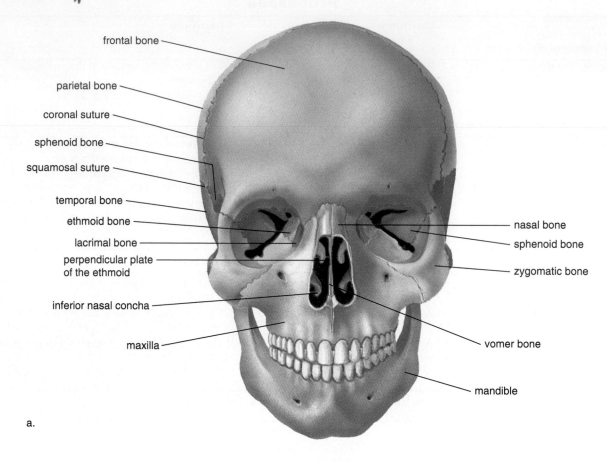

frontal bone

parietal bone

coronal suture

sphenoid bone

squamosal suture

temporal bone

ethmoid bone

lacrimal bone

perpendicular plate
of the ethmoid

inferior nasal concha

maxilla

nasal bone

sphenoid bone

zygomatic bone

vomer bone

mandible

a.

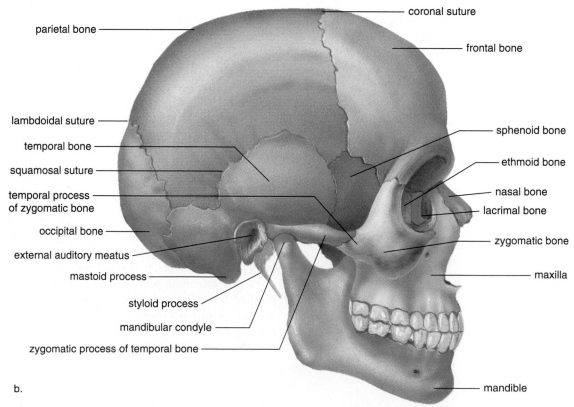

parietal bone

coronal suture

frontal bone

lambdoidal suture

temporal bone

squamosal suture

temporal process
of zygomatic bone

occipital bone

external auditory meatus

mastoid process

styloid process

mandibular condyle

zygomatic process of temporal bone

sphenoid bone

ethmoid bone

nasal bone

lacrimal bone

zygomatic bone

maxilla

mandible

b.

figure 6.7 Skull anatomy continued. **a.** Inferior view. **b.** Superior view.

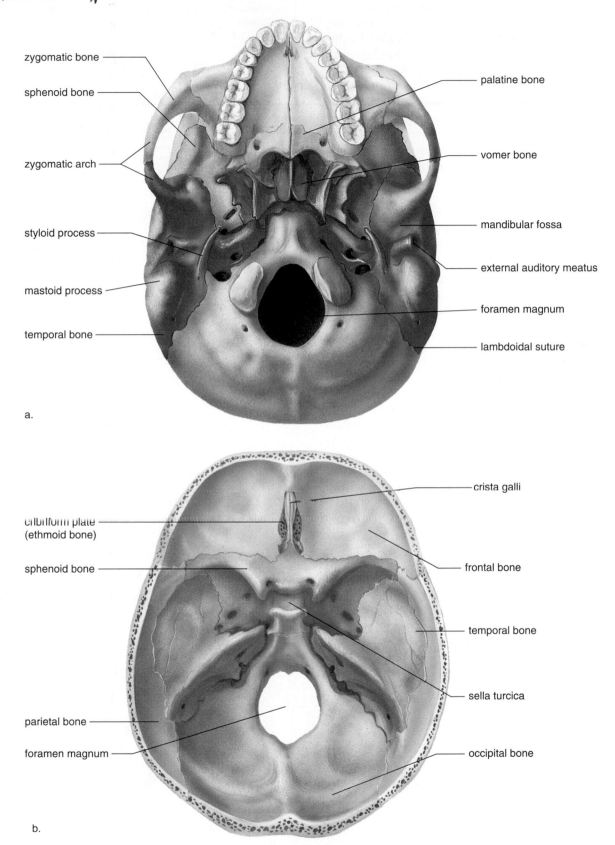

zygomatic bone

sphenoid bone

zygomatic arch

styloid process

mastoid process

temporal bone

palatine bone

vomer bone

mandibular fossa

external auditory meatus

foramen magnum

lambdoidal suture

a.

cribriform plate
(ethmoid bone)

sphenoid bone

parietal bone

foramen magnum

crista galli

frontal bone

temporal bone

sella turcica

occipital bone

b.

Vertebral Column (Spine)

The **vertebral column** (fig. 6.8) extends from the skull to the pelvis. It consists of a series of separate bones, the **vertebrae,** separated by pads of fibrocartilage called the **intervertebral disks.** The vertebral column is located in the middorsal region and forms the vertical axis. The skull rests on the superior end of the vertebral column, which also supports the thoracic cage and serves as a point of attachment for the pelvic girdle. The vertebral column also protects the spinal cord, which passes through a vertebral canal formed by the vertebrae. Thirty-three vertebrae are named according to their location: seven *cervical* (neck) *vertebrae,* twelve *thoracic* (chest) *vertebrae,* five *lumbar* (lower back) *vertebrae,* five *sacral vertebrae* fused to form the sacrum, and three to five *coccygeal vertebrae* fused into one coccyx.

When viewed from the side, the vertebral column has four normal curvatures named for their location. The cervical and lumbar curvatures are convex anteriorly, and the thoracic and sacral curvatures are concave anteriorly. In the fetus, the vertebral column has but one curve—concave anteriorly. The cervical curve develops three to four months after birth, when the child begins to hold the head up. The lumbar curvature develops when a child begins to stand and walk, at about one year of age. The curvatures of the vertebral column provide more support than a straight column would, and they also provide the balance needed to walk upright.

The curvatures of the vertebral column are subject to abnormalities. An abnormally exaggerated lumbar curvature is called **lordosis** (lor-do'sis), or "swayback." People who are balancing a heavy midsection, such as pregnant women or men with "potbellies," have swayback. An increased roundness of the thoracic curvature is **kyphosis** (ki-fo'sis), or "hunchback." This abnormality sometimes develops in older people. An abnormal lateral (side-to-side) curvature is called **scoliosis** (sko"le-o'sis), or "twisted disease." Occurring most often in the thoracic region, scoliosis is usually first seen during late childhood.

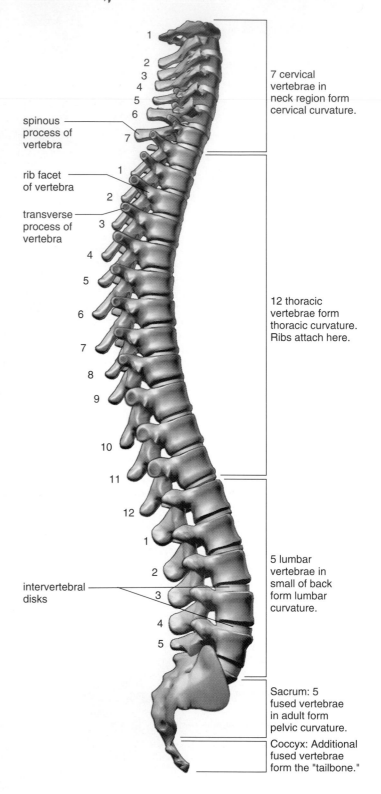

figure 6.8 Curvatures of the spine. The vertebrae are named for their location in the body. Note the presence of the coccyx, also called the tailbone.

spinous process of vertebra

rib facet of vertebra

transverse process of vertebra

intervertebral disks

7 cervical vertebrae in neck region form cervical curvature.

12 thoracic vertebrae form thoracic curvature. Ribs attach here.

5 lumbar vertebrae in small of back form lumbar curvature.

Sacrum: 5 fused vertebrae in adult form pelvic curvature.

Coccyx: Additional fused vertebrae form the "tailbone."

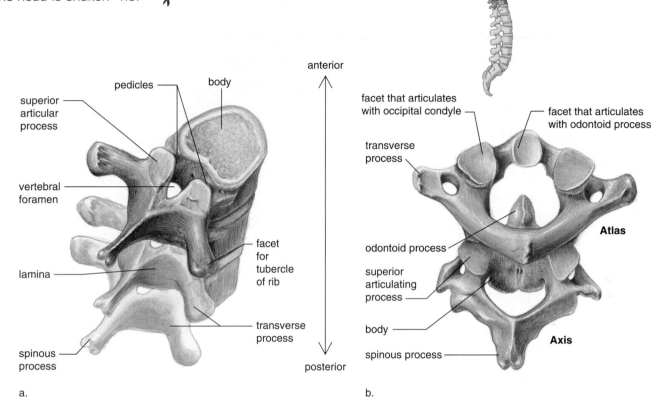

figure 6.9 Vertebrae. **a.** Typical vertebra in articular position. The vertebral canal where the spinal cord is found is formed by adjacent vertebral foramina. **b.** Atlas and axis, showing how they articulate with one another. The odontoid process of the axis is the pivot around which the atlas turns, such as when the head is shaken "no."

Labels for figure a: pedicles; body; superior articular process; vertebral foramen; lamina; spinous process; facet for tubercle of rib; transverse process; anterior; posterior; a.

Labels for figure b: facet that articulates with occipital condyle; facet that articulates with odontoid process; transverse process; odontoid process; superior articulating process; body; spinous process; **Atlas**; **Axis**; b.

Vertebrae

Figure 6.9*a* shows that a typical vertebra has an anteriorly placed body and a posteriorly placed vertebral arch. The vertebral arches form the walls of the *vertebral foramen* (pl., *foramina*). The foramina form a canal through which the spinal cord passes.

There is a posteriorly placed *spinous process,* where the two thin plates of bone called *laminae* meet, and two laterally placed transverse processes, where the pedicle joins the laminae. They serve for the attachment of muscles and ligaments. There are also articular processes (superior and inferior), which serve for the joining of vertebrae.

The vertebrae have regional differences. For example, as the vertebral column descends, the bodies get bigger and are better able to carry more weight. The spinous processes are short and tend to have a dip in the cervical region; the thoracic spines are long and slender and project downward; the lumbar spines are massive and square and project

posteriorly. The transverse processes of thoracic vertebrae have articular facets for ribs.

Atlas and Axis The first two cervical vertebrae are not typical (fig. 6.9*b*). The **atlas** supports and balances the head. It has two depressions that articulate with the occipital condyles, allowing movement of the head forward and back. The **axis** has an *odontoid process* that projects into the ring of the atlas. When the head moves from side to side, the atlas pivots around the odontoid process.

Sacrum and Coccyx The five sacral vertebrae are fused to form the **sacrum.** It articulates with the pelvic girdle and forms the posterior wall of the pelvic cavity (see fig. 6.18*a*). The **coccyx,** or tailbone, is the last part of the vertebral column. It is formed from a fusion of three to five vertebrae.

figure 6.10 The thoracic cage. This structure includes the thoracic vertebrae, the ribs, and the sternum. The three bones that make up the sternum are the manubrium, body, and xiphoid process. The ribs numbered 1–7 are true ribs, those numbered 8–12 are false ribs.

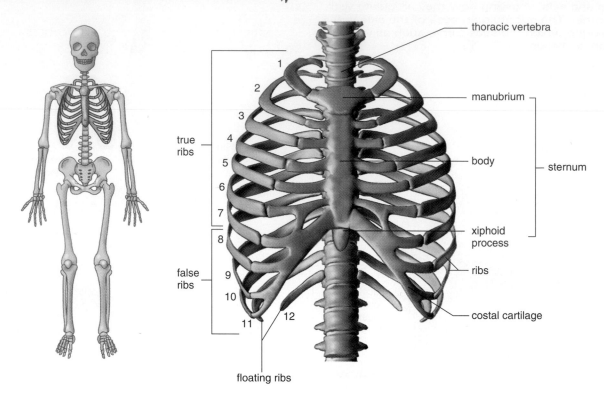

Intervertebral Disks

The fibrocartilage disks located between the vertebrae act as a cushion. They prevent the vertebrae from grinding against one another and absorb shock caused by such movements as running, jumping, and even walking. The disks also allow motion between the vertebrae so that a person can bend forward, backward, and from side to side. Unfortunately, these disks become weakened with age, and can slip or even rupture (called a **herniated disk**). The damaged disk pressing against the spinal cord and/or spinal nerves causes pain. The body may heal itself, or the disk can be removed surgically. If surgery is required, the vertebrae are fused together, limiting the body's flexibility.

> The vertebral column contains the cervical, thoracic, lumbar, sacral, and coccygeal vertebrae, which are separated by intervertebral disks, and has four curvatures.

Thoracic Cage

The **thoracic cage** (fig. 6.10) protects the heart and lungs, plays a role in breathing, and supports the bones of the shoulders.

Ribs

There are 12 pairs of **ribs.** All 12 pairs connect directly to the thoracic vertebrae in the back. A rib articulates with the body and transverse process of its corresponding thoracic vertebra. Each rib curves outward and then forward and downward.

The upper seven pairs of ribs connect directly to the sternum by means of costal cartilages. These are called "true ribs." The lower five pairs of ribs do not connect directly to the sternum, and they are called "false ribs." Three pairs of false ribs attach to the sternum by means of a common cartilage. The other two pairs are "floating ribs" because they do not attach to the sternum at all.

Sternum

The **sternum** contains three parts: the *manubrium, body,* and *xiphoid process.* The ribs articulate with the manubrium and body of the sternum. In addition, the manubrium articulates with the clavicles.

> The thoracic cage contains the thoracic vertebrae, ribs, and sternum.

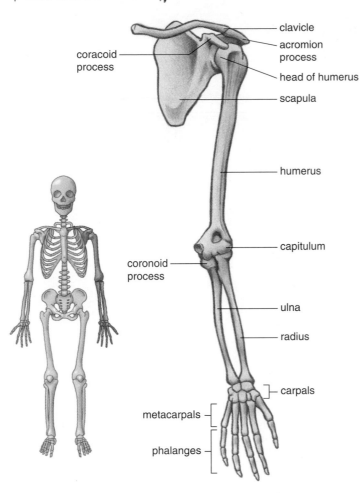

figure 6.11 The bones of the pectoral girdle, the arm, and the hand.

- clavicle
- acromion process
- coracoid process
- head of humerus
- scapula
- humerus
- capitulum
- coronoid process
- ulna
- radius
- carpals
- metacarpals
- phalanges

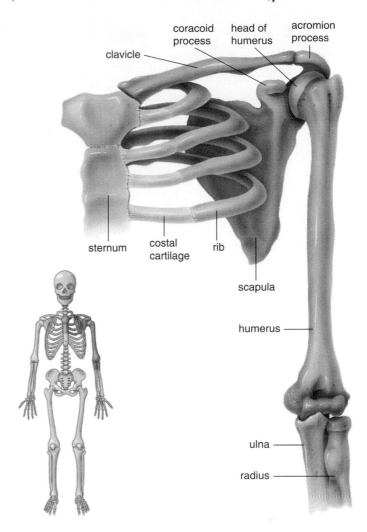

figure 6.12 Articulation of the left humerus with the left portion of the pectoral girdle.

- coracoid process
- head of humerus
- acromion process
- clavicle
- sternum
- costal cartilage
- rib
- scapula
- humerus
- ulna
- radius

Appendicular Skeleton

The **appendicular skeleton** contains the bones of the pectoral girdle, upper limbs (arms), pelvic girdle, and lower limbs (legs).

Pectoral Girdle

The **pectoral girdle** (shoulder girdle) contains four bones: two clavicles and two scapulae (figs. 6.11 and 6.12). It supports the arms and serves as a place of attachment for muscles that move the arms. The bones of this girdle are not held tightly together and are weakly attached and held in place by ligaments and muscles. This arrangement allows great flexibility but means that the pectoral girdle is easily dislocated.

Clavicles

The **clavicles** (collarbones) are slender and S-shaped. Each clavicle articulates medially with the manubrium of the sternum. This is the only place of attachment of the pectoral girdle to the axial skeleton.

Each clavicle also articulates with a scapula. The clavicle serves as a brace for the scapula and helps stabilize the shoulder. It is structurally weak, however, and if undue force is applied to the shoulder, the clavicle will fracture.

Scapulae

The **scapulae** (shoulder blades) are broad bones that somewhat resemble triangles (fig. 6.13). One reason for the pectoral girdle's flexibility is that the scapulae are not joined to each other (see fig. 6.4).

Each scapula has a spine and also the following features:

acromion process, which articulates with a clavicle and provides a place of attachment for arm and chest muscles;

coracoid process, which serves as a place of attachment for arm and chest muscles;

figure 6.13 Scapula. **a.** Posterior surface. **b.** Lateral view. **c.** Anterior surface.

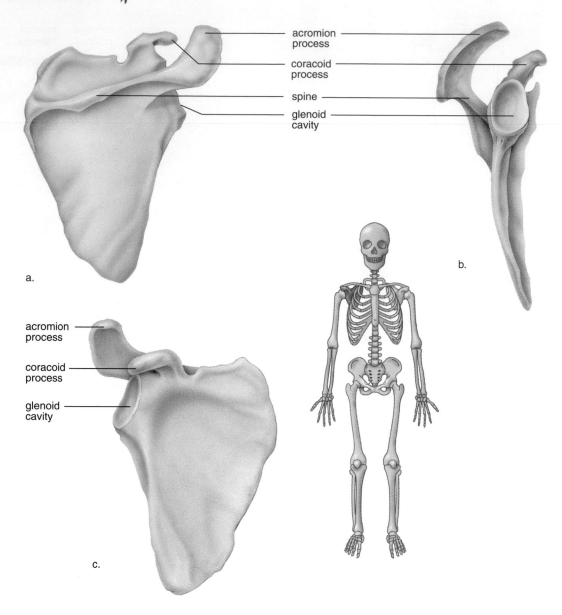

acromion process

coracoid process

spine

glenoid cavity

a.

b.

acromion process

coracoid process

glenoid cavity

c.

glenoid cavity, which articulates with the head of the upper arm bone (humerus). The pectoral girdle's flexibility is also a result of the glenoid cavity being smaller than the head of the humerus.

The pectoral (shoulder) girdle contains two clavicles and two scapulae.

Upper Limb (Arm)

The upper limb (see fig. 6.11) includes the bones of the upper arm (humerus), the forearm (radius and ulna), and the hand (carpals, metacarpals, and phalanges).

Humerus

The **humerus** (fig. 6.14) is the bone of the upper arm. It is a long bone with the following features at the proximal end:

head, which fits into the glenoid cavity of the scapula;

greater and **lesser tubercles,** which provide attachments for muscles that move arm and shoulder;

intertubercular groove, which holds the tendon from the biceps brachii, a muscle of the upper arm;

deltoid tuberosity, which provides an attachment for the deltoid, a muscle that covers the shoulder joint.

figure 6.14 Humerus of the left arm. **a.** Posterior surface. **b.** Anterior surface.

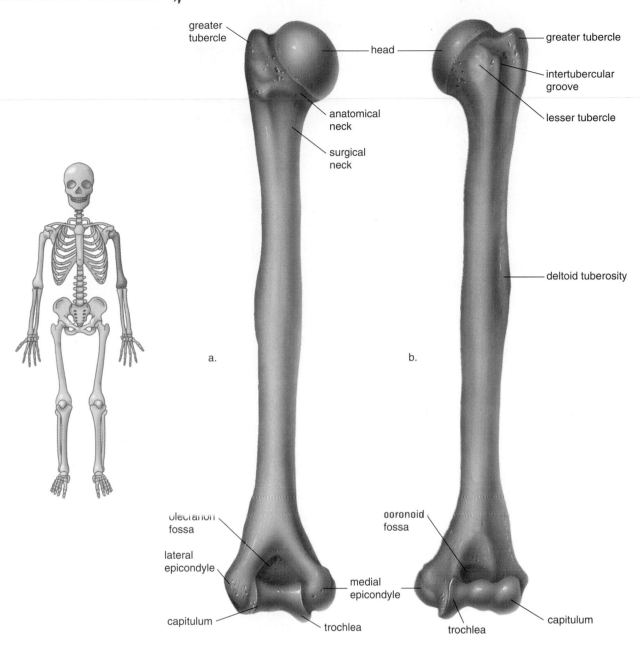

The humerus has the following features at the distal end:

capitulum, a lateral condyle that articulates with the radius;

trochlea, a spool-shaped condyle that articulates with the ulna;

coronoid fossa, a depression for a process of the ulna when the elbow is bent;

olecranon fossa, a depression for a process of the ulna when the elbow is extended.

Radius

The **radius** and **ulna** (see figs. 6.11 and 6.15) are the bones of the forearm (lower arm). The radius is on the thumb side when the palm faces forward, but crosses over the ulna when the hand is turned so that the palm faces backward. Proximally, the radius has the following features:

head, which articulates with the capitulum of the humerus and fits into the radial notch of the ulna;

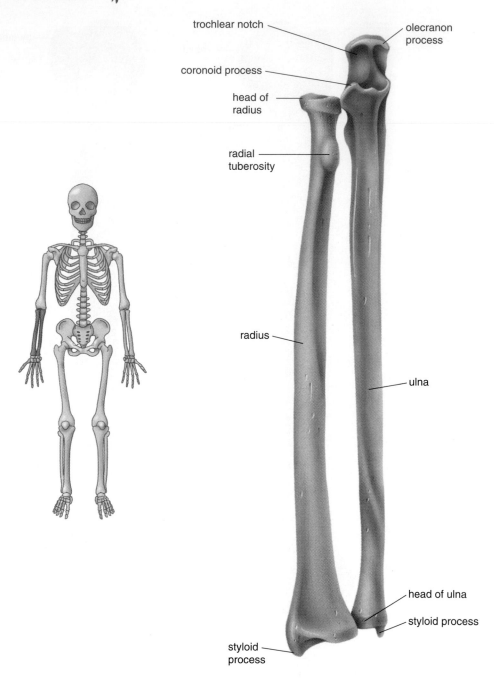

radial tuberosity

coronoid process
head of radius
trochlear notch
olecranon process

radius

ulna

head of ulna
styloid process

styloid process

radial tuberosity, which serves as a place of attachment for a tendon from the biceps brachii.

Ulna

The ulna is the longer bone of the forearm. Proximally, the ulna has the following features:

coronoid process, which articulates with the coronoid fossa of the humerus when the elbow is bent;

olecranon process, which articulates with the olecranon fossa of the humerus when the elbow is extended;

trochlear notch, which articulates with the trochlea of the humerus.

figure 6.16 Posterior view of the right hand.

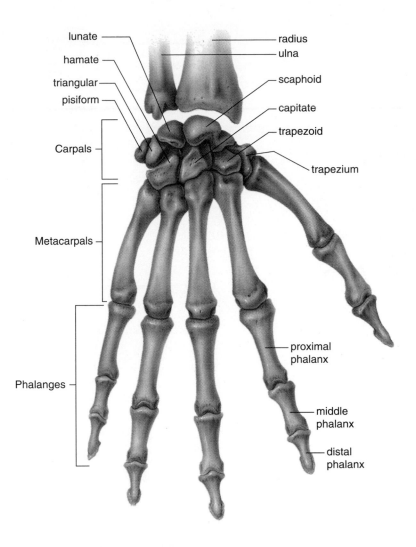

lunate

hamate

triangular

pisiform

Carpals

Metacarpals

Phalanges

radius

ulna

scaphoid

capitate

trapezoid

trapezium

proximal
phalanx

middle
phalanx

distal
phalanx

Hand

Each hand (see figs. 6.11 and 6.16) has a wrist, a palm, and five fingers. The wrist contains eight small **carpal bones** (tightly bound by ligaments in two rows of four each). The palm has five **metacarpal bones** that form the knuckles when a fist is made. The fingers contain the **phalanges**. The thumb has only two phalanges, but the other fingers have three each.

> The upper limb contains the humerus, the radius, the ulna, and the bones of the hand (the carpals, metacarpals, and phalanges).

Pelvic Girdle

The **pelvic girdle**, or **pelvis**, contains two coxal bones (hip-bones), as well as the sacrum and coccyx (figs. 6.17 and 6.18). The strong bones of the pelvic girdle are firmly attached to one another and bear the weight of the body. The pelvis also serves as the place of attachment for the legs and protects the urinary bladder, the internal reproductive organs, and a portion of the large intestine.

Coxal Bones

Each **coxal bone** has the following three parts (fig. 6.18b):

1. *ilium.* The **ilium** is the largest part of a coxal bone and flares outward to give the hip prominence. The margin of the ilium is called the iliac crest. Each ilium connects posteriorly with the sacrum at a **sacroiliac joint.**
2. *ischium.* The ischium is the most inferior part of a coxal bone and has a posterior region, the *ischial tuberosity,* that allows a person to sit. Near the junction of the ilium and ischium is the **ischial spine,**

figure 6.17 The bones of the pelvic girdle, the leg, and the foot.

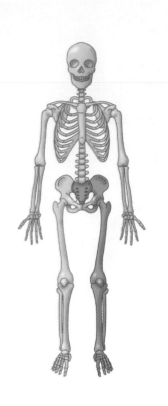

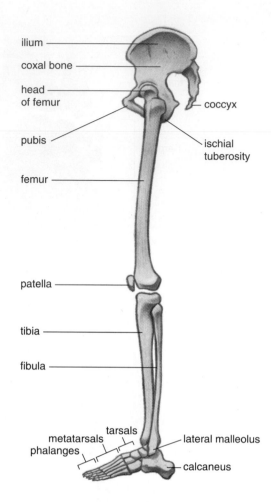

ilium

coxal bone

head of femur

pubis

femur

patella

tibia

fibula

coccyx

ischial tuberosity

tarsals

metatarsals

phalanges

lateral malleolus

calcaneus

which projects into the pelvic cavity. The distance between the ischial spines tells the size of the pelvic cavity. The **greater sciatic notch** is the site where blood vessels and the large sciatic nerve pass posteriorly into the lower leg.

3. *pubis.* The pubis is the anterior part of a coxal bone. The two pubic bones join together at the *pubic symphysis.* Posterior to where the pubis and the ischium join together is a large opening, the *obturator foramen,* through which blood vessels and nerves pass anteriorly into the lower leg.

Where the three bones meet, there is a depression called the **acetabulum,** which receives the rounded head of the femur.

> The pelvic girdle contains two coxal bones, as well as the sacrum and coccyx.

False and True Pelvises

The false pelvis is bounded laterally on the other side by the flared parts of the ilium. This space is much larger than that of the true pelvis. The true pelvis is inferior to the false pelvis and is the ring formed by the sacrum, lower ilium, ischium, and pubic bones. The true pelvis is said to have an upper inlet and a lower outlet. The dimensions of these outlets are important for females because the outlets must be large enough to allow a baby to pass through during the birth process.

figure 6.18 Pelvic girdle. **a.** Female pelvis (*top*) is wider than the male pelvis (*bottom*). **b.** Lateral surface of the left coxal bone.

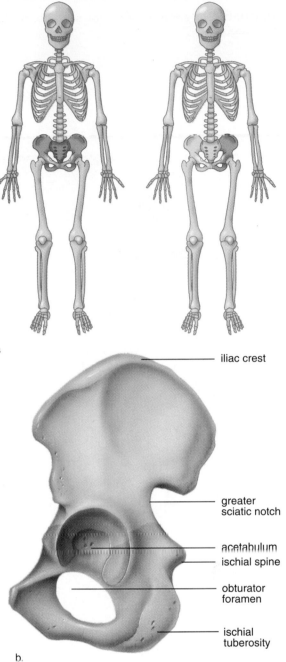

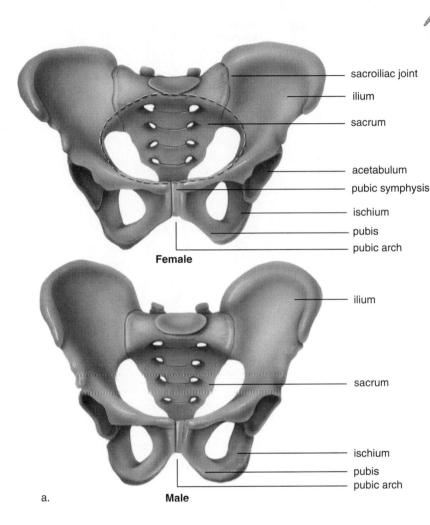

sacroiliac joint

ilium

sacrum

acetabulum

pubic symphysis

ischium

pubis

pubic arch

Female

ilium

sacrum

ischium

pubis

pubic arch

a.   **Male**

iliac crest

greater sciatic notch

acetabulum

ischial spine

obturator foramen

ischial tuberosity

b.

Sex Differences

Female and male pelvises (see fig. 6.18*a*) usually differ in several ways, including the following:

1. Female iliac bones are more flared than those of the male; therefore, the female has broader hips.
2. The female pelvis is wider between the ischial spines and the ischial tuberosities.
3. The female inlet and outlet of the true pelvis are wider.
4. The female pelvic cavity is more shallow, while the male pelvic cavity is more funnel-shaped.
5. Female bones are lighter and thinner.
6. The female pubic arch (angle at the pubic symphysis) is wider.

In addition to these differences in pelvic structure, male pelvic bones are larger and heavier, the articular ends are thicker, and the points of muscle attachment may be larger.

The female pelvis is generally wider and more shallow than the male pelvis.

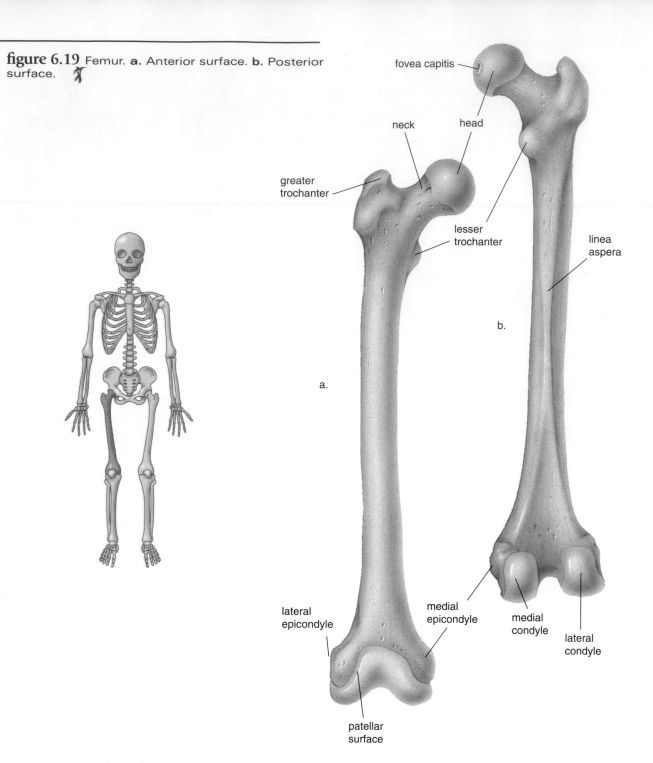

figure 6.19 Femur. **a.** Anterior surface. **b.** Posterior surface.

fovea capitis

neck

head

greater trochanter

lesser trochanter

linea aspera

b.

a.

lateral epicondyle

medial epicondyle

medial condyle

lateral condyle

patellar surface

Lower Limb (Leg)

The lower limb includes the bones of the thigh (femur), the lower leg (tibia and fibula), and those of the foot (tarsals, metatarsals, and phalanges) (see fig. 6.17).

Femur

The **femur** (fig. 6.19), or thighbone, is the longest and strongest bone in the body. Proximally, the femur has the following:

head, which fits into the acetabulum of the coxal bone;

greater and **lesser trochanters,** which provide a place of attachment for the muscles of the legs and buttocks;

linea aspera, a crest that serves as a place of attachment for several muscles.

Distally, the femur articulates with the **patella** (kneecap) and has **lateral** and **medial condyles** that articulate with the tibia.

figure 6.20 Tibia and fibula, showing how they articulate.

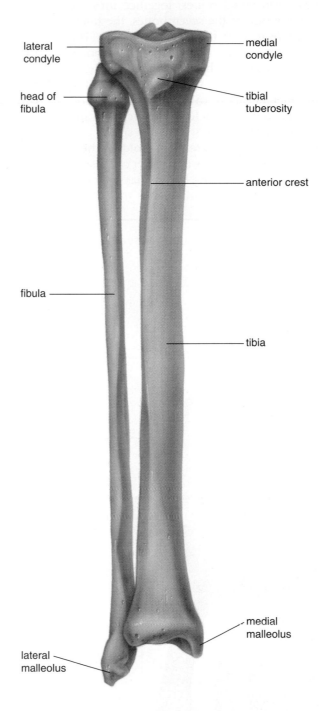

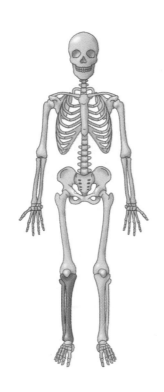

Tibia

The **tibia** and **fibula** (fig. 6.20) are the bones of the lower leg. The tibia, or shinbone, is medial to the fibula and has the following:

medial and **lateral condyles,** which articulate with the femur;

tibial tuberosity, where the patellar (kneecap) ligaments attach;

anterior crest, commonly called the shin;

medial malleolus, the bulge of the inner ankle.

Fibula

The fibula is lateral to the tibia and is more slender. It has a head that articulates with the tibia just below the lateral condyle and a distal **lateral malleolus** that forms the outer part of the ankle.

Foot

Each foot (fig. 6.21) has an ankle, an instep, and five toes.

The ankle has seven **tarsal bones**; together, they are called the tarsus. Only one of the seven bones, the **talus**, can move freely where it joins the tibia and fibula. The largest of the ankle bones is the **calcaneus,** or heel bone. Along with the talus, it supports the weight of the body.

The instep has five elongated **metatarsal bones.** The distal ends of the metatarsals form the ball of the foot. Along with the tarsals, these bones form the arches of the foot (longitudinal and transverse), which give spring to a person's step. If the ligaments and tendons holding these bones together weaken, fallen arches or flat feet can result.

Toes contain the **phalanges.** The big toe has only two phalanges, but the other toes have three each.

figure 6.21 Anterior view of the right foot.

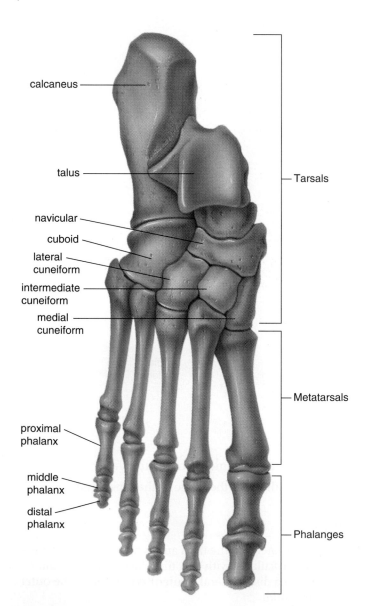

- calcaneus
- talus
- navicular
- cuboid
- lateral cuneiform
- intermediate cuneiform
- medial cuneiform
- proximal phalanx
- middle phalanx
- distal phalanx
- Tarsals
- Metatarsals
- Phalanges

The lower limb contains the femur, the tibia, the fibula, and the bones of the foot (the tarsals, metatarsals, and phalanges).

Joints (Articulations)

Bones articulate at the **joints,** which are often classified according to the amount of movement they allow.

Classification

Some bones, such as those that make up the cranium, are sutured together by a thin layer of fibrous connective tissue and are *immovable.* Such joints are called **synarthroses.** Review figures 6.6 and 6.7, and note the following immovable sutures:

> **coronal suture** occurs between the parietal bones and the frontal bone;
>
> **lambdoidal suture** occurs between the parietal bones and the occipital bone;
>
> **squamosal suture** occurs between each parietal bone and each temporal bone;
>
> **sagittal suture** occurs between the parietal bones (not shown).

Other joints are *slightly movable,* connected by hyaline cartilage or fibrocartilage, and are referred to as **amphiarthroses.** For example, the vertebrae are separated by intervertebral disks (see fig. 6.8) that increase vertebrae flexibility. Also, the ribs are joined to the sternum by costal cartilage (see fig. 6.10), and the **pubic symphysis** occurs

figure 6.22 Generalized freely movable joint. Notice the cavity between the bones that is lined by synovial membrane.

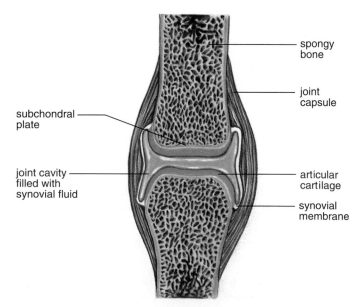

- spongy bone
- joint capsule
- subchondral plate
- joint cavity filled with synovial fluid
- articular cartilage
- synovial membrane

between the pubic bones (see fig. 6.18). Due to hormonal changes, this joint becomes more flexible during late pregnancy, which allows the pelvis to expand during childbirth.

Most joints are *freely movable* **synovial joints,** known as **diarthroses,** in which the two bones are separated by a cavity (fig. 6.22). **Ligaments** composed of fibrous connective tissue bind the two bones to one another, holding them in place as they form a capsule. In a double-jointed individual, the ligaments are unusually loose. **Tendons** are cords of fibrous connective tissue that connect muscles to bones and help stabilize joints. The joint capsule is lined by a **synovial** (sĭ-no've-al) **membrane,** which produces **synovial fluid,** a lubricant for the joint.

The knee is an example of a synovial joint (fig. 6.23). In the knee, as in other freely movable joints, the bones are capped by cartilage. In addition, the knee contains **menisci** (sing., *meniscus*), crescent-shaped pieces of cartilage between the bones. These give added stability, helping to support the weight placed on the knee joint. Unfortunately, athletes often suffer an injury to the menisci, known as torn cartilage. The knee joint contains 13 fluid-filled sacs called **bursae** (bur'se), which ease friction between tendons and ligaments, and between tendons and bones. Inflammation of the bursae is called **bursitis.** Tennis elbow is a form of bursitis.

figure 6.23 Knee Joint. The knee joint is a synovial joint. Notice the cavity between the bones, which is encased by ligaments and lined by synovial membrane. The patella (kneecap) serves to protect the joint and guide the quadriceps tendon over the joint when flexion or extension occurs.

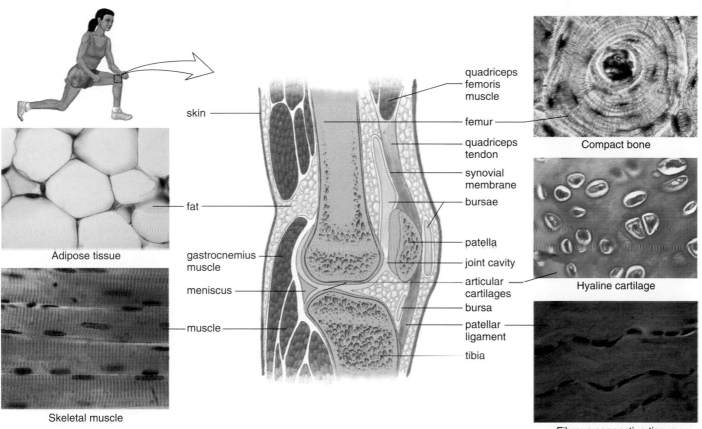

Adipose tissue

Skeletal muscle

skin

fat

gastrocnemius muscle

meniscus

muscle

quadriceps femoris muscle

femur

quadriceps tendon

synovial membrane

bursae

patella

joint cavity

articular cartilages

bursa

patellar ligament

tibia

Compact bone

Hyaline cartilage

Fibrous connective tissue

figure 6.24 Types of synovial joints.

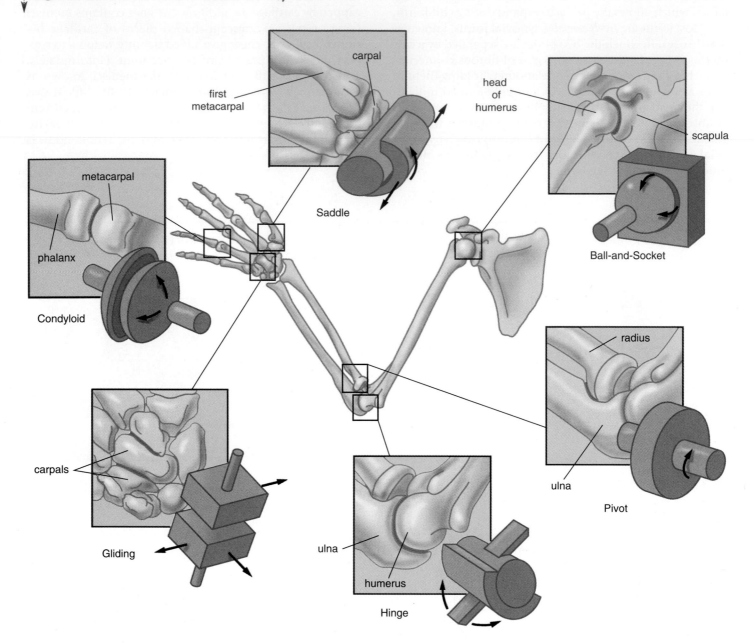

first metacarpal

carpal

Saddle

head of humerus

scapula

Ball-and-Socket

metacarpal

phalanx

Condyloid

radius

ulna

Pivot

carpals

Gliding

ulna

humerus

Hinge

Different types of freely movable synovial joints are listed here and depicted in figure 6.24:

Saddle joint, in which each bone is saddle-shaped and fits into the complementary regions of the other. A variety of movements is possible. Example: the joint between the carpal and metacarpal bones of the thumb.

Ball-and-socket joint, in which the ball-shaped head of one bone fits into the cup-shaped socket of another. Movement in all planes and rotation is possible. Example: the shoulder and hip joint.

Pivot joint, in which a small, cylindrical projection of one bone pivots within a ring formed of bone and

MedAlert

Arthritic Joints

*a*t the joints, the ends of bones are covered with articular cartilage, which normally reduces friction. When a person has osteoarthritis (OA), the articular cartilage softens, cracks, and wears away entirely in some areas. If diseased cartilage wears away, the joint becomes swollen and warm to the touch as body enzymes break down the debris. As the disease progresses, the exposed bone thickens and forms spurs that cause the bone ends to enlarge and joint movement to be restricted. The fingers, spine, hips, and knees are most often affected.

The cause of OA is not known, but a combination of genetic and mechanical factors is believed to be involved. It appears that arthritis begins in joints that have been overworked. Compression and abrasion may repeatedly damage the articular cartilage, which then must be broken down and replaced more often. Apparently, in persons with OA, cartilage breakdown occurs at a faster rate than replacement.

Weight loss is one recommendation for combating OA. Taking off 3 pounds can reduce the load on a hip or knee joint by 9–15 pounds. In addition, activities that cause finger pain should be avoided by using such items as lightweight pots and pans, Velcro fasteners, and electric can openers. A sensible exercise program can be designed to build up muscles and stabilize joints; low-impact activities like biking and swimming are best. Recently, acetaminophen (present in Tylenol) has been found to be as effective in controlling pain as a nonsteroidal anti-inflammatory drug such as ibuprofen, which is in Advil, Nuprin, and Motrin. Proper use of crutches or a cane can also assist mobility.

If all else fails, joint surgery may be recommended. During arthroscopy, surgeons use an arthroscope, a tiny tubular instrument that not only bears a lens and light source but also a camera to view the diseased joint on a television screen. Probes and surgical instruments can then be guided into the joint to remove debris and smooth the ends of bones. A diseased knee or hip can be replaced with a **prosthesis** (pros-the′-sis) (artificial substitute). Figure 6A shows that half of the artificial joint is anchored by a metal shank driven into the marrow of the femur, and the other half is attached to the coxal bone (hip) or tibia (knee). Previously, the metal was always fastened to the bone with cement, but now, an alternative method is available. Some metal implants have a porous surface, and bone tissue can grow into the crevices to fasten the prosthesis. Since it takes time for bone to grow, patients are on their feet and can walk faster when cement is used. Artificial knees and hips do not have the flexibility and durability of natural ones; jumping, twisting, and heavy lifting are discouraged, as is excessive walking.

Questions

Why would you expect:

1. osteoarthritis to be present in an older rather than a younger person?

2. an artificial hip to dislocate faster than a natural hip?

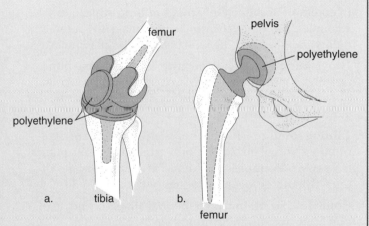

figure 6.A Artificial joints in which polyethylene replaces articular cartilage. **a.** Knee. **b.** Hip.

figure 6.25 Joint movement.

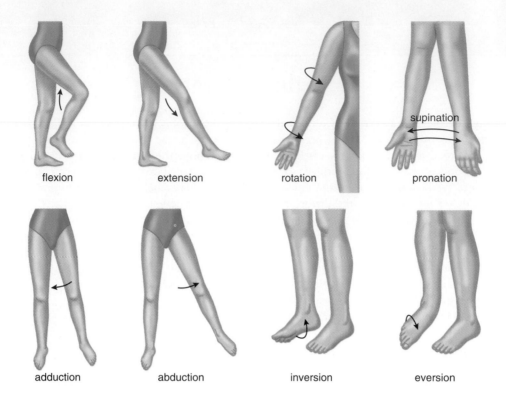

flexion   extension   rotation   supination pronation

adduction   abduction   inversion   eversion

ligament of another. Only rotation is possible. Examples: the joint between the proximal ends of the radius and ulna, and the joint between the atlas and axis.

Hinge joint, in which the convex surface of one bone articulates with the concave surface of another. Up-and-down motion in one plane is possible. Example: the elbow and knee joint.

Gliding joint, in which flat or slightly curved surfaces of bones articulate. Sliding or twisting in various planes is possible. Examples: the joints between the bones of the wrist and between the bones of the ankle.

Condyloid joint, in which the oval-shaped condyle of one bone fits into the elliptical cavity of another. Movement in different planes is possible, but rotation is not. Examples: the joints between the metacarpals and phalanges.

Synovial joints are subject to various disorders. A sudden movement that twists or wrenches a joint can stretch or tear a ligament, which is called a *sprain*. *Arthritis* is a more serious disorder. In **rheumatoid arthritis,** the synovial membrane becomes inflamed and grows thicker. Degenerative changes take place that make the joint almost immovable and painful to use. Evidence indicates that these effects are brought on by an autoimmune reaction. In **osteoarthritis,** the cartilage at the ends of the bones disintegrates so that the two bones become rough and irregular. The MedAlert reading on page 103 discusses osteoarthritis.

Gout, or gouty arthritis, is caused by an excessive buildup of uric acid (a metabolic waste) in the blood. Rather than being excreted in the urine, it becomes deposited as crystals in the joints, where it causes inflammation and pain.

> Joints are regions of articulations between bones. They are classified according to their degree of movement. Some joints are immovable, some are slightly movable, and some are freely movable.

Movements Permitted by Synovial Joints

Intact skeletal muscles are attached to bones by tendons that span joints. When a muscle contracts, one bone moves in relation to another bone.

Angular movements increase or decrease the joint angle between the bones of a joint (fig. 6.25):

Flexion decreases the joint angle. Flexion of the elbow moves the forearm toward the upper arm; flexion of the knee moves the lower leg toward the upper leg. *Dorsiflexion* is flexion of the foot upward, as when you stand on your heels; *plantar flexion* is flexion of the foot downward, as when you stand on your toes.

Extension increases the joint angle. Extension of the flexed elbow straightens the arm so that there is a $180°$ angle at the elbow. Hyperextension occurs when a portion of the body parts are extended beyond $180°$. It is possible to hyperextend the head and the trunk of the body.

Abduction is the movement of a body part laterally away from the midline. Abduction of the arms or legs moves them to the side, away from the body.

Adduction is the opposite of abduction. It is the movement of a body part toward the midline. For example, adduction of the arms or legs moves them back to the sides, toward the body.

Circular movements occur at ball-and-socket joints (see fig. 6.25):

Rotation is the movement of a body part around its own axis, as when the head is turned to answer "no" or when the arm is twisted one way and then the other.

Supination is the rotation of the lower arm so that the palm is upward; **pronation** is the opposite—the movement of the lower arm so that the palm is downward.

Circumduction is the movement of a body part in a wide circle, as when a person makes arm circles. If the motion is observed carefully, one can see that, because the proximal end of the arm is stationary, the shape outlined by the arm is actually a cone.

Inversion and **eversion** are terms that apply only to the feet. Inversion is turning the foot so that the sole is inward, and eversion is turning the foot so that the sole is outward.

Elevation and **depression** are the lifting up and down, respectively, of a body part, such as when you shrug your shoulders.

> Movements at joints are broadly classified as angular and circular.

Effects of Aging

Both cartilage and bone tend to deteriorate as a person ages. The chemical nature of cartilage changes, and the bluish color typical of young cartilage changes to an opaque, yellowish color. The chondrocytes die, and reabsorption occurs as the cartilage undergoes calcification, becoming hard and brittle. Calcification interferes with the ready diffusion of nutrients and waste products through the matrix. The articular cartilage may no longer function properly, and the symptoms of osteoarthritis can appear.

Beginning at about age 30 in both men and women, the bone resorbed by osteoclasts exceeds the amount of new bone synthesized by osteoblasts. **Osteoporosis** is present when weak and thin bones cause aches and pains, and tend to fracture easily. The likelihood of osteoporosis increases in women due to a reduction in estrogen levels after menopause. Although it is not known how estrogen acts on bone maintenance, it seems to play a role in calcium metabolism. Behavior is also important, since a lack of exercise and too little calcium in the diet help to trigger osteoporosis. The bones of an older person are porous, but the matrix that remains is harder, which causes the bones to be brittle and more easily broken. A fracture that results from osteoporosis takes longer to heal because bone is serviced by a reduced number of blood vessels.

Working Together

The accompanying illustration shows how the skeletal system works with other organ systems of the body to maintain homeostasis. All the organ systems of the body are interrelated.

> **BodyWorks CD-ROM** 💿
> The module accompanying chapter 6 is Skeletal System.

working together Skeletal System

Integumentary System

Bones provide support for skin.

Skin protects bones; helps provide vitamin D for Ca^{2+} absorption.

Muscular System

Bones provide attachment sites for muscles; store Ca^{2+} for muscle function.

Muscular contraction causes bones to move joints; muscles help protect bones.

Nervous System

Bones protect sense organs, brain, and spinal cord; store Ca^{2+} for nerve function.

Receptors send sensory input from bones to joints.

Endocrine System

Bones provide protection for glands; store Ca^{2+} used as second messenger.

Growth hormone regulates bone development; parathyroid hormone and calcitonin regulate Ca^{2+} content.

Circulatory System

Rib cage protects heart; red bone marrow produces blood cells; bones store Ca^{2+} for blood clotting.

Blood vessels deliver nutrients and oxygen to bones, carry away wastes.

How the Skeletal System works with other body systems

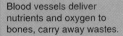

Lymphatic System/Immunity

Red bone marrow produces white blood cells involved in immunity.

Lymphatic vessels pick up excess tissue fluid; immune system protects against infections.

Respiratory System

Rib cage protects lungs and assists breathing; bones provide attachment sites for muscles involved in breathing.

Gas exchange in lungs provides oxygen and rids body of carbon dioxide.

Digestive System

Jaws contain teeth that chew food; hyoid bone assists swallowing.

Digestive tract provides Ca^{2+} and other nutrients for bone growth and repair.

Urinary System

Bones provide support and protection.

Kidneys provide active vitamin D for Ca^{2+} absorption and help maintain blood level of Ca^{2+}, needed for bone growth and repair.

Reproductive System

Bones provide support and protection of reproductive organs.

Sex hormones influence bone growth and density in males and females.

Selected New Terms

Basic Key Terms

abduction (ab-duk'shun), p. 105
adduction (ah-duk'shun), p. 105
appendicular skeleton (ap"en-dik'u-lar skel'ĕ-ton), p. 91
articular cartilage (ar-tik'u-lar kar'tĭ-lij), p. 79
articulation (ar-tik"u-la'shun), p. 79
axial skeleton (ak'se-al skel'ĕ-ton), p. 82
bursa (bur'sah), p. 101
circumduction (ser"kum-duk'shun), p. 105
compact bone (kom'pakt bōn), p. 81
diaphysis (di-af'ĭ-sis), p. 79
diarthrosis (di"ar-thro'sis), p. 101
epiphyseal disk (ep"ĭ-fiz'e-al disk), p. 81
epiphysis (ĕ'-pif'ĭ-sis), p. 79
eversion (e-ver'zhun), p. 105
extension (ek-sten'shun), p. 105
flexion (flek'shun), p. 104
fontanel (fon"tah-nel), p. 82
hematopoiesis (hem"ah-to-poi-e'sis), p. 81
intervertebral disk (in"ter-ver'tĕ-bral disk), p. 88
inversion (in-ver'zhun), p. 105
joint, p. 100
ligament (lig'ah-ment), p. 101
medullary cavity (med'u-lār"e kav'ĭ-te), p. 79
meniscus (mĕ-nis'kus), p. 101
osteoblast (os'te-o-blast"), p. 81
osteoclast (os'te-o-klast"), p. 81
osteocyte (os'te-o-sīt), p. 81
pectoral girdle (pek'tor-al ger'dl), p. 91
pelvic girdle (pel'vik ger'dl), p. 95
periosteum (per"e-os'te-um), p. 79
pronation (pro-na'shun), p. 105
red marrow (red mār'o), p. 81
rotation (ro-ta'shun), p. 105
sinus (si'nus), p. 82
spongy bone (spunj'e bōn), p. 81
supination (soo"pĭ-na'shun), p. 105
suture (soo'cher), p. 82
synovial membrane (si-no've-al mem'bran), p. 101
synovial fluid (si-no've-al floo'id), p. 101
synovial joint (si-no've-al joint), p. 101
tendon (ten'dun), p. 101

Clinical Key Terms

bursitis (bur-si'tis), p. 101
fracture (frak'chur), p. 82
herniated disk (her'ne-a-ted disk), p. 90
kyphosis (ki-fo'sis), p. 88
lordosis (lor-do'sis), p. 88
mastoiditis (mas"toi-di'tis), p. 82
osteoarthritis (os"te-o-ar-thri'tis), p. 104
osteoporosis (os"te-o-po-ro'sis), p. 105
prosthesis (pros-the'sis), p. 103
rheumatoid arthritis (roo'mah-toid ar-thri'tis), p. 104
scoliosis (sko"le-o'sis), p. 88

Summary

I. **Skeleton: Overview**
   A. The skeleton supports and protects the body, produces red blood cells, serves as a storehouse for inorganic calcium and phosphorus salts, and permits flexible movement.
   B. A long bone has a shaft (diaphysis) and two ends (epiphyses), which are covered by articular cartilage. The diaphysis contains a medullary cavity with yellow marrow and is bounded by compact bone. The epiphyses contain spongy bone with red bone marrow that produces red blood cells.
   C. Bone is a living tissue and is always being remodeled. Osteoblasts produce bone; osteoclasts break down bone.
   D. Fractures are of various types, but repair requires four steps: (1) hematoma, (2) fibrocartilage callus, (3) bony callus, and (4) remodeling.

II. **Axial Skeleton**
   The axial skeleton lies in the midline of the body and consists of the skull, the hyoid bone, the vertebral column, and the thoracic cage.
   A. The skull is formed by the cranium and the facial bones. The cranium includes the frontal bone, two parietal bones, one occipital bone, two temporal bones, one sphenoid bone, and one ethmoid bone. The facial bones include two maxillae, two palatine bones, two zygomatic bones, two lacrimal bones, two nasal bones, the vomer bone, two inferior nasal conchae, and the mandible.
   B. The U-shaped hyoid bone is located in the neck. It anchors the tongue and does not articulate with any other bone.
   C. The typical vertebra has a body, a vertebral arch surrounding the vertebral foramen, and a spinous

process. The first two vertebrae are the atlas and axis. The vertebral column contains the cervical, thoracic, lumbar, sacral, and coccygeal vertebrae, which are separated by intervertebral disks, and has four curvatures.

D. The thoracic cage contains the thoracic vertebrae, ribs, and sternum.

III. **Appendicular Skeleton**

The appendicular skeleton consists of the bones of the pectoral girdle, upper limbs (arms), pelvic girdle, and lower limbs (legs).

A. The pectoral (shoulder) girdle contains two clavicles and two scapulae.

B. The upper limb contains the humerus, the radius, the ulna, and the bones of the hand (the carpals, metacarpals, and phalanges).

C. The pelvic girdle contains two coxal bones, as well as the sacrum and coccyx. The female pelvis is generally wider and more shallow than the male pelvis.

D. The lower limb contains the femur, the tibia, the fibula, and the bones of the foot (the tarsals, metatarsals, and phalanges).

IV. **Joints (Articulations)**

A. Joints are regions of articulations between bones. They are classified according to their degree of

movement. Some joints are immovable, some are slightly movable, and some are freely movable. The different kinds of freely movable synovial joints are ball-and-socket, hinge, condyloid, pivot, gliding, and saddle.

B. Movements at joints are broadly classified as angular (flexion, extension, abduction, adduction) and circular (rotation, supination, pronation, circumduction, inversion, eversion, elevation, and depression).

Study Questions

1. What are five functions of the skeleton? (p. 79)
2. What are five major categories of bones based on their shapes? (p. 79)
3. What are the parts of a long bone? (pp. 79–80) What are some differences between compact bone and spongy bone? (p. 81)
4. How does bone grow in children, and how is it remodeled in all age groups? (pp. 81–82)
5. What are the various types of fractures? What four steps are required for fracture repair? (p. 82)
6. List the bones of the axial and appendicular skeletons. (p. 83)
7. What are the bones of the cranium and the face? What are the special

features of the temporal bones, sphenoid bone, and ethmoid bone? (pp. 82–87)

8. What are the parts of the vertebral column, and what are its curvatures? Distinguish between the atlas, axis, sacrum, and coccyx. (pp. 88–89)
9. What are the bones of the thoracic cage, and what are several of its functions? (p. 90)
10. What are the bones of the pectoral girdle? Give examples to demonstrate the flexibility of the pectoral girdle. What are the special features of each scapula? (pp. 91–92)
11. What are the bones of the upper limb, and what are their special features? (pp. 92–95)

12. What are the bones of the pelvic girdle, and what are their functions? Give examples to demonstrate the strength and stability of the pelvic girdle. (pp. 95–96)
13. What are the false and true pelvises, and what are several differences between male and female pelvises? (pp. 96–97)
14. What are the bones of the lower limb? Describe the special features of these bones. (pp. 98–100)
15. How are joints classified? Give examples of each type of joint. (pp. 100–102, 104)
16. How are joint movements classified? Give an example of each type. (pp. 104–105)

Objective Questions

I. **Match the items in the key to the bones given in questions 1–6.**
Key:
- a. forehead
- b. chin
- c. cheekbone
- d. elbow
- e. shoulder blade
- f. hip
- g. ankle

1. temporal and zygomatic
2. tibia and fibula
3. frontal bone
4. ulna
5. coxal bone
6. scapula

II. **Match the items in the key to the bones listed in questions 7–13.**

Key:
- a. external auditory meatus
- b. cribriform plate
- c. xiphoid process
- d. glenoid cavity
- e. olecranon process
- f. acetabulum
- g. greater and lesser trochanter

7. scapula
8. sternum
9. femur
10. temporal
11. coxal bone
12. ethmoid
13. ulna

III. **Fill in the blanks.**

14. Long bones are _____ than they are wide.

15. The epiphysis of a long bone contains _____ bone, where red blood cells are produced.

16. The _____ are the air-filled spaces in the cranium.

17. The sacrum is a part of the _____ , and the sternum is a part of the _____ .

18. The pectoral girdle is specialized for _____ , while the pelvic girdle is specialized for _____ .

19. The term *phalanges* is used for the bones of both the _____ and the _____ .

20. The knee is a freely movable synovial joint of the _____ type.

Medical Terminology Reinforcement Exercise

Consult Appendix B for help in pronouncing and analyzing the meaning of the terms that follow.

1. chondromalacia (kon″dro-mah-la′she-ah)
2. osteomyelitis (os″te-o-mi″e-li′tis)
3. craniosynostosis (kra″ne-o-sin″os-to′sis)
4. myelography (mi″ĕ-log′rah-fe)
5. acrocyanosis (ak″ro-si″ah-no′sis)
6. syndactylism (sin-dak′tĭ-lizm)
7. orthopedist (or″tho-pe′dist)
8. prognathism (prog′nah-thizm)
9. micropodia (mi″kro-po′de-ah)
10. arthroscopic surgery (ar″thro-skop′ik)

Website Link

For a listing of the most current Websites related to this chapter, please visit the Mader home page at:

www.mhhe.com/maderap

chapter 7

The Muscular System

Chapter Outline and Learning Objectives

After you have studied this chapter, you should be able to:

Types and Functions of Muscles (p. 111)

- Describe the three types of muscles, and indicate whether each type is voluntary/involuntary, striated/nonstriated.
- Name and discuss four functions of muscles.

Skeletal Muscle Structure and Contraction (p. 111)

- Describe the anatomy of a whole muscle and a muscle fiber.
- Describe the manner in which a muscle fiber contracts.
- Describe a muscle twitch, summation, and tetanus.
- Describe how ATP is made available for muscle contraction.
- Discuss how muscles work together to achieve movement.
- Explain muscle tone and the effect of contraction on the size of a muscle.

Skeletal Muscles of the Body (p. 117)

- Name the superficial muscles of the head, neck, and trunk; shoulder and upper limb (arm); and thigh and lower limb (leg). Indicate their origins and insertions, and give their functions.

Effects of Aging (p. 129)

- Anatomical and physiological changes occur in the muscular system as we age.

Working Together (p. 130)

- The muscular system works with other systems of the body to maintain homeostasis.

Visual Focus

Contraction of a Muscle (p. 113)

MedAlert

Benefits of Exercise (p. 116)

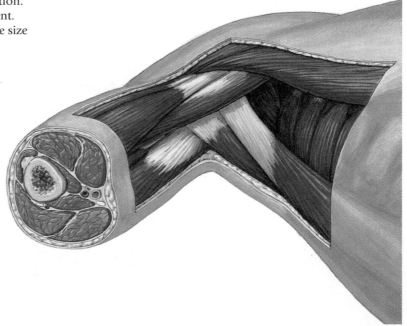

Cross section of the upper arm shows how the muscles are arranged in relation to bone.

Types and Functions of Muscles

Types of Muscles

Humans have three types of muscle tissue (table 7.1): smooth, cardiac, and skeletal. The contractile cells of these tissues are called **muscle fibers.**

Smooth muscle fibers are spindle-shaped cells, each with a single nucleus (uninucleated). The cells are usually arranged in parallel lines, forming sheets. The **striations** (bands of light and dark) seen in cardiac and skeletal muscle are not found in this tissue. Smooth muscle is located in the walls of hollow internal organs. Contraction of smooth muscle is involuntary, occurring without conscious control. Although smooth muscle is slower to contract than skeletal muscle, it can sustain prolonged contractions and does not fatigue easily.

Cardiac muscle fibers are uninucleated, striated, cylindrical, and branched, which allows the fibers to interlock at intercalated disks. Intercalated disks permit contractions to spread quickly throughout the heart. Cardiac fibers relax completely between contractions, which prevents fatigue. Contraction of cardiac muscle is rhythmical; it occurs without outside nervous stimulation and without conscious control. Thus, cardiac muscle contraction is involuntary.

Skeletal muscle fibers are cylindrical, multinucleated, and striated. They make up the skeletal muscles attached to the skeleton. They run the length of the muscle and can be quite long. Skeletal muscle is voluntary because its contraction can be consciously stimulated and controlled by the nervous system.

Uninucleated, spindle-shaped fibers make up the involuntary smooth muscle found in the walls of hollow internal organs. Uninucleated, branched, and striated fibers with intercalated disks make up the involuntary cardiac muscle of the heart. Multinucleated, striated, and cylindrical fibers make up the voluntary skeletal muscle attached to the skeleton.

Functions of Muscles

Muscles have at least four functions:

1. The **tendons** of skeletal muscles often extend across joints, thereby stabilizing the joints. The skeleton would fall apart if joints were not stabilized by tendons and ligaments.
2. Muscle contraction opposes the force of gravity and allows us to remain upright. Muscle contraction refinements allow us to assume different positions.
3. Muscle contraction produces movements. Skeletal muscle contraction accounts for body movements, eye movements, and facial expressions, and also contributes to breathing. Smooth muscle contraction moves food in the digestive tract and reduces the size of blood vessels. Cardiac muscle contraction pumps the blood.
4. Muscle contraction produces heat, which is used to maintain body temperature.

Muscle tendons extend across joints, stabilizing them. Muscle contraction allows us to remain upright, produces movement of the body and internal organs, and creates heat that warms the body.

Skeletal Muscle Structure and Contraction

This chapter is concerned with skeletal muscles—those muscles that make up the bulk of the human body. Skeletal muscles are attached to the skeleton, and their contraction causes the movement of bones. Nerve impulses originating in the brain and spinal cord innervate skeletal muscles by way of nerves and, through a series of steps, bring about contraction. Blood vessels serve the muscles, bringing to muscle cell mitochondria the oxygen and nutrients needed to produce a supply of ATP for muscle contraction.

Skeletal muscles make up the bulk of the body, and their contraction accounts for the movement of bones. Nerves innervate skeletal muscles, and blood vessels bring skeletal muscles oxygen and nutrients.

table 7.1	Types of Muscle Tissue			
	Feature	Smooth	Cardiac	Skeletal
	Location	Internal organs	Heart	Attached to skeleton
	Fiber Appearance	Spindle-shaped	Cylindrical, but branching with intercalated disks	Cylindrical
	Number of Nuclei	Uninucleated	Uninucleated	Multinucleated
	Striations	Nonstriated	Striated	Striated
	Control	Involuntary	Involuntary	Voluntary

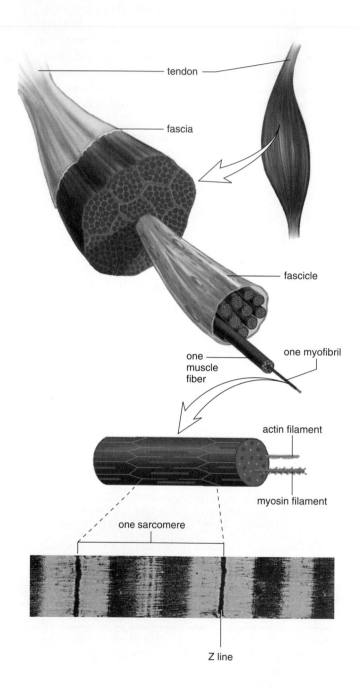

tendon

fascia

fascicle

one muscle fiber

one myofibril

actin filament

myosin filament

one sarcomere

Z line

Anatomy of a Skeletal Muscle

Muscles are covered by several layers of connective tissue called the **fascia** (fash'e-ah), which extends beyond the muscle to become its tendon (fig. 7.1). Fascia also surrounds bundles of muscle fibers called fascicles and it separates the muscle fibers within a fascicle.

Muscle Fiber

A muscle fiber can contract because it contains hundreds and sometimes even thousands of contractile elements, called **myofibrils.** Myofibrils run the length of a muscle fiber, and the striations seen in observations of muscular tissue are due to the placement of filaments within sarcomeres, the units of a myofibril. There are thin **(actin)** filaments and thick **(myosin)** filaments. When a myofibril contracts, the actin filaments slide past the myosin filaments, causing the myofibril, and therefore the muscle fiber, to shorten and thicken. This is the sliding filament theory of muscle contraction.

> Myofibrils are the contractile elements of muscle fibers. The placement of the myofilaments actin and myosin accounts for the striations of skeletal muscle.

🎞 Innervation of Muscle

Muscles are innervated; that is, they are supplied with nerves, and nerve impulses cause muscles to contract. A motor neuron arising from the spinal cord branches to several muscle fibers; collectively, this is called a **motor unit.** A **neuromuscular junction** occurs where a motor neuron fiber meets a muscle fiber (fig. 7.2).

In muscle fibers, the plasma membrane, called the *sarcolemma*, forms tubules that penetrate and dip down into the cell. These tubules, which comprise the *T system* (T for transverse), come near to expanded portions of the endoplasmic reticulum (ER). The endoplasmic reticulum is called the *sarcoplasmic reticulum* and the expanded portions are *calcium storage sacs*, where calcium ions (Ca^{2+}) are stored.

After nerve impulses arrive at a neuromuscular junction, they move down the T system, prompting the release of Ca^{2+} from the calcium storage sacs. If ATP is available as an energy source, the actin filaments now slide past myosin filaments, and the sarcomere contracts.

> Motor neurons meet muscle fibers at neuromuscular junctions. Impulses travel down the tubules of the T system and cause calcium to be released from the sarcoplasmic reticulum. Contraction occurs due to the presence of calcium and ATP in the muscle cells, which prompts actin filaments to slide past myosin filaments, shortening the length of the sarcomere.

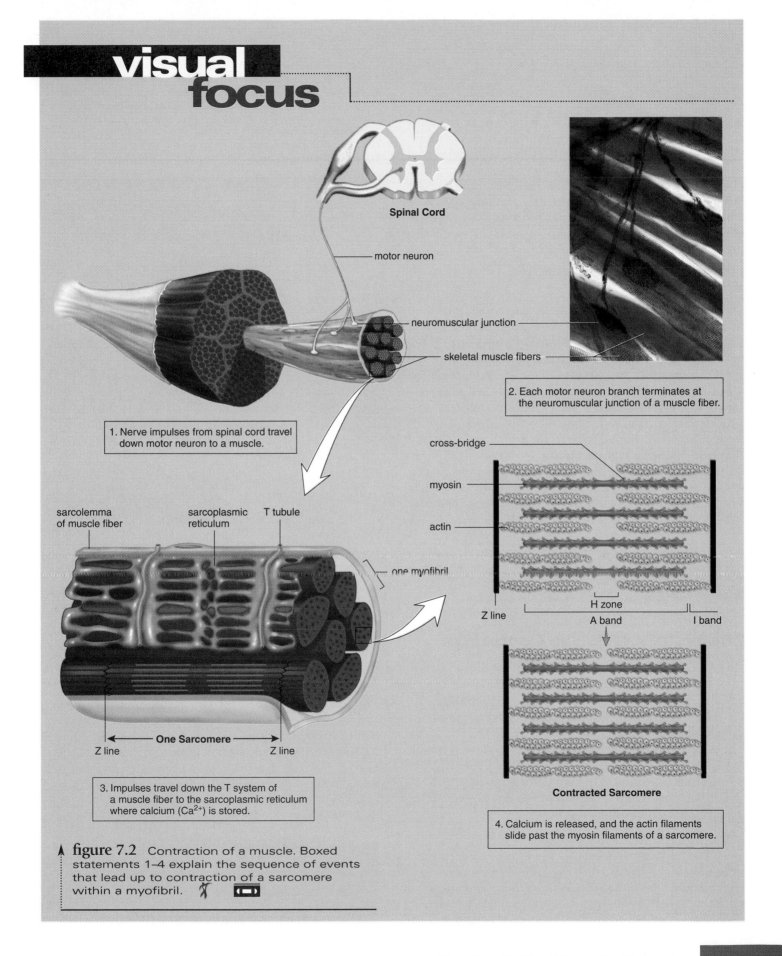

Spinal Cord

motor neuron

neuromuscular junction

skeletal muscle fibers

1. Nerve impulses from spinal cord travel down motor neuron to a muscle.

2. Each motor neuron branch terminates at the neuromuscular junction of a muscle fiber.

sarcolemma of muscle fiber

sarcoplasmic reticulum

T tubule

one myofibril

One Sarcomere

Z line

Z line

3. Impulses travel down the T system of a muscle fiber to the sarcoplasmic reticulum where calcium (Ca^{2+}) is stored.

cross-bridge

myosin

actin

Z line

H zone

A band

I band

Contracted Sarcomere

4. Calcium is released, and the actin filaments slide past the myosin filaments of a sarcomere.

figure 7.2 Contraction of a muscle. Boxed statements 1–4 explain the sequence of events that lead up to contraction of a sarcomere within a myofibril.

Physiology of Muscle Contraction

Researchers have noted the following aspects of muscle contraction in the laboratory:

All-or-None Law

When a muscle fiber is stimulated, it behaves in an all-or-none manner; it either contracts completely or not at all. On the other hand, the strength of the contraction of a whole muscle can increase according to how many muscle fibers are contracted. In other words, an entire muscle does not obey the **all-or-none law** because the total amount of contraction depends on how many muscle fibers are contracted at that time.

Muscle Twitch, Summation, and Tetanus

A single stimulus causes a muscle to contract and then relax, resulting in a muscle **twitch**. Figure 7.3*a* shows that a muscle twitch can be divided into the latent period (the time between stimulus and initiation of contraction), the contraction period, and the relaxation period. Normally in the body, a muscle receives many impulses in rapid succession. Because of this, tension summates until maximal sustained contraction called **tetanus** is achieved (fig. 7.3*b*). If the muscle is not allowed to rest, **fatigue** will set in. Fatigue is apparent when a muscle relaxes, even though stimulation is continued. When a muscle fiber fatigues (gives out), it has run out of ATP. The muscles of long-distance runners have been known to fatigue when most of the fibers have run out of ATP, causing the runner to collapse.

> Muscle fibers obey the all-or-none law, but whole muscles do not. The occurrence of a muscle twitch, summation, or tetanus depends on the frequency with which a muscle is stimulated.

Chemistry of Muscle Contraction

Muscle contraction requires a plentiful supply of ATP. ATP can be made available in three ways:

1. Muscle cells are generously supplied with mitochondria, within which ATP is formed by means of aerobic cellular respiration.
2. Muscle cells contain *creatine phosphate,* which is used as a storage supply of high-energy phosphate. Creatine phosphate does not participate directly in muscle contraction. Instead, it is used to regenerate ATP by the following reaction:

$$\text{creatine} \sim \textcircled{P} + ADP \rightarrow ATP + \text{creatine}$$

3. When the creatine phosphate supply has been depleted, a muscle cell can still produce ATP anaerobically. Anaerobic respiration occurs when the cells are not being supplied with oxygen quickly enough to make aerobic respiration possible. This would occur, for

figure 7.3 Physiology of a skeletal muscle contraction. These are myograms, visual representations of the contraction of a muscle that has been dissected from an animal. **a.** A simple muscle twitch is composed of three periods: latent, contraction, and relaxation. **b.** Summation and tetanus. When a muscle is not allowed to relax completely between stimuli, the contraction gradually increases in intensity. The muscle becomes maximally contracted until it fatigues.

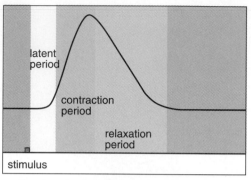

a. **Simple muscle twitch**

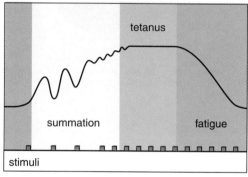

b. **Summation, tetanus, fatigue**

example, during times of strenuous exercise. In practice, anaerobic respiration can supply ATP for only a short time because a waste product called lactic acid builds up, resulting in muscular aching and fatigue.

Oxygen Debt

We have all experienced the fact that, following strenuous exercise, we continue to breathe deeply and pant, even while resting. This continued intake of oxygen is required to complete the metabolism of lactic acid that has accumulated during exercise and represents an **oxygen debt** that the body must pay to rid itself of lactic acid. The lactic acid is transported to the liver, where it is completely broken down to carbon dioxide and water.

> ATP is required for muscle contraction and can be generated by way of aerobic or anaerobic cellular respiration and from creatine phosphate. During anaerobic cellular respiration, a waste product called lactic acid is produced. Lactic acid represents an oxygen debt, since oxygen is required to metabolize this product.

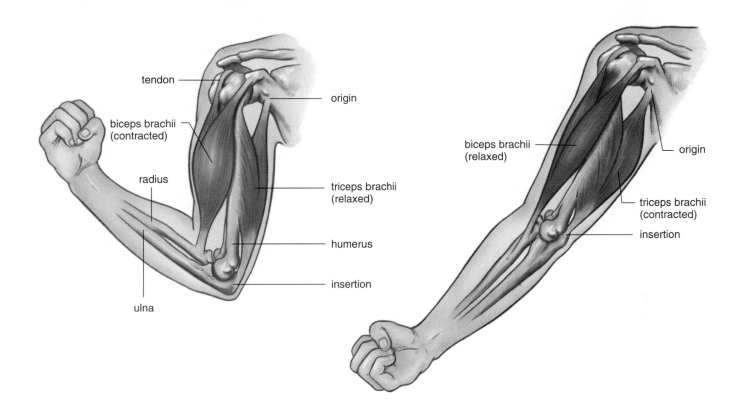

figure 7.4 Attachment of the skeletal muscles. The origin of a muscle is on a bone that remains stationary, and the insertion of a muscle is on a bone that moves when the muscle contracts. The muscles in this drawing are antagonistic. When the biceps brachii contracts, the lower arm flexes, and when the triceps brachii contracts, the lower arm extends.

Aspects of Muscle Contraction

When muscle contracts, one bone remains fairly stationary, and the other one moves. The **origin** of a muscle is on the stationary bone, and the **insertion** of a muscle is on the bone that moves.

Frequently, a body part is moved by a group of muscles working together. Even so, one muscle does most of the work and is called the **prime mover.** The assisting muscles are called the **synergists.** When muscles contract, they shorten; therefore, muscles can only pull; they cannot push.

Muscles have **antagonists,** and antagonistic pairs work opposite one another to bring about movement in opposite directions. For example, the biceps brachii and the triceps brachii are antagonists; one flexes the forearm, and the other extends the forearm (fig. 7.4). The discussion that follows will point out other antagonistic pairs.

> When muscles cooperate to achieve movement, some act as prime movers, others as synergists, and still others as antagonists.

Isotonic Versus Isometric Contraction

Ordinarily, when muscles contract, they shorten and a movement occurs. This is called **isotonic contraction.** For example, when a barbell weight is moved upward, the biceps contracts isotonically. On occasion, muscles contract but do not shorten or produce movement. This is called an **isometric contraction.** When a stationary bar is pulled on, the biceps contracts isometrically, and no movement occurs.

Muscle Tone

Whole skeletal muscles have **tone,** a condition in which some fibers are always contracted. Muscle tone is particularly important in maintaining posture. If the muscles of the neck, trunk, and legs suddenly relax, the body collapses.

Maintenance of the right amount of tone requires the use of special sense receptors called **muscle spindles.** A muscle spindle consists of a bundle of modified muscle fibers, with sensory nerve fibers wrapped around a short, specialized region. A spindle contracts along with muscle

MedAlert

Benefits of Exercise

*e*xercise programs improve muscular strength, muscular endurance, and flexibility. Muscular strength is defined as the force a muscle group (or muscle) can exert against a resistance in one maximal effort. Muscular endurance is judged by the ability of a muscle group (or muscle) to contract repeatedly or to sustain a contraction for an extended period. Flexibility is tested by observing the range of motion about a joint.

As muscular strength improves, the overall size of the muscle, as well as the number of muscle fibers and myofibrils in the muscle, increases. The total amount of protein, the number of capillaries, and the amounts of connective tissue, including tissue found in tendons and ligaments, also increase. Physical training with weights can improve muscular strength and endurance in all adults, regardless of their age.

A surprising finding, however, is that health benefits also accompany less strenuous programs, such as those described in table 7.A. A study of 12,000 men by Dr. Arthur Leon at the University of Minnesota showed that even moderate exercise lowered the risk of a heart attack by one-third. People with arthritis reported much less pain, swelling, fatigue, and depression after only four months of attending a twice-weekly, low-impact aerobics class. Increasing daily activity by walking to the corner store instead of driving and by taking the stairs instead of the elevator can improve a person's health.

The benefits of exercise are most apparent with regard to cardiovascular health. Brisk walking for 2.5–4 hours a week can raise the blood level of high-density lipoprotein (HDL), a chemical that promotes healthy blood vessels (see chapter 12).

table 7.A	A Checklist for Staying Fit	
	Children, Ages 7–12	**Teenagers, Ages 13–18**
	Engage in vigorous activity 1–2 hours daily.	Engage in vigorous activity three to five times a week.
	Participate in free play.	Build muscle with calisthenics.
	Build motor skills through team sports, dance, swimming.	Plan aerobic exercise to control buildup of fat cells.
	Exercise outside of physical education classes.	Pursue tennis, swimming, horseback riding—sports that can be enjoyed for a lifetime.
	Be involved in family outings: bowling, boating, camping, hiking.	Continue team sports, dancing, hiking, swimming.

fibers, but thereafter, it sends stimuli to the central nervous system, which then regulates muscle contraction so that tone is maintained.

Effect of Contraction on Size of Muscle

Exercise greatly benefits the health of all people, as is discussed in the MedAlert reading above. Forceful muscular activity over a prolonged period causes muscles to increase in size. This increase, called **hypertrophy,** occurs only if the muscle contracts to at least 75% of its maximum tension. However, only a few minutes of forceful exercise several times a week are required for hypertrophy to occur. The muscle fibers show a gain in metabolic potential, as well as an increase in their number of myo-

fibrils. Contrary to former belief, the number of muscle fibers may also increase, resulting in muscles that can work longer before they fatigue. Some athletes take steroids, either testosterone or related chemicals, to promote muscle growth. This practice has many side effects, as discussed in chapter 10.

Muscles that are not used or that are used for only very weak contractions decrease in size, or **atrophy.** Atrophy can occur when a limb is placed in a cast or when the nerve serving a muscle is damaged. If nerve stimulation is not restored, muscle fibers are gradually replaced by fat and fibrous tissue. Unfortunately, atrophy causes muscle fibers to shorten progressively, leaving body parts contracted in contorted positions.

Exercise also helps prevent osteoporosis, a condition in which the bones are weak and tend to break. The stronger the bones are when a person is young, the less chance of osteoporosis as a person ages. Exercise promotes the activity of osteoblasts (as opposed to osteocytes) in young people, as well as older people. An increased activity level can also keep off unwanted pounds, which is a worthwhile goal since added body weight aggravates numerous conditions, such as the likelihood of developing type II diabetes. In addition, increased muscle activity causes glucose to be transported into muscle cells and makes the body less dependent on the presence of insulin.

Cancer prevention and early detection involve eating properly, not smoking, avoiding cancer-causing chemicals and radiation, undergoing appropriate medical screening tests, and knowing the early warning signs of cancer. However, evidence indicates that exercise also helps prevent certain kinds of cancer. Studies show that people who exercise are less likely to develop colon, breast, cervical, uterine, and ovarian cancers.

Questions

1. Regular bowel movements seem to prevent colon cancer. What effect do you predict exercise has on promoting regular bowel movements?
2. Fat intake and its assimilation in the body seem to promote the development of breast cancer. What effect do you predict that exercise has on body fat?
3. Physical training lowers the resting heart rate. Why might this be beneficial to longevity?

Adults, Ages 19–55	Seniors, Age 55 and Up
Engage in vigorous activity for a half hour, three times a week.	Engage in moderate exercise three times a week.
Exercise to prevent lower back pain: aerobics, stretching, yoga.	Plan a daily walk.
Take active vacations: hike, bicycle, cross-country ski.	Do stretching exercises daily.
Find exercise partners: join a running club, bicycle club, outing group.	Learn a new sport or activity: golf, fishing, ballroom dancing.
	Try low-impact aerobics.

Muscles contract isotonically and isometrically. They have tone. Forceful contraction can cause them to increase in size, and lack of contraction weakens them.

Skeletal Muscles of the Body

Naming Muscles

The names of the various skeletal muscles (figs. 7.5 and 7.6) may indicate the following characteristics of the muscle:

1. *Size.* For example, the gluteus maximus is the largest muscle that makes up the buttocks.
2. *Shape.* For example, the deltoid is shaped like a delta, or triangle.
3. *Direction of fibers.* For example, the rectus abdominis is a longitudinal muscle of the abdomen (*rectus* means straight).
4. *Location.* For example, the frontalis overlies the frontal bone.
5. *Number of attachments.* For example, the biceps brachii has two attachments, or origins.
6. *Action.* For example, the extensor digitorum extends the fingers or digits.

The names of muscles often include information about the muscles' size, shape, direction of fibers, location, number of attachments, and action.

figure 7.5 Anterior view of superficial skeletal muscles.

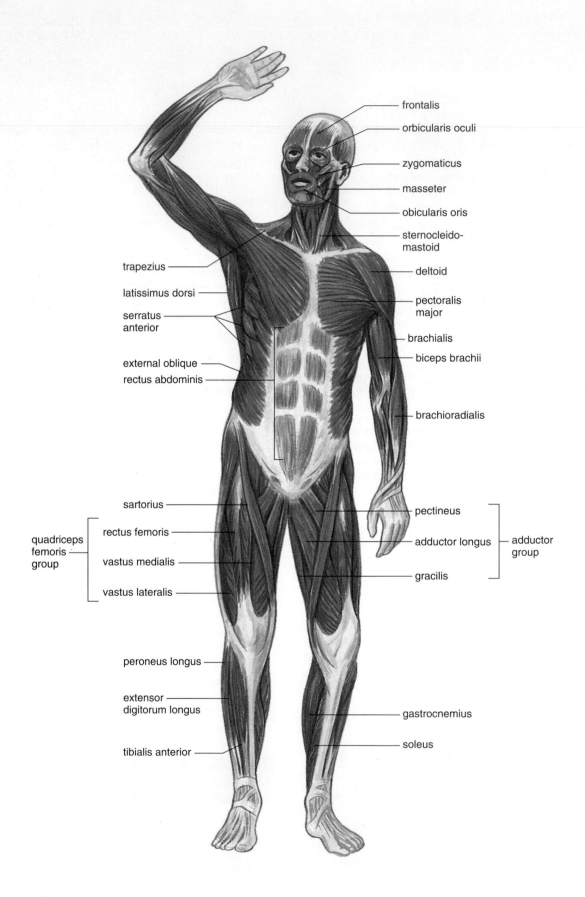

frontalis
orbicularis oculi
zygomaticus
masseter
obicularis oris
sternocleido-mastoid
deltoid
pectoralis major
brachialis
biceps brachii
brachioradialis
pectineus
adductor longus
gracilis
adductor group
gastrocnemius
soleus

trapezius
latissimus dorsi
serratus anterior
external oblique
rectus abdominis
quadriceps femoris group
sartorius
rectus femoris
vastus medialis
vastus lateralis
peroneus longus
extensor digitorum longus
tibialis anterior

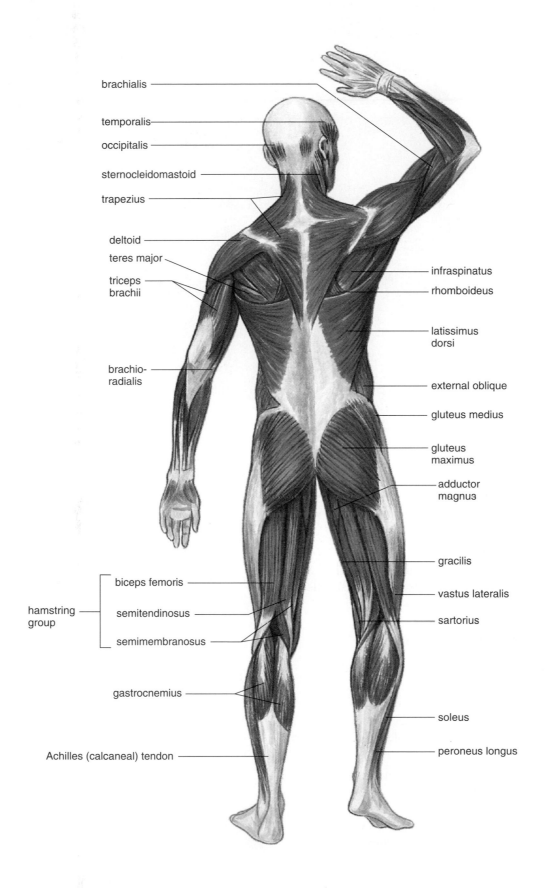

figure 7.6 Posterior view of superficial skeletal muscles.

brachialis

temporalis

occipitalis

sternocleidomastoid

trapezius

deltoid

teres major

triceps brachii

brachio-radialis

infraspinatus

rhomboideus

latissimus dorsi

external oblique

gluteus medius

gluteus maximus

adductor magnus

gracilis

vastus lateralis

sartorius

hamstring group

biceps femoris

semitendinosus

semimembranosus

gastrocnemius

soleus

peroneus longus

Achilles (calcaneal) tendon

figure 7.7 Lateral view of superficial facial muscles involved in facial expression and mastication. The cranial aponeurosis is a broad sheet of connective tissue that lies over the cranium like a cap.

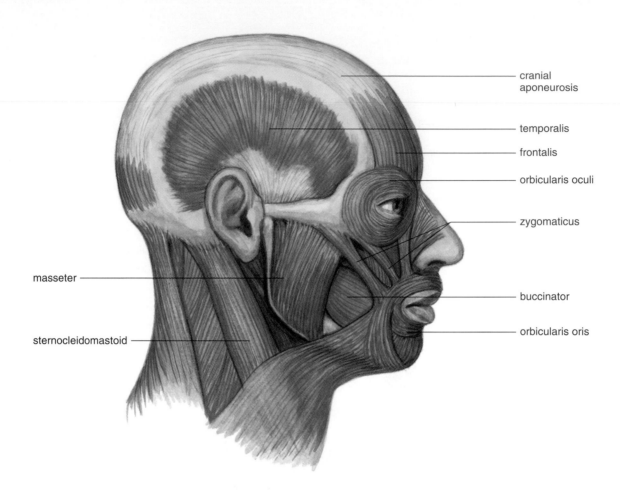

Muscles of the Head

The muscles of the head (fig. 7.7) are divided into the muscles of facial expression and the muscles of mastication. These muscles are listed in table 7.2.

Muscles of Facial Expression

Frontalis lies over the frontal bone and raises the eyebrows and wrinkles the brow.

Orbicularis oculi is a ringlike band of muscle that encircles (forms an orbit about) the eye. It causes the eye to close or blink, and is responsible for crow's feet at the eye corners.

Orbicularis oris encircles the mouth and is used to pucker the lips, as in forming a kiss.

Buccinator is located in the cheek area, and when it contracts, the cheek is compressed, such as when a person whistles or blows out air. (This muscle is also

called the trumpeter's muscle.) Important to everyday life, this muscle helps hold food in contact with the teeth during chewing.

Zygomaticus extends from the zygomatic arch (cheekbone) to the corners of the mouth. It raises the corners of the mouth when a person smiles.

Muscles of Mastication

The muscles of mastication are used when we chew food.

Masseter extends from the zygomatic arch to the mandible. It is a muscle of mastication (chewing) because it raises the mandible.

Temporalis is a fan-shaped muscle that overlies the temporal bone. It acts as a synergist to the masseter.

The muscles of the head are divided into those for facial expression and those for mastication (chewing).

table 7.2

Muscles of the Head

Name	Function	Origin/Insertion
Muscles of Facial Expression		
Frontalis	Raises eyebrows	Cranial aponeurosis / skin and muscles around eye
Orbicularis oculi	Closes eye	Maxillary and frontal bones / skin around eye
Orbicularis oris	Closes and protrudes lips	Muscles near the mouth / skin around mouth
Buccinator	Compresses cheeks inward	Outer surfaces of maxilla and mandible / orbicularis oris
Zygomaticus	Raises corner of mouth	Zygomatic bone / skin and muscle around mouth
Muscles of Mastication		
Masseter	Closes jaw	Zygomatic arch / mandible
Temporalis	Closes jaw	Temporal bone / mandibular coronoid process

table 7.3

Muscles of the Neck and Trunk

Name	Function	Origin/Insertion
Muscles That Move the Head		
Sternocleidomastoid	Flexes neck and rotates head	Sternum and clavicle / mastoid process of temporal bone
Trapezius	Extends neck and adducts scapula	Occipital bone and all cervical and thoracic vertebrae / spine of scapula and clavicle
Muscles of the Trunk		
External intercostals	Pull ribs up and out	Superior rib / inferior rib
Internal intercostals	Pull ribs down and in	Inferior rib / superior rib
External oblique	Tenses abdominal wall and lateral rotation	Lower eight ribs / iliac crest
Internal oblique	Tenses abdominal wall and lateral rotation	Iliac crest / lower three ribs
Transversus abdominis	Tenses abdominal wall	Lower six ribs / crest of pubis
Rectus abdominis	Flexes vertebral column	Crest of pubis, pubic symphysis / xiphoid process of sternum, fifth to seventh costal cartilages

Muscles of the Neck and Trunk

Anterior muscles of the neck and trunk are shown in figure 7.8 and listed in table 7.3. Certain of these muscles are discussed here.

Muscles That Move the Head

Sternocleidomastoid muscles occur in the sides of the neck and extend from the sternum to the mastoid process (they are named for their attachments). When both of these muscles contract, the neck is flexed and the head is bent toward the chest. When only one contracts, the head turns to the opposite side.

Each **trapezius** muscle is triangular, but together, they take on a diamond or trapezoid shape. A trapezius muscle extends from the base of the skull down to the end of the thoracic vertebrae and inserts on a scapula laterally. The trapezius muscles extend the neck and help move the scapulae when the shoulders are shrugged or pulled back.

The sternocleidomastoid muscles and trapezius muscles are antagonistic because the sternocleidomastoid muscles flex the neck, and the trapezius muscles extend it.

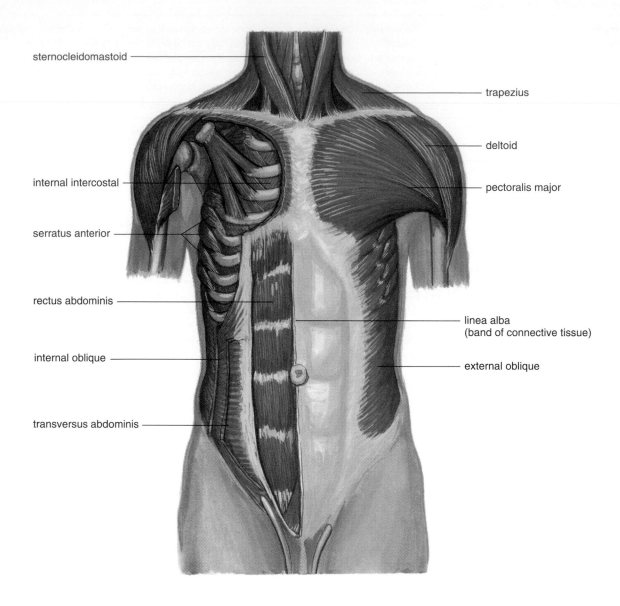

sternocleidomastoid

trapezius

deltoid

internal intercostal

pectoralis major

serratus anterior

rectus abdominis

linea alba
(band of connective tissue)

internal oblique

external oblique

transversus abdominis

Muscles of the Trunk

External and **internal obliques,** and the **transversus abdominis** are located in the abdominal wall. The external and internal obliques occur on a slant and are at right angles to one another between the lower ribs and the pelvic girdle (see fig. 7.8). Below the obliques, the transversus abdominis extends horizontally across the abdomen. As with plywood, the abdominal wall is strengthened by having muscle fibers that run in different directions. All of these muscles tense and support the abdominal wall.

Rectus abdominis has a straplike appearance but takes its name from the fact that it runs straight (*rectus* means straight) up from the pubic bones to the ribs and sternum. It is the outermost muscle that compresses the contents of the abdominal cavity, but it also helps flex the vertebral column.

The obliques (external and internal), the transversus abdominis, and the rectus abdominis all function to provide a sturdy abdominal wall.

figure 7.9 Muscles of the posterior neck, shoulder, and trunk.

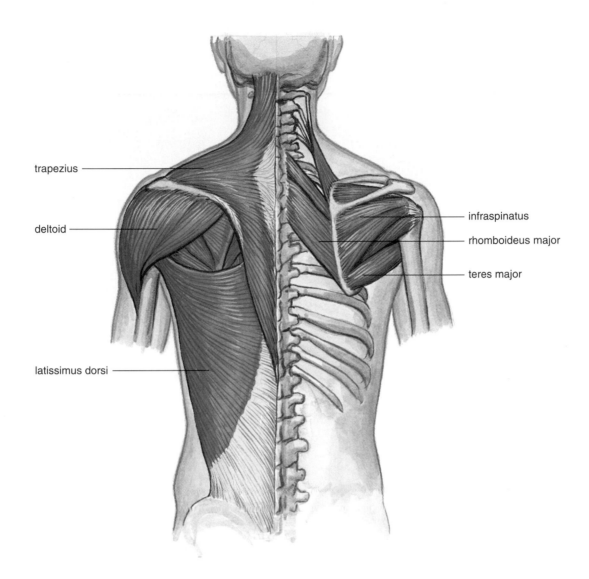

trapezius

deltoid

latissimus dorsi

infraspinatus

rhomboideus major

teres major

Muscles of the Shoulder and Upper Limb (Arm)

Muscles of the shoulder and upper limb (arm) are shown in figures 7.9 and 7.10 and listed in table 7.4. Certain of these muscles are discussed here.

Muscles That Move the Pectoral Girdle and Arm

Deltoid is a large, fleshy, triangular muscle (*deltoid* in Greek means triangular) that covers the shoulder and causes the bulge of the upper arm. It runs from both the clavicle and the scapula to the humerus. This muscle abducts the arm (raises the arm laterally) to the horizontal position. (The term *arm* is used in this section to refer to the entire upper limb.)

Pectoralis major is a large (major) anterior muscle of the upper chest. It originates from the clavicles of the pectoral girdle, but also from the sternum and ribs. It inserts on the humerus. If you press your hands together forcefully, you can feel these muscles contract isometrically. The pectoralis major flexes the arm (raises it anteriorly) and adducts the arm, pulling it across the chest.

Latissimus dorsi (located both laterally and dorsally) is a wide, triangular muscle of the lower back. This muscle originates from the lower spine and sweeps

figure 7.10 Muscles of the shoulder and arm. **a.** Muscles of the anterior shoulder and cross section of upper arm. **b.** Muscles of the posterior shoulder and upper arm. **c.** Muscles of the posterior arm.

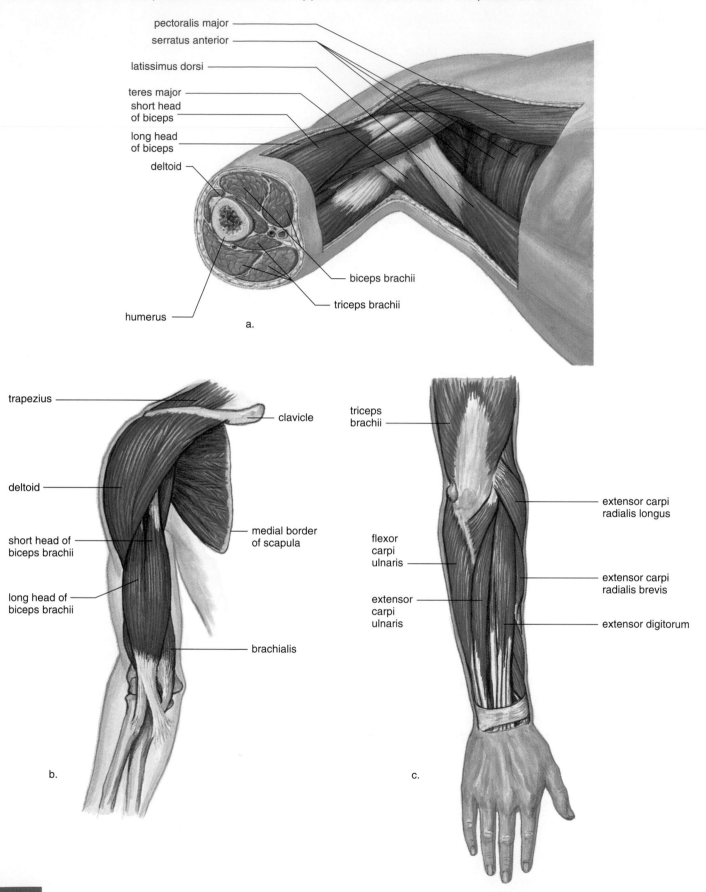

pectoralis major

serratus anterior

latissimus dorsi

teres major

short head of biceps

long head of biceps

deltoid

biceps brachii

triceps brachii

humerus

a.

trapezius

clavicle

deltoid

medial border of scapula

short head of biceps brachii

long head of biceps brachii

brachialis

b.

triceps brachii

flexor carpi ulnaris

extensor carpi ulnaris

extensor carpi radialis longus

extensor carpi radialis brevis

extensor digitorum

c.

table 7.4

Muscles of the Shoulder and Upper Limb (Arm)

Name	Function	Origin/Insertion
Muscles That Move the Pectoral Girdle and Arm		
Rhomboideus major	Raises and adducts scapula	Spines of upper thoracic vertebrae / medial border of scapula
Infraspinatus	Rotates arm laterally	Posterior surface of scapula / greater tubercle of humerus
Serratus anterior	Pulls scapula downward and forward	Upper nine ribs / vertebral border of scapula
Deltoid	Abducts arm	Acromion process, spine of scapula, and the clavicle / deltoid tuberosity of humerus
Pectoralis major	Flexes and adducts arm	Clavicle, sternum, second to sixth costal cartilages / intertubular groove of humerus
Latissimus dorsi	Extends or adducts arm	Iliac crest / intertubular groove of humerus
Teres major	Extends humerus or adducts and rotates arm medially	Lateral border of scapula / intertubular groove of humerus
Muscles That Move the Forearm		
Biceps brachii	Flexes forearm and supinates hand	Scapula / radial tuberosity
Triceps brachii	Extends forearm	Scapula, proximal humerous / olecranon process of ulna
Brachialis	Flexes forearm	Anterior humerus / coronoid process of ulna
Muscles That Move the Hand		
Extensor carpi and flexor carpi	Move wrist and hand	Humerus / carpals and metacarpals
Extensor digitorum and flexor digitorum	Move fingers	Humerus, radius, ulna / phalanges

upward to insert on the humerus. The latissimus dorsi extends and adducts the arm (brings it down from a raised position). This muscle is very important for swimming, rowing, and climbing a rope.

Serratus anterior is located below the axilla (armpit) on the side of the chest. It runs between the upper ribs and the scapula. It pulls the scapula downward and forward, as when we push something. It also helps to raise the arm above the horizontal level.

The deltoid, pectoralis major, latissimus dorsi, and serratus anterior muscles all function to move the humerus and, therefore, the arm, in relation to the trunk.

Muscles That Move the Forearm

Biceps brachii is a muscle of the anterior upper arm that is familiar because it bulges when the forearm is flexed. It also supinates the hand when a doorknob is turned or the cap of a jar is unscrewed. The name of the muscle refers to its two heads that attach to the scapula, where it originates. The biceps brachii inserts on the radius.

Triceps brachii is the only muscle of the posterior upper arm. It has three heads that attach to the scapula and humerus, and it inserts on the ulna. The triceps extends the forearm when something is pushed. It is sometimes called the boxer's muscle because it straightens the elbow when a punch is thrown.

The biceps and triceps brachii are antagonistic muscles because the biceps flexes the forearm, and the triceps extends it.

Muscles That Move the Hand

Flexor carpi and **extensor carpi** muscles (see fig. 7.10) originate on the bones of the forearm and insert on the bones of the hand. They move the wrist and hand.

table 7.5

Muscles of the Thigh and Lower Limb (Leg)

Name	Function	Origin/Insertion
Muscles That Move the Thigh		
Iliopsoas	Flexes thigh	Lumbar vertebrae, ilium / lesser trochanter of femur
Gluteus maximus	Extends thigh	Posterior ilium, sacrum / proximal femur
Gluteus medius	Abducts thigh	Ilium / greater trochanter of femur
Adductor group	Adducts thigh	Pubis, ischium / femur and tibia
Muscles That Move the Lower Limb (Leg)		
Quadriceps femoris group	Extends lower leg	Ilium, femur / patella tendon that continues as a ligament to tibial tuberosity
Hamstring group	Flexes lower leg and extends hip	Ischial tuberosity / lateral and medial tibia
Sartorius	Flexes, abducts, and rotates leg	Ilium / medial tibia
Muscles That Move the Ankle and Foot		
Gastrocnemius	Plantar flexion and eversion of foot	Condyles of femur / calcaneus by way of Achilles tendon
Tibialis anterior	Dorsiflexion and inversion of foot	Condyles of tibia / tarsal and metatarsal bones
Peroneus group	Plantar flexion and eversion of foot	Fibula / tarsal and metatarsal bones
Soleus	Plantar flexes foot	Tibia, fibula / calcaneus
Flexor and extensor digitorum longus	Moves toes	Tibia, fibula / phalanges

Flexor digitorum and **extensor digitorum** muscles also originate on the bones of the forearm and insert on the bones of the hand. They move the fingers.

> The muscles that move the hand and fingers span the wrist.

Muscles of the Thigh and Lower Limb (Leg)

Muscles of the thigh and lower limb (leg) tend to be large and heavy because they are used to move the entire weight of the body and to resist the force of gravity. Therefore, they are important for movement and balance. These muscles are shown in figures 7.11–7.14 and listed in table 7.5. Certain of these muscles are discussed here.

Muscles That Move the Thigh

Muscles that move the thigh are shown in figures 7.11 and 7.12.

Iliopsoas (includes psoas major and iliacus) originates from the ilium and the bodies of the lumbar vertebrae, and inserts on the femur anteriorly. This muscle flexes the thigh (raises the leg anteriorly) and is important to the process of walking. (The term *leg* is used in this section and the next to refer to the entire lower limb.)

It also helps prevent the trunk from falling backward when a person is standing erect.

Gluteus maximus is the largest muscle in the body and covers a large part of the buttock (*gluteus* means buttocks in Greek). It originates at the ilium and sacrum, and inserts on the femur. The gluteus maximus acts to straighten the leg at the hip (and, in that way, to extend the thigh) when a person is walking, climbing stairs, or jumping from a crouched position.

> The iliopsoas and the gluteus maximus are antagonistic muscles because the iliopsoas flexes the thigh, and the gluteus maximus extends it.

Gluteus medius lies partly behind the gluteus maximus. It runs between the ilium and the femur, and functions to abduct the thigh (raise the leg sideways to a horizontal position).

Adductor muscles (pectineus, adductor longus, adductor magnus, gracilis) are located on the medial part of the thigh. They originate from the pubic bone and ischium, and insert on the femur. Adductor muscles adduct the thigh (bring the leg down from a horizontal position) and press the thighs together.

> The gluteus medius and adductor muscles are antagonistic because the gluteus medius acts to abduct the thigh, and the adductor muscles act to adduct it.

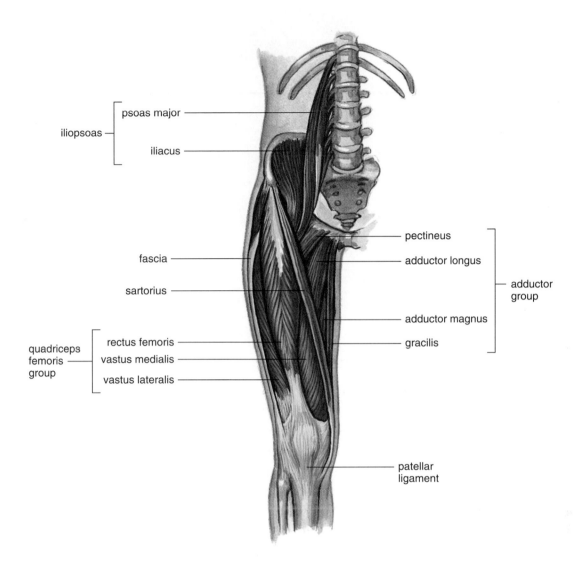

figure 7.11 Muscles of the anterior right thigh.

iliopsoas — {
psoas major
iliacus
}

pectineus

adductor longus

adductor group

fascia

sartorius

adductor magnus

gracilis

quadriceps femoris group — {
rectus femoris
vastus medialis
vastus lateralis
}

patellar ligament

Muscles That Move the Lower Limb (Leg)

Muscles that move the lower limb (leg) are shown in figures 7.11 and 7.12.

Quadriceps femoris group (rectus femoris, vastus lateralis, vastus medialis, vastus intermedius) includes muscles found in the front and on the sides of the thigh. The rectus femoris, which originates from the ilium, lies in front of the vastus intermedius, and therefore this muscle is not shown in figure 7.11. The vastus muscles originate from the femur. The vastus muscles are the primary extensors of the lower leg, such as when a ball is kicked.

Hamstring group (biceps femoris, semimembranosus, semitendinosus) is composed of several muscles found at the back of the thigh. They all have origins on the ischium and insert on the tibia. Their strong tendons can be felt behind the knee. These same tendons are present in hogs and were used by butchers as strings to hang up hams for smoking—hence, the name. The hamstrings help flex the lower leg and extend the hip.

The quadriceps femoris group of muscles and the hamstring group are antagonistic because the quadriceps femoris group extends the lower leg, and the hamstring group flexes it.

figure 7.12 Muscles of the posterior right thigh.

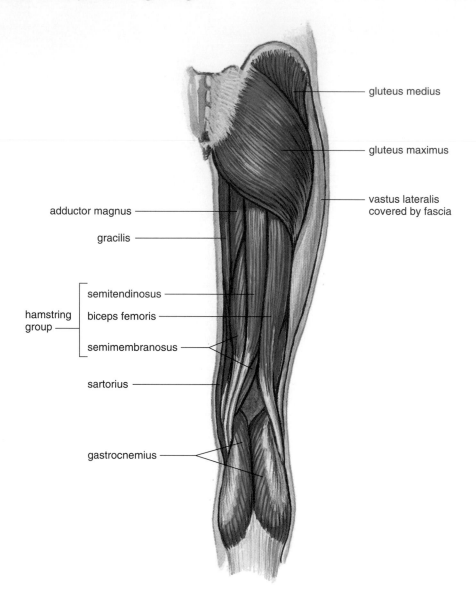

gluteus medius

gluteus maximus

vastus lateralis covered by fascia

adductor magnus

gracilis

semitendinosus

hamstring group — biceps femoris

semimembranosus

sartorius

gastrocnemius

Sartorius is a long, straplike muscle that begins at the iliac spine and then passes inward across the front of the thigh to descend over the medial side of the knee. It flexes the leg and is used to sit cross-legged, as tailors were accustomed to do in another era. Therefore, it is sometimes called the tailor's muscle, and, in fact, *sartor* means tailor in Latin.

Muscles That Move the Ankle and Foot

Muscles that move the ankle and foot are shown in figures 7.13 and 7.14.

Gastrocnemius is located on the back of the leg, where it forms a large part of the calf. It arises from the femur; distally, the muscle joins the strong Achilles tendon,

which attaches behind the calcaneus bone (heel). The gastrocnemius is a powerful plantar flexor of the foot that aids in pushing the body forward during walking or running. It is sometimes called the toe dancer's muscle because it allows a person to stand on tiptoe.

Tibialis anterior is a long, spindle-shaped muscle located on the front of the lower leg. It arises from the surface of the tibia and attaches to the bones of the ankle and foot. Contraction of this muscle causes dorsiflexion and inversion of the foot.

The gastrocnemius and the tibialis anterior are antagonistic muscles because the gastrocnemius brings about plantar flexion, and the tibialis anterior dorsiflexion, of the foot.

figure 7.13 Muscles of the anterior right lower leg.

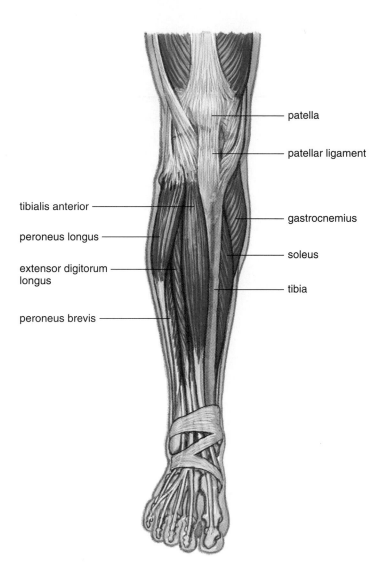

patella

patellar ligament

tibialis anterior

gastrocnemius

peroneus longus

extensor digitorum longus

soleus

peroneus brevis

tibia

Peroneus muscles (peroneus longus, peroneus brevis) are found on the lateral side of the leg, connecting the fibula to the metatarsal bones of the foot. These muscles evert the foot and also help bring about plantar flexion.

Flexor and **extensor digitorum longus** muscles are found on the lateral and posterior portion of the leg. They arise mostly from the tibia and insert on the toes. They flex and extend the toes, and assist in other movements of the feet.

> The peroneus muscles and the flexor and extensor digitorum muscles move the feet and the toes.

Effects of Aging

Muscle mass and strength tend to decrease as people age. How much of this is due to lack of exercise and a poor diet has yet to be determined. Deteriorated muscle elements are replaced initially by connective tissue and, eventually, by fat. With age, degenerative changes take place in the mitochondria, and endurance decreases. Also, changes in the nervous and circulatory systems adversely affect the structure and function of muscles.

Muscle mass and strength can improve remarkably if elderly people undergo a training program. Exercise at any age appears to stimulate muscle buildup. As discussed

figure 7.14 Muscles of the lateral right lower leg.

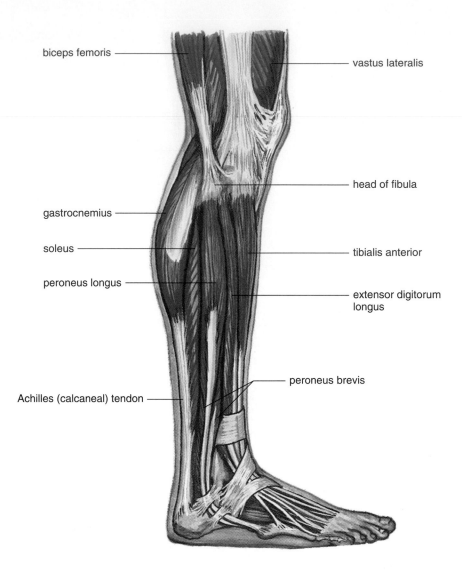

biceps femoris

vastus lateralis

head of fibula

gastrocnemius

soleus

tibialis anterior

peroneus longus

extensor digitorum longus

peroneus brevis

Achilles (calcaneal) tendon

in the MedAlert reading for this chapter, the benefits of exercise exceed this advantage. For example, exercise improves the cardiovascular system and reduces the risk of diabetes and glycation. During glycation, excess glucose molecules stick to body proteins so that the proteins no longer have their normal structure and cannot function properly. Exercise burns glucose and, in this way, helps prevent muscle deterioration.

Working Together

The accompanying illustration shows how the muscular system works with other organ systems of the body to maintain homeostasis. All the organ systems of the body are interrelated.

BodyWorks CD-ROM
The module accompanying chapter 7 is Muscular System.

working together   Muscular System

Integumentary System

Muscle contraction provides heat to warm skin.

Skin protects muscles; rids the body of heat produced by muscle contraction.

How the Muscular System works with other body systems

Lymphatic System/Immunity

Skeletal muscle contraction moves lymph; physical exercise enhances immunity.

Lymphatic vessels pick up excess tissue fluid; immune system protects against infections.

Skeletal System

Muscle contraction causes bones to move joints; muscles help protect bones.

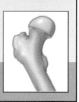

Bones provide attachment sites for muscles; store Ca^{2+} for muscle function.

Respiratory System

Muscle contraction assists breathing; physical exercise increases respiratory capacity.

Lungs provide oxygen for, and rid the body of, carbon dioxide from contracting muscles.

Nervous System

Muscle contraction moves eyes, permits speech, creates facial expressions.

Brain controls nerves that innervate muscles; receptors send sensory input from muscles to brain.

Digestive System

Smooth muscle contraction accounts for peristalsis; skeletal muscles support and help protect abdominal organs.

Digestive tract provides glucose for muscle activity; liver metabolizes lactic acid following anaerobic muscle activity.

Endocrine System

Muscles help protect glands.

Androgens promote growth of skeletal muscle; epinephrine stimulates heart and constricts blood vessels.

Urinary System

Smooth muscle contraction assists voiding of urine; skeletal muscles support and help protect urinary organs.

Kidneys maintain blood levels of Na^+, K^+, and Ca^{2+}, which are needed for muscle innervation, and eliminate creatinine, a muscle waste.

Circulatory System

Muscle contraction keeps blood moving in heart and blood vessels.

Blood vessels deliver nutrients and oxygen to muscles, carry away wastes.

Reproductive System

Muscle contraction occurs during orgasm and moves gametes; abdominal and uterine muscle contraction occurs during childbirth.

Androgens promote growth of skeletal muscle.

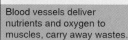

Selected New Terms

Basic Key Terms

actin (ak'tin), p. 112
all-or-none law, p. 114
cardiac muscle (kar'de-ak mus'el), p. 111
fascia (fash'e-ah), p. 112
fatigue (fah-tĕg'), p. 114
insertion (in-ser'shun), p. 115
isometric contraction (i"so-met'rik kon-trak'shun), p. 115
isotonic contraction (i"so-ton'ik kon-trak'shun), p. 115
motor unit (mo'tor u'nit), p. 112
muscle fiber (mus'el fi'ber), p. 111
myofibril (mi"o-fi'bril), p. 112
myosin (mi'o-sin), p. 112

neuromuscular junction (nu"ro-mus'ku-lar junk'shun), p. 112
origin (or'ĭ-jin), p. 115
prime mover (prīm mo͞ov'er), p. 115
skeletal muscle (skel'ĕ-tal mus'el), p. 111
smooth muscle (smo͞oth mus'el), p. 111
synergist (sin'er-jist), p. 115
tendon (ten'don), p. 111
tetanus (tet'ah-nus), p. 114
tone (tōn), p. 115

Clinical Key Terms

atrophy (at'ro-fe), p. 116
hypertrophy (hi-per'tro-fe), p. 116

Summary

I. Types and Functions of Muscles

A. Uninucleated, spindle-shaped fibers make up the involuntary smooth muscle found in the walls of hollow internal organs. Uninucleated, branched, and striated fibers with intercalated disks make up the involuntary cardiac muscle of the heart. Multinucleated, striated, and cylindrical fibers make up the voluntary skeletal muscle attached to the skeleton.

B. Muscle tendons extend across joints, stabilizing them. Muscle contraction allows us to remain upright, produces movement of the body and internal organs, and creates heat that warms the body.

II. Skeletal Muscle Structure and Contraction

Skeletal muscles make up the bulk of the body, and their contraction accounts for the movement of bones. Nerves innervate skeletal muscles, and blood vessels bring skeletal muscles oxygen and nutrients.

A. Anatomy of a skeletal muscle. Myofibrils are the contractile elements of muscle fibers. The placement of the myofilaments actin and myosin accounts for the striations of skeletal muscle.

B. Physiology of muscle contraction.

1. Motor neurons meet muscle fibers at neuromuscular junctions. Impulses travel down the tubules of the T system and cause calcium to be released from the sarcoplasmic reticulum. Contraction occurs due to the presence of calcium and ATP in the muscle cells, which prompts actin myofilaments to slide past myosin myofilaments, shortening the length of the sarcomere.

2. Muscle fibers obey the all-or-none law, but whole muscles do not. The occurrence of a muscle twitch, summation, or tetanus depends on the frequency with which a muscle is stimulated.

3. ATP is required for muscle contraction and can be generated by way of aerobic or anaerobic cellular respiration and from creatine phosphate. During anaerobic cellular respiration, a waste product called lactic acid is produced. Lactic acid represents an oxygen debt, since oxygen is required to metabolize this product.

4. When muscles cooperate to achieve movement, some act as prime movers, others as synergists, and still others as antagonists.

5. Muscles contract isotonically and isometrically. They have tone. Forceful contraction can cause them to increase in size, and lack of contraction weakens them.

III. Skeletal Muscles of the Body

A. The names of muscles often include information about the muscles' size, shape, direction of fibers, location, number of attachments, and action.

B. Muscles of the head. These muscles are divided into those for facial expression and those for mastication (chewing).

C. Muscles of the neck and trunk.

1. Muscles that move the head. The sternocleidomastoid muscles and trapezius muscles are antagonistic because the sternocleidomastoid flexes the neck, and the trapezius muscles extend it.

2. Muscles of the trunk. The obliques (external and internal), the transversus abdominis, and the rectus abdominis all function to provide a sturdy abdominal wall.
D. Muscles of the shoulder and upper limb (arm).
  1. Muscles that move the pectoral girdle and arm. The deltoid, pectoralis major, latissimus dorsi, and serratus anterior muscles all function to move the humerus and, therefore, the arm in relation to the trunk.
  2. Muscles that move the forearm. The biceps and triceps brachii are antagonistic muscles because the biceps flexes the forearm, and the triceps extends it.
  3. Muscles that move the hand. The muscles that move the hand and fingers span the wrist.
E. Muscles of the thigh and lower limb (leg).
  1. Muscles that move the thigh. The iliopsoas and the gluteus maximus are antagonistic muscles because the iliopsoas flexes the thigh, and the gluteus maximus extends it. The gluteus medius and adductor muscles are antagonistic because the gluteus medius acts to abduct the thigh, and the adductor muscles act to adduct it.
  2. Muscles that move the lower limb (leg). The quadriceps femoris group of muscles and the hamstring group are antagonistic because the quadriceps femoris group extends the lower leg, and the hamstring group flexes it.
  3. Muscles that move the ankle and foot. The gastrocnemius and the tibialis anterior are antagonistic muscles because the gastrocnemius brings about plantar flexion, and the tibialis anterior dorsiflexion, of the foot. The peroneus muscles and the flexor and extensor digitorum muscles move the feet and the toes.

Study Questions

1. Name and describe the three types of muscles, and give a general location for each type. (p. 111)
2. List and discuss four functions of muscles. (p. 111)
3. Describe the anatomy of a muscle, from the whole muscle to the myofilaments within a sarcomere. Name the layers of fascia that cover a skeletal muscle and divide the muscle interior. (p. 112)
4. List the sequential events that occur when a nerve impulse reaches a muscle. (pp. 112–13)
5. How is ATP supplied to muscles? What is oxygen debt? (p. 114)
6. What is the all-or-none law? What is the difference between a single muscle twitch, summation, and tetanus? (p. 114)

7. Describe how muscles are attached to bones. Define the terms *prime mover, synergist,* and *antagonist.* (p. 115)
8. What is the difference between isotonic and isometric contraction? What is muscle tone? How does muscle contraction affect muscle size? (p. 115)
9. How do muscles get their names? Give an example for each method of naming a muscle. (p. 117)
10. Which of the head muscles are used for facial expression? Which are used for chewing? (p. 120)
11. Which muscles of the neck and trunk flex and extend the head? (p. 121)
12. What are the muscles of the abdominal wall? (p. 122)

13. Name the muscles that move the humerus, and give their actions. (pp. 123, 125)
14. Which of the muscles of the upper limb move the forearm, and what are their actions? Name the muscles that move the hand and fingers. (pp. 125–26)
15. Which of the muscles of the lower limb move the thigh, and what are their actions? Which of the muscles of the lower limb move the lower leg, and what are their actions? Which of the muscles of the lower limb move the feet? (pp. 126–29)

I. Fill in the blanks.

1. _____ muscle is uninucleated, nonstriated, and located in the walls of internal organs.

2. The fascia called _____ separates muscle fibers from one another within a fascicle.

3. When a muscle fiber contracts, an _____ myofilament slides past a myosin myofilament.

4. The energy molecule _____ is needed for muscle fiber contraction.

5. Whole muscles have _____, a condition in which some fibers are always contracted.

6. When muscles contract, the _____ does most of the work, but the _____ help.

7. The _____ is a muscle in the upper arm that has two origins.

8. The _____ acts as the origin of the latissimus dorsi, and the _____ acts as the insertion during most activities.

II. For questions 9–12, name the muscle indicated by the combination of origin and insertion shown.

	Origin	Insertion
9.	temporal bone	mandibular coronoid process
10.	scapula, clavicle	humerus
11.	scapula, proximal humerus	olecranon process of ulna
12.	posterior ilium, sacrum	proximal femur

III. Match the muscles in the key to the actions listed in questions 13–20.

Key:
   a. orbicularis oculi
   b. zygomaticus
   c. deltoid
   d. serratus anterior
   e. rectus abdominis
   f. iliopsoas
   g. gluteus maximus
   h. gastrocnemius

13. allows a person to stand on tiptoe
14. tenses abdominal wall
15. abducts arm
16. flexes thigh
17. raises corner of mouth
18. closes eyes
19. extends leg
20. pulls scapula downward and forward

Medical Terminology Reinforcement Exercise

Consult Appendix B for help in pronouncing and analyzing the meaning of the terms that follow.

1. hyperkinesis (hi″per-ke-ne′sis)
2. dystrophy (dis′tro-fe)
3. electromyogram (e-lek″tro-mi′-o-gram)
4. menisectomy (men″ĭ-sek′to-me)
5. tenorrhaphy (ten-or′ah-fe)
6. myatrophy (mi-at′ro-fe)
7. leiomyoma (li″o-mi-o′mah)
8. kinesiotherapy (kĭ-ne″se-o-ther′ah-pe)
9. myocardiopathy (mi″o-kar″de-op′ ah-the)
10. myasthenia (mi″as-the′ne-ah)

Website Links

For a listing of the most current Websites related to this chapter, please visit the Mader home page at:

www.mhhe.com/maderap

chapter 8

The Nervous System

◼ Chapter Outline and Learning Objectives

After you have studied this chapter, you should be able to:

Nervous System (p. 136)
- ◾ Describe the three functions of the nervous system.
- ◾ Describe the structure and function of the three types of neurons and four types of neuroglial cells.
- ◾ Explain how a nerve impulse is conducted along a nerve and across a synapse.
- ◾ Describe the structure of a nerve and the differences between the three different types of nerves.
- ◾ Describe the structure of a reflex arc and the function of a reflex.

Central Nervous System (p. 144)
- ◾ Describe the major parts of the brain and the lobes of the cerebral cortex. State functions for each structure.
- ◾ Describe in detail the structure of the spinal cord, and state its functions.
- ◾ Describe the three layers of meninges, and state the function of each.
- ◾ Describe how cerebrospinal fluid is formed and circulates.

Peripheral Nervous System (p. 152)
- ◾ Name the twelve pairs of cranial nerves, and give a function for each.
- ◾ Describe the structure and function of spinal nerves.
- ◾ Define and describe the autonomic nervous system.
- ◾ Distinguish between the sympathetic and parasympathetic divisions in four ways, and give examples of their respective effects on specific organs.

Effects of Aging (p. 158)
- ◾ Anatomical and physiological changes occur in the nervous system as we age.

Working Together (p. 158)
- ◾ The nervous system works with other systems of the body to maintain homeostasis.

Visual Focus
Synapse Structure and Function (p. 141)

Medical Focus
The Left and Right Brain (p. 146)
EEG (p. 149)
Spinal Cord Injuries (p. 150)

MedAlert
Alzheimer Disease (p. 160)

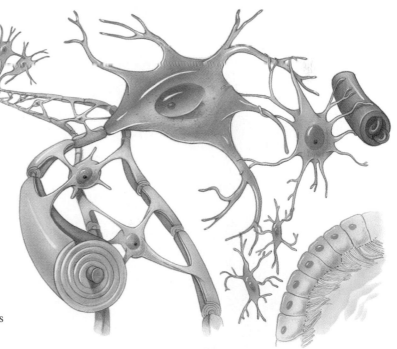

Cells found in the central nervous system (brain and spinal cord).

Part III

Nervous System

The nervous system and the endocrine system are responsible for regulation and coordination of body parts. Together, they maintain homeostasis. The nervous system permits a quick response to external and internal stimuli, while the endocrine system is slower to act but provides longer-lasting effects. This chapter examines the nervous system; the endocrine system is discussed in chapter 10.

Divisions of the Nervous System

The nervous system has two major divisions: the central nervous system and the peripheral nervous system (fig. 8.1). As mentioned earlier, the **central nervous system (CNS)** includes the brain and spinal cord, which have a *central* location—they lie in the midline of the body. The **peripheral nervous system (PNS),** which is further divided into the somatic division and the autonomic division, includes all the cranial and spinal nerves.

figure 8.1 Overall organization of the human nervous system. **a.** The central nervous system (CNS) is composed of the brain and spinal cord. The peripheral nervous system (PNS) contains nerves. **b.** The CNS communicates with the PNS. In the somatic system, nerves send sensory impulses from sensory receptors to the CNS and send motor impulses from the CNS to the skeletal muscles. In the autonomic system, consisting of the sympathetic and parasympathetic divisions, motor impulses travel to smooth muscle, cardiac muscle, and the glands.

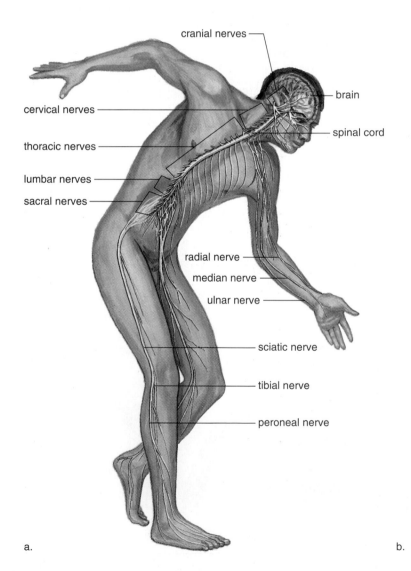

a.

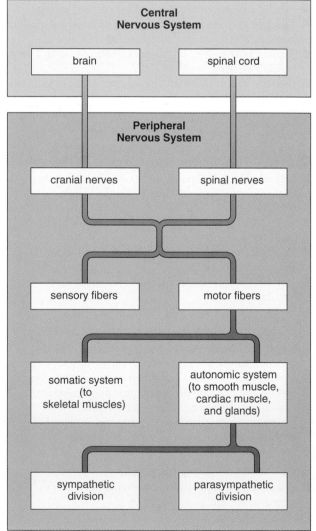

b.

Nerves have a *peripheral* location in the body—they project out from the central nervous system. The division between the central nervous system and the peripheral nervous system is arbitrary; the two systems work together and are connected to one another, as is illustrated in a reflex arc (discussed later in the chapter).

> The nervous system is divided into the central nervous system and the peripheral nervous system. The CNS lies in the midline of the body, and the PNS is located peripherally to the CNS.

Functions of the Nervous System

The nervous system has three specific functions:

1. *Permits sensory input.* Receptors present in skin and organs respond to external and internal stimuli by generating nerve impulses that travel to the central nervous system (CNS), consisting of the brain and spinal cord.
2. *Performs integration.* The CNS integrates (sums up) the input it receives from all over the body. Integration in the brain allows us to make decisions about the body in general. The CNS then sends out nerve impulses to effectors.
3. *Stimulates motor output.* Nerve impulses from the CNS go to the muscles and glands. Muscle contractions and gland secretions are responses to stimuli received by receptors.

Cells of Nervous Tissue

There are two types of cells in nervous tissue: **neurons,** which conduct nerve impulses, and **neuroglial cells,** which support, protect, and nourish the neurons.

Neurons

Neurons vary in size and shape, but they all have three parts: the dendrite(s), the cell body, and the axon (fig. 8.2). The **dendrites** receive information from other neurons and conduct signals typically toward the cell body. The **axon,** on the other hand, conducts nerve impulses typically away from the cell body. The **cell body** contains the nucleus and other organelles typically found in cells.

Two types of neurons are shown in figure 8.2. **Motor neurons,** which have short dendrites and a long axon, conduct nerve impulses from the central nervous system to effector organs—for example, either muscles or glands. Muscle tissue is located in the walls of internal organs and in the skeletal muscles. Glands are located throughout the body. Because motor neurons cause muscle fibers to contract and glands to secrete, they are said to innervate these structures.

Sensory neurons (fig. 8.2b) conduct nerve impulses from peripheral body parts to the CNS. Sensory neurons begin at specialized endings called receptors, which are sensitive to either external and/or internal stimuli. When a receptor is stimulated, a nerve impulse begins that travels along the sensory neuron to the CNS.

Sometimes, a sensory neuron is referred to as an afferent neuron, and a motor neuron is called an efferent neuron. These words, which are derived from Latin, mean "running to" and "running away from," respectively. Obviously, they refer to the relationship of these neurons to the CNS.

A third type of neuron, called an **interneuron,** or association neuron, is only found within the CNS. An interneuron conducts nerve impulses between various parts of the central nervous system, such as from one side of the brain or spinal cord to the other, or from the brain to the cord, and vice versa. An interneuron has short dendrites and either a long axon or a short axon. Table 8.1 lists the three types of neurons and their functions.

table 8.1	Neurons	
	Type of Neuron	**Function**
	Sensory (afferent)	Conducts nerve impulses from periphery to CNS*
	Motor (efferent)	Conducts nerve impulses from CNS to periphery
	Interneuron	Conducts nerve impulses within CNS

*CNS = central nervous system

figure 8.2 Neuron anatomy and function. All neurons have dendrites, a cell body, and an axon. **a.** A motor neuron conducts nerve impulses away from the CNS. **b.** A sensory neuron conducts nerve impulses to the CNS.

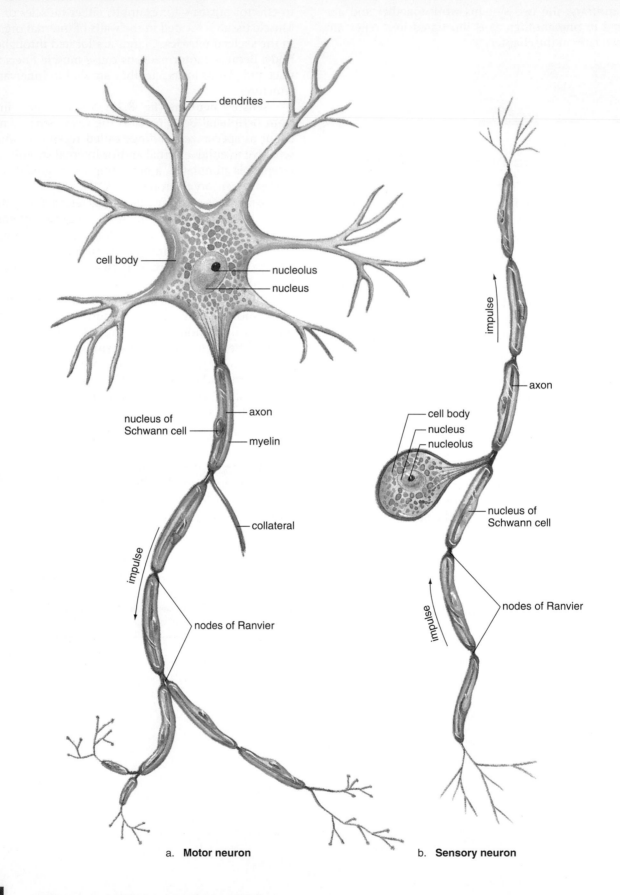

a. **Motor neuron**

b. **Sensory neuron**

figure 8.3 Myelin and neurilemmal sheaths. The myelin sheath forms when Schwann cells wrap themselves around a nerve fiber in the manner shown. The neurilemmal sheath contains the cytoplasm, nucleus, and outer plasma membrane of the Schwann cell. ▼

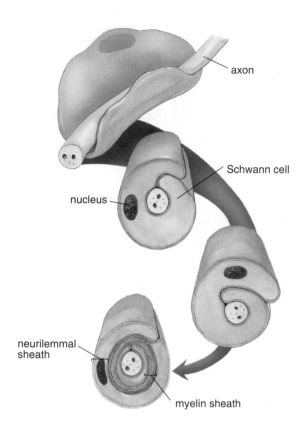

figure 8.4 Neuroglial cells in the central nervous system. ▼

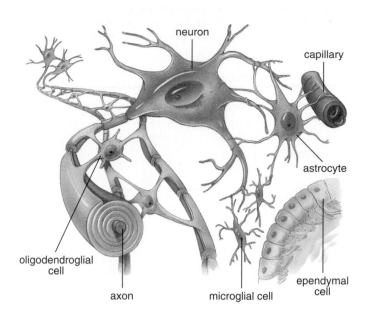

Neuroglial Cells

Schwann cells are a type of neuroglial cell found outside the CNS. Microglial cells, astrocytes, and oligodendroglial cells and ependymal cells are among the neuroglia found inside the CNS (fig. 8.4). *Microglial cells* are small, phagocytic cells that migrate to the site of an injury, where they engulf microbes and clean away debris. *Astrocytes* have cytoplasmic processes that provide structural support by forming bridges between neurons and capillaries. They may have a nutritive function involving the transport of molecules between capillaries and neurons. *Oligodendroglial cells* produce the myelin sheath that envelops neurons in the CNS. *Ependymal cells* line cavities in the CNS and assist in the production and circulation of the cerebrospinal fluid that fills the cavities of the CNS.

> Nervous tissue contains neurons and neuroglial cells. Each type of neuron has three parts (dendrite[s], cell body, and axon) but is specific as to function. Neuroglial cells support, protect, and nourish the neurons.

The dendrites and axons of neurons are sometimes called processes or neuron **fibers.** Outside the CNS, long fibers are covered by a white **myelin sheath** formed by **Schwann cells,** which wrap themselves tightly around these fibers. The portions of the Schwann cells that contain cytoplasm and nuclei form a **neurilemmal sheath** (fig. 8.3). The neurilemmal sheath promotes regeneration of a nerve fiber if it is injured. Gaps between Schwann cells form the nodes of Ranvier and are important in nerve cell conduction (see fig. 8.2).

Multiple sclerosis (MS) is a disease of the myelin sheath. Lesions develop that soon become hardened scleroses, or scars, that interfere with normal conduction of nerve impulses. The effects are widespread because of the number of fibers covered by a myelin sheath. The disease is chronic and tends to worsen with time.

figure 8.5 The resting and action potential. **a.** Resting potential. The outside of the membrane is positive and the inside is negative due to the presence of large negative organic ions inside the cytoplasm. Note also the distribution of Na⁺ and K⁺ across the membrane. **b.** Action potential. First, depolarization occurs as Na⁺ moves into the neuron (the inside of the cell becomes positive). Then repolarization occurs, and K⁺ moves out of the cell (the inside of the cell is negative once again). A nerve impulse consists of a wave of depolarization immediately followed by a wave of repolarization.

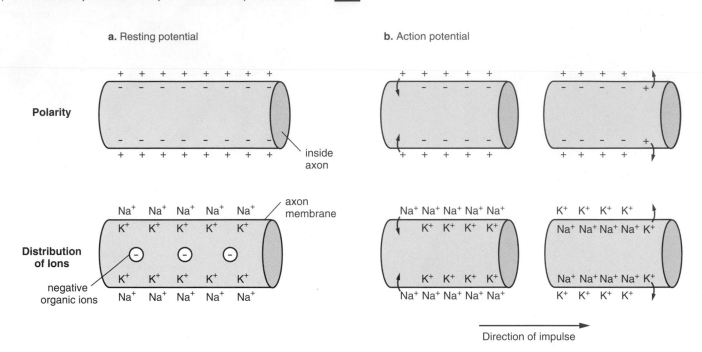

Nerve Impulses

When an axon is not conducting a nerve impulse, the axon membrane is polarized; that is, the outside is positive compared to the inside, which is negative. A protein carrier in the membrane called the sodium-potassium pump pumps sodium (Na^+) out of the axon and potassium (K^+) into the axon. Another factor that causes the inside of the axon to be negative compared to the outside is the presence of large, negatively charged protein ions inside an axon.

The difference in charge across an axon that is not conducting impulses is called the *resting potential* (fig. 8.5). When the nerve fiber is conducting a **nerve impulse,** which is also called the **action potential,** a change in polarity flows along the axon's membrane. As the nerve impulse passes by, the inside of an axon first becomes positive compared to the outside (this is called **depolarization**), and then the inside becomes negative again (this is called **repolarization**). During depolarization, Na^+ ions move to the inside of the axon, and during repolarization, K^+ ions move to the outside. In myelinated fibers, the nerve impulse skips from one node of Ranvier to the next, which increases the speed with which a fiber can conduct nerve impulses.

All neurons transmit the same type of nerve impulse: a change in polarity that flows along the membrane of a nerve fiber.

The Synapse

A **synapse** (sin'aps) is a region where an axon ending, called a synaptic knob, on one neuron meets but does not touch a dendrite on another neuron (fig. 8.6). The axon membrane is called the *presynaptic membrane,* and the dendrite membrane is the *postsynaptic membrane.* The small gap between the synaptic knob and the dendrite is called a **synaptic cleft.** Chemicals called **neurotransmitters** transmit nerve impulses across a synapse. When nerve impulses reach the end of an axon, a neurotransmitter diffuses across the synaptic cleft and binds to receptors in the membrane of the receiving dendrite (or cell body). Now, action potentials (nerve impulses) begin in the next neuron.

Acetylcholine (as"ĕ-til-ko'lēn) **(ACh)** is a well-known neurotransmitter that is active in all parts of the nervous system. Once acetylcholine has been received, it is broken down by an enzyme called **acetylcholinesterase** (as"ĕ-til-ko"lin-es'ter-ăs) **(AChE).** Norepinephrine (NE) is another well-known neurotransmitter. ACh and NE can be stimulatory or inhibitory, depending on the type of receptor receiving them. Stimulatory neurotransmitters increase the likelihood of a neuron conducting a nerve impulse, while inhibitory ones decrease the likelihood of a neuron conducting a nerve impulse. Serotonin and dopamine are neurotransmitters associated with behavioral states, such as mood, tension, learning, and memory.

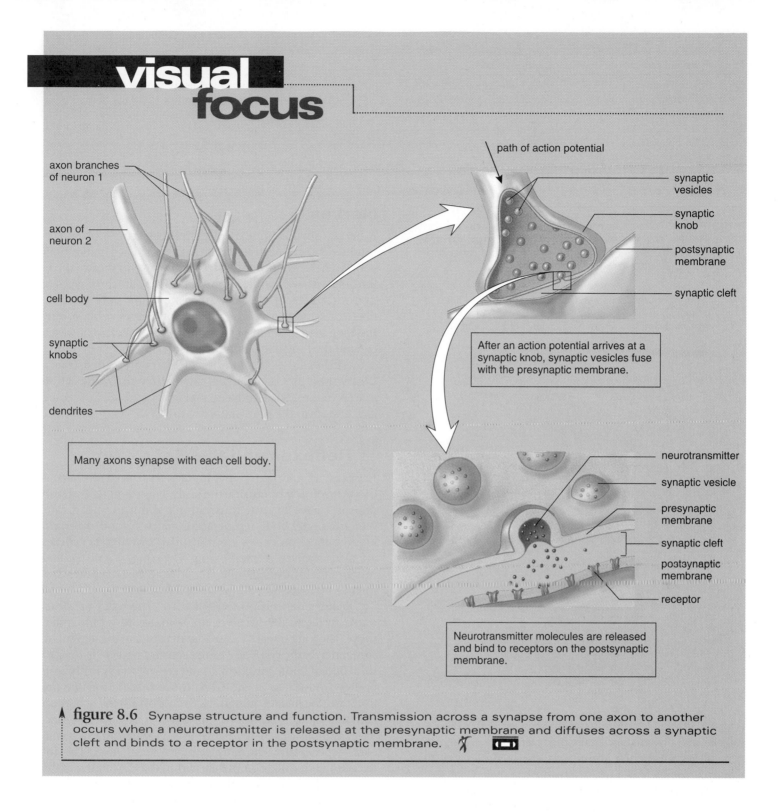

axon branches of neuron 1

axon of neuron 2

cell body

synaptic knobs

dendrites

Many axons synapse with each cell body.

path of action potential

synaptic vesicles

synaptic knob

postsynaptic membrane

synaptic cleft

After an action potential arrives at a synaptic knob, synaptic vesicles fuse with the presynaptic membrane.

neurotransmitter

synaptic vesicle

presynaptic membrane

synaptic cleft

postsynaptic membrane

receptor

Neurotransmitter molecules are released and bind to receptors on the postsynaptic membrane.

figure 8.6 Synapse structure and function. Transmission across a synapse from one axon to another occurs when a neurotransmitter is released at the presynaptic membrane and diffuses across a synaptic cleft and binds to a receptor in the postsynaptic membrane.

Neurotransmitters and Neurological Disorders

Several neurological illnesses, such as **Parkinson disease** and **Huntington disease**, are due to an imbalance in neurotransmitters within the brain. Parkinson disease is characterized by a wide-eyed, unblinking expression, an involuntary tremor of the fingers and thumbs, muscular rigidity, and a shuffling gait. All of these symptoms are due to a deficiency of dopamine. Huntington disease is characterized by a progressive deterioration of the individual's nervous system, which eventually leads to constant thrashing and writhing movements, and finally, to insanity and death. The problem is believed to be due to a malfunction of GABA, another neurotransmitter of the brain. As

table 8.2	Nerves		
	Type of Nerve	**Consists of**	**Function**
	Sensory nerves	Long fibers of sensory neurons	Conduct nerve impulses from receptors to CNS
	Motor nerves	Long fibers of motor neurons	Conduct nerve impulses from CNS to effectors
	Mixed nerves	Long fibers of sensory neurons and of motor neurons	Conduct nerve impulses to and away from CNS

Note: Compare this table to table 8.1.

figure 8.7 Cross section of a nerve, showing that it contains several fascicles, which, in turn, contain many nerve fibers. The axon, the fascicle, and the nerve have connective tissue coverings.

Endoneurium: connective tissue sheath of a nerve fiber
Fascicle: bundle of nerve fibers
Perineurium: connective tissue sheath of a fascicle
Nerve: bundle of fascicles
Epineurium: tough, fibrous sheath of nerve

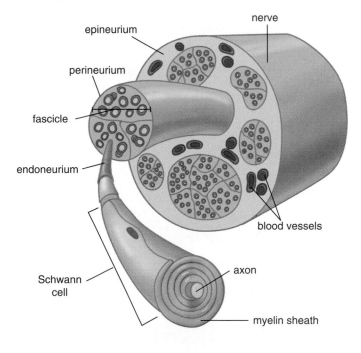

discussed in the MedAlert reading on page 160, **Alzheimer** (altz'hi-mer) **disease,** a brain disorder characterized by a general loss of mental abilities, is due to the loss of neurons that use ACh as a neurotransmitter.

Treatment of individuals with brain disorders has focused on trying to restore the proper balance of neurotransmitter substances. More recently, however, researchers have been exploring the possibility of transplanting tissue that produces the missing neurotransmitter.

Transmission of a nerve impulse across a synapse is dependent on the release of a neurotransmitter substance into a synaptic cleft.

Nerves

A **nerve** contains bundles of peripheral neuron fibers surrounded by connective tissue (fig. 8.7). Each nerve fiber within a nerve is surrounded by a white myelin sheath; therefore, nerves have a white appearance. There are three types of nerves (table 8.2). **Sensory nerves** contain only the long fibers of sensory neurons, and **motor nerves** contain only the long axons of motor neurons. **Mixed nerves,** however, contain the long fibers of both sensory and motor neurons.

Within the CNS, bundles of fibers are called **tracts.** When these fibers are myelinated, they form the **white matter** of the CNS. Unmyelinated fibers and cell bodies are gray and form the **gray matter** of the CNS.

Reflexes and the Reflex Arc

Reflexes are automatic, involuntary responses to changes occurring inside or outside the body. For example, a person does not have to consciously think about the heart rate, breathing rate, body temperature, or even food digestion because these processes are controlled by reflexes. Reflexes are also involved in swallowing, sneezing, vomiting, and urination. However, people are more aware of reflexes that involve reactions to painful stimuli (fig. 8.8).

Reflexes depend on the reflex arc. The reflex arc illustrated in figure 8.8 involves the spinal cord and a spinal nerve. The long fibers of a sensory neuron are located in the spinal nerve; the cell body of the sensory neuron is located in a **dorsal-root ganglion** (gang'gle-on). (Typically, cell bodies are located in the CNS; those that lie outside the CNS are located in **ganglia.**) Each dorsal-root ganglion contains the cell bodies of hundreds of sensory neurons. Axon endings of sensory neurons are located in the spinal cord—specifically, the gray matter's *dorsal horn.* The interneuron is located completely within the gray matter of the spinal cord (CNS). The cell body and dendrites of the motor neuron are located in the spinal cord—specifically, within the gray matter's *ventral horn.* The axon of the motor neuron is located in the spinal nerve. Spinal nerves are called *mixed nerves* because they contain sensory and motor fibers.

A **sensory receptor** is a structure specialized to receive information from the environment and to generate nerve impulses. As shown in figure 8.8, when a sensory receptor at the end of a sensory neuron is stimulated—in this case,

figure 8.8 Cross section of a spinal cord and spinal nerve that illustrates a reflex arc.

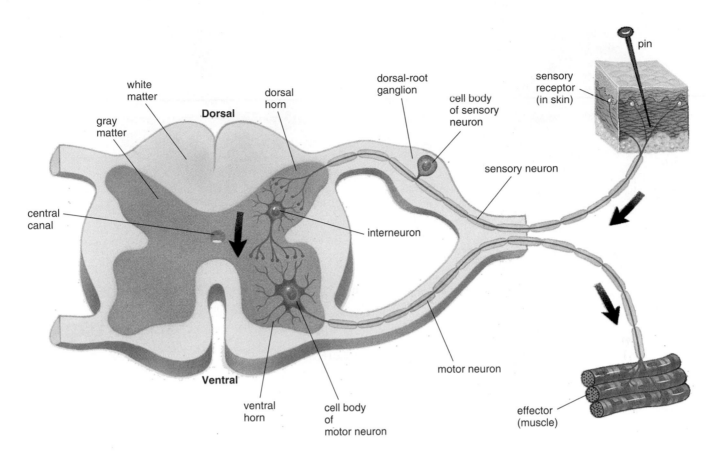

by injury—it generates nerve impulses that are conducted along the sensory neuron toward the cell body in a dorsal-root ganglion. From the cell body, the impulses are conducted along the axon of the sensory neuron. They are then transmitted from the sensory neuron to the interneuron, which passes them on to the motor neuron. The axon of the motor neuron conducts the nerve impulses to muscle fibers, which then contract so that the hand is withdrawn from the injury-causing object. The response to nerve impulses transmitted along a reflex arc is called a reflex.

Various other reactions usually accompany a reflex response: The person may look in the direction of the object, jump back, and utter appropriate exclamations. These responses are a result of the sensory neuron stimulating several interneurons, which take impulses to all parts of the central nervous system, including those responsible for consciousness.

Reflexes are essential to homeostasis. They keep the internal organs functioning within normal bounds and protect the body from external harm. Reflexes can also be used to determine if the nervous system is reacting properly. Two of these types of reflexes are:

Knee-jerk reflex (patellar reflex): This reflex is initiated by striking the patellar ligament just below the patella. The response is contraction of the quadriceps femoris muscles, which causes the lower leg to extend.

Ankle-jerk reflex: This reflex is initiated by tapping the Achilles tendon just above its insertion on the calcaneus. The response is plantar flexion due to contraction of the gastrocnemius and soleus muscles.

Some reflexes are important for avoiding injury, but the knee-jerk and ankle-jerk reflexes are important for normal physiological functions. For example, the knee-jerk reflex helps a person stand erect. If the knee begins to bend slightly when a person stands still, the quadriceps femoris is stretched, and the leg straightens.

Reflexes (automatic reactions to internal and external stimuli) depend on the reflex arc. Some reflexes are important for avoiding injury, and others are necessary for normal physiological functions.

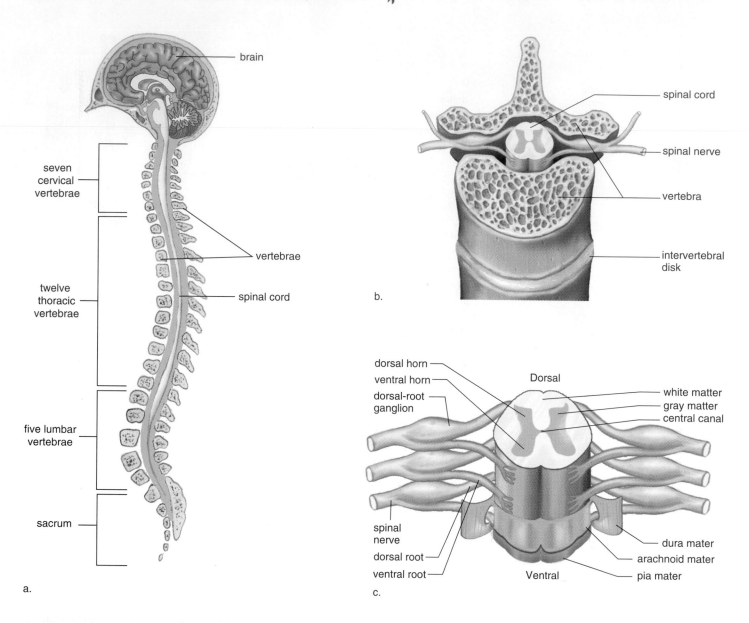

figure 8.9 Central nervous system. **a.** The brain and spinal cord are protected by bone. **b.** The spinal cord is protected by vertebrae. **c.** Anatomy of the spinal cord.

a.

brain

seven cervical vertebrae

twelve thoracic vertebrae

vertebrae

spinal cord

five lumbar vertebrae

sacrum

b.

spinal cord

spinal nerve

vertebra

intervertebral disk

c.

dorsal horn

ventral horn

dorsal-root ganglion

Dorsal

white matter

gray matter

central canal

spinal nerve

dorsal root

ventral root

Ventral

dura mater

arachnoid mater

pia mater

Central Nervous System

The central nervous system (CNS) consists of the brain and the spinal cord. As figure 8.9 illustrates, the CNS is protected by bone: The brain occupies the cranial cavity, and the spinal cord occupies the vertebral canal.

Brain

The human brain is divided into the medulla oblongata, pons, midbrain, cerebellum, hypothalamus, thalamus, and cerebrum (fig. 8.10*a*). The brain has four cavities, called **ventricles:** two lateral ventricles, the third ventricle, and the fourth ventricle (fig. 8.10*b*). *Interventricular foramina* connect the lateral ventricles with the third ventricle. The third and fourth ventricles are connected by a small canal called the *cerebral aqueduct* (aqueduct of Sylvius).

Brain Stem

The medulla oblongata, pons, and midbrain lie in a portion of the brain known as the *brain stem.* The **medulla oblongata** (mĕ-dul'ah ob"long-gah'tah) lies between the spinal cord and the pons and is anterior to the cerebellum. It contains a number of vital centers for regulating heartbeat, breathing, and blood pressure. It also contains the reflex centers for vomiting, coughing, sneezing, hiccoughing, and swallowing. The medulla contains tracts that ascend or descend between the spinal cord and the brain's higher centers.

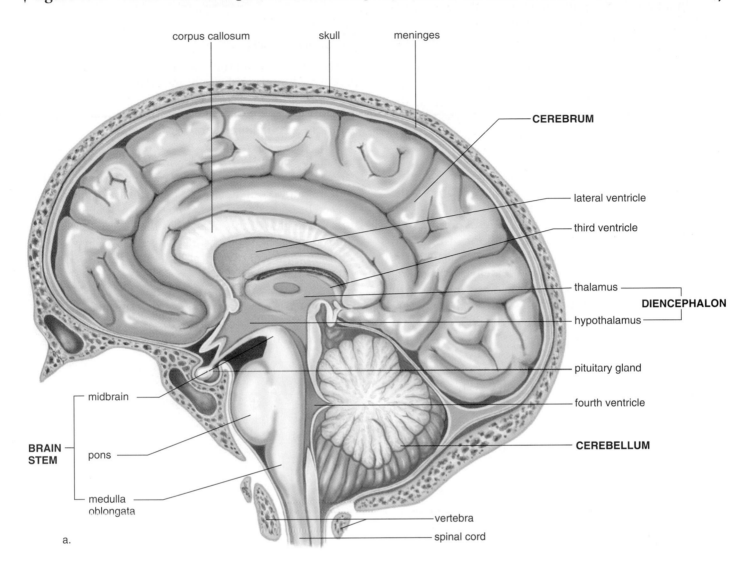

corpus callosum   skull   meninges

CEREBRUM

lateral ventricle

third ventricle

thalamus

DIENCEPHALON

hypothalamus

pituitary gland

fourth ventricle

CEREBELLUM

midbrain

BRAIN STEM

pons

medulla oblongata

vertebra

spinal cord

a.

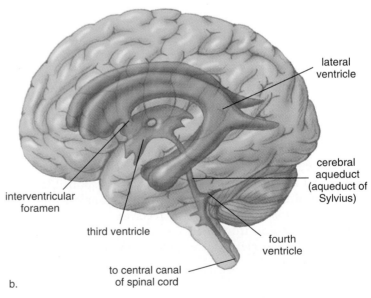

lateral ventricle

cerebral aqueduct (aqueduct of Sylvius)

interventricular foramen

third ventricle

fourth ventricle

to central canal of spinal cord

b.

The **pons** is a "bridge;" it contains bundles of axons traveling between the cerebellum and the rest of the CNS. In addition, the pons functions with the medulla to regulate breathing rate and has reflex centers concerned with head movements in response to visual and auditory stimuli.

The **midbrain** encloses the cerebral aqueduct. Aside from acting as a relay station for tracts passing between the cerebrum and the spinal cord or cerebellum, the midbrain has reflex centers for visual, auditory, and tactile responses.

Diencephalon

The hypothalamus and thalamus are in a portion of the brain known as the **diencephalon** (di"en-sef'ah-lon), where the third ventricle is located. The **hypothalamus** (hi"po-thal'ah-mus), which forms the floor of the third ventricle, maintains homeostasis, or the constancy of the internal environment, and contains centers for regulating hunger, sleep, thirst, body temperature, water balance, and blood pressure. The hypothalamus controls the pituitary gland and thereby serves as a link between the nervous and endocrine systems.

The **thalamus,** in the lateral walls of the third ventricle, is the last portion of the brain for sensory input before the cerebrum. It serves as a central relay station for sensory impulses traveling upward from other parts of the body and brain to the cerebrum. It receives all sensory impulses (except those associated with the sense of smell) and channels them to appropriate regions of the cortex for interpretation.

> The brain stem and the diencephalon contain reflex centers that are involved in controlling internal organs, and tracts that channel impulses into the brain or between brain regions. The hypothalamus maintains homeostasis and controls the secretions of the pituitary gland.

Cerebellum

The **cerebellum** (ser"ĕ-bel'um), which lies below the posterior portion of the cerebrum, is separated from the brain stem by the fourth ventricle. The cerebellum has two parts called hemispheres that are joined by a constricted median portion. The surface of the cerebellum is gray matter, and the interior is largely white matter. The cerebellum functions in muscle coordination, integrating impulses received from higher centers to ensure that all of the skeletal muscles work together to produce smooth and graceful motion. The cerebellum is also responsible for maintaining normal muscle tone and transmitting impulses that maintain posture. It receives information about body position from the inner ear and then sends impulses to the muscles, whose contraction maintains or restores balance.

> The cerebellum controls balance and complex muscular movements.

Cerebrum

The **cerebrum** (ser'ĕ-brum) is the largest and most superior part of the brain. It is the only area of the brain responsible for consciousness. The cerebrum is divided into halves known as the right and left **cerebral hemispheres;** each hemisphere contains a lateral ventricle.

The outer layer of the cerebrum, called the *cortex,* is gray and contains cell bodies and short fibers. The cortex has convolutions known as **gyri,** which are separated by shallow grooves called *sulci* and deep grooves called *fissures.* The cerebrum is almost divided by a deep, longitudinal fissure. At the base of this fissure lies the **corpus callosum,** a bridge of myelinated fibers that joins the two hemispheres. Left-brain and right-brain abilities are examined in the Medical Focus reading on this page.

Each cerebral hemisphere has four lobes: **frontal, parietal** (pah-ri'ĕ-tal), **temporal,** and **occipital** (ok-sip'ĭ-tal) (fig. 8.11), which are named for the bones that cover them. Each lobe has particular functions (table 8.3).

The Left and Right Brain

Some years ago, Roger W. Sperry and Michael Gazzaniga severed the corpus callosum in some of their patients who suffered from epilepsy. The corpus callosum connects the left and right sides of the brain. From these procedures and further experimentation, they learned that the left brain and right brain have different abilities.

Sperry and Gazzaniga found that the left brain contains centers for speech and is responsible for language ability. Therefore, patients could report what was seen by the left half of each eye but not what was seen by the right half of each eye. In contrast, patients were unable to report verbally on left-hand activities because the left hand is controlled by the right half of the brain.

Sperry and Gazzaniga also determined that the right brain is far superior to the left brain in dealing with spatial relationships. For example, the left hand, not the right hand, is better able to recognize and remember objects by their shape. In addition to spatial relationships, the right brain also appears to be involved with musical and artistic activities and the expression of emotions.

figure 8.11 Cerebral cortex. The convoluted cortex of the cerebrum is divided into four lobes: frontal, temporal, parietal, and occipital. Each lobe has the functions noted in table 8.3.

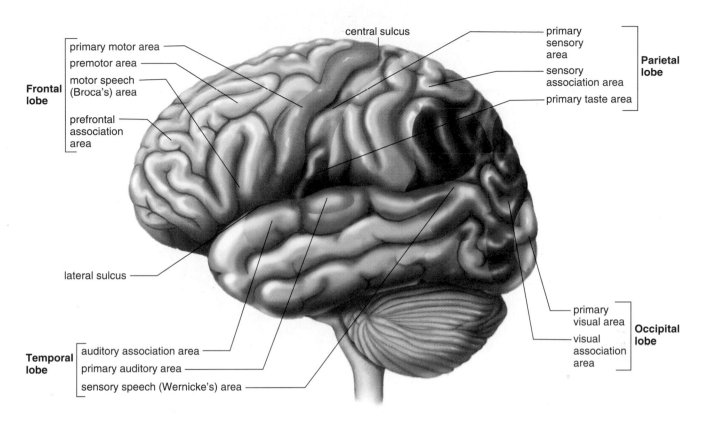

table 8.3	**Functions of the Cerebral Lobes**			
	Lobe	**Functions**	**Lobe**	**Functions**
	Frontal lobes	Primary motor area controls movements of voluntary skeletal muscles.	Temporal lobes	Primary auditory area is responsible for hearing.
		Prefrontal association area carries on higher intellectual processes, such as those required for concentrating, planning, complex problem solving, and judging the consequences of behavior.		Auditory association area interprets sensory experiences and remembers visual scenes, music, and other complex sensory patterns.
	Parietal lobes	Primary sensory area is responsible for the sensations of temperature, touch, pressure, and pain from the skin.	Occipital lobes	Primary visual area is responsible for vision.
		Sensory association area functions in understanding speech and in using words to express thoughts and feelings.		Visual association area combines visual images with other sensory experiences.

Association areas are believed to contain areas for intelligence, artistic and creative ability, and learning. The *primary sensory area* receives nerve impulses from the sense organs and produces what are termed sensations. The particular sensation produced is related to the area of the brain that is stimulated, since the nerve impulse itself always has the same nature (as described previously). The *primary motor area* of the cerebrum initiates nerve impulses that control muscle fibers.

The primary sensory area of the cerebral cortex lies in the parietal lobe just posterior to the central sulcus; the primary motor area lies just anterior to the central sulcus.

figure 8.12 Portions of the body that are controlled by the primary motor area and the primary sensory area of the cerebrum. Notice that the size of the body part in the diagram reflects the amount of cerebral cortex devoted to that body part.

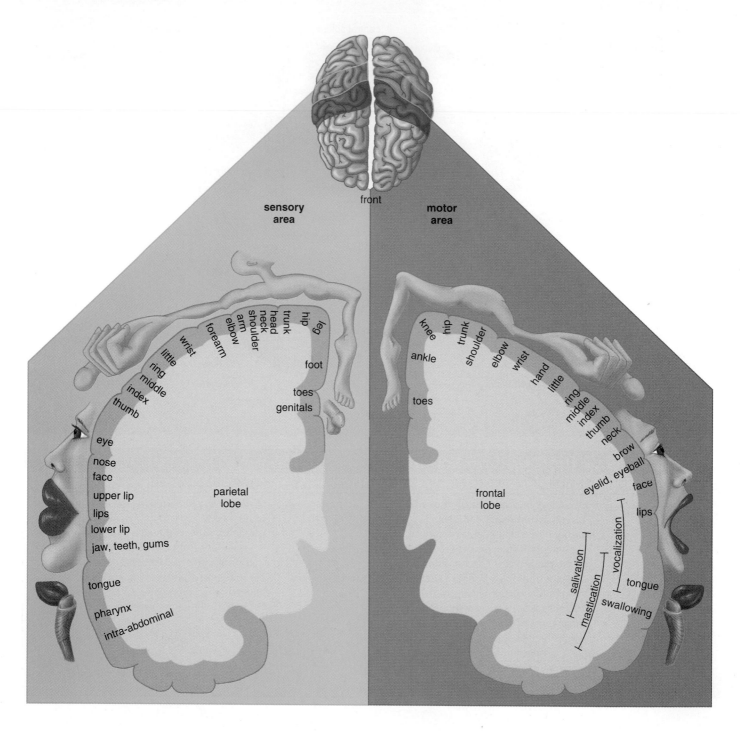

These two areas in particular demonstrate how it is possible to map the cerebrum by associating particular parts of the cerebrum with particular parts of the body (fig. 8.12). A momentary lack of oxygen during birth can damage the motor areas of the cerebral cortex so that the individual de- velops the symptoms of **cerebral palsy,** a condition charac- terized by a spastic weakness of the arms and legs.

The motor area for speech, called *Broca's area,* is lo- cated near the base of the primary motor area (see fig. 8.11). Broca's area is usually only found in the left cerebral

EEG

*t*he electrical activity of the brain can be recorded in the form of an **electroencephalogram** (e-lek″tro-in-sef′lah-gram) **(EEG).** Electrodes are taped to different parts of the scalp, and an instrument called an electroencephalograph records the brain waves.

When the subject is awake, two types of waves are recorded: (1) *alpha waves*, with a frequency of about six to thirteen per second and a potential of about 45 microvolts, which predominate when the eyes are closed; and (2) *beta waves*, with higher frequencies but lower voltage, which predominate when the eyes are open.

During an 8-hour sleep, there are usually five periods when the brain waves become slower and larger than alpha waves. During each of these periods, there are irregular episodes of the eyes moving back and forth rapidly. When subjects are awakened during this **REM** (rapid eye movement) **sleep,** they always report that they were dreaming. The significance of REM sleep is still being debated, but some studies indicate that REM sleep is needed for memory to occur.

The EEG is a good diagnostic tool; for example, an irregular brain wave pattern can signify epilepsy or a brain tumor. A flat EEG signifies lack of electrical activity of the brain, or brain death, and thus may be used to determine the precise time of death.

Consciousness is under the control of the cerebrum, the most highly developed portion of the brain. It is responsible for higher mental processes, including the interpretation of sensory input and the initiation of voluntary muscular movements.

Limbic System   The **limbic system** involves portions of both the unconscious and conscious brain. It lies just beneath the cerebral cortex and contains neural pathways that connect portions of the frontal lobes, the temporal lobes, the thalamus, and the hypothalamus (fig. 8.13).

Stimulation of different areas of the limbic system causes the subject to experience rage, pain, pleasure, or sorrow. By causing pleasant or unpleasant feelings about experiences, the limbic system apparently guides the individual into behavior that is likely to increase the chance of survival.

The limbic system is also involved in learning and memory. Learning requires memory, and memory is stored in the sensory regions of the cerebrum, but just what permits memory development is not definitely known. The involvement of the limbic system in memory explains why emotionally charged events result in our most vivid memories. The fact that the limbic system communicates with the sensory areas for touch, smell, vision, and so forth accounts for the ability of any particular sensory stimulus to awaken a complex memory.

The limbic system is involved in learning and memory, and in causing the emotions that guide behavior.

figure 8.13   The limbic system includes portions of the cerebrum, the thalamus, and the hypothalamus. This system joins higher mental functions like reasoning with more primitive feelings like fear and pleasure.

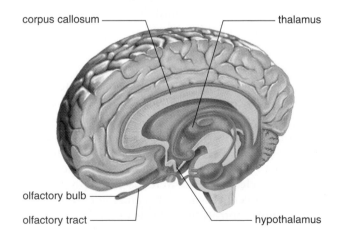

corpus callosum — — thalamus

olfactory bulb —

olfactory tract — — hypothalamus

hemisphere. Damage to this area can interfere with a person's ability to understand words (written or spoken) and to communicate with others. This explains why a stroke is often accompanied by speech problems. **Strokes** (p. 235) occur when the blood supply to the brain is temporarily halted due to a burst blood vessel or to a blood clot in an artery. Electroencephalograms (EEGs), which record the brain's electrical activity, are examined in the Medical Focus reading on this page.

Basal Nuclei   The basal nuclei constitute several masses of gray matter that lie deep within the white matter of the cerebrum. The precise function of the basal nuclei is not known, but they are a part of the limbic system discussed next. Also they must have some control over voluntary muscle action because when they are diseased, Parkinson disease may develop and there may be spastic movement of limbs.

Spinal Cord Injuries

Spinal cord injuries are an ever-present danger from accidents and trauma. The damage may result in transection (complete cut across) or partial section of the cord. The location and extent of the damage produce a variety of effects because of the partial or complete stoppage of impulses passing up and down the spinal cord. If the spinal cord is completely transected, no sensations or somatic motor impulses traveling in the cord will be able to pass this point. If the injury is between the first thoracic vertebra (T1) and the second lumbar vertebra (L2), paralysis of the lower body and legs occurs. This condition is known as **paraplegia** (par-ah-ple′je-ah). If the injury is between the fourth cervical vertebra (C4) and the first thoracic vertebra (T1), the entire body and all four limbs are usually affected. This condition is called **quadriplegia** (kwah-drah-ple′je-ah). If the injury is a unilateral hemisection (half cut), motor loss will occur on the same side as the injury because motor neuron crossover occurs in the medulla oblongata. At the same time, loss of sensation will vary, and the pattern and type of such loss can be analyzed to locate the lesion.

Spinal Cord

Structure

The spinal cord extends from the base of the brain through the foramen magnum of the skull into the vertebral canal formed by the vertebrae. The spinal cord terminates between the first and second lumbar vertebrae.

A cross section of the spinal cord reveals the central canal and distinct white and gray areas (see fig. 8.9c). The *gray matter* is centrally located and shaped like the letter H. It is gray because it contains cell bodies and short, unmyelinated fibers. The two posterior projections of the H are the dorsal (posterior) horns, and the two anterior projections of the H are the ventral (anterior) horns. Sensory neurons enter the spinal cord through the dorsal roots; motor neurons exit the spinal cord through the ventral roots. The dorsal and ventral roots join to form the spinal nerves.

The *white matter* fills in the areas around the gray matter. The white matter contains bundles of myelinated fibers within tracts that form columns. *Ascending tracts* take nerve impulses up through the spinal cord and the brain, and *descending tracts* take nerve impulses down through the brain and the spinal cord.

Functions

The spinal cord has two main functions. First, it is the center for thousands of reflex arcs. Figure 8.8 shows that a nerve impulse generated by a receptor passes along a sensory neuron to an interneuron in the spinal cord. The interneuron passes the nerve impulse to a motor neuron that brings about a reflex. Withdrawal reflexes, the knee-jerk reflex, and the ankle-jerk reflex were discussed on page 143. The tracts cross over in the medulla, and therefore, the right side of the brain controls the left side of the body, and vice versa. Reflexes allow quick response to a stimulus because they do not require conscious thought.

The spinal cord's second function is that it provides a means of communication between the brain and the peripheral nerves that leave the cord. Sensory impulses travel up the cord to the brain in ascending tracts, and motor impulses travel down the cord from the brain in descending tracts. Sensory impulses that reach the cerebrum result in sensation, and motor impulses that travel from the cerebrum allow us to voluntarily control our limbs and other body parts. If the spinal cord is injured, we suffer a loss of sensation and a loss of voluntary control—that is, we suffer a *paralysis* (see the Medical Focus reading on this page).

> The spinal cord extends from the base of the brain into the vertebral canal formed by the vertebrae. It is a center for reflex action and allows communication between the brain and the peripheral nerves leaving the spinal cord.

Meninges and Cerebrospinal Fluid

Both the brain and spinal cord are wrapped in three protective membranes known as **meninges** (mĕ-nin′jēz). The outer meningeal layer, called the **dura mater,** is tough, white, fibrous connective tissue that lies next to the skull and vertebrae. In some areas, it splits into two layers, forming channels (called dural sinuses) that collect venous blood before it is returned to the circulatory system. Bleeding due to a head injury is called an **epidural hematoma.** When there is blood between the dura mater and bone, the injury is called a **subdural hematoma.** The middle meningeal layer is the **arachnoid membrane,** a weblike connective tissue covering with thin strands that attach to the **pia mater,** the inner and third meningeal layer. Beneath the arachnoid membrane is the subarachnoid space,

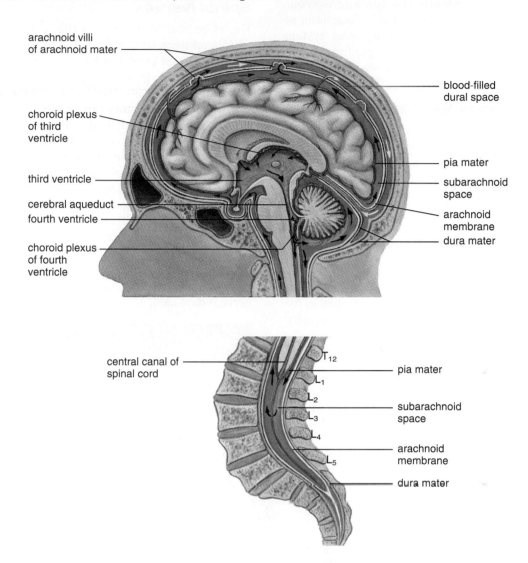

which is filled with cerebrospinal fluid. The pia mater is very thin and closely follows the contours of the brain and spinal cord (figs. 8.9*c* and 8.14).

Cerebrospinal Fluid

Cerebrospinal fluid (CSF) is a clear tissue fluid that forms a protective cushion around and within the CNS. It also supplies the CNS with nutrients that have been filtered from the blood and collects wastes that are returned to the blood. CSF is a circulating fluid. CSF is produced by and filters out of specialized masses of capillaries called choroid plexuses into the lateral ventricles. Ependymal cells cover the choroid plexuses and aid in the production of CSF. CSF moves from the lateral ventricles into the third ventricle and then flows through the cerebral aqueduct

into the fourth ventricle. From the fourth ventricle, it moves into the **central canal** of the spinal cord and then out into the subarachnoid spaces. After that, it returns to the veins of the brain.

Normally, the CSF within the brain maintains a constant pressure because it steadily drains into the circulatory system by way of the dural sinuses. However, if there is a blockage in an infant whose cranial sutures have not yet closed, the brain can enlarge due to fluid accumulation. This condition is called **hydrocephalus**, or "water on the brain." If fluid collects in an adult, the brain cannot enlarge, and instead, it is pushed against the skull, possibly causing injury.

The CNS is protected by the meninges and the cerebrospinal fluid.

Peripheral Nervous System

The peripheral nervous system (PNS) contains the somatic and autonomic nervous systems. The **somatic nervous system (SNS)** is primarily concerned with reactions to outside stimuli, while the **autonomic nervous system (ANS)** is primarily concerned with the proper functioning of the internal organs to maintain homeostasis.

Somatic Nervous System

The somatic nervous system includes all of the nerves that serve the musculoskeletal system and the exterior sense organs, including those of the skin. Exterior sense organs have *receptors* that receive environmental stimuli and then initiate nerve impulses. On the other hand, muscle fibers are *effectors* that bring about a reaction to the stimulus. Muscle effectors were studied in chapter 7, and receptors are discussed in chapter 9.

Cranial Nerves

Humans have twelve pairs of **cranial nerves** attached to the brain (fig. 8.15 and table 8.4). Some of these are sensory, some are motor, and still others are mixed. Although the brain is a part of the CNS, the cranial nerves are a part of the PNS. All cranial nerves, except the vagus, are concerned with the head, neck, and facial regions of the body. The vagus nerve has many branches to serve the internal organs.

Spinal Nerves

The 31 pairs of spinal nerves (fig. 8.16) are a part of the peripheral nervous system. Some of the major peripheral nerves are listed in table 8.5.

Each **spinal nerve** emerges from the spinal cord (see fig. 8.9c) by two short branches, called the dorsal and ventral roots. The dorsal root can be identified by the presence of an enlargement called the dorsal-root ganglion. This ganglion contains the cell bodies of the sensory neurons whose fibers conduct impulses toward the spinal cord. The ventral root of each spinal nerve contains the axons of motor neurons that conduct impulses away from the spinal cord. These two roots join just before the spinal nerve leaves the vertebral column. Therefore, all spinal nerves are mixed nerves with many sensory dendrites and motor axons.

The segments of skin supplied by a particular spinal nerve are called **dermatomes**. Abnormal skin sensations in a particular dermatome can be used to detect which spinal nerve is damaged.

> Cranial nerves take impulses to and/or from the brain.
> Spinal nerves take impulses to and from the spinal cord.

figure 8.15 Ventral surface of the brain, showing attachment of cranial nerves. The 12 pairs of cranial nerves and their functions are listed in table 8.4.

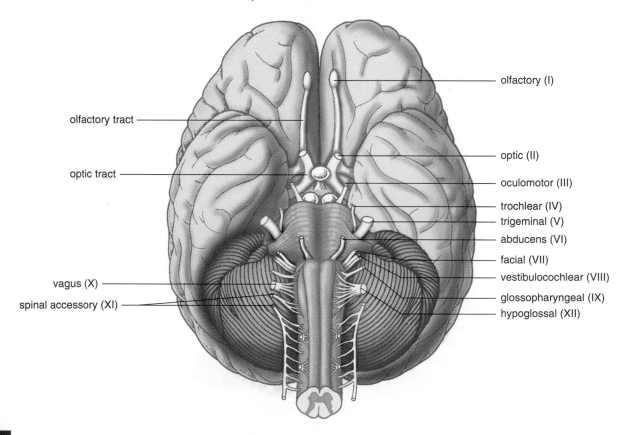

olfactory tract

optic tract

vagus (X)

spinal accessory (XI)

olfactory (I)

optic (II)

oculomotor (III)

trochlear (IV)

trigeminal (V)

abducens (VI)

facial (VII)

vestibulocochlear (VIII)

glossopharyngeal (IX)

hypoglossal (XII)

table 8.4 — Cranial Nerves

Nerve	Type		Brain Location	Transmits Nerve Impulses to (Motor) or from (Sensory)
Olfactory (I)	Sensory		Olfactory bulb	Olfactory receptors for sense of smell
Optic (II)	Sensory		Thalamus	Retina for sense of light
Oculomotor (III)	Motor		Midbrain	Eye muscles (including eyelids and lens); pupil (parasympathetic division)
Trochlear (IV)	Motor		Midbrain	Eye muscles
Trigeminal (V)	Mixed	Sensory	Pons	Teeth, eyes, skin, and tongue
		Motor		Jaw muscles (chewing)
Abducens (VI)	Motor		Pons	Eye muscles
Facial (VII)	Mixed	Sensory	Pons	Taste buds of anterior tongue
		Motor		Facial muscles (facial expression) and glands (tear and salivary)
Vestibulocochlear (VIII)	Sensory		Pons	Inner ear for sense of balance and hearing
Glossopharyngeal (IX)	Mixed	Sensory	Medulla oblongata	Pharynx
		Motor		Pharyngeal muscles (swallowing)
Vagus (X)	Sensory		Medulla oblongata	Internal organs
	Motor			Internal organs (parasympathetic division)
Spinal accessory (XI)	Motor		Medulla oblongata	Neck and back muscles
Hypoglossal (XII)	Motor		Medulla oblongata	Tongue muscles

table 8.5 — Major Peripheral Nerves

Name	Spinal Nerves Involved*	Function
Musculocutaneous nerves	C5–T1	Supply muscles of the arms on the anterior sides, and skin of the forearms
Radial nerves	C5–T1	Supply muscles of the arms on the posterior sides, and skin of the forearms and hands
Median nerves	C5–T1	Supply muscles of the forearms, and muscles and skin of the hands
Ulnar nerves	C5–T1	Supply muscles of the forearms and hands, and skin of the hands
Phrenic nerves	C3–C5	Supply the diaphragm
Intercostal nerves	T2–T12	Supply intercostal muscles, abdominal muscles, and skin of the trunk
Femoral nerves	L2–L4	Supply muscles and skin of the thighs and legs
Sciatic nerves	L4–S3	Supply muscles and skin of the thighs, legs, and feet

*C = cervical; T = thoracic; L = lumbar

figure 8.16 Spinal nerves. The number and kinds of spinal nerves are given on the right. The location of major peripheral nerves is given on the left. Table 8.5 lists the functions of these nerves.

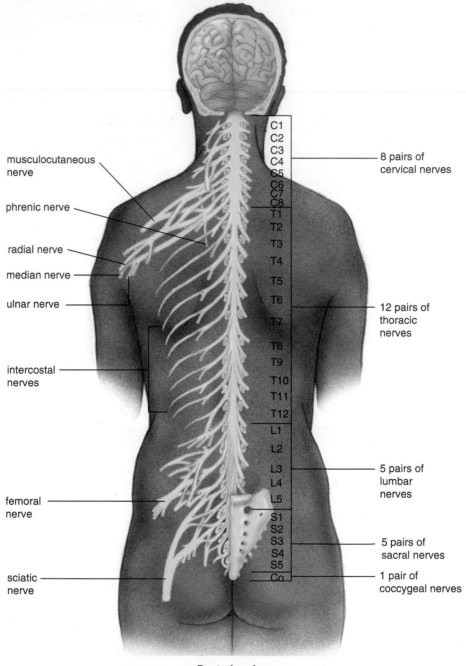

Posterior view

Part III Integration and Coordination

figure 8.17 Somatic versus autonomic nerves. In the somatic pathway, there is only one motor neuron between the CNS and a skeletal muscle. In the autonomic pathway, there are two motor neurons between the CNS and an internal organ. The cell body for the first motor neuron is in the CNS, and the cell body for the second motor neuron is in a ganglion.

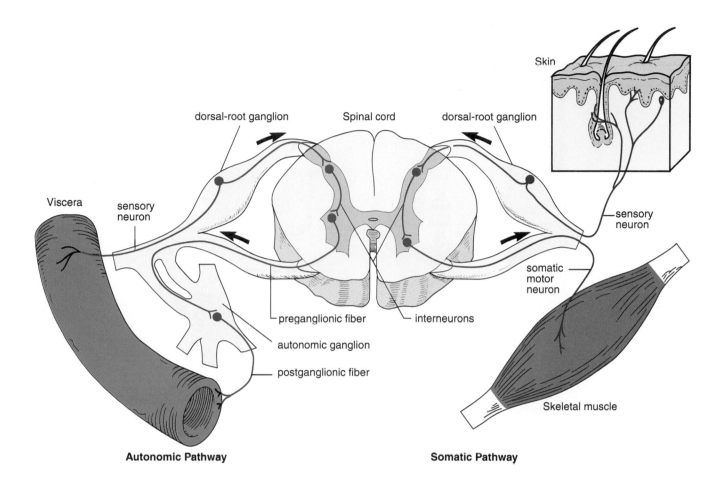

Autonomic Pathway

Somatic Pathway

Autonomic Nervous System

The autonomic nervous system (ANS) is made up of motor neurons that innervate smooth muscle, cardiac muscle, and glands automatically and usually without the need for conscious intervention. For each impulse there is one autonomic ganglion and two motor neurons (fig. 8.17). The first neuron has a cell body within the central nervous system and a **preganglionic fiber.** The second neuron has a cell body within the ganglion and a **postganglionic fiber.**

The ANS has two divisions: the sympathetic division (fig. 8.18) and the parasympathetic division (fig. 8.19). Both of these function automatically and usually subconsciously in an involuntary manner.

The divisions of the autonomic nervous system: (1) function automatically and usually subconsciously in an involuntary manner, (2) innervate all internal organs, and (3) utilize two motor neurons and one ganglion for each impulse.

Sympathetic Division

The preganglionic fibers of the **sympathetic division** arise from the thoracic and lumbar (thoracolumbar) levels of the spinal cord and almost immediately terminate in ganglia that lie near the spinal cord (see fig. 8.18). Thus, in this system, the preganglionic fibers are short, but the postganglionic fibers that make contact with the organs are long.

The sympathetic division is especially important during emergency situations and is associated with the "fight-or-flight" response. For example, it inhibits the digestive tract, but dilates the pupils, accelerates the heartbeat, and increases the breathing rate. Postganglionic axons release the neurotransmitter norepinephrine (NE).

The sympathetic division brings about the responses associated with the "fight-or-flight" response.

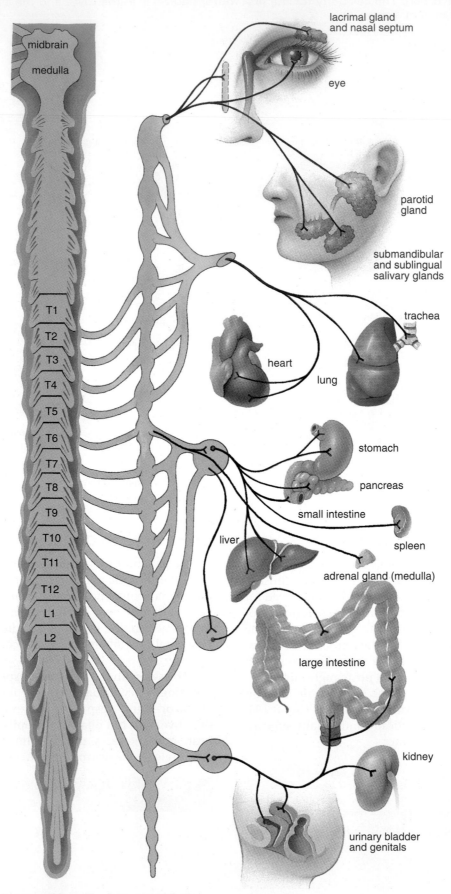

figure 8.18 Sympathetic division. The sympathetic fibers arise from the thoracic and lumbar portions of the cord. Note that this division regulates the same organs as the parasympathetic division (see fig. 8.19).

Part III Integration and Coordination

figure 8.19 Parasympathetic division. The parasympathetic fibers arise from the brain and sacral portions of the cord. Note that this division regulates the same organs as the sympathetic division (see fig. 8.18).

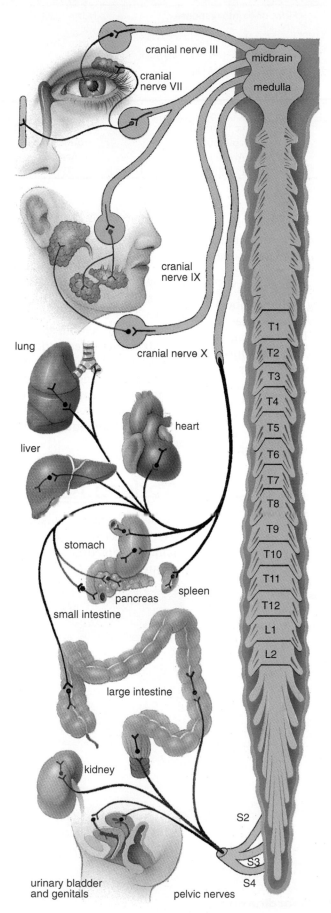

table 8.6

Sympathetic Versus Parasympathetic Divisions

	Sympathetic	Parasympathetic
Function	"Fight-or-flight" response	Normal activity
Neurotransmitter	Norepinephrine	Acetylcholine
Fiber Length	Postganglionic fiber is longer than preganglionic fiber.	Preganglionic fiber is longer than postganglionic fiber.
Fiber Origin	Preganglionic fiber arises from thoracolumbar portion of ANS.	Preganglionic fiber arises from craniosacral portion of ANS.

Parasympathetic Division

Cranial nerves, including the vagus nerve, and fibers that arise from the sacral levels of the spinal cord form the **parasympathetic division** (see fig. 8.19). Therefore, this system is often called the craniosacral portion of the ANS. In the parasympathetic division, the preganglionic fibers are long, and the postganglionic fibers are short because the ganglia lie near or within the organ. The parasympathetic division promotes all of the internal responses associated with a relaxed state. For example, it causes the pupils to contract, promotes digestion of food, and retards the heartbeat. The parasympathetic division utilizes the neurotransmitter acetylcholine.

Table 8.6 lists differences between the sympathetic and parasympathetic divisions. Figures 8.18 and 8.19 contrast the two divisions.

> The parasympathetic division brings about the responses associated with normally restful activities.

Effects of Aging

After age 60, the brain begins to lose thousands of neurons a day. When these cells die, they are not replaced. By age 80, the brain weighs about 10% less than when the person was a young adult. The cerebral cortex shrinks more than other areas of the brain, losing as much as 45% of its cells. Therefore, such mental activities as learning, memory, and reasoning decline.

Neurotransmitter production also decreases, resulting in slower synaptic transmission. As a person ages, thought processing and translating a thought into action take longer. This explains why younger athletes tend to outshine older athletes in sports.

Alzheimer disease, discussed in the MedAlert reading in this chapter, is usually associated with age. The progressive loss of memory and other intellectual functions in Alzheimer patients is accompanied by abnormalities in neuron structure. The neurons develop neurofibrillary tangles and amyloid plaques, perhaps as a result of a genetic defect.

Working Together

The accompanying illustration shows how the nervous system works with other organ systems of the body to maintain homeostasis. All the organ systems of the body are interrelated.

> **BodyWorks CD-ROM**
> The module accompanying chapter 8 is Nervous System.

working together Nervous System

Integumentary System

Brain controls nerves that regulate size of cutaneous blood vessels, activate sweat glands and arrector pili muscles.

Skin protects nerves, helps regulate body temperature; skin receptors send sensory input to brain.

Skeletal System

Receptors send sensory input from bones and joints to brain.

Bones protect sense organs, brain, and spinal cord; store Ca^{2+} for nerve function.

Muscular System

Brain controls nerves that innervate muscles; receptors send sensory input from muscles to brain.
Muscle contraction moves eyes, permits speech, and creates facial expressions.

Endocrine System

Hypothalamus is part of endocrine system; nerves innervate certain glands of secretion.

Sex hormones affect development of brain.

Circulatory System

Brain controls nerves that regulate the heart and dilation of blood vessels.

Blood vessels deliver nutrients and oxygen to neurons, carry away wastes.

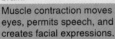
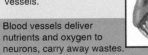

How the Nervous System works with other body systems

Lymphatic System/Immunity

Microglial cells engulf and destroy pathogens.

Lymphatic vessels pick up excess tissue fluid; immune system protects against infections of nerves.

Respiratory System

Respiratory centers in brain regulate breathing rate.

Lungs provide oxygen for neurons and rid the body of carbon dioxide produced by neurons.

Digestive System

Brain controls nerves that innervate smooth muscle and permit digestive tract movements.
Digestive tract provides nutrients for growth, maintenance, and repair of neurons and neuroglial cells.

Urinary System

Brain controls nerves that innervate muscles that permit urination.

Kidneys maintain blood levels of Na^+, K^+, and Ca^{2+}, which are needed for nerve conduction.

Reproductive System

Brain controls onset of puberty; nerves are involved in erection of penis and clitoris, contraction of ducts that carry gametes, and contraction of uterus.
Sex hormones masculinize or feminize the brain, exert feedback control over the hypothalamus, and influence sexual behavior.

Alzheimer Disease

Alzheimer disease (AD) is a disorder characterized by a gradual loss of reason that begins with memory lapses and ends with an inability to perform any type of daily activity. Personality changes signal the onset of AD. A normal 50- to 60-year-old adult might forget the name of a friend not seen for years. People with AD, however, forget the name of a neighbor who visits daily. With time, they have trouble traveling and cannot perform simple errands. People afflicted with AD become confused and tend to repeat the same question. Signs of mental disturbances eventually appear, and patients gradually become bedridden and die of a complication, such as pneumonia.

A normal neuron (nerve cell), and a neuron damaged by Alzheimer disease (AD), are shown in figure 8A. The AD neuron has two abnormalities not seen in the normal neuron: (1) Bundles of fibrous protein, called neurofibrillary tangles, surround the nucleus in the cell, and (2) protein-rich accumulations, called amyloid plaques, envelop the axon branches. These abnormal neurons are especially seen in the portions of the brain that are involved in reason and memory (frontal lobe and limbic system). To see the abnormal brain neurons, brain tissue must be examined microscopically after the patient dies.

A chemical test can be used to check brain tissue for the presence of a protein called Alzheimer disease associated protein (ADAP), which is believed to be the protein contained in the neurofibrillary tangles. If ADAP is proven to be the protein involved in AD, individuals could be tested for this protein by obtaining a spinal tap of cerebrospinal fluid.

Over a life span of 100 years, the likelihood of developing AD is 16% for people with no family history of AD, and 24% for those having first-degree relatives with AD. This difference in susceptibility suggests that AD might have a genetic basis. Researchers have discovered that in some families whose members have a 50% chance of AD, a genetic defect exists on chromosome 21. This is of extreme interest because Down syndrome (p. 381) results from the inheritance of three copies of chromosome 21, and people with Down syndrome tend to develop AD. Further, the genetic defect affects the normal production of amyloid precursor protein (APP), which may be the cause of the amyloid plaques.

Acetylcholine is a chemical that stimulates neurons to carry nerve impulses, and it appears that this chemical may be in short supply in the brains of patients with AD. Drugs that enhance acetylcholine production are currently being tested in AD patients. Experimental drugs that prevent neuron degeneration are also being tested. For example, it is possible that nerve growth factor, a substance that is made by the body and that promotes the growth of neurons, will one day be available to AD patients.

Questions

1. Why are drugs that enhance acetylcholine production being tested in AD patients?
2. What evidence suggests that AD might have a genetic basis?
3. How does the AD neuron differ from a normal neuron?

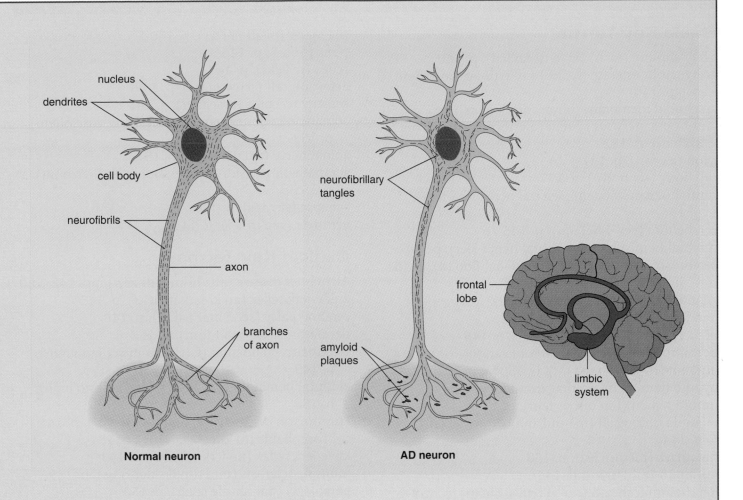

figure 8A Alzheimer disease (AD). An AD neuron has neurofibrillary tangles and amyloid plaques. AD neurons are particularly present in the frontal lobe and limbic system. This accounts for the development of symptoms of Alzheimer disease.

Selected New Terms

Basic Key Terms

acetylcholine (ACh) (as"ĕ-til-ko'lēn), p. 140

acetylcholinesterase (AChE) (as"ĕ-til-ko"lin-es'ter-ās), p. 140

arachnoid membrane (ah-rak'noid mem'brān), p. 150

autonomic nervous system (aw"to-nom'ik ner'vus sis'tem), p. 152

axon (ak'son), p. 137

cell body (sel bod'e), p. 137

central nervous system (sen'tral ner'vus sis'tem), p. 136

cerebellum (ser"ĕ-bel'um), p. 146

cerebral hemisphere (ser'ĕ-bral hem'ĭ-sfēr), p. 146

cerebrospinal fluid (ser"e-bro-spi'nal floo'id), p. 151

cerebrum (ser'ĕ-brum), p. 146

cranial nerve (kra'ne-al nerv), p. 152

dendrite (den'drīt), p. 137

diencephalon (di"en-sef'ah-lon), p. 146

dorsal-root ganglion (dor'sal root gang'gle-on), p. 142

dura mater (du'rah ma'ter), p. 150

hypothalamus (hi"po-thal'ah-mus), p. 146

limbic system (lim'bik sis'tem), p. 149

medulla oblongata (mĕ-dul'ah ob"long-ga'tah), p. 144

meninges (mĕ-nin'jēz), p. 150

midbrain (mid'brān), p. 146

neurilemmal sheath (nu"rĭ-lem'al shēth), p. 139

neuron (nu'ron), p. 137

neurotransmitter (nu"ro-trans'mit-er), p. 140

parasympathetic division (par"ah-sim"pah-thet'ik dĭ-vizh'un), p. 158

peripheral nervous system (pĕ-rif'er-al ner'vus sis'tem), p. 136

pia mater (pi'ah ma'ter), p. 150

pons (ponz), p. 146

reflex (re'fleks), p. 142

Schwann cell (schwon sel), p. 139

sensory receptor (ri-sep'ter), p. 142

somatic nervous system (so-mat'ik ner'vus sis'tem), p. 152

spinal nerve (spi'nal nerv), p. 152

sympathetic division (sim"pah-thet'ik dĭ-vizh'un), p. 155

synapse (sin'aps), p. 140

ventricle (ven'trĭ-k'l), p. 144

Clinical Key Terms

Alzheimer disease (altz'hi-mer dĭ-zēz'), pp. 142, 160

ankle-jerk reflex (an'kl-jerk re'fleks), p. 143

cerebral palsy (ser'ĕ-bral pal'ze), p. 148

dermatome (der'mah-tōm), p. 152

electroencephalogram (e-lek"tro-in-sef'lah-gram), p. 149

epidural hematoma (ep"ĭ-du'ral he"mah-to'mah), p. 150

Huntington disease (hun'ting-tun dĭ-zēz'), p. 141

hydrocephalus (hi"dro-sĕ'fah-lus), p. 151

knee-jerk reflex (ne-jerk re'fleks), p. 143

multiple sclerosis (mul'tĭ-pul skler-o'sis), p. 139

paraplegia (par-ah-ple'je-ah), p. 150

Parkinson disease (par'kin-sun dĭ-zēz'), p. 141

quadriplegia (kwah-drah-ple'je-ah), p. 150

stroke (strōk), p. 149

subdural hematoma (sub"du'ral he"mah-to'mah), p. 150

Summary

I. **Nervous System**
   A. Divisions of the nervous system. The nervous system is divided into the central nervous system (brain and spinal cord) and the peripheral nervous system (somatic and autonomic nervous systems). The CNS lies in the midline of the body, and the PNS is located peripherally to the CNS.
   B. Functions of the nervous system. The nervous system permits sensory input, performs integration, and stimulates motor output.
   C. Cells of nervous tissue. Nervous tissue contains neurons and neuroglial cells. Each type of neuron has three parts (dendrites, cell body, and axon) but is specific as to function. Neuroglial cells support, protect, and nourish the neurons.
   D. Nerve impulses. All neurons transmit the same type of nerve impulse: a change in polarity that flows along the membrane of a nerve fiber.
   E. Synapse. Transmission of a nerve impulse across a synapse is dependent on the release of a neurotransmitter into a synaptic cleft.

F. Nerves. A nerve contains bundles of long fibers covered by fibrous, connective tissue layers. In the CNS, bundles of long fibers are found in tracts. White matter is composed of myelinated fibers, and gray matter is composed of cell bodies and unmyelinated fibers.

G. Reflexes and the reflex arc. Reflexes (automatic reactions to internal and external stimuli) depend on the reflex arc. Some reflexes are important for avoiding injury, and others are necessary for normal physiological functions.

II. Central Nervous System

A. Ventricles of the brain. The brain has four ventricles. The lateral ventricles are found in the left and right cerebral hemispheres. The third ventricle is found in the diencephalon. The fourth ventricle is found in the brain stem.

B. Brain stem. The brain stem contains the medulla oblongata, pons, and midbrain. The medulla oblongata contains vital centers for regulating heartbeat, breathing, and blood pressure. The pons assists the medulla oblongata in regulating the breathing rate. The midbrain contains tracts that conduct impulses to and from the higher parts of the brain.

C. Diencephalon. The hypothalamus helps control the functioning of most internal organs and controls the secretions of the pituitary gland. The thalamus receives sensory impulses from all parts of the body and channels them to the cerebrum.

D. Cerebellum. The cerebellum controls balance and complex muscular movements.

E. Cerebrum. Consciousness is under the control of the cerebrum, the most highly developed portion of the brain. It is responsible for higher mental processes, including the interpretation of sensory input and the initiation of voluntary muscular movements.

F. Limbic system. The limbic system includes portions of the cerebrum, the thalamus, and the hypothalamus. It is involved in learning and memory and in causing the emotions that guide behavior.

G. Spinal cord. The spinal cord is located in the vertebral column in cross sections composed of white matter and gray matter. White matter contains bundles of nerve fibers, called tracts, that conduct nerve impulses to and from the higher centers of the brain. Gray matter is mainly made up of short fibers and cell bodies. The spinal

cord is a center for reflex action and allows communication between the brain and the peripheral nerves leaving the spinal cord.

H. Meninges and cerebrospinal fluid. The CNS is protected by the meninges and the cerebrospinal fluid.

III. Peripheral Nervous System

A. Somatic nervous system. Cranial nerves take impulses to and/or from the brain. Spinal nerves take impulses to and from the spinal cord.

B. Autonomic nervous system. The ANS controls the functioning of internal organs without need of conscious control.
   1. The divisions of the autonomic nervous system: (1) function automatically and usually subconsciously in an involuntary manner, (2) innervate all internal organs, and (3) utilize two motor neurons and one ganglion for each impulse.
   2. The sympathetic division brings about the responses associated with the "fight-or-flight" response.
   3. The parasympathetic division brings about the responses associated with normally restful activities.

Study Questions

1. What are the two main divisions of the nervous system? How are these divisions subdivided? (pp. 136–37)
2. What are the types of neurons and neuroglial cells? How are they similar, and how are they different? (pp. 137–39)
3. What does the term *nerve impulse* mean, and how is a nerve impulse brought about? (p. 140)
4. What is a neurotransmitter? Where is it stored, and how does it function? How is it destroyed? Name several well-known neurotransmitters. (pp. 140–41)
5. Describe the structure of a nerve, and state the location of nerves and tracts. (p. 142)

6. What is the path of a spinal reflex that involves three neurons? What is the function of reflexes? (pp. 142–43)
7. Where are the ventricles of the brain located? (p. 144)
8. Name the various parts of the brain, state where the parts are located, and give their functions. (pp. 144–48)
9. What does it mean to say that the cerebral cortex can be mapped? Discuss this in relation to the primary motor areas and the primary sensory areas. (pp. 146–49)
10. Describe the anatomy of the spinal cord. What are the functions of the gray and white matter in the spinal cord? (p. 150)

11. What are the three different meninges, and what is their function? (pp. 150–51)
12. What is cerebrospinal fluid? Where is it made, and how does it circulate? (p. 151)
13. What are the different cranial nerves, and what is the function of each? (pp. 152–53)
14. What are the structure and function of the spinal nerves? (p. 152)
15. What is the autonomic nervous system, and what are its two major divisions? Describe several similarities and differences between these divisions. (pp. 155–58)

Objective Questions

Fill in the blanks.

1. A(n) _____ carries nerve impulses away from the cell body.
2. During the depolarization portion of an action potential, _____ ions are moving to the _____ of the nerve fiber.
3. The space between the axon ending of one neuron and the dendrite of another is called the _____ .
4. ACh is broken down by the enzyme _____ after it has initiated an action potential on a neighboring neuron.
5. Motor nerves stimulate _____ .
6. In a reflex arc, only the _____ is completely within the CNS.
7. The _____ is the part of the brain responsible for coordination of body movements.
8. The _____ is the part of the brain that is responsible for consciousness.
9. The brain and spinal cord are covered by protective layers called _____ .
10. The vagus nerve is a _____ nerve that controls _____ .
11. Whereas the central nervous system is composed of the _____ and _____, the peripheral nervous system is composed of the _____ .
12. The limbic system includes portions of the _____, _____, and _____ .
13. Whereas the _____ division of the autonomic nervous system brings about organ responses that are part of the "fight-or-flight" response, the _____ division brings about responses associated with normal restful conditions.
14. The electrical activity of the brain can be recorded in the form of a(n) _____ .

Medical Terminology Reinforcement Exercise

Consult Appendix B for help in pronouncing and analyzing the meaning of the terms that follow.

1. neuropathogenesis (nu″ro-path″o-jen′ĕ-sis)
2. anesthesia (an″es-the′ze-ah)
3. encephalomyeloneuropathy (en-sef″ah-lo-mi″ĕ-lo-nu-rop′ah-the)
4. hemiplegia (hem″ĭ-ple′je-ah)
5. glioblastoma (gli″o-blas-to′mah)
6. subdural hemorrhage (sub-du′ral hem′or-ij)
7. cephalometer (sef″ah-lom′ĕ-ter)
8. pneumoencephalography (nu″mo-en-sef″ah-log′rah-fe)
9. meningoencephalocele (me-ning″go-en-sef″ah-lo-sēl″)
10. neurorrhaphy (nu-rōr′ah-fe)
11. ataxiaphasia (ah-tak″se-ah-fa′ze-ah)
12. dysphagia (dis-fa′je-ah)

Website Links

For a listing of the most current Websites related to this chapter, please visit the Mader home page at:

www.mhhe.com/maderap

c h a p t e r 9

The Sensory System

Chapter Outline and Learning Objectives

After you have studied this chapter, you should be able to:

General Receptors (p. 166)
- Categorize sensory receptors according to the system used in the text.
- Name the four senses of the skin, and state the location of their receptors.
- Discuss the function of visceral receptors.
- Discuss the function of proprioceptors.

Chemoreceptors (p. 168)
- Name the chemoreceptors, and state their location, anatomy, and mechanism of action.

Photoreceptors (p. 169)
- Describe the anatomy and function of the accessory organs of the eye.
- Describe the anatomy of the eye and the function of each part.
- Describe the sensory receptors for sight, their mechanism of action, and the mechanism for stereoscopic vision.
- Describe common disorders of sight discussed in the text.

Mechanoreceptors (p. 179)
- Describe the anatomy of the ear and the function of each part.
- Describe the sensory receptors for balance and hearing, and their mechanism of action.

Effects of Aging (p. 183)
- Anatomical and physiological changes occur in the sensory system as we age.

Medical Focus
Myasthenia Gravis (p. 171)
Corrective Lenses (p. 178)
Hearing Damage and Deafness (p. 182)

MedAlert
Age-Related Macular Degeneration (p. 174)

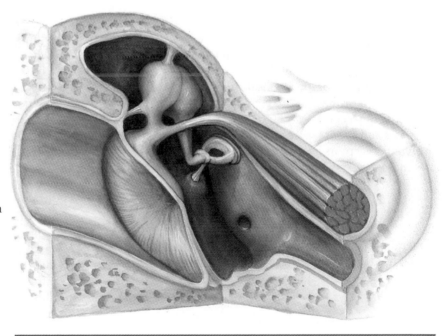

Sound waves cause the bones of the middle ear to vibrate, which eventually results in the ability to hear.

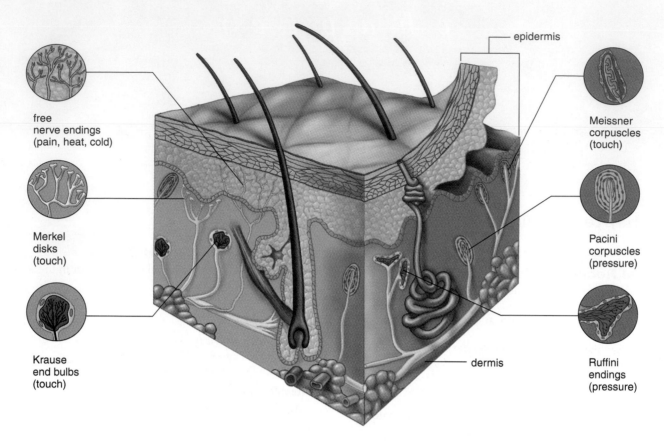

figure 9.1 Sensory receptors in human skin. Each receptor has the function(s) indicated.

free nerve endings (pain, heat, cold)

Merkel disks (touch)

Krause end bulbs (touch)

epidermis

Meissner corpuscles (touch)

Pacini corpuscles (pressure)

dermis

Ruffini endings (pressure)

Sense perception is dependent upon sensory receptors. When a receptor is stimulated, it generates nerve impulses that are transmitted to the spinal cord and/or brain, but a person is conscious of a sensation only if the impulses reach the cerebrum.

Sensory receptors can be divided in two types: general receptors and special receptors. The special receptors include the chemoreceptors in the mouth and nose, the photoreceptors in the eyes, and the mechanoreceptors in the ears.

General Receptors

General receptors are sensory receptors generally present throughout the body. These receptors are present in the skin, visceral organs, and muscles and joints.

Skin

The skin contains sensory receptors of touch, pressure, pain, and temperature (fig. 9.1). It is a mosaic of these tiny receptors, as can be determined by passing a metal probe slowly over the skin. At certain points, there will be a feeling of pressure, and at others, a feeling of hot or cold (depending on the probe's temperature).

Three types of receptors are responsive to touch: the Meissner corpuscle, the Merkel disk, and the Krause end bulb. The Meissner corpuscle is concentrated in the fingertips, the palms, the lips, the tongue, the nipples, the penis, and the clitoris. Their prevalence provides these regions with special sensitivity. The Pacinian corpuscle and Ruffini ending are two different types of pressure receptors. Pacinian corpuscles are onion-shaped sense organs that lie deep inside the dermis. Ruffini endings are encapsulated by sheaths of connective tissue and contain lacy networks of nerve fibers. Temperature and pain receptors are free nerve endings in the epidermis. Some free nerve endings are responsive to cold; others are responsive to heat. Cold receptors are far more numerous than heat receptors, but there are no known structural differences between the two.

Part III   Integration and Coordination

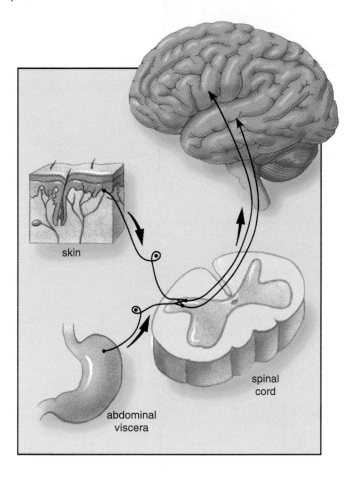

figure 9.2 Referred pain. Sensory receptors in the abdominal organs send impulses to a connector neuron that also receives impulses from the skin. Thereafter, the brain mistakenly senses pain in the skin.

skin

abdominal viscera

spinal cord

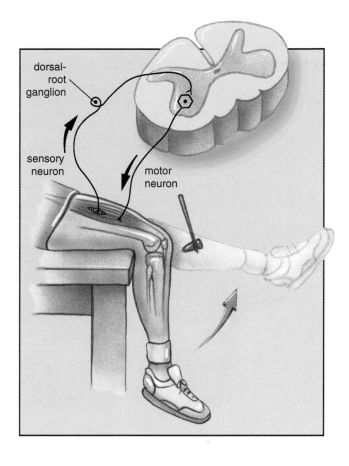

figure 9.3 Proprioceptors in muscles and joints. Muscle spindle fibers are stretch receptors, as can be demonstrated by the knee-jerk reflex.

dorsal-root ganglion

sensory neuron

motor neuron

Visceral Organs

The visceral (internal) organs have receptors that aid in maintaining homeostasis. For example, stretch receptors in the lungs respond to lung expansion, the aortic and carotid bodies are sensitive to pH and oxygen levels of the blood, and osmoreceptors in the hypothalamus detect blood osmolarity, which is dependent on the concentration of solutes in the blood.

Like the skin, many internal organs have pain receptors. Sometimes, this pain is felt as pain from the skin. This is called **referred pain** (fig. 9.2). Different internal organs have a referred pain relationship with certain areas of the skin. For example, pain arising from the intestine is located in the skin of the back, groin, and abdomen; pain from the heart is felt in the left shoulder and arm. This happens because nerve stimuli from the pain receptors of the internal organs travel to the spinal cord, where they make contact with neurons also receiving messages from the skin. The brain interprets this as pain in the skin.

Muscles and Joints

The sense of position and movement of limbs is dependent upon receptors termed **proprioceptors** (pro″pre-o-sep′-torz). Proprioceptors are located in the joints and in associated ligaments and tendons that respond to stretching, pressure, and pain. Nerve impulses from these receptors are integrated within the cerebellum with those received from other types of receptors so that the person can monitor the position of body parts.

A muscle spindle consists of a bundle of modified muscle fibers with sensory nerve fibers wrapped around a short, specialized region somewhere near the middle of the spindle's length. Stretching of associated muscle fibers causes muscle spindles to increase the rate at which they fire, and for this reason, they are sometimes called *stretch receptors*. The knee-jerk reflex illustrates how muscle spindles act as stretch receptors (fig. 9.3). When the legs are crossed at the knee and the tendon at the knee is tapped, both the tendon and a muscle in the thigh are stretched.

Chapter 9 The Sensory System 167

table 9.1	Special Sense Organs			
	Sense Organ	Type of Receptor	Specific Receptor	Senses
	Taste buds	Chemoreceptor	Taste cells	Taste
	Nose	Chemoreceptor	Olfactory cells	Smell
	Eye	Photoreceptor	Rods and cones in retina	Vision
	Ear	Mechanoreceptor	Hair cells in utricle, saccule, and semicircular canals	Equilibrium
			Hair cells in organ of Corti	Hearing

figure 9.4 Taste buds. **a.** Papillae on the tongue contain taste buds that are sensitive to sweet, sour, salty, and bitter tastes as indicated. **b.** Taste buds occur along the walls of the papillae. **c.** Taste cells end in microvilli. When molecules bind to the microvilli, nerve impulses are generated that go to the brain where the sensation of taste occurs.

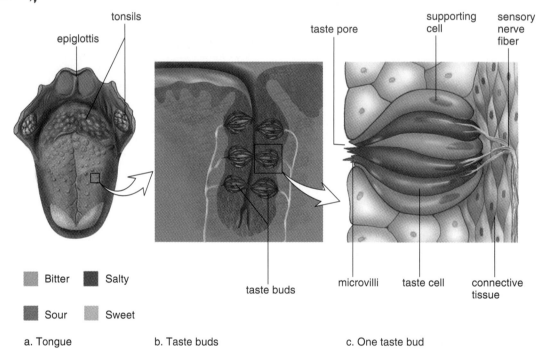

Bitter   Salty

Sour   Sweet

a. Tongue   b. Taste buds   c. One taste bud

Stimulated by the stretching, muscle spindles transmit impulses to the spinal cord, and thereafter, muscle contraction occurs, which causes the lower leg to jerk upward in a kicking motion. Normally, the sensory input from muscle spindles allows the central nervous system to regulate muscle contraction so that tone is maintained.

General receptors are located in the skin, the visceral organs, and the muscles and joints. General receptors in the skin respond to touch, pressure, pain, and temperature; others help to maintain homeostasis and allow us to monitor the position of body parts.

Chemoreceptors

Taste buds and the nose are *special sense organs*. Special sense organs contain groups of specialized sensory receptors (table 9.1). Taste and smell are called the *chemical senses* because the receptors for these senses are sensitive to certain chemical substances in food and in the air.

There are four types of tastes (bitter, sour, salty, and sweet), and the **taste buds** for each are concentrated on the tongue in particular regions (fig. 9.4). Most taste buds lie along the walls of papillae, the small elevations visible to the naked eye. The taste buds are pockets of taste cells in the tongue epithelium with microvilli that project out of a pore. Nerve impulses, generated when chemicals bind to the microvilli, go to the brain where the sensation of taste occurs.

figure 9.5 Olfactory cell location and anatomy. **a.** The olfactory epithelium in humans is located in the nasal cavity. **b.** Olfactory cells end in cilia that bear receptor proteins for specific odor molecules. The cilia of each olfactory cell can bind to only one type of odor molecule signified here by color. If a rose causes olfactory cells sensitive to purple and green odor molecules to be stimulated, then interneurons, designated by purple and green, in the olfactory bulb are activated. The primary olfactory area of the cerebral cortex interprets the pattern of interneurons stimulated as the scent of a rose.

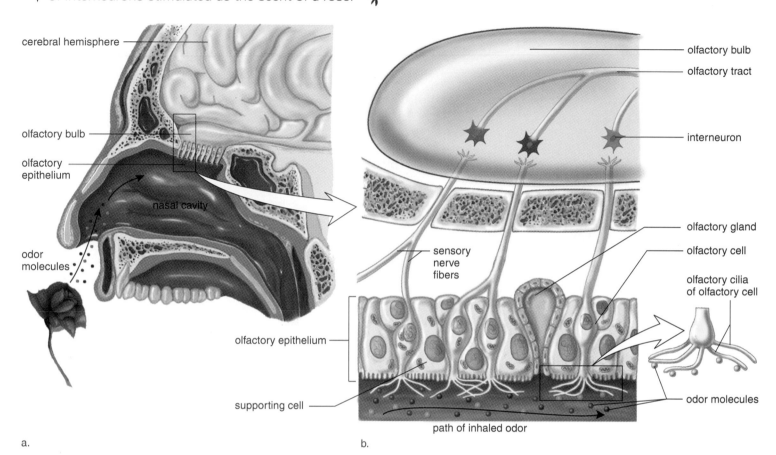

cerebral hemisphere

olfactory bulb

olfactory epithelium

nasal cavity

odor molecules

a.

olfactory bulb

olfactory tract

interneuron

sensory nerve fibers

olfactory epithelium

supporting cell

olfactory gland

olfactory cell

olfactory cilia of olfactory cell

odor molecules

path of inhaled odor

b.

The **olfactory cells** (fig. 9.5) are in the superior aspect of a nasal cavity. These cells, which are specialized endings of the fibers that make up the olfactory nerve, lie among supporting epithelial cells. It is likely that there are at least 1,000 different types of olfactory cells. Each cell ends in a tuft of six to eight cilia, and when chemicals bind to these cilia, nerve impulses are generated. The particular aroma that a person perceives results from the combination of olfactory cells that have been stimulated. For example, a rose may stimulate one combination of olfactory cells, while a daffodil stimulates another. The olfactory cells stimulate interneurons in the olfactory bulb, which pass this information via the olfactory tract to the olfactory areas of the cerebral cortex where the perception of the odor occurs.

The senses of taste and smell supplement each other, creating a combined effect when interpreted by the cerebral cortex. For example, when a person has a cold, food seems tasteless, but actually, the person has temporarily lost the ability to sense the food's smell. When a person smells something, some of the molecules move from the nose down into the mouth region and stimulate the taste buds. Therefore, part of what is referred to as smell may actually be taste.

The sensory receptors for taste (taste buds) and for smell (olfactory cells) work together to produce the senses of taste and smell.

Photoreceptors

The photoreceptors are in the eyes. The eyes are located in orbits formed by seven of the skull's bones (frontal, lacrimal, ethmoid, zygomatic, maxilla, sphenoid, and palatine). The supraorbital ridge protects the eye from blows, and the eyebrow diverts sweat around the eye. The eye's accessory organs are discussed first.

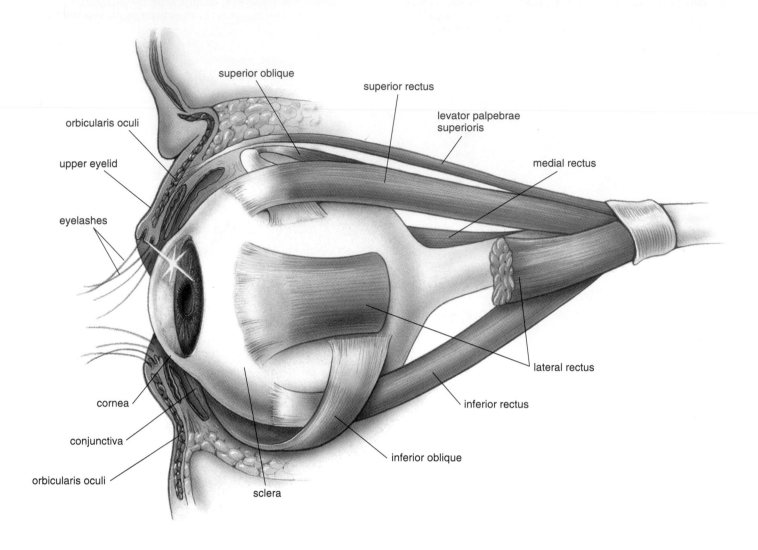

Accessory Organs of the Eye
Extrinsic Muscles

Within an orbit, the eye is anchored in place by the **extrinsic muscles,** whose contractions move the eyes. There are three pairs of antagonistic extrinsic muscles (fig. 9.6):

1. **Superior rectus**   Rolls eye upward
   Inferior rectus   Rolls eye downward
2. **Lateral rectus**   Turns eye outward, away from midline
   Medial rectus   Turns eye inward, toward midline
3. **Superior oblique**   Rotates eye counterclockwise
   Inferior oblique   Rotates eye clockwise

Three cranial nerves—the oculomotor, abducens, and trochlear—control these muscles. The oculomotor nerve innervates the superior, inferior, and medial rectus muscles, as well as the inferior oblique muscles; the abducens nerve innervates the lateral rectus muscle; and the trochlear nerve innervates the superior oblique muscle.

Eyelids and Eyelashes

The thin skin of an eyelid covers muscle and connective tissue (see fig. 9.6). Lining the inner surface is a transparent mucous membrane, called the conjunctiva. The conjunctiva folds back to cover the anterior of the eye, except for the cornea, and prevents tears from entering the orbits. The eyelids have eyelashes that can trap a tiny particle of grit, causing the eyes to close immediately. Sebaceous glands associated with each eyelash produce an oily secretion that lubricates the eye. Inflammation of one of the glands is called a **sty.**

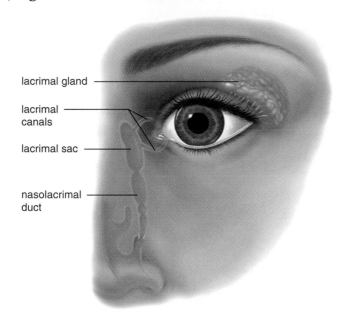

figure 9.7 Lacrimal apparatus.

lacrimal gland

lacrimal canals

lacrimal sac

nasolacrimal duct

The eyelids are operated by the orbicularis oculi, which closes the lid, and the levator palpebrae superioris, which raises the lid. As explained in the Medical Focus reading on this page, a person with *myasthenia gravis* has muscle weakness due to an inability to respond to acetylcholine, and the eyelids often have to be taped open.

Lacrimal Apparatus

A **lacrimal apparatus** consists of the lacrimal gland and the lacrimal sac with its ducts (fig. 9.7). The lacrimal gland, which lies in the orbit above the eye, produces tears that flow over the eye when the eyelids are blinked. The tears, collected by two small ducts, pass into the lacrimal sac before draining into the nose by way of the nasolacrimal duct.

> Accessory structures of the eye include the extrinsic muscles, which move the eye; the eyelids and eyelashes; and the lacrimal apparatus, which produces tears.

Medical Focus

Myasthenia Gravis

*m*yasthenia gravis (mi″as-the′ne-ah grah′vis) is a disease involving an interruption of normal skeletal muscle contraction because the immune system of the affected individual mistakenly produces antibodies that destroy acetylcholine receptors. The disease is characterized by muscle weakness that especially affects the muscles of the eyelids, face, neck, and extremities. Respiratory muscles are also affected.

Myasthenia gravis may affect any age group, with more females affected in their twenties and thirties and more men affected in their fifties and sixties. Overall, the incidence of the disease shows a female-to-male ratio of 3:2. In many cases, ocular muscle weakness occurs first, resulting in a drooping of the eyelid and double vision. Muscles of the face, throat, extremities, and neck also become weak. Respiratory muscle weakness may cause difficulty in breathing. Characteristically, the weakness fluctuates, even during the course of one day. Illness, exercise, and emotional stress worsen the condition.

Myasthenia gravis is not normally life threatening, although a crisis involving respiratory muscle weakness may require assisted ventilation. Methods of treatment include drugs that are antagonistic to the enzyme cholinesterase, drugs that suppress the immune system, and surgical removal of the thymus.

figure 9.8 Sectioned eyeball.

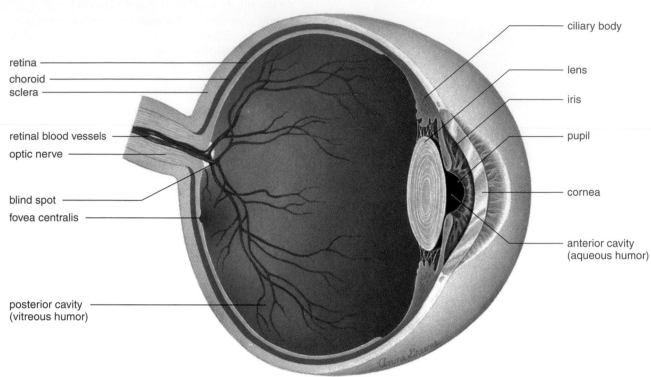

retina

choroid

sclera

retinal blood vessels

optic nerve

blind spot

fovea centralis

posterior cavity
(vitreous humor)

ciliary body

lens

iris

pupil

cornea

anterior cavity
(aqueous humor)

Structure of the Eye

The eye (fig. 9.8 and table 9.2), an elongated sphere about 1 inch in diameter, has three layers, or coats: the outer sclera, the middle choroid, and the inner retina. The outer **sclera** is a white, fibrous layer except for the transparent **cornea,** the window of the eye. The middle, thin, dark brown layer, the **choroid,** contains many blood vessels and absorbs stray light rays. Toward the front, the choroid thickens and forms the ring-shaped ciliary body containing the **ciliary muscle,** which controls the shape of the lens for near and far vision. Finally, the choroid becomes a thin, circular, muscular, and pigmented diaphragm, the **iris,** which regulates the size of the **pupil,** the hole through which light enters the eyeball. The **lens,** attached to the ciliary body by ligaments, divides the cavity of the eye into two smaller cavities. The posterior cavity behind the lens is filled with **vitreous humor** (vit're-us hu'mor), a viscous, gelatinous material. The anterior cavity between the cornea and the lens is filled with **aqueous humor** (a'kwe-us hu'-mor), a watery solution secreted by the ciliary body. (The anterior cavity contains two chambers: an anterior chamber between the cornea and iris, and the posterior chamber between the iris and the lens.) A small amount of aqueous humor is continually produced each day. Normally, it leaves the anterior cavity by way of tiny ducts located where the iris meets the cornea.

table 9.2

Function of Parts of the Eye

Part	Function
Lens	Refracts and focuses light rays
Iris	Regulates light entrance
Pupil	Admits light
Choroid	Absorbs stray light
Sclera	Protects and supports eyeball
Cornea	Refracts light rays
Humors	Refract light rays
Ciliary body	Holds lens in place, accommodation
Retina	Contains sight receptors
Rods	Make black-and-white vision possible
Cones	Make color vision possible
Optic nerve	Transmits impulse to brain
Fovea centralis	Makes acute vision possible

When a person has **glaucoma,** these drainage ducts are blocked, and aqueous humor builds up. If glaucoma is not treated, the resulting pressure compresses the arteries that serve the nerve fibers of the retina, where the sight receptors are located. The nerve fibers begin to die due to lack of nutrients, and the person becomes partially blind. Over time, total blindness can result.

figure 9.9 Cells of the retina. Rods and cones are located at the back of the retina, followed by the bipolar and ganglionic cells, whose fibers become the optic nerve. Notice that rods share bipolar cells, but cones do not.

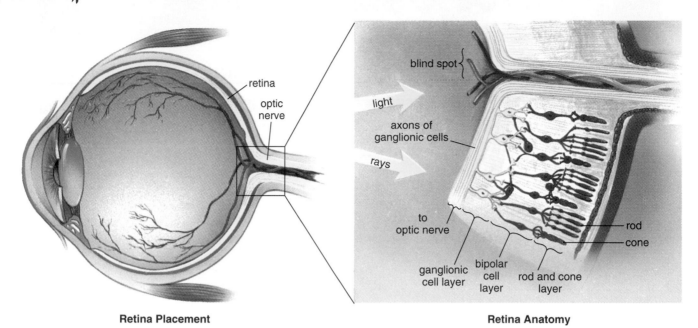

Retina Placement

Retina Anatomy

Retina

The inner layer of the eye, the **retina**, has three layers of cells (fig. 9.9). The layer closest to the choroid contains the sense receptors for sight, the **rods** and **cones;** the middle layer contains bipolar cells; and the innermost layer contains ganglionic cells whose fibers become the **optic nerve.** Only the rods and cones contain light-sensitive pigments, and therefore, light must penetrate to the back of the retina before nerve impulses are generated.

Nerve impulses initiated by the rods and cones are sent to the bipolar cells, which, in turn, send them to the ganglionic cells. The axons of the ganglionic cells pass in front of the retina, forming the optic nerve, which turns to pierce the layers of the eye. As shown in figure 9.9, rods and cones greatly outnumber ganglionic cells. In fact, the retina has as many as 150 million rods but only 1 million ganglionic cells and optic nerve fibers. This means that there is con-

siderable mixing of messages and a certain amount of integration before nerve impulses are sent to the thalamus and then on to the occipital lobe of the cerebrum. There are no rods or cones where the optic nerve passes through the retina; therefore, the optic disk is known as a **blind spot,** where vision is impossible.

The retina contains a very special region called the *macula lutea,* an oval, yellowish area with a depression called the **fovea centralis.** In this region, vision is most acute because there is a great concentration of cone cells. The MedAlert reading on page 174 concerns age-related macular degeneration, which is the most frequent cause of legal blindness in the United States.

> The retina contains the rods and cones, which are the sensory receptors for sight. When either is stimulated, nerve impulses begin and are transmitted via the optic nerve to the brain.

MedAlert

Age-Related Macular Degeneration

*t*he macula lutea, a yellowish area in the central region of the retina, contains a concentration of cones, especially in the fovea centralis. Vision is most acute and colors are detected in the macula lutea. **Macular degeneration** in the elderly (or age-related macular degeneration) is now the most frequent cause of legal blindness in the United States. Individuals with this condition have a distorted visual field: Blurriness or a blind spot is present, straight lines may look wavy, objects may appear to be larger or smaller than they are, and colors may look faded (fig. 9A).

There are two main forms of age-related macular degeneration. "Wet" macular degeneration means that abnormal growth of new blood vessels is evident in the region of the macula. The blood vessels leak serum and blood, and the retina becomes distorted, leading to severe scarring that completely destroys the macula. "Dry" macular degeneration is not accompanied by the growth of blood vessels, and visual loss is less dramatic.

Heredity plays a role in the development of age-related macular degeneration: 15% of people with a family history of the condition develop the disease after age 60. Also, light-eyed people tend to be afflicted more frequently than dark-eyed people. Smoking, hypertension, and excessive sun exposure are possible contributing factors.

A yearly eye examination assists in the early detection of many eye diseases, including macular degeneration, cataracts, and glaucoma. When an ophthalmologist presents an Amsler grid (a crosshatched pattern of straight lines) to someone with macular degeneration, the grid looks blurred, distorted, or discolored. Signs of the "wet" form can be detected by an examination of the retina and confirmed by a fluorescein angiogram. In this test, a number of pictures are taken of the macula lutea after an orange dye has been injected into a vein in the patient's arm.

Currently, the only treatment for the "dry" form of macular degeneration is the use of vitamin and mineral supplements, which may help stem the disease. For example, research indicates that consumption of zinc may prevent further loss of vision. On the other hand, when the "wet" form of the disease is diagnosed early, laser treatment can sometimes stop the growth of blood vessels.

Although people with age-related macular degeneration are classified as blind, they still have normal peripheral vision (outside the macula), which they can learn to use effectively. Because the periphery of the retina contains a high concentration of rods, vision here is less acute, and colors are not detected. But high-powered eyeglasses, magnifying devices, closed-circuit television, and special lamps can help patients see details more clearly.

Accumulating evidence suggests that both macular degeneration and cataracts, which tend to occur in the elderly, are caused by long-term exposure to the ultraviolet rays of the sun. Therefore, everyone—especially those who live in sunny climates or work outdoors—should wear sunglasses that absorb ultraviolet light. Large lenses worn close to the eyes offer further protection. The Sunglass Association of America has

normal

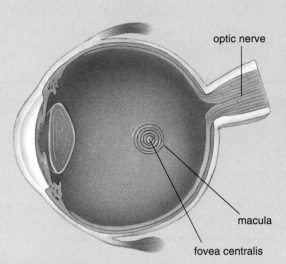

optic nerve

macula

fovea centralis

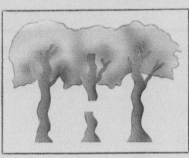

macular degeneration

figure 9A Macular degeneration. When a person with macular degeneration looks at a clump of trees, the trees may appear larger or smaller than they really are, the trunks may look wavy, details may be absent, and the colors may be dim.

devised the system that follows for categorizing sunglasses. (Note: UV-B rays are the cancer-causing ultraviolet rays. UV-A rays are believed to increase the effect of UV-B rays.)

- Cosmetic lenses absorb at least 70% of UV-B rays, 20% of UV-A rays, and 60% of visible light. Such lenses are worn for comfort rather than protection.

- General-purpose lenses absorb at least 95% of UV-B rays, 60% of UV-A rays, and 60–92% of visible light. They are good for outdoor activities in temperate regions.

- Special-purpose lenses block at least 99% of UV-B rays, 60% of UV-A rays, and 20–97% of visible light. They are good for bright sun combined with sand, snow, or water.

Questions

Why would you expect a person with age-related macular degeneration to:

1. see a tree limb but not the bark on the limb?
2. be color blind?
3. be able to move about the house day or night without bumping into objects?

figure 9.10 Structure and function of rods and cones. The outer segment of rods and cones is composed of stacks of membranous disks, which contain visual pigments. In rods, the membrane of each disk contains rhodopsin, a complex molecule containing the protein opsin and the pigment retinal. When retinal absorbs energy, it changes shape and disengages from opsin. This leads to nerve impulses in bipolar cells (see fig. 9.9).

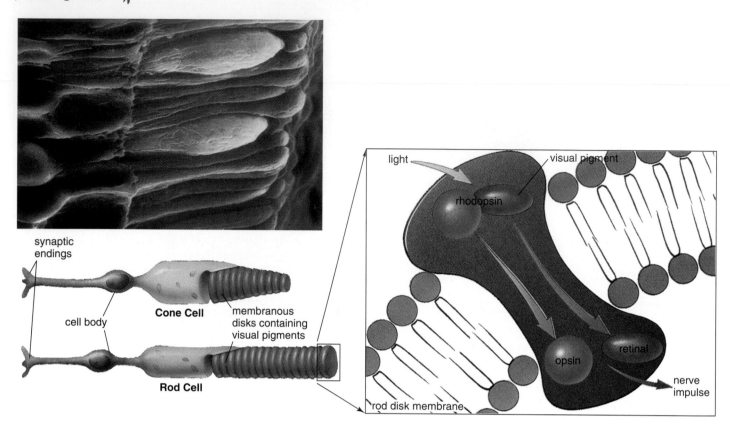

Rods

In dim light, the iris causes the pupil to enlarge so that more light rays can enter the eye. As the faint light rays enter, they strike the rods and cones, but only the 150 million rods located in the periphery of the eyes are sensitive to faint light. The rods do not detect fine detail or color, so at night, for example, all objects appear to be blurred and have a shade of gray. However, because of their abundance and position in the eyes, rods do detect even the slightest motion.

The rods contain *rhodopsin*, a molecule that contains the protein opsin and the pigment retinal. When light strikes rhodopsin, rhodopsin breaks down to its components, and this generates nerve impulses (fig. 9.10). The more rhodopsin present in the rods, the more sensitive the eyes are to dim light. Therefore, during the time required for adjustment to dim light, when it is difficult to see, rhodopsin is being formed in the rods. Retinal is a derivative of vitamin A, which is abundant in carrots, so the suggestion that eating carrots helps vision is not without foundation.

> The rods are responsible for vision in dim light. They do not see fine detail or color, but they do detect motion.

Cones

The cones, located primarily in the fovea centralis, function in bright light to detect fine detail and color. To perceive depth, as well as to see color, we turn our eyes so that reflected light from the object strikes the fovea centralis. Color vision depends on three kinds of cones, one kind for each of three colors: blue, green, and red. The colors we see depend on which of these cones are activated.

Complete **color blindness** is extremely rare. In most instances, a particular type of cone is lacking or deficient in number. The lack of either red or green cones is the most common type of color blindness, affecting about 5% of the American population. If the eye lacks red cones, the green colors become accentuated, and vice versa.

> The cones are responsible for vision in bright light. They detect fine detail and color.

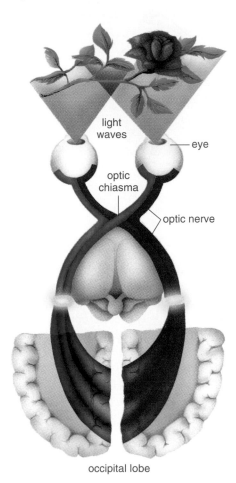

figure 9.11 Pathway of signals from the retina through the optic chiasma to the brain, resulting in stereoscopic vision.

light waves

eye

optic chiasma

optic nerve

occipital lobe

Stereoscopic Vision

Each eye sends its own information to the brain about the placement of an object because each forms an image from a slightly different angle. These data are pooled to produce **stereoscopic vision** by a two-step process. First, because the optic nerves cross at the optic chiasma (fig. 9.11), one-half of the brain receives information from both eyes about the same part of an object. Later, the two halves of the brain communicate to arrive at a complete, three-dimensional interpretation of the whole object.

Lens

When we look at an object, light rays are bent (refracted) and focused on the retina. The cornea, vitreous humor, and lens all help in this process. Although the lens remains flat when we view distant objects, it rounds up when we view close objects, a process called **accommodation** (fig. 9.12). Lens shape is controlled by the ciliary muscle within the ciliary body. When we view a distant object, the ciliary muscle is relaxed, causing the suspensory ligaments attached to the ciliary body to be taut; therefore, the lens remains relatively flat. When we view a near object, the ciliary muscle contracts, releasing the tension on the ligaments; the lens then rounds up due to its natural elasticity (table 9.3). Close work, which requires contraction of the ciliary muscle, often causes eyestrain.

When the eyeball is too long or too short, accommodation by the lens may not be sufficient to bring an object into focus. Also, after age 40, the lens loses some of its elasticity and is unable to accommodate as well. Then a person needs glasses, as described in the Medical Focus reading on page 178.

With age, the lens also is subject to cataracts. A **cataract** occurs when the lens becomes opaque and incapable of transmitting light rays. Recent research suggests that cataracts develop when crystalline proteins within the lens oxidize. Knowing this may help researchers find a way to treat cataracts medically. At present, surgery is the only viable cataract treatment. First, a surgeon opens the eye near the rim of the cornea. Then the enzyme zonulysin may be used to digest the ligaments holding the lens in place. A plastic lens is then implanted in the eye, and the patient does not need to wear thick glasses or contact lenses.

The lens focuses light rays on the retina. The lens is flat for distant vision and rounds up for near vision.

table 9.3

Accommodation Adjustment to Distances

Focus	Ciliary Muscle	Lens
Distant	Ciliary muscle relaxes; ligaments are taut.	Lens is flattened.
Near	Ciliary muscle contracts; ligaments are relaxed.	Lens is more rounded.

figure 9.12 Accommodation. **a.** With distant vision, the lens is flat. **b.** With close vision, the lens is rounded.

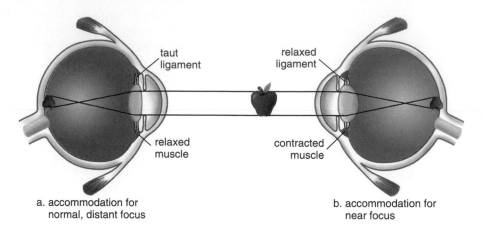

taut
ligament

relaxed
muscle

a. accommodation for
normal, distant focus

relaxed
ligament

contracted
muscle

b. accommodation for
near focus

Medical Focus

Corrective Lenses

*m*ost people can see what is designated as a size "20" letter 20 feet away and, therefore, are said to have 20/20 vision. People who can see close objects but cannot see the letters from this distance are said to be nearsighted. They often have an elongated eyeball, and when they attempt to look at a far object, the image is brought to focus in front of the retina. This condition is called **myopia** (mi-o′pe-ah) (fig. 9B). Nearsighted people need to wear concave lenses that diverge the light rays so that the image of a distant object can be focused on the retina.

Persons who can easily see the optometrist's chart but cannot see close objects are said to be farsighted. They often have a shortened eyeball, and when they try to look at a near object, the image is focused behind the retina. This condition is called **hyperopia** (hi″pĕ-ro′pe-ah) (fig. 9B). Farsighted people need to wear a convex lens to increase the bending of light rays so that the image of a close object will be focused on the retina.

With aging, the lens loses some of its elasticity and is unable to accommodate as well to bring objects into focus. This necessitates the wearing of **bifocals,** lenses that contain an upper part for distant vision and a lower part for near vision.

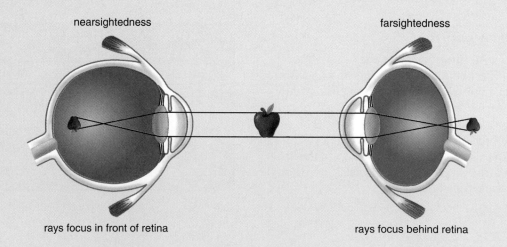

nearsightedness

farsightedness

rays focus in front of retina

rays focus behind retina

figure 9.B Nearsightedness (myopia) and farsightedness (hyperopia)

Mechanoreceptors

The mechanoreceptors are in the ears. The ear accomplishes two sensory functions: equilibrium (balance) and hearing. The receptors for both of these functions are located in the inner ear and consist of hair cells with cilia that respond to mechanical stimulation. Each hair cell has from 30 to 150 cilia. When the cilia of any particular hair cell are displaced in a certain direction, the cell generates nerve impulses, which are sent along the eighth cranial nerve to the brain.

Table 9.4 lists the parts of the ear, and figure 9.13 shows its structure. The ear has three divisions: external, middle, and inner.

figure 9.13 Human ear. **a.** Structure of the human ear. **b.** Close-up of the inner ear.

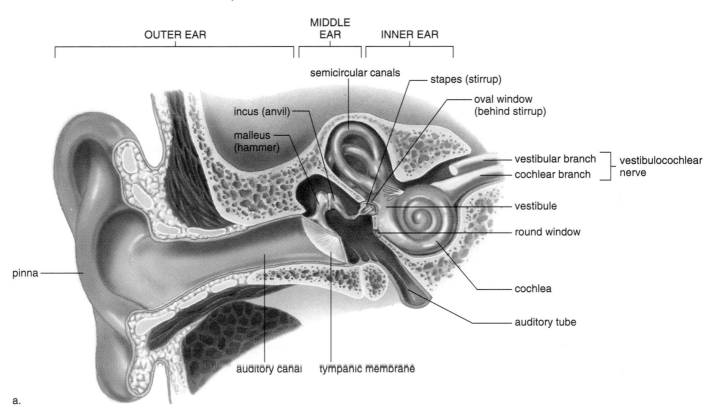

a.

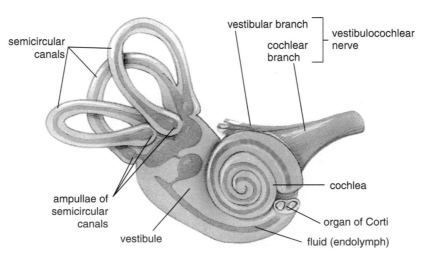

b.   **Inner ear**

table 9.4

Function of Parts of the Ear

Part	Function
External Ear	
Pinna	Collects sound waves
Auditory canal	Filters air
Middle Ear	
Tympanic membrane and ossicles	Amplify sound waves
Auditory tube	Equalizes air pressure
Inner Ear	
Vestibule (contains utricle and saccule)	Helps to maintain balance (static equilibrium)
Semicircular canals	Helps to maintain balance (dynamic equilibrium)
Cochlea	Transmits pressure waves that cause the organ of Corti to generate nerve impulses, resulting in hearing

External Ear

The **external ear** consists of the **pinna** (external flap) and external **auditory canal.** The opening of the auditory canal is lined with fine hairs and sweat glands. In the upper wall are ceruminous glands, modified sweat glands that secrete earwax, which helps guard the ear against the entrance of foreign materials, such as air pollutants.

Middle Ear

The **middle ear** begins at the **tympanic membrane** (eardrum) and ends at a bony wall with two small openings covered by membranes. These openings are called the **oval window** and the **round window.** Between the tympanic membrane and the oval window are three small bones called the **ossicles.** The ossicles are the **malleus** (mal′e-us), also called the hammer, **incus** (ing′kus), also called the anvil, and **stapes** (sta′pēz), also called the stirrup (fig. 9.13). The ossicles are named for the objects they resemble. The malleus adheres to the tympanic membrane, and the stapes touches the oval window. The posterior wall of the middle ear also has an opening that leads to mastoid sinuses of the skull.

The **auditory (eustachian) tubes** (fig. 9.13), which extend from the middle ear to the nasopharynx, permit equalization of air pressure. Chewing gum, yawning, and swallowing help move air through the auditory tubes during ascent and descent in elevators and airplanes.

Infections of the middle ear (*otitis media*) can occur frequently during childhood. As a precautionary measure to prevent perforation of the tympanic membrane, children with frequent infections may need to have an incision of

the tympanic membrane, called **myringotomy** (mir″in-got′o-me), followed by the insertion of a tiny tube into the membrane. The tube ensures that the pressure is equal on each side of the tympanic membrane, regardless of whether the auditory tube is blocked by the pus that results from the infection. With time, the tubes are sloughed out of the ears or surgically removed.

Inner Ear

The **inner ear** (fig. 9.13*b*), anatomically speaking, has three areas: the first two—a vestibule and the semicircular canals—are concerned with balance; the third—the cochlea—is concerned with hearing.

The **semicircular canals** are arranged so that there is one canal in each dimension of space. The base of each canal, called an **ampulla,** is slightly enlarged. Within the ampullae (fig. 9.13*b*), the cilia of small hair cells insert into a gelatinous medium.

A **vestibule,** or chamber, lies between the semicircular canals and the cochlea. It contains two small sacs called the **utricle** and the **saccule.** Within both of these are the cilia of small hair cells that protrude into a gelatinous substance. **Otoliths,** calcium carbonate granules, rest on this gelatinous material.

The spirals of the **cochlea** (kok′le-ah) resemble a snail shell. The cochlea contains the **organ of Corti** (kor′tē), which sends nerve impulses to the brain stem by way of the eighth cranial nerve. Eventually, the nerve impulses are relayed to the temporal lobe of the cerebrum, where they are interpreted as sound.

> The external ear, middle ear, and cochlea are necessary for hearing. The vestibule and semicircular canals are concerned with the sense of balance.

Hearing

The process of hearing begins when sound waves enter the auditory canal. Just as ripples travel across the surface of a pond, sound travels by the successive vibrations of molecules. Ordinarily, sound waves do not carry much energy, but when a large number of waves strike the eardrum, it moves back and forth (vibrates) slightly. The ossicles receive the vibrations from the eardrum and transmit them to the oval window, having amplified them about 20 times. Vibrations of the oval window cause pressure waves in the fluid of the cochlea.

The tubular cochlea has three canals: the vestibular canal, the **cochlear canal,** and the tympanic canal. Along the length of the basilar membrane, which forms the lower wall of the cochlear canal, are little hair cells whose cilia are embedded in another membrane, called the **tectorial membrane.** The hair cells of the cochlear canal compose the organ of Corti (fig. 9.14*a*).

figure 9.14 Organ of Corti. **a.** Enlarged cross section through the organ of Corti, showing the receptor hair cells from the side. **b.** Cochlea unwound, showing the placement of the organ of Corti along its length. The arrows represent the pressure waves that move from the oval window to the round window, causing the basilar membrane to vibrate and the cilia of a portion of the at least 20,000 hair cells to bend against the tectorial membrane so that they generate nerve impulses.

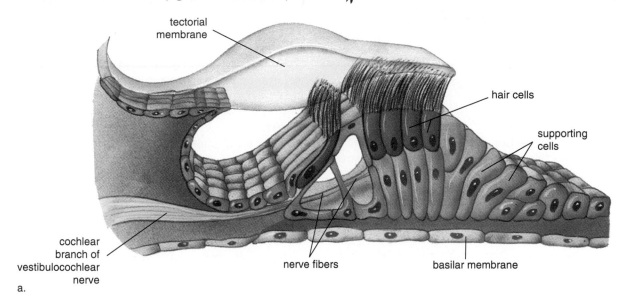

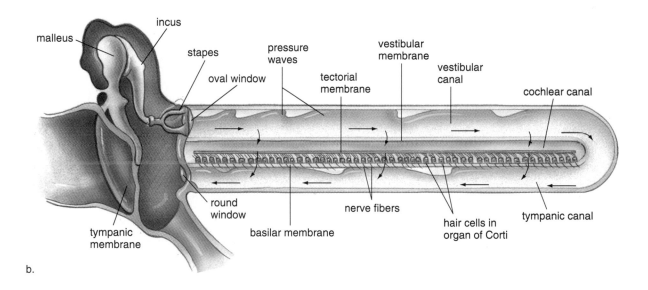

If the cochlea is unwound, as shown in figure 9.14*b*, the vestibular canal is seen to connect with the tympanic canal; therefore, as the figure indicates, pressure waves move from one canal to the other toward the round window, a membrane that can bulge to absorb the pressure. As a result of the movement of the fluid within the cochlea, the basilar membrane moves up and down, and the cilia of a portion of the at least 20,000 hair cells are bent. This bending of the cilia initiates nerve impulses that pass by way of the cochlear branch of the vestibulocochlear nerve to the brain stem. Eventually, the impulses reach the temporal lobe of the cerebrum, where they are interpreted as a sound. As discussed in the Medical Focus reading on page 182, deafness occurs when the process of hearing is interrupted.

The sense receptors for sound are hair cells on the basilar membrane (the organ of Corti). When the basilar membrane vibrates, the delicate hairs that are embedded in the tectorial membrane bend, and nerve impulses begin in the cochlear nerve and are transmitted to the brain.

Hearing Damage and Deafness

*t*wo major types of deafness are conduction deafness and nerve deafness. In **conduction deafness,** the ossicles tend to fuse together, restricting their ability to magnify sound waves. Conduction deafness can be caused by a congenital defect, particularly when a pregnant woman contracts German measles (rubella) during the first trimester of pregnancy. (For this reason, every female should be immunized against rubella before the childbearing years.) Conduction deafness can also be due to repeated infections or otosclerosis. With **otosclerosis** (o"to-sklĕ-ro'sis), the normal bone of the middle ear is replaced by vascular spongy bone.

Nerve deafness most often occurs when cilia on the receptors within the cochlea have worn away. Since this may happen with normal aging, older people are more likely to have trouble hearing. However, studies also suggest that age-associated hearing loss can be prevented if ears are protected from loud noises, starting even during infancy. Hospitals are now aware of the problem and are taking steps to ensure that neonatal intensive care units and nurseries are as quiet as possible.

In today's society, exposure to the types of noises listed in table 9A is common. Noise is measured in decibels, and any noise above 80 decibels could damage the hair cells of the organ of Corti. Since even listening to city traffic for extended periods can damage hearing, frequent attendance at rock concerts and constant loud music from a stereo are obviously dangerous. Noisy indoor or outdoor equipment, such as a rug-cleaning machine or a chain saw, are also troublesome. Even motorcycles and recreational vehicles, such as snowmobiles and motocross bikes, can contribute to a gradual hearing loss. Exposure to intense sounds of short duration, such as a burst of gunfire, can result in an immediate hearing loss. Since the butt of a rifle offers some protection, hunters may have a significant hearing reduction in the ear opposite the shoulder on which they carry their gun.

The first hint of a problem could be temporary hearing loss, a "full" feeling in the ears, muffled hearing, or tinnitus (ringing in the ears). If you have any of these symptoms, modify your listening habits immediately to prevent further damage. If exposure to noise is unavoidable, use specially designed noise-reduction earmuffs or purchase earplugs made from a compressible, spongelike material at a drugstore or sporting goods store. These earplugs are not the same as those worn for swimming, and they should not be used interchangeably.

Finally, people need to be aware that some medicines are **ototoxic** (damaging to any of the elements of hearing or balance). Anticancer drugs—most notably, cisplatin—and certain antibiotics (for example, streptomycin, kanamycin, and gentamicin) make ears especially susceptible to a hearing loss. Anyone taking such medications should protect ears from any loud noises.

Cochlear implants that directly stimulate the auditory nerve are available for persons with nerve deafness. However, they are costly, and people wearing these electronic devices report that the speech they hear is like that of a robot.

table 9A — Sound Intensity and Hearing Damage

Type of Noise	Sound Level (Decibels)	Effect
Rock concert, shotgun, jet engine	Over 125	Beyond threshold of pain; potential for hearing loss is high.
Nightclub, boom box, thunderclap	Over 120	Hearing loss is likely.
Chain saw, pneumatic drill, jackhammer, symphony orchestra, snowmobile, garbage truck, cement mixer	100–200	Regular exposure of longer than 1 minute risks permanent hearing loss.
Farm tractor, newspaper press, subway, motorcycle	90–100	Fifteen minutes of unprotected exposure is potentially harmful.
Lawnmower, food blender	85–90	Continuous daily exposure for more than 8 hours can cause hearing damage.
Diesel truck, average city traffic noise	80–85	Annoying; constant exposure may cause hearing damage.

Source: National Institute on Deafness and Other Communication Disorders, National Institutes of Health, January 1990.

figure 9.15 Organs of equilibrium. **a.** Rotational movement displaces the cilia of hair cells in ampullae of semicular canals. **b.** Bending the head displaces the cilia in the utricle and saccule.

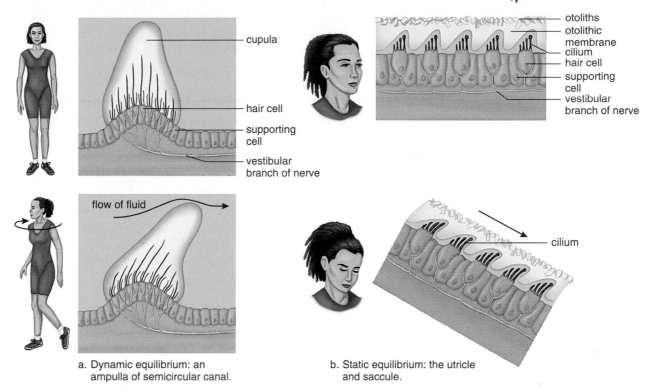

a. Dynamic equilibrium: an ampulla of semicircular canal.

b. Static equilibrium: the utricle and saccule.

Balance (Equilibrium)

Balance can be divided into two senses: *dynamic equilibrium,* requiring a knowledge of angular and/or rotational movement, and *static equilibrium,* requiring a knowledge of movement in one place, either vertical or horizontal.

Dynamic equilibrium is required when the body is moving (fig. 9.15*a*). During movement, the fluid within the semicircular canals flows over and displaces the gelatinous cupula within the ampullae. This causes the cilia of the hair cells to bend and to initiate nerve impulses that travel to the brain. Continuous movement of the fluid in the semicircular canals causes one form of motion sickness.

When the body is still, the otoliths in the utricle and the saccule rest on the otolithic membrane above the hair cells (fig. 9.15*b*). Static equilibrium is required when the body moves horizontally or vertically. At that time, the otoliths are displaced, and the otolithic membrane sags, bending the cilia of the hair cells beneath it. The hair cells then generate nerve impulses that travel to the brain via the vestibular branch of the vestibulocochlear nerve.

> Movement of fluid within the semicircular canals contributes to the sense of dynamic equilibrium. Movement of the otoliths within the utricle and the saccule is important for static equilibrium.

Effects of Aging

The need for eyeglasses increases with age. Also, three serious visual disorders are seen more frequently in older persons: (1) The lens of the eye does not accommodate as well and may also develop a cataract, a clouding of the lens that impairs vision; (2) age-related macular degeneration (see the MedAlert reading on page 174) is sometimes seen in older people; and (3) glaucoma is more likely to develop because of a reduction in the size of the anterior cavity of the eye (see fig. 9.8).

The need for a hearing aid also increases with age. *Otosclerosis,* an overgrowth of bone that causes the stapes to adhere to the oval window, is the most frequent cause of conduction deafness in adults. The condition actually begins during youth but may not become evident until later in life. Dizziness and the inability to maintain balance may be due to changes in the inner ear. With age, atrophy of the organ of Corti can lead to **presbycusis** (pres′be cus″ is age-related hearing decline). First, there is a loss in the ability to detect high-frequency tones, and then the lower tones are affected. Eventually, speech can be heard, but words cannot be detected.

> BodyWorks CD-ROM
> The module accompanying chapter 9 is Sensory Organs.

Basic Key Terms

accommodation (ah-kom″o-da′shun), p. 177
ampulla (am-pul′lah), p. 180
aqueous humor (a′kwe-us hu′mor), p. 172
auditory canal (aw′dĭ-to″re kah-nal), p. 180
auditory tube (aw′dĭ-to″re toob), p. 180
blind spot (blīnd spot), p. 173
choroid (ko′roid), p. 172
ciliary muscle (sil′e-er″e mus′el), p. 172
cochlea (kōk′le-ah), p. 180
cochlear canal (kōk′le-ar kah-nal′), p. 180
cone (kōn), p. 173
cornea (kor′ne-ah), p. 172
incus (ing′kus), p. 180
iris (i′ris), p. 172
lacrimal apparatus (lak′rĭ-mal ap″ah-ră′tus), p. 171
lens (lenz), p. 172
malleus (mal′e-us), p. 180
olfactory cell (ol-fak′to-re sel), p. 169
optic nerve (op′tik nerv), p. 173
organ of Corti (or′gan ov kor′tē), p. 180
ossicle (os′sĭ-k′l), p. 180
otolith (ō-′tō-lith), p. 180
pinna (pin′nah), p. 180
proprioceptor (pro″pre-o-sep′tor), p. 167
pupil (pu′pil), p. 172
retina (ret′ĭ-nah), p. 173
rod (rod), p. 173

saccule (sak′ūl), p. 180
sclera (skler′ah), p. 172
semicircular canal (sem″e-ser′ku-lar kah-nal′), p. 180
stapes (sta′pēz), p. 180
taste bud (tāst bud), p. 168
tympanic membrane (tim-pan′ik mem′brān), p. 180
utricle (u′trĭ-k′l), p. 180
vitreous humor (vit′re-us hu′mor), p. 172

Clinical Key Terms

bifocals (bi′fo-klz), p. 178
cataract (kat′ah-rakt), p. 177
cochlear implant (kōk′le-er im′plant), p. 182
color blindness (kul′er blind′nes), p. 176
conduction deafness (kon-duk′shun def′nes), p. 182
glaucoma (glau-ko′mah), p. 172
hyperopia (hi″pĕ-ro′pe-ah), p. 178
macular degeneration (mă′kyu-lar de″jen-er-a′shun), p. 174
myasthenia gravis (mi″as-the′ne-ah grah′vis), p. 171
myopia (mi-o′pe-ah), p. 178
myringotomy (mir″in-got′o-me), p. 180
nerve deafness (nerv def′nes), p. 182
otosclerosis (ō″tō-sklĕ-ro′sis), p. 182
ototoxic (ō″tō-tok′sik), p. 182
presbycusis (pres″bĭ-ku′sis), p. 183
sty (sti′), p. 170

Summary

I. General Receptors
General receptors are located in the skin, the visceral organs, and the muscles and joints.
  A. Skin. Sensory receptors in the human skin respond to temperature, touch, pressure, and pain.
  B. Visceral organs. Sensory receptors in the visceral organs aid in maintaining homeostasis. Many internal organs also have pain receptors.
  C. Muscles and joints. Sensory receptors in the muscles and joints, called proprioceptors, relay information on the position of body parts.

II. Chemoreceptors
The sensory receptors for taste (taste buds) and for smell (olfactory cells) work together to produce the senses of taste and smell.

III. Photoreceptors
  A. Accessory organs of the eye include the extrinsic muscles, which move the eye; the eyelids and eyelashes; and the lacrimal apparatus, which produces tears.
  B. The eye has three layers: the outer sclera, the middle choroid, and the inner retina.
  C. The retina contains the rods and cones, which are the sensory receptors for sight. When either is stimulated, nerve impulses begin and are transmitted via the optic nerve to the brain.
  D. The rods are responsible for vision in dim light. They do not detect fine detail or color, but they do detect motion.
  E. The cones are responsible for vision in bright light. They detect fine detail and color.
  F. The lens focuses light rays on the retina. The lens is flat for distant vision and rounds up for near vision. Inability of the lens to

accommodate as a person ages may necessitate the use of corrective lenses.

IV. **Mechanoreceptors**
   A. The external ear, middle ear, and cochlea are necessary for hearing. The vestibule and semicircular canals are concerned with the sense of balance.
   B. Movement of fluid within the semicircular canals contributes to the sense of dynamic equilibrium. Movement of the otoliths within the utricle and saccule is important for static equilibrium.
   C. The sensory receptors for sound are hair cells on the basilar membrane (the organ of Corti).

When the basilar membrane vibrates, the delicate hairs that touch the tectorial membrane bend, and nerve impulses begin in the cochlear nerve and are transmitted to the brain.

Study Questions

1. What type of sensory receptors are categorized as general, and what type are categorized as special? (p. 166)
2. Discuss the sensory receptors of the skin, visceral organs, and muscles and joints. (pp. 166–68)
3. Discuss the structure and function of the chemoreceptors. (pp. 168–69)
4. Describe the anatomy of the eye. (pp. 170–73)
5. Describe sight in dim light. What chemical reaction is responsible for vision in dim light? Explain color vision. (p. 176)
6. Explain focusing and accommodation. (pp. 177–78)
7. Describe the anatomy of the ear and how a person hears. (pp. 179–81)
8. Discuss the two major causes of deafness, including why young people frequently suffer loss of hearing. (p. 182)
9. Describe the role of the utricle, saccule, and semicircular canals in balance. (p. 183)

Objective Questions

Fill in the blanks.

1. The sensory organs for position and movement are called _____ .
2. Taste buds and olfactory cells are termed _____ because they are sensitive to chemicals in the air and food.
3. The sensory receptors for sight, the _____ and _____ , are located in the _____ , the inner layer of the eye.
4. The cones give us _____ vision and work best in _____ light.
5. The lens _____ for viewing close objects.
6. People who are nearsighted cannot see objects that are _____ . A _____ lens will restore this ability.
7. The ossicles are the _____ , _____ , and _____ .
8. The semicircular canals are involved in the sense of _____ .
9. The organ of Corti is located in the _____ canal of the _____ .
10. Vision, hearing, taste, and smell do not occur unless nerve impulses reach the proper portion of the _____ .

Medical Terminology Reinforcement Exercise

Consult Appendix B for help in pronouncing and analyzing the meaning of the terms that follow.

1. ophthalmologist (of″thal-mol′o-jist)
2. presbyopia (pres″be-o′pe-ah)
3. blepharoptosis (blef″ah-ro-to′sis)
4. keratoplasty (ker′ah-to-plas″te)
5. optometrist (op-tom′ĕ-trist)
6. lacrimator (lak′rĭ-ma″tor)
7. otitis media (o-ti′tis me′de-ah)
8. myringotomy (mir″in-got′o-me)
9. tympanocentesis (tim″pah-no-sen-te′sis)
10. microtia (mi″kro′she-ah)

Website Link

For a listing of the most current Websites related to this chapter, please visit the Mader home page at:

www.mhhe.com/maderap

chapter 10

The Endocrine System

Chapter Outline and Learning Objectives

After you have studied this chapter, you should be able to:

Endocrine System (p. 187)
- Define a hormone, and explain the mechanism of hormone action.
- Name the major endocrine glands, and identify their locations.
- Discuss control of glandular secretion by negative feedback.

Hypothalamus and Pituitary Gland (p. 190)
- Explain the anatomical and functional relationships between the hypothalamus and the pituitary gland.
- Name and discuss the two hormones produced by the hypothalamus that are secreted by the posterior pituitary.
- Discuss the physiological action of growth hormone (GH).
- Name six hormones produced by the anterior pituitary, and indicate which of these control other endocrine glands.
- Draw and explain the negative feedback mechanism.

Thyroid and Parathyroid Glands (p. 193)
- Discuss the anatomy of the thyroid gland, and the chemistry and physiological function of thyroxine.
- Discuss the function of parathyroid hormone (PTH) and calcitonin.

Adrenal Glands (p. 194)
- Describe the relationship between the adrenal medulla and the adrenal cortex.
- Discuss the function of the adrenal medulla and its relationship to the nervous system.
- Name three categories of hormones produced by the adrenal cortex, give an example of each category, and discuss their physiological action.

Pancreas (p. 197)
- Describe the microscopic anatomy of the pancreas.
- Name two hormones produced by the pancreas, and discuss their function.
- Discuss the two types of diabetes mellitus, and contrast hypoglycemia with hyperglycemia.

Testes and Ovaries (p. 199)
- Name the most important male and female hormones. Discuss their functions.

Other Endocrine Glands and Hormones (p. 199)
- State the location and function of the pineal gland and the thymus gland.
- Discuss atrial natriuretic hormone, growth factors, and prostaglandins as hormones not produced by glands.

Effects of Aging (p. 202)
- Anatomical and physiological changes occur in the endocrine systems as we age.

Working Together (p. 202)
- The endocrine system works with other systems of the body to maintain homeostasis.

Visual Focus
Hypothalamus and the Pituitary (p. 191)

Medical Focus
Glucocorticoid Therapy (p. 195)

MedAlert
Side Effects of Anabolic Steroids (p. 200)

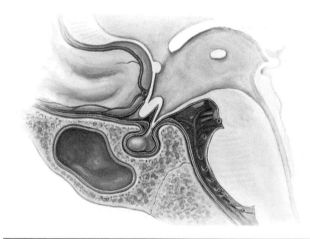

The pituitary gland is a pea-sized organ at the base of the brain whose secretions have a powerful effect on the body.

figure 10.1 Cellular activity of hormones. **a.** Peptide hormones combine with receptors located on the plasma membrane. This promotes the production of cyclic AMP, which, in turn, leads to activation of particular enzymes. **b.** Steroid hormones pass through the plasma membrane to bind with receptors; the hormone receptor complex moves into the nucleus and activates certain genes, leading to protein synthesis. ▣

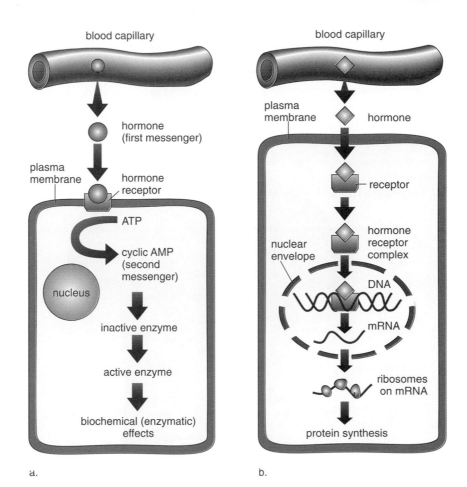

a.

b.

Endocrine System

The endocrine system consists of a number of glands located throughout the body that produce and secrete substances called hormones into the blood. A **hormone** is a chemical that influences or controls the activity of a specific tissue or organ. Like the nervous system, the endocrine system coordinates the functioning of body parts. The nervous system is fast acting and utilizes nerve impulses that travel rapidly along nerve fibers. The endocrine system is slower acting because it takes time to deliver hormones to target organs, but the action of hormones is longer lasting.

Hormone Action

Hormones can be divided into two categories: (1) peptides (used here to include amino acids, polypeptides, and protein hormones) and (2) steroids (complex rings of carbon and hydrogen atoms) (see fig. 2.10). Peptide hormones act as messengers that stimulate cells by binding to specific receptors in the plasma membrane (fig. 10.1a). The hormone receptor complex activates an enzyme that produces cyclic AMP (cAMP), a compound derived from ATP. Cyclic AMP, serving as a so-called second messenger, then activates other enzymes that carry out various cellular activities. Other peptide hormones utilize calcium (Ca^{2+}) as a second messenger. In contrast to a two-messenger system, steroid hormones (which are lipid-soluble) pass through the plasma membrane and bind to receptors present in the cytoplasm (fig. 10.1b). This hormone receptor complex moves into the nucleus, where the complex activates certain genes involved in protein synthesis.

Endocrine Glands

All hormones are carried throughout the body by the blood, but each one affects only a specific body part or parts, appropriately termed the target organ(s). The hormones discussed here are secreted by the endocrine glands

figure 10.2 Location of major endocrine glands in the body. Female sex organs are shown in the inset.

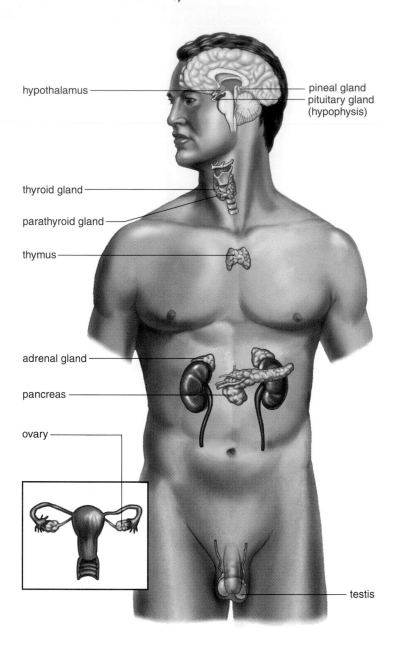

hypothalamus

pineal gland
pituitary gland
(hypophysis)

thyroid gland

parathyroid gland

thymus

adrenal gland

pancreas

ovary

testis

pictured in figure 10.2 and listed in table 10.1. Endocrine glands are ductless glands that secrete hormones directly into the blood. Exocrine glands, on the other hand, secrete their products into ducts that transport them to their destination. For example, the salivary glands secrete saliva into ducts that enter the mouth.

Hormones affect metabolism, appearance, and behavior. They help maintain homeostasis by keeping the fluid, electrolyte, and glucose balance of the blood within normal limits. Hormone secretion by endocrine glands is often regulated by negative feedback.

Negative Feedback

In a self-regulating negative feedback mechanism, an adaptive response dampens or even cancels the stimulus that brought about the response (see fig. 1.9). An endocrine gland can be sensitive to either the condition it is regulating or to the blood level of the hormone it is producing. For example, when blood is concentrated, the hypothalamus produces a hormone that causes blood dilution, which then is a signal to the hypothalamus to stop producing the hormone. On the other hand, the pituitary gland produces a hormone that stimulates the thyroid gland.

Part III Integration and Coordination

table 10.1

Principal Endocrine Glands and Their Hormones

Endocrine Gland	Hormone Released	Target Tissues/Organ	Chief Function of Hormone
Hypothalamus	Hypothalamic-releasing and release-inhibiting hormones	Anterior pituitary	Regulate anterior pituitary hormones
Posterior pituitary (storage of hypothalamic hormones)	Antidiuretic hormone (ADH), also known as vasopressin	Kidneys	Stimulates water reabsorption by kidneys
	Oxytocin	Uterus, mammary glands	Stimulates uterine muscle contraction and release of milk by mammary glands
Anterior pituitary	Growth hormone (GH), also known as somatotropin	Soft tissues, bones	Stimulates cell division, protein synthesis, and bone growth
	Prolactin (PRL)	Mammary glands	Stimulates milk production and secretion
	Thyroid-stimulating hormone (TSH)	Thyroid	Stimulates thyroid
	Adrenocorticotropic hormone (ACTH)	Adrenal cortex	Stimulates adrenal cortex
	Gonadotropic hormones	Gonads (testes and ovaries)	Control gamete and sex hormone production
Thyroid	Thyroxine	All tissues	Increases metabolic rate; helps to regulate growth and development
	Calcitonin	Bones, kidneys, intestine	Lowers blood calcium level
Parathyroids	Parathyroid hormone (PTH)	Bones, kidneys, intestine	Raises blood calcium level
Adrenal medulla	Epinephrine and norepinephrine	Cardiac and other muscles	Stimulate "fight-or-flight" reactions; raise blood glucose level
Adrenal cortex	Glucocorticoids* (e.g., cortisol)	All tissues	Raise blood glucose level
	Mineralocorticoids* (e.g., aldosterone)	Kidneys	Stimulate kidneys to reabsorb sodium and to excrete potassium
	Sex hormones	Sex organs, skin, muscles, bones	Stimulate development of secondary sex characteristics
Pancreas	Insulin	Liver, muscles, adipose tissue	Lowers blood glucose level
	Glucagon	Liver, muscles, adipose tissue	Raises blood glucose level
Testes	Androgens* (e.g., testosterone)	Sex organs, skin, muscles	Stimulate spermatogenesis; develop and maintain primary and secondary male sex characteristics
Ovaries	Estrogen and progesterone*	Sex organs, skin, muscles, bones	Stimulate oogenesis; develop and maintain primary and secondary female sex characteristics
Thymus	Thymosins	T lymphocytes	Stimulate maturation of T lymphocytes
Pineal gland	Melatonin	Various tissues	Involved in daily rhythms; possibly involved in maturation of sex organs

*Steroid hormones

When the blood level of a hormone produced by the thyroid rises, the pituitary gland no longer stimulates the thyroid gland.

The endocrine glands secrete their hormones into the bloodstream for transport to target organ cells, where the hormones are received by receptors. Hormone secretion by endocrine glands is often regulated by negative feedback.

Hypothalamus and Pituitary Gland

The **hypothalamus,** located beneath the thalamus in the lower walls and floor of the third ventricle of the brain, helps regulate the body's internal environment. For example, the hypothalamus helps control heart rate, body temperature, and water balance, as well as the activity of the pituitary gland.

The pituitary gland is small—about 1 centimeter in diameter—and lies just inferior to the hypothalamus (see fig. 10.2). It has two portions: (1) the **anterior pituitary,** or hypophysis, and (2) the **posterior pituitary.**

Posterior Pituitary

The posterior pituitary is connected to the hypothalamus by means of a stalklike structure. The hormones released by the posterior pituitary are made by neurosecretory cells in the hypothalamus. The hormones then migrate through axons that terminate in the posterior pituitary (fig. 10.3).

Antidiuretic (an″tǐ-di″u-ret′ik) **hormone (ADH),** also called *vasopressin,* promotes the reabsorption of water from the kidneys, thereby preventing dehydration. The hypothalamus is believed to contain cells that are sensitive to blood solute concentrations. When these cells detect that the blood lacks sufficient water, ADH is produced by special neurosecretory cells and is transported by their fibers to the posterior pituitary, where it is released (fig. 10.3). As the blood becomes more dilute, the hormone ceases to be produced and released.

Inability to produce ADH causes **diabetes insipidus** (watery urine), in which a person produces copious amounts of urine with a resultant loss of electrolytes from the blood. The condition can be corrected by the administration of ADH.

Oxytocin is another hormone made in the hypothalamus and released by the posterior pituitary. Oxytocin causes the uterus to contract and can be used to artificially induce labor. It also stimulates the release of milk from the breast when a baby is nursing.

figure 10.4 Effect of growth hormone. The amount of growth hormone production during childhood affects the height of an individual. Excessive growth hormone results in a tall stature and even giants. Little growth hormone results in limited stature and even pituitary dwarfism.

The hormones of the posterior pituitary—ADH and oxytocin—are produced in the hypothalamus.

Anterior Pituitary

The hypothalamus controls the anterior pituitary by producing hypothalamic-releasing and release-inhibiting hormones, which are transported to the anterior pituitary by the blood within a portal system. Each of these hypothalamic hormones causes the anterior pituitary either to secrete or to stop secreting a specific hormone. The anterior pituitary produces several different hormones (fig. 10.3).

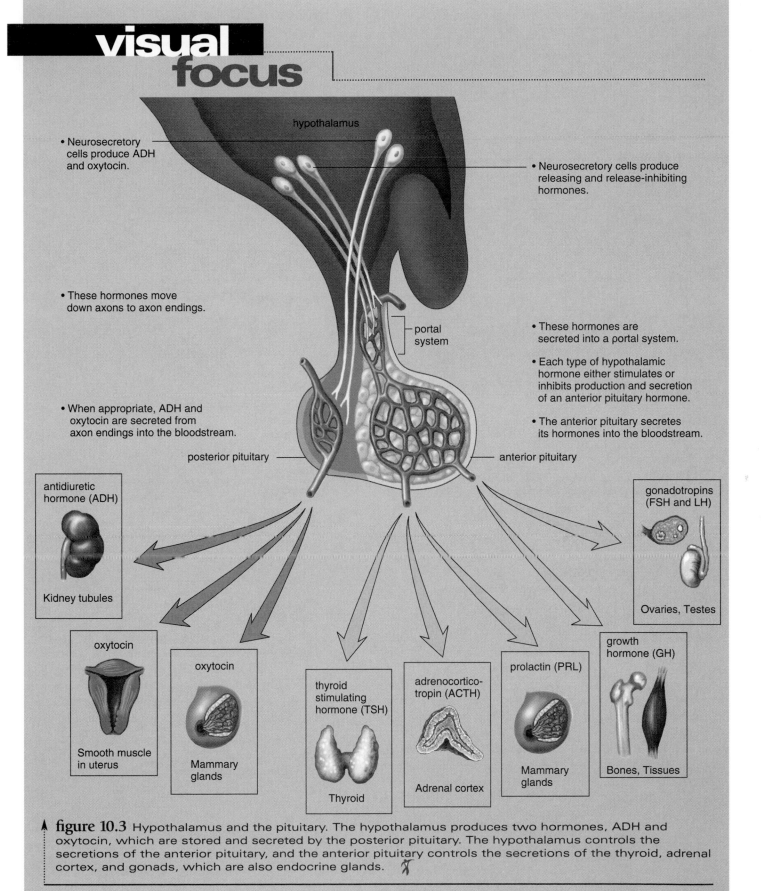

hypothalamus

- Neurosecretory cells produce ADH and oxytocin.

- Neurosecretory cells produce releasing and release-inhibiting hormones.

- These hormones move down axons to axon endings.

portal system

- These hormones are secreted into a portal system.

- Each type of hypothalamic hormone either stimulates or inhibits production and secretion of an anterior pituitary hormone.

- When appropriate, ADH and oxytocin are secreted from axon endings into the bloodstream.

- The anterior pituitary secretes its hormones into the bloodstream.

posterior pituitary

anterior pituitary

antidiuretic hormone (ADH)

Kidney tubules

gonadotropins (FSH and LH)

Ovaries, Testes

oxytocin

Smooth muscle in uterus

oxytocin

Mammary glands

thyroid stimulating hormone (TSH)

Thyroid

adrenocortico-tropin (ACTH)

Adrenal cortex

prolactin (PRL)

Mammary glands

growth hormone (GH)

Bones, Tissues

figure 10.3 Hypothalamus and the pituitary. The hypothalamus produces two hormones, ADH and oxytocin, which are stored and secreted by the posterior pituitary. The hypothalamus controls the secretions of the anterior pituitary, and the anterior pituitary controls the secretions of the thyroid, adrenal cortex, and gonads, which are also endocrine glands.

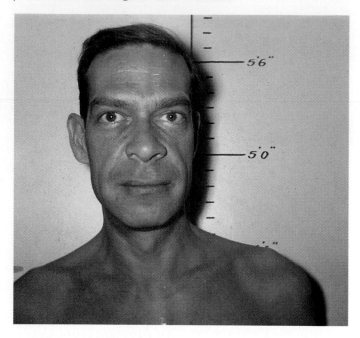

figure 10.5 Acromegaly. This condition is caused by overproduction of growth hormone in the adult. It is characterized by an enlargement of the bones in the face, the fingers, and the toes.

figure 10.6 Negative feedback in the regulation of the hypothalamus, anterior pituitary, and thyroid. **a.** The level of thyroid-stimulating hormone (TSH) exerts feedback control over the hypothalamus. **b.** The level of thyroxine exerts feedback control over the anterior pituitary. **c.** The level of thyroxine exerts feedback control over the hypothalamus. In this way, thyroxine controls its own secretion. Cortisol and sex hormone levels are controlled in similar ways.

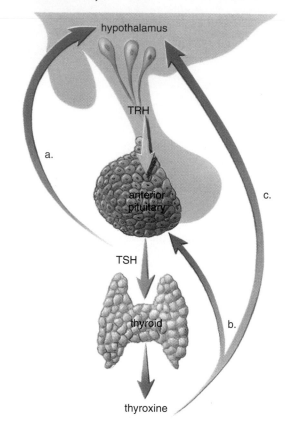

Growth hormone (GH), or somatotropin, produced by the anterior pituitary, affects the physical appearance dramatically since it determines an individual's size and height. If little or no GH is secreted by the anterior pituitary during childhood, a person can become a **pituitary dwarf,** characterized by perfect proportions but small stature (fig. 10.4). If too much GH is secreted, a person can become a giant. Giants usually have poor health, primarily because GH has a secondary effect on the blood sugar level, promoting an illness called diabetes mellitus.

GH is produced in greatest quantities during childhood and adolescence, when most body growth is occurring, but is still produced (though in lower quantities) in adults to aid in continued protein synthesis and normal cell division and replacement. If GH production increases in an adult after full height has been obtained, only the bones of the jaw, eyebrow ridges, nose, fingers, and toes respond. When these bones begin to grow, the person acquires a slightly grotesque look, with huge fingers and toes. This condition is called **acromegaly** (fig. 10.5).

Prolactin (PRL) is produced by the anterior pituitary only after childbirth. It causes the mammary glands in the breasts to develop and produce milk.

The anterior pituitary also secretes the hormones that follow. Since these hormones have an effect on other endocrine glands, the anterior pituitary is sometimes called the master gland.

1. **Thyroid-stimulating hormone (TSH),** which stimulates the thyroid to produce thyroxine
2. **Adrenocorticotropic** (ad-re"no-kor"te-ko-trop'ik) **hormone (ACTH),** which stimulates the adrenal cortex to produce and secrete hormones
3. **Gonadotropic** (gon"ah-do-trōp'ik) **hormones,** which stimulate the gonads—the testes in males and the ovaries in females—to secrete sex hormones

A three-tiered relationship exists between the hypothalamus, anterior pituitary, and other endocrine glands. The hypothalamus produces releasing hormones that control the anterior pituitary, which produces hormones that control the thyroid, adrenal cortex, and gonads. Figure 10.6 illustrates the negative feedback mechanism that controls the activity of these glands.

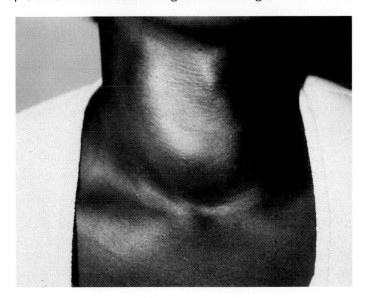

figure 10.7 Simple goiter. An enlarged thyroid gland is often caused by a lack of iodine in the diet. Without iodine, the thyroid is unable to produce thyroxine, and continued anterior pituitary stimulation causes the gland to enlarge.

figure 10.8 Exophthalmic goiter (Graves' disease). In hyperthyroidism, the eyes protrude because of edema in the tissues of the eye sockets.

Thyroid and Parathyroid Glands

The thyroid gland (see fig. 10.2) is located in the neck and is attached to the trachea just inferior to the larynx. The parathyroid glands are embedded in the posterior surface of the thyroid gland.

Thyroid Gland

The thyroid gland is composed of a large number of follicles, each a small, spherical structure made of thyroid cells and filled with stored thyroxine. **Thyroxine,** one hormone produced by the thyroid gland, occurs in two forms. Thyroxine is usually secreted as T_4 (tetraiodothyronine), which contains four iodine atoms, but eventually this form is converted to T_3 (triiodothyronine), the active form of the hormone. Iodine, which is required for thyroxine production, is actively transported into the thyroid gland, where its concentration may be as much as 25 times that found in the blood. If iodine is lacking in the diet, the thyroid gland enlarges, producing a **simple goiter** (fig. 10.7). The use of iodized salt helps prevent such a condition.

Figure 10.6 clearly demonstrates the cause of thyroid enlargement. When the level of thyroxine in the blood is low (called hypothyroidism), the anterior pituitary produces TSH, which stimulates the thyroid. If iodine is not present in the thyroid, the level of thyroxine will remain the same, and TSH will continue to stimulate the thyroid. The result is hypertrophy of the gland, called a simple goiter.

Thyroxine increases the metabolic rate. It does not have one target organ; instead, it stimulates most of the cells of the body to metabolize at a faster rate. For example, it causes more glucose to be broken down.

Failure of the thyroid to develop properly results in a condition called **cretinism** (kre'tĭ-nizm). Cretins are short and stocky, and have had extreme hypothyroidism since childhood or infancy. Thyroxine therapy can initiate growth, but unless treatment is begun within the first two months of birth, mental retardation results. Hypothyroidism in adults produces the condition known as **myxedema** (mik"sĕ-de'mah), which is characterized by lethargy, weight gain, loss of hair, slower pulse rate, lowered body temperature, and thickness and puffiness of the skin. The administration of adequate doses of thyroxin restores normal function and appearance.

In the case of hyperthyroidism, or **Graves'** (grāvz) **disease,** the thyroid gland is enlarged and overactive, causing a goiter to form and the eyes to protrude because of edema in eye socket tissues and swelling of extrinsic eye muscles. This type of goiter is called **exophthalmic** (ek"sof-thal'mik) **goiter** (fig. 10.8). The patient usually becomes hyperactive, nervous, and irritable, and suffers from insomnia. Hyperthyroidism can also be caused by a thyroid tumor, which is usually detected as a lump during physical examination. The treatment for hyperthyroidism is surgery in combination with administration of radioactive iodine. The prognosis for most patients is excellent.

figure 10.9 Control of calcium (Ca^{2+}) levels. The opposing actions of parathyroid hormone and calcitonin on bone breakdown maintain the homeostatic level of calcium in the blood. See figure 10.2 for the location of the thyroid gland in the body.

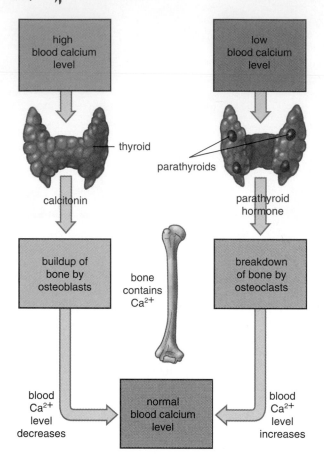

Calcitonin

In addition to thyroxine, the thyroid gland also produces the hormone **calcitonin** (kal″sĭ-to′nin), which helps regulate the calcium level in the blood and opposes the action of parathyroid hormone. (The interaction of these two hormones is discussed in the next section.) Calcitonin lowers blood calcium by increasing the buildup of bone (fig. 10.9).

The anterior pituitary produces TSH, a hormone that promotes the production of thyroxine by the thyroid. Thyroxine, which increases metabolism, can affect the entire body, as exemplified by cretinism and myxedema. The thyroid also produces calcitonin, which lowers the blood calcium level.

Parathyroid Glands

Many years ago, the four, small parathyroid glands were sometimes mistakenly removed during thyroid surgery. **Parathyroid hormone (PTH),** the hormone produced by the parathyroid glands, causes the calcium (Ca^{2+}) level in the blood to increase and the phosphate (HPO_4^{2-}) level to decrease. PTH promotes bone breakdown and calcium retention by the kidneys, and activates vitamin D, which, in turn, stimulates the absorption of calcium from the intestine. It also promotes the kidneys' excretion of phosphate in urine.

Parathyroid hormone inhibits the activity of osteoblasts and promotes the activity of osteoclasts in bone, thereby raising the blood calcium level. Calcitonin has the opposite effect, and therefore, the homeostatic balance of calcium in the blood is achieved through the action of both hormones (fig. 10.9).

If insufficient parathyroid hormone is produced, the blood calcium level drops, resulting in **tetany** (tet′ah-ne). In tetany, the body shakes from continuous muscle contraction. The effect is actually brought about by increased excitability of the nerves, which fire spontaneously and without rest. Calcium plays an important role in both nervous conduction and muscle contraction. It is also necessary to blood clotting.

PTH maintains a high blood calcium level by promoting calcium absorption in the intestine, calcium retention by the kidneys, and bone breakdown. These actions are opposed by calcitonin produced by the thyroid gland.

Adrenal Glands

The adrenal glands, as their name implies (*ad* means near; *renal* means kidneys), lie atop the kidneys (fig. 10.10). Each consists of an outer portion, called the *cortex*, and an inner portion, called the *medulla*. These portions, like the anterior and posterior pituitaries, have no connection with one another.

The hypothalamus exerts control over the activity of both portions of the adrenal glands. It can initiate nerve impulses that travel by way of the brain stem, spinal cord, and sympathetic nerve fibers to the adrenal medulla, which then secretes its hormones. The hypothalamus, by means of corticotrophin-releasing hormone, controls the anterior pituitary's secretion of adrenocorticotropic hormone (ACTH), which, in turn, stimulates the adrenal cortex. Stress of all types, including both emotional and physical trauma, prompts the hypothalamus to stimulate the adrenal glands to release hormones.

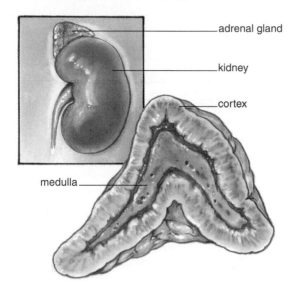

figure 10.10 Location and structure of adrenal glands. See figure 10.2 for the location of the adrenal glands in the body.

- adrenal gland
- kidney
- cortex
- medulla

Adrenal Medulla

The adrenal medulla, which is under the control of the sympathetic division of the autonomic nervous system, produces **epinephrine** (ep"ĭ-nef'rin), also called adrenaline, and **norepinephrine** (nor"ep-ĭ-nef'rin), also called noradrenaline. These hormones are responsible for the "fight-or-flight" reaction that occurs in times of emergency. Epinephrine and norepinephrine bring about these reactions:

- Blood glucose level rises, and the metabolic rate increases.
- Bronchioles dilate, and breathing rate increases.
- Blood vessels to the digestive tract constrict; those to the skeletal muscles dilate.
- Cardiac muscle contraction is more forceful, and heart rate increases.

The adrenal medulla releases epinephrine and norepinephrine into the bloodstream, helping the body cope with situations that seem to threaten survival.

Adrenal Cortex

The adrenal cortex produces two major types of hormones: (1) the glucocorticoids, which help regulate the level of glucose in the blood; and (2) mineralocorticoids, which help regulate the level of minerals in the blood. It also secretes a small amount of male sex hormones and a small amount of female sex hormones in both sexes; that is, the adrenal cortex produces both male and female sex hormones in males, and both male and female sex hormones in females.

Glucocorticoids

Cortisol is responsible for the greatest amount of glucocorticoid activity. Cortisol promotes the hydrolysis of muscle protein to amino acids, which enter the blood. This leads to a higher blood glucose level when the liver converts these amino acids to glucose. Cortisol also favors metabolism of fatty acids rather than carbohydrates. In opposition to insulin (a pancreatic hormone to be discussed shortly), cortisol raises the blood glucose level. Cortisol also counteracts the inflammatory response that leads to joint pain and swelling in arthritis and bursitis. The administration of cortisol aids these conditions because it reduces inflammation (see the Medical Focus reading on this page).

Cortisol, a glucocorticoid secreted by the adrenal cortex, raises the blood glucose level.

Medical Focus

Glucocorticoid Therapy

Glucocorticoid preparations have been used for allergies, severe asthma, rheumatoid arthritis, lupus erythematosus, ulcerative colitis, acute rheumatic fever, organ transplant rejection, and brain edema, among other illnesses. However, long-term administration of glucocorticoids for therapeutic purposes will cause some degree of Cushing syndrome. In addition, sudden withdrawal from glucocorticoid therapy causes symptoms of diminished secretory activity by the adrenal cortex. This occurs because glucocorticoids suppress the release of adrenocorticotropic hormone (ACTH) by the anterior pituitary and lead to a decrease in glucocorticoid production by the adrenal cortex. Therefore, withdrawal of glucocorticoids following long-term use must be tapered. During an alternate-day schedule, the dosage is gradually reduced and then finally discontinued, as the patient's adrenal cortex resumes activity.

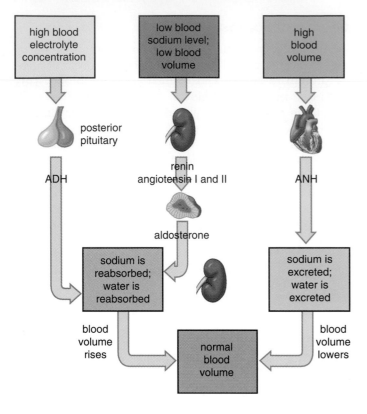

figure 10.11 Blood volume maintenance. Normal blood volume is maintained by ADH (antidiuretic hormone) and aldosterone, whose actions raise blood volume, and by ANH (atrial natriuretic hormone), whose action lowers blood volume.

Mineralocorticoids

Aldosterone (al″dos′ter-ōn) is the most important of the mineralocorticoids. These hormones maintain the electrolyte (ion) concentration in blood and, therefore, other body fluids. Aldosterone's primary target organ is the kidney, where it promotes renal absorption of sodium (Na^+) and renal excretion of potassium (K^+). The levels of sodium and potassium in the blood are critical for nerve conduction and muscle contraction; in fact, cardiac failure may result from too low a level of potassium.

The secretion of mineralocorticoids is not under the control of the anterior pituitary. When the blood volume and blood sodium level is low, the kidneys secrete the enzyme **renin**. Renin converts the plasma protein angiotensinogen to angiotensin I, which is changed to angiotensin II by a converting enzyme found in the lungs. Angiotensin II stimulates the adrenal cortex to release aldosterone. The effect of this system, called the renin-angiotensin-aldosterone system, is to raise the blood volume and pressure in two ways. First, angiotensin II con-

stricts the arterioles directly, and second, aldosterone causes the kidneys to reabsorb sodium. When the blood sodium level rises, water is reabsorbed, and blood volume and pressure are maintained.

Two other hormones play a role in the homeostatic maintenance of blood volume. As discussed earlier, antidiuretic hormone (ADH) helps increase blood volume by causing the kidney to reabsorb water. Also, when the atria of the heart are stretched due to increased blood volume, cardiac cells release a hormone called **atrial natriuretic** (a′trē-al na″trē-u-ret′ik) **hormone (ANH)**, which inhibits renin secretion by the kidneys and aldosterone secretion from the adrenal cortex. The effect of ANH, therefore, is to cause sodium excretion—that is, *natriuresis*. When sodium is excreted, so is water, and therefore, blood volume and blood pressure decrease (fig. 10.11).

> Aldosterone, a mineralocorticoid secreted by the adrenal cortex, and ADH raise blood volume by causing the kidneys to reabsorb Na^+ and water. Their action is opposed by ANH, and in this way, normal blood volume is maintained.

Disorders of the Adrenal Cortex

Addison Disease A person with a low level of adrenal cortex hormones due to hyposecretion develops **Addison disease.** Typically, symptoms include a peculiar bronzing of the skin. Because the lack of cortisol results in a potentially severe drop in blood glucose level, the individual is highly susceptible to any kind of stress due to an insufficient energy supply. Even a mild infection can cause death. Due to the lack of aldosterone, the blood sodium level is low, and the person experiences low blood pressure and possibly severe dehydration. Left untreated, Addison disease can be fatal.

Cushing Syndrome A person with a high level of adrenal cortex hormones due to hypersecretion develops **Cushing syndrome.** Excess cortisol causes a tendency toward diabetes mellitus, a decrease in muscular protein, and an increase in subcutaneous fat. Because of these effects, the person usually has an obese trunk, while the arms and legs remain normal. Due to the high level of sodium in the blood, the blood is basic (pH greater than normal), hypertension occurs, and there is edema of the face, which gives it a moonlike shape. Masculinization may occur in women due to oversecretion of the adrenal male sex hormone.

> Addison disease is due to adrenal cortex hyposecretion. Cushing syndrome is due to adrenal cortex hypersecretion.

figure 10.12 Gross and microscopic anatomy of the pancreas. The pancreas lies in the abdomen, extending from the C-shaped curvature of the duodenum to the left. As an exocrine gland, it secretes digestive enzymes that enter the duodenum by way of a duct utilized also by bile. As an endocrine gland, the pancreatic islets (of Langerhans) secrete insulin and glucagon into the bloodstream.

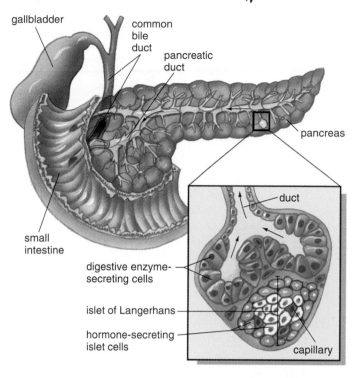

Pancreas

The pancreas is a long, soft organ that lies transversely in the abdomen (fig. 10.12) between the kidneys and near the duodenum of the small intestine. It is composed of two types of tissues: One produces and secretes the digestive juices that are carried by the ducts to the small intestine, and the other, called the **pancreatic islets (of Langerhans)**, produces and secretes the hormones insulin and glucagon directly into the blood. Insulin is secreted by beta cells, and glucagon is secreted by alpha cells.

Insulin is secreted when the blood glucose level is high, which usually occurs immediately after eating. Insulin has three different actions: (1) It stimulates all cells, and in particular fat, liver, and muscle cells, to absorb and metabolize glucose; (2) it stimulates the liver and muscles to store glucose as glycogen; and (3) it promotes the buildup of fats and proteins, and inhibits their use as an energy source so that they will be available during leaner times. As a result of its activities, insulin lowers the blood glucose level.

Glucagon is secreted from the pancreas between meals, and its effects are opposite those of insulin. Glucagon stimulates the breakdown of glycogen and raises the blood glucose level (fig. 10.13).

> The pancreas produces and secretes the hormones insulin and glucagon. Insulin lowers the blood glucose level, while glucagon raises the blood glucose level.

Diabetes Mellitus

Diabetes mellitus (sugar diabetes) is the most common illness due to hormonal imbalance. Symptoms include: sugar in the urine; frequent, copious urination; abnormal thirst; **polyphagia** (pŏ"le-fa'je-ah), which is excessive eating; rapid weight loss; general weakness; drowsiness and fatigue; itching of the genitals and skin; visual disturbances and blurring; and skin disorders, such as boils, carbuncles, and infection.

Many of these symptoms develop because cells are not metabolizing sugar. The liver fails to store glucose as glycogen, and all of the cells fail to utilize glucose as an energy source. This means that the blood glucose level is very high after eating, causing glucose to be excreted in the urine. Because more water than usual is excreted, the diabetic is extremely thirsty.

Since carbohydrates are not being metabolized, the body turns to the breakdown of proteins and fat for energy. Unfortunately, the breakdown of these molecules leads to the buildup of ketones. **Ketones** are acidic

figure 10.13 Regulation of blood glucose level. When the blood glucose level is high, the pancreas secretes insulin. Insulin promotes the storage of glucose as glycogen and the synthesis of proteins and fats (as opposed to their use as energy sources). Therefore, insulin lowers the blood glucose level to normal. When the blood glucose level is low, the pancreas secretes glucagon. Glucagon acts opposite to insulin; therefore, glucagon raises the blood glucose level to normal.

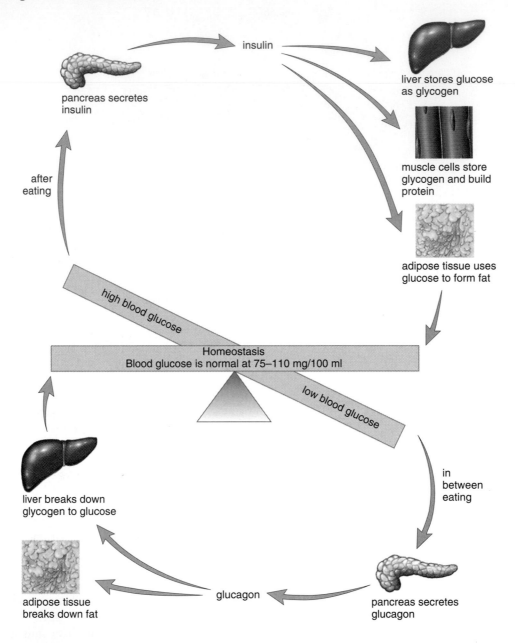

insulin

pancreas secretes insulin

liver stores glucose as glycogen

muscle cells store glycogen and build protein

adipose tissue uses glucose to form fat

after eating

high blood glucose

Homeostasis
Blood glucose is normal at 75–110 mg/100 ml

low blood glucose

in between eating

liver breaks down glycogen to glucose

adipose tissue breaks down fat

glucagon

pancreas secretes glucagon

molecules, and their buildup in the body causes an increase in acid levels (**acidosis**). The reduction in blood volume due to production of copious urine and the acidosis due to ketones can eventually lead to coma and death. The symptoms of **hyperglycemia** (high blood sugar) develop slowly, and intervention can reverse symptoms.

There are two types of diabetes mellitus. In *type I diabetes mellitus* (insulin-dependent diabetes mellitus [IDDM]), the pancreas does not produce insulin, requiring the diabetic to have daily insulin injections. These injections control the symptoms but can still cause inconveniences. Either an overdose of insulin or the absence of regular eating can induce symptoms of **hypoglycemia** (low blood sugar) because the blood sugar level is below normal. Since the brain requires a constant supply of sugar, unconsciousness results. The cure for low blood sugar is quite simple: An immediate source of sugar, such as a sugar cube or fruit juice, can counteract hypoglycemia immediately. Obviously, insulin injections are not the same as a fully functioning pancreas that responds on demand to a high glucose level by supplying insulin. For this reason, some doctors advocate a pancreatic islet transplant for type I diabetes.

Of the over 12 million people in the United States who now have diabetes, at least 11 million have *type II - diabetes mellitus* (non-insulin-dependent diabetes mellitus [NIDDM]). Risk of developing type II diabetes has a genetic component, and obesity greatly increases the risk. In type II diabetes, the pancreas produces insulin, but the cells do not respond to it. At first, the cells lack the receptors necessary to detect the presence of insulin; later, the cells are incapable of taking up glucose. If type II diabetes is left untreated, the results can be as serious as those for type I diabetes. Diabetics are prone to blindness, kidney disease, and circulatory disorders, including strokes. Pregnancy carries an increased risk of diabetic coma, and the child of a diabetic is somewhat more likely to be stillborn or to die shortly after birth. For these reasons, type II diabetes should be prevented or at least controlled by diet and regular exercise. If diet and exercise fail to control it, oral drugs are available that make the cells more sensitive to the effects of insulin or that stimulate the pancreas to produce more insulin.

Gestational diabetes mellitus (GDM) is diabetes that sometimes occurs during pregnancy. It is probably due to the presence of placental hormones and is expected to disappear once the child is born.

> The most common illness due to hormonal imbalance is diabetes mellitus, caused by a lack of insulin or an insensitivity of cells to insulin.

Testes and Ovaries

The sex organs are the testes, located in the male scrotum, and the ovaries, located in the female pelvic cavity. As will be discussed again in chapter 17, the testes produce the **androgens** (for example, testosterone), which are the male sex hormones, and the ovaries produce estrogen and progesterone, the female sex hormones. The hypothalamus and pituitary gland control the hormonal secretions of these organs in the same manner as described for the thyroid gland in figure 10.6.

Testosterone

The male sex hormone, **testosterone,** is essential for the normal development and functioning of the male sex organs. It is also necessary for the maturation of sperm.

Greatly increased testosterone secretion at the time of puberty stimulates the growth of the penis and the testes. Testosterone also brings about and maintains the male secondary sex characteristics that develop at the time of puberty, such as a beard, axillary (underarm) hair, and pubic hair. It prompts the larynx and vocal cords to enlarge, causing the voice to change. Testosterone is responsible for the greater muscular strength of males, which is why some athletes take supplemental amounts of *anabolic steroids,* which are either testosterone or similar chemicals. But testosterone also promotes closure of the epiphyses of long bones and therefore stops growth. Other side effects of taking anabolic steroids are listed in the MedAlert reading on page 200. Testosterone also causes sebaceous and sweat glands in the skin to secrete; therefore, it is largely responsible for acne and body odor. Another effect of testosterone activity is **pattern baldness.** Genes for baldness probably are inherited by both sexes, but baldness is seen more often in males because of the presence of testosterone.

Testosterone is believed to be largely responsible for the sex drive. It may even contribute to the supposed aggressiveness of males.

Estrogen and Progesterone

The female sex hormones, **estrogen** and **progesterone,** have many effects on the body. In particular, estrogen secreted at the time of puberty stimulates the growth of the uterus and the vagina. Estrogen is necessary for egg maturation and is largely responsible for female secondary sex characteristics, such as female body hair and fat distribution. In general, females have a more rounded appearance than males because of a greater accumulation of fat beneath the skin. Also, the pelvic girdle enlarges in females so that the pelvic cavity has a larger relative size compared to males; this means that females have wider hips. Estrogen also promotes closure of the epiphyses of long bones and therefore stops growth. Both estrogen and progesterone are also required for breast development and regulation of the uterine cycle, which includes monthly *menstruation* (discharge of blood from the uterus).

> The androgens, primarily testosterone, are the male sex hormones produced by the testes. Estrogen and progesterone are the female sex hormones produced by the ovaries. The sex hormones maintain the sex organs and the secondary sex characteristics.

Other Endocrine Glands and Hormones

Other glands in the body also produce hormones, and we will discuss two of these. The **pineal** (pīn′e-al) **gland** is a cone-shaped gland located in the roof of the brain's third ventricle (see fig. 10.2). It is smaller than the pituitary gland and decreases in size as a person ages. In the adult, it becomes a thickened strand of fibrous tissue.

The pineal gland secretes the hormone **melatonin,** particularly at night, which is believed to regulate daily rhythms, such as a person's sleep pattern. An injection of melatonin can induce sleep. It also inhibits the secretion of the gonadotropic hormones FSH and LH; therefore, excessive

MedAlert

Side Effects of Anabolic Steroids

*a*nabolic steroids are synthetic forms of the male sex hormone testosterone. These drugs were developed in the 1930s for medical reasons. They prevent muscular atrophy in patients with debilitating illness, they speed recovery in surgery and burn patients, and they are helpful in rare forms of anemia and breast cancer.

Trainers may have been the first to acquire anabolic steroids for weight lifters, bodybuilders, and other athletes, such as professional football players. When taken in large doses (10 to 100 times the amount prescribed by doctors for illnesses), anabolic steroids promote larger muscles when the abuser also exercises. The end result in muscle cells is increased amounts of the proteins actin and myosin. Occasionally, steroid abuse makes the news because an Olympic winner tests positive for the drug and must relinquish a medal. Steroid use has been outlawed by the International Olympic Committee.

Being a steroid user can have serious detrimental effects. Men often experience decreased sperm counts and a decrease of sexual desire due to atrophy of the testicles. Some develop an enlargement of the prostate gland or grow breasts. On the other hand, women can develop male sexual characteristics. Some cease ovulating or menstruating, sometimes permanently. They grow hair on their chests and faces, and lose hair from their heads; many experience abnormal enlargement of the clitoris.

It appears that steroids upset the body's hormonal balance of testosterone based on negative feedback. If the level of testosterone is low, the hypothalamus signals the anterior pituitary gland to trigger increased production. If the level of testosterone is high, as it is in the case of steroid abusers, the hypothalamus signals the pituitary to stop production. After athletes stop taking steroids, the hypothalamus fails to start the system again.

There are several other health risks associated with steriod use. Steroids have even been linked to heart disease in both sexes and implicated in the deaths of young athletes from liver cancer and a type of kidney tumor. Steroids can cause the body to retain fluid, which results in increased blood pressure. Users then try to get rid of "steroid bloat" by taking large doses of diuretics. A young California weight lifter had a fatal heart attack after using steroids, and the postmortem showed a lack of electrolytes, chemicals that help regulate the heart, from taking diuretics.

The Federal Drug Administration bans most steroids, but they are brought into the United States illegally and sold through the mail or in gyms and health clubs. According to federal officials, 1–3 million Americans now take anabolic steroids. Of great concern is their increased use by teenagers wishing to build bulk quickly, possibly due to society's emphasis on physical appearance and adolescents' need to feel better about how they look.

The many and varied harmful effects of anabolic steroids are listed in figure 10A. Some researchers predict that two or three months of high-dosage use of anabolic steroids as a teen can cause death by age 30 or 40. Unfortunately, these drugs also increase aggression and make a person feel invincible. One abuser even had his friend videotape him as he drove his car at 40 miles an hour into a tree!

Questions

1. In what way does steroid abuse decrease the chances of a male having children?
2. In what way does steroid abuse have a masculinizing effect in women regarding distribution of hair?

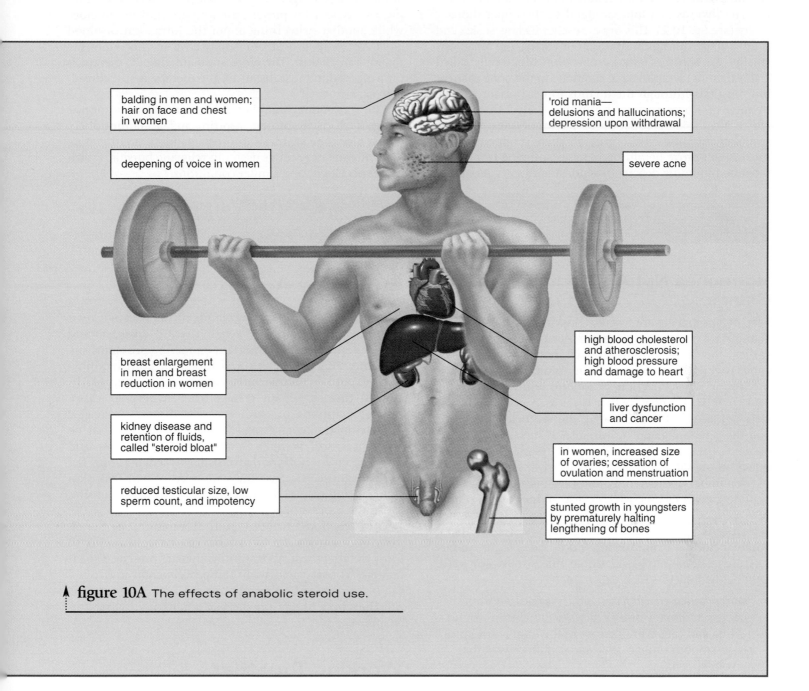

balding in men and women; hair on face and chest in women

deepening of voice in women

'roid mania— delusions and hallucinations; depression upon withdrawal

severe acne

breast enlargement in men and breast reduction in women

kidney disease and retention of fluids, called "steroid bloat"

reduced testicular size, low sperm count, and impotency

high blood cholesterol and atherosclerosis; high blood pressure and damage to heart

liver dysfunction and cancer

in women, increased size of ovaries; cessation of ovulation and menstruation

stunted growth in youngsters by prematurely halting lengthening of bones

figure 10A The effects of anabolic steroid use.

amounts of melatonin inhibit the ovarian and uterine cycles.

The **thymus** is a lobular gland in the upper thoracic cavity (see fig. 10.2). This organ reaches its largest size and is most active during childhood. With aging, the organ gets smaller and becomes fatty. Certain white blood cells, called T (for thymus) lymphocytes, originate in the bone marrow but must pass through the thymus to reach maturity. The thymus produces various hormones called **thymosins,** which aid the differentiation of T lymphocytes and may stimulate immune cells in general. There is hope that thymosins will prove helpful in patients suffering from AIDS (acquired immune deficiency syndrome).

> The pineal gland secretes melatonin, which is believed to regulate daily rhythms. The thymus gland secretes thymosins necessary to immunity.

Hormones Not Associated with Glands

Even organs that are not usually considered to be endocrine glands have been found to secrete hormones. As discussed earlier, the heart produces atrial natriuretic hormone (ANH), which helps regulate blood volume and pressure by promoting renal excretion of sodium and water (see fig. 10.11). ANH is a peptide that is released not only by the atria but also by the aortic arch, the ventricles, the lungs, and the pituitary gland in response to increases in blood pressure. The stomach and the small intestine produce peptide hormones that help regulate digestive secretions.

A number of different types of organs and cells secrete peptide growth factors, which cause an increase in certain cells. Peptide growth factors are like hormones in that they act on cells that have specific receptors to receive them. Some, including lymphokines and blood cell growth factors, are released into blood; others diffuse to nearby cells. Other growth factors include:

Platelet-derived growth factor, which is released from platelets and many other cell types. It helps in wound healing and causes an increase in the number of fibroblasts, smooth muscle cells, and certain cells of the nervous system.

Epidermal growth factor and *nerve growth factor,* which stimulate the cells indicated by their names as well as many others.

Tumor angiogenesis factor, which stimulates the formation of capillary networks and is released by tumor cells. One treatment for cancer is to prevent the activity of this growth factor.

Prostaglandins (PG) are produced by cells and act on tissues or cells in the immediate vicinity. They are active in very small quantities and have diverse actions that affect such processes as nervous system function, blood flow in the kidneys, pregnancy, and the inflammation of arthritis.

Sometimes, prostaglandins have contrary effects. For example, one type helps prevent the formation of blood clots, while another helps bring about the formation of blood clots. Also, a large dose of PG may have an effect opposite that of a small dose. Therefore, standardizing PG therapy is difficult, and in most instances, the therapy is still considered experimental.

Prostaglandins stimulate the inflammatory response. Drugs such as ibuprofen, aspirin, and acetaminophen, which block the synthesis of prostaglandins, are therefore anti-inflammatory drugs useful in relieving bursitis, arthritis, tennis elbow, and similar conditions.

> Various growth factors stimulate cell production. Prostaglandins are only active locally and have many varied effects.

Effects of Aging

Although hormone-secreting glands shrink with age, their performance is often unaffected. Thyroid disorders and diabetes are the most important endocrine problems significantly affecting health and function. Both hypothyroidism and hyperthyroidism are seen in the elderly. Graves' disease, which results from hyperthyroidism, causes symptoms of cardiovascular disease, increased body temperature, and apathy. In addition, there may be a weight loss of as much as 20 pounds, depression, and mental confusion. Hypothyroidism (myxedema) may fail to be diagnosed because the symptoms of hair loss, skin changes, and mental deterioration may be attributed simply to the process of aging.

The true incidence of type II diabetes among the elderly is unknown. Its symptoms can be confused with those of other medical conditions that are present. Type II diabetes is associated with being overweight and often can be controlled by a proper diet.

The effect of age on the sex organs is discussed in chapter 17.

Working Together

The accompanying illustration shows how the endocrine system works with other organ systems of the body to maintain homeostasis. All the organ systems of the body are interrelated.

> BodyWorks CD-ROM ◆
> The module accompanying chapter 10 is Endocrine System.

working together

Endocrine System

Integumentary System

Androgens activate sebaceous glands and help regulate hair growth.

Skin provides sensory input that results in the activation of certain endocrine glands.

Skeletal System

Growth hormone regulates bone development; parathyroid hormone and calcitonin regulate Ca^{2+} content.

Bones provide protection for glands; store Ca^{2+} used as second messenger.

Muscular System

Androgens promote growth of skeletal muscle; epinephrine stimulates heart and constricts blood vessels.

Muscles help protect glands.

Nervous System

Sex hormones affect development of brain.

Hypothalamus is part of endocrine system; nerves innervate glands of secretion.

Circulatory System

Epinephrine increases blood pressure; ADH, aldosterone, and atrial natriuretic hormone help regulate blood volume; growth factors control blood cell formation.

Blood vessels transport hormones from glands; blood services glands; heart produces atrial natriuretic hormones.

How the Endocrine System works with other body systems

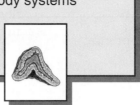

in female

Lymphatic System/Immunity

Thymus is necessary for maturity of T lymphocytes.

Lymphatic vessels pick up excess tissue fluid; immune system protects against infections.

Respiratory System

Epinephrine promotes ventilation by dilating bronchioles; growth factors control production of red blood cells that carry oxygen.

Gas exchange in lungs provides oxygen and rids body of carbon dioxide.

Digestive System

Hormones help control secretion of digestive glands and accessory organs; insulin and glucagon regulate glucose storage in liver.

Stomach and small intestine produce hormones.

Urinary System

ADH, aldosterone, and atrial natriuretic hormone regulate reabsorption of water and Na^+ by kidneys.

Kidneys keep blood values within normal limits so that transport of hormones continues.

Reproductive System

Hypothalamic, pituitary, and sex hormones control sex characteristics and regulate reproductive processes.

Gonads produce sex hormones.

Basic Key Terms

adrenocorticotropic hormone (ACTH) (ad-re″no-kor″te-ko-trop′ik hor′mōn), p. 192

aldosterone (al″dos′ter-ōn), p. 196

androgen (an′dro-jen), p. 199

anterior pituitary (an-ter′e-or pĭ-tu′ĭ-tār″e), p. 190

antidiuretic hormone (an″tĭ-di″u-ret′ik hor′mōn), p. 190

atrial natriuretic hormone (a′trē-al na″trē-u-ret′ik hor′mōn), p. 196

calcitonin (kal″sĭ-to′nin), p. 194

cortisol (kor′tĭ-sol), p. 195

epinephrine (ep″ĭ-nef′rin), p. 195

estrogen (es′tro-jen), p. 199

glucagon (gloo′kah-gon), p. 197

gonadotropic hormone (gon″ah-do-trōp′ik hor′mōn), p. 192

growth hormone (grōth hor′mōn), p. 192

hormone (hor′mōn), p. 187

insulin (in′su-lin), p. 197

melatonin (mel″ah-to′nin), p. 199

norepinephrine (nor″ep-ĭ-nef′rin), p. 195

pancreatic islets (of Langerhans) (pan″kre-at′ik i′lets ov lahng′er-hanz), p. 197

parathyroid hormone (PTH) (par″ah-thi′roid hor′mōn), p. 194

pineal gland (pīn′e-al gland), p. 199

posterior pituitary (pos-tēr′e-or pĭ-tu′ĭ-tār″e), p. 190

progesterone (pro-jes′tĕ-rōn), p. 199

prolactin (pro-lak′tin), p. 192

prostaglandins (PG) (pros″tah-glan′dinz), p. 202

renin (re′nin), p. 196

testosterone (tes-tos′tĕ-rōn), p. 199

thymosin (thi′mo-sin), p. 202

thymus (thi′mus), p. 202

thyroid-stimulating hormone (TSH) (thi′roid stim′u-lāt-ing hor′mōn), p. 192

thyroxine (thi-rok′sin), p. 193

Clinical Key Terms

acidosis (as″ĭ-do′sis), p. 198

acromegaly (ak″ro-meg′ah-le), p. 192

Addison disease (ă′dah-son dĭ-zēz′), p. 196

cretinism (kre′tĭ-nizm), p. 193

Cushing syndrome (koosh′ing sin′drōm), p. 196

diabetes insipidus (di″ah-be′tēz in-sip′ĭ-dus), p. 190

diabetes mellitus (di″ah-be′tēz mel-li′tus), p. 197

exophthalmic goiter (ek″sof-thal′mik goi′ter), p. 193

gestational diabetes mellitus (jes-ta′shun′l di″ah-be′tēz mĕ-li′tus), p. 199

Graves' disease (grāvz dĭ-zēz′), p. 193

hyperglycemia (hi″per-gli-se′me-ah), p. 198

hypoglycemia (hi″po-gli-se′me-ah), p. 198

myxedema (mik″sĕ-de′mah), p. 193

pattern baldness (pă′tern bŏld′nes), p. 199

pituitary dwarf (pĭ-tu′ĭ-tār″e dwarf), p. 192

polyphagia (pŏ″le-fa′je-ah), p. 197

simple goiter (sim′p′l goi′ter), p. 193

tetany (tet′ah-ne), p. 194

Summary

I. Endocrine System
   A. The endocrine glands secrete their hormones into the bloodstream for transport to target organ cells, where the hormones are received by receptors.
   B. Peptide hormones combine with receptors located on the plasma membrane, while steroid hormones pass through the plasma membrane to bind with receptors in the cytoplasm.
   C. Hormone secretion by endocrine glands is often regulated by negative feedback.

II. Hypothalamus and Pituitary Gland
   The pituitary gland has two parts: the posterior pituitary and the anterior pituitary.
   A. Posterior pituitary. The hormones of the posterior pituitary—antidiuretic hormone (ADH) and oxytocin—are produced in the hypothalamus. ADH stimulates water reabsorption by the kidneys; oxytocin stimulates uterine contractions and milk release from the mammary glands.
   B. Anterior pituitary.
      1. Growth hormone (GH) and prolactin (PRL) are two

hormones of the anterior pituitary that do not affect other glands. GH influences the height of children and can cause a condition called acromegaly in adults. PRL promotes milk production after childbirth.
      2. The anterior pituitary also secretes thyroid-stimulating hormone (TSH), adrenocorticotropic hormone (ACTH), and gonadotropic hormones, all of which affect other endocrine glands.
      3. The hypothalamus, anterior pituitary, and the other

endocrine glands controlled by the anterior pituitary are all involved in a self-regulating negative feedback loop.

III. **Thyroid and Parathyroid Glands**
   A. Thyroid gland. The anterior pituitary produces thyroid-stimulating hormone (TSH), a hormone that promotes the production of thyroxine by the thyroid. Thyroxine, which increases metabolism, can affect the entire body, as exemplified by cretinism and myxedema. The thyroid also produces calcitonin, which lowers the blood calcium level.
   B. Parathyroid glands. Parathyroid hormone (PTH) maintains a high blood calcium level by promoting calcium absorption in the intestine, calcium retention by the kidneys, and bone breakdown. These actions are opposed by calcitonin produced by the thyroid gland.

IV. **Adrenal Glands**
   The adrenal glands have two parts: an outer cortex and an inner medulla. The adrenal medulla is under nervous control, and the adrenal cortex is under the hormonal control of adrenocorticotropic hormone (ACTH), an anterior pituitary hormone.
   A. Adrenal medulla. The adrenal medulla releases epinephrine and norepinephrine into the bloodstream, helping the body cope with situations that seem to threaten survival.
   B. Adrenal cortex.
      1. Cortisol, a glucocorticoid secreted by the adrenal cortex, raises the blood glucose level.
      2. Aldosterone, a mineralocorticoid secreted by the adrenal cortex, and antidiuretic hormone (ADH) raise blood volume by causing the kidneys to reabsorb sodium and water. Their action is opposed by atrial natriuretic hormone (ANH), and in this way, normal blood volume is maintained.
      3. Addison disease is due to adrenal cortex hyposecretion. Cushing syndrome is due to adrenal cortex hypersecretion.

V. **Pancreas**
   A. The pancreas produces and secretes the hormones insulin and glucagon. Insulin lowers the blood glucose level by causing cells to take up glucose and the liver to convert glucose to glycogen. Glucagon opposes the actions of insulin and raises the blood glucose level.
   B. The most common illness due to hormonal imbalance is diabetes mellitus, caused by a lack of insulin or insensitivity of cells to insulin.

VI. **Testes and Ovaries**
   A. The androgens, primarily testosterone, are the male sex hormones produced by the testes. Estrogen and progesterone are the female sex hormones produced by the ovaries. The sex hormones maintain the sex organs and the secondary sex characteristics.

VII. **Other Endocrine Glands and Hormones**
   A. The pineal gland secretes melatonin, which is believed to regulate daily rhythms.
   B. The thymus gland secretes thymosins necessary to immunity.
   C. Various growth factors stimulate cell production. Prostaglandins are only active locally and have many varied effects.

Study Questions

1. What is a hormone, and how do hormones work? (p. 187)
2. Define *endocrine gland* and *target organ.* (pp. 187–88)
3. Give the anatomical location of each of the endocrine glands listed in question 4. (p. 188)
4. For each of the following endocrine glands, name the hormone(s) secreted, the effect of the hormone(s), and the medical illnesses, if any, that result from too much or too little of each hormone: posterior pituitary, thyroid, parathyroids, adrenal medulla, adrenal cortex, pancreas. (p. 189)
5. How does the hypothalamus control the posterior pituitary? How does it control the anterior pituitary? (pp. 190–91)
6. Describe the functioning of the hormone secreted by the anterior pituitary that has its effect on the body proper, rather than on other glands. (p. 190)
7. Explain the three-tier relationship between the hypothalamus, anterior pituitary, and thyroid gland. (p. 192)
8. How do calcitonin and parathyroid hormone control normal blood calcium level? (p. 194)
9. What role is played by ADH, aldosterone, and ANH in the maintenance of normal blood volume? (p. 196)
10. Draw a diagram to describe the action and control of insulin and glucagon. (p. 198)
11. Discuss the action of atrial natriuretic hormone, growth factors, and prostaglandins. (p. 202)

Fill in the blanks.

1. Generally, hormone production is self-regulated by a _____ mechanism.
2. The hypothalamus _____ the hormones _____ and _____ , released by the posterior pituitary.
3. The _____ secreted by the hypothalamus control the anterior pituitary.
4. Growth hormone is produced by the _____ pituitary.
5. Simple goiter occurs when the thyroid is producing _____ (too much or too little) _____ .

6. Parathyroid hormone increases the level of _____ in the blood.
7. Adrenocorticotropic hormone (ACTH), produced by the anterior pituitary, stimulates the _____ of the adrenal glands.
8. An overproductive adrenal cortex results in the condition called _____ .
9. Type I diabetes mellitus is due to a malfunctioning _____ , but type II diabetes is due to malfunctioning _____ .
10. Prostaglandins are not carried in the _____ , as are hormones

that are secreted by the endocrine glands.
11. Whereas _____ hormones are lipid soluble and bind to receptor proteins within the cytoplasm of target cells, _____ hormones bind to membrane-bound receptors, thereby activating second messengers.
12. Whereas the adrenal _____ is under the control of the autonomic nervous system, the adrenal _____ secretes its hormones in response to _____ from the anterior pituitary gland.

Medical Terminology Reinforcement Exercise

Consult Appendix B for help in pronouncing and analyzing the meaning of the terms that follow.

1. antidiuretic (an″tĭ-di″u-ret′ik)
2. hypophysectomy (hi-pof″ĭ-sek′to-me)
3. gonadotropic (gon″ah-do-trop′ik)
4. hyperglycemia (hi″per-gli-se′me-ah)
5. hypokalemia (hi″po-kal″e′me-ah)
6. acromegaly (ak″ro-meg′ah-le)
7. lactogenic (lak″to-jen′ik)
8. adrenopathy (ad″ren-op′ah-the)
9. adenomalacia (ad″ĕ-no-mah-la′she-ah)
10. parathyroidectomy (par″ah-thi″roi-dek′to-me)

Website Link

For a listing of the most current Websites related to this chapter, please visit the Mader home page at:

www.mhhe.com/maderap

chapter 11

Blood

Chapter Outline and Learning Objectives

After you have studied this chapter, you should be able to:

Composition of Blood (p. 208)
- Describe, in general, the composition of blood.
- Describe the structure and function of red blood cells, white blood cells, and platelets.
- Explain the hematopoietic role of stem cells in the red bone marrow.
- Describe, in general, the composition of plasma.

Functions of Blood (p. 214)
- Describe the functions of blood.
- Discuss the transport function of blood, and describe capillary exchange within the tissues.
- Describe the blood clotting process and how it is associated with thromboembolism.

Blood Groups and Typing (p. 216)
- Explain the ABO and Rh systems of blood typing.
- Describe how each person's blood type is determined for transfusion purposes.

Effects of Aging (p. 218)
- Blood disorders increase in frequency as we age.

Visual Focus
Hematopoiesis (p. 213)

Medical Focus
Abnormal Red and White Blood Cell Counts (p. 210)
Who Should Not Give Blood (p. 218)

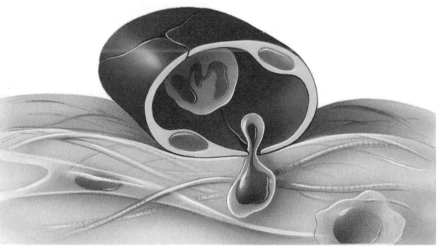

Some white blood cells can squeeze through capillary walls to enter the tissues, where they fight infection.

Part IV

Composition of Blood

Blood is classified as a connective tissue because it consists of cells separated by a matrix, the liquid called **plasma** (see chapter 4). An adult male of average size (70 kilograms or 154 pounds) has a blood volume of about 5 liters (5.2 quarts).

A blood sample that is transferred to a test tube and prevented from clotting separates into two layers. The upper layer, plasma, contains a variety of inorganic and organic substances dissolved or suspended in water. Plasma accounts for about 55% of the volume of whole blood. The lower layer comprises about 45% of the volume of whole blood and is composed mainly of red blood cells; this percentage is called the **hematocrit (HCT).**

Red blood cells (erythrocytes) are one of the **formed elements** in blood. The other formed elements—white blood cells (leukocytes) and blood platelets (thrombocytes)—are found in a narrow layer between the plasma and the red blood cells.

> Blood is a liquid connective tissue. The liquid portion is termed plasma, and the solid portion consists of the formed elements.

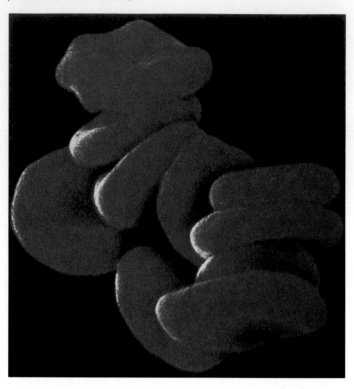

figure 11.1 Red blood cells, as seen by scanning electron microscope.

Red Blood Cells (Erythrocytes)

The **red blood cells (RBCs),** or erythrocytes, are small, biconcave, disk-shaped cells without nuclei (fig. 11.1 and fig. 11.2). They contain **hemoglobin** (he″mo-glo′bin), a protein carrier for oxygen. Hemoglobin binds with oxygen in the cool, neutral conditions of the lungs and readily gives up oxygen under the warm, more acidic conditions of the tissues. **Oxyhemoglobin** (hemoglobin plus oxygen) is bright red. **Deoxyhemoglobin** (hemoglobin minus oxygen) is dark purple. The presence or absence of oxygen attached to hemoglobin accounts for the difference in color between arteries and veins.

The number of red blood cells in a cubic millimeter (mm^3) of blood is called the **red blood cell count.** Males usually have 4.6 to 6.2 million cells per cubic millimeter, and females have 4.2 to 5.4 million cells per cubic millimeter. The Medical Focus reading on page 210 examines disorders resulting from abnormal red and white blood cell counts. Each cell contains about 200 million hemoglobin molecules. If this much hemoglobin were suspended within the plasma, rather than being enclosed within the cells, the blood would be so thick that the heart would have difficulty pumping it.

Red blood cells live only about 120 days and are destroyed chiefly in the liver and spleen, where they are engulfed by large phagocytic cells. When red blood cells are broken down, the hemoglobin is released and broken down into **heme** (hēm), the iron-containing portion, and **globin** (glo′bin), a protein. The iron is recovered and re-

turned to the red bone marrow for reuse. The rest of the heme undergoes further chemical degradation and is excreted by the liver in the bile as bile pigments, which are primarily responsible for the color of feces.

> Red blood cells do not have nuclei, but they do have hemoglobin, which carries oxygen.

White Blood Cells (Leukocytes)

White blood cells (WBCs), or leukocytes, differ from red blood cells in that they are usually larger, have a nucleus, lack hemoglobin, and without staining, appear to be white. White blood cells are not as numerous as red blood cells. The **white blood cell count** is usually 5,000–11,000 per cubic millimeter of blood. White blood cells fight infection in ways discussed at greater length in chapter 13, which concerns immunity.

Red blood cells are confined to the blood, but white blood cells are also found in lymph and tissue fluid. They are able to squeeze through pores in the capillary wall (see figure on p. 207). Some are always found in small numbers in the tissues, but their number greatly increases when there is an infection. Many white blood cells live only a few days—they probably die while engaging pathogens. Others live months or even years.

figure 11.2 Composition of blood. When blood is transferred to a test tube and is prevented from clotting, it forms two layers. The transparent, yellow, top layer is plasma, the liquid portion of blood. The formed elements are in the bottom layer. The tables describe these components in detail.

FORMED ELEMENTS	Function and Description	Source
Red Blood Cells (erythrocytes)  4 million–6 million per mm³ blood	Transport O_2 and help transport CO_2 7–8 μm in diameter Bright-red to dark-purple biconcave disks without nuclei	Red bone marrow
White Blood Cells (leukocytes) 4,000–11,000 per mm³ blood	Fight infection	Red bone marrow
Granular leukocytes		
• Basophil  20–50 per mm³ blood	10–12 μm in diameter Spherical cells with lobed nuclei; large, irregularly shaped, deep-blue granules in cytoplasm	
• Eosinophil 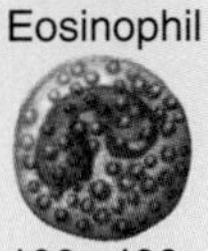 100–400 per mm³ blood	10–14 μm in diameter Spherical cells with bilobed nuclei; coarse, deep-red, uniformly sized granules in cytoplasm	
• Neutrophil 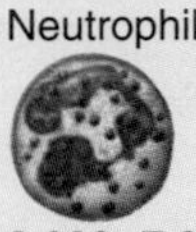 3,000–7,000 per mm³ blood	10–14 μm in diameter Spherical cells with multilobed nuclei; fine, pink granules in cytoplasm	
Agranular leukocytes		
• Lymphocyte  1,500–3,000 per mm³ blood	5–17 μm in diameter (average 9–10 μm) Spherical cells with large round nuclei	
• Monocyte 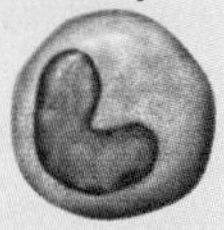 100–700 per mm³ blood	10–24 μm in diameter Large spherical cells with kidney-shaped, round, or lobed nuclei	
• **Platelets** (thrombocytes)  150,000–300,000 per mm³ blood	Aid clotting 2–4 μm in diameter Disk-shaped cell fragments with no nuclei; purple granules in cytoplasm	Red bone marrow

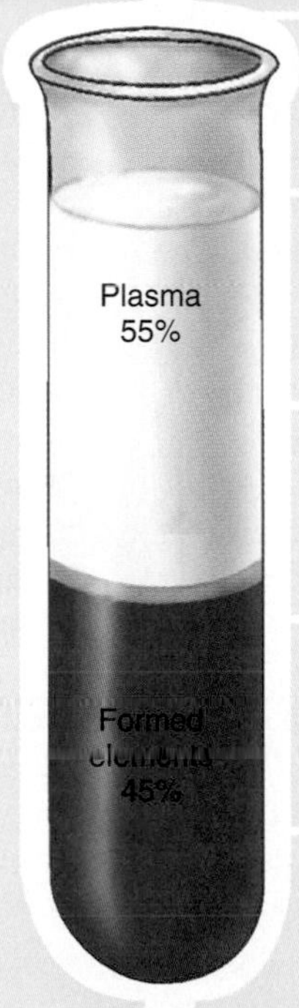

Plasma 55%

Formed elements 45%

PLASMA	Function	Source
Water (90–92% of plasma)	Maintains blood volume; transports molecules	Absorbed from intestine
Plasma proteins (7–8% of plasma)	Maintain blood osmotic pressure and pH	Liver
Albumin	Maintain blood volume and pressure	
Globulins	Transport; fight infection	
Fibrinogen	Clotting	
Salts (less than 1% of plasma)	Maintain blood osmotic pressure and pH; aid metabolism	Absorbed from intestine
Gases		
Oxygen	Cellular respiration	Lungs
Carbon dioxide	End product of metabolism	Tissues
Nutrients	Food for cells	Absorbed from intestine
Fats Glucose Amino acids		
Nitrogenous waste	Excretion by kidneys	Liver
Uric acid Urea		
Other		
Hormones, vitamins, etc.	Aid metabolism	Varied

• with Wright's stain

Types of White Blood Cells

White blood cells are classified as **granular leukocytes** or **agranular leukocytes.** Both types of cells have granules in the cytoplasm surrounding the nucleus, but the granules are more prominent in granular leukocytes. The granules contain various enzymes and antibiotic-like proteins that help white blood cells defend the body. There are three types of granular leukocytes and two types of agranular leukocytes. They differ somewhat by the size of the cell and the shape of the nucleus.

Abnormal Red and White Blood Cell Counts

Polycythemia (pol″e-si-the′me-ah) is a disorder in which an excessive number of red blood cells makes the blood so thick it is unable to flow properly. There is also an increased risk of clot formation associated with this condition.

Anemia is characterized by an insufficient number of red cells, or the cells do not have enough hemoglobin. Normally, the blood hemoglobin level is 12 to 17 grams per 100 milliliters. In **iron deficiency anemia,** a common type of anemia, the hemoglobin count is low, and the individual feels tired and run-down. The person's diet may not contain enough iron. Certain foods, such as raisins and liver, are rich in iron, and the inclusion of these in the diet can help prevent this type of anemia.

In another type of anemia, called **pernicious** (per-ni′shus) **anemia,** the digestive tract is unable to absorb enough vitamin B_{12}, which is essential to the proper formation of red cells. Without it, large numbers of immature red cells tend to accumulate in the bone marrow. A special diet and injections of vitamin B_{12} are effective treatments for pernicious anemia.

In **aplastic** (a-plas′tik) **anemia,** the red bone marrow has been damaged due to radiation or chemicals, and not enough red blood cells are produced. In **hemolytic** (he-mah-lĭtik) **anemia,** there is an increased rate of red blood cell destruction.

Sickle-cell disease is a hereditary condition in which the individual has sickle-shaped red blood cells (fig. 11A). Such cells tend to rupture and wear out easily as they pass through the narrow capillaries, leading to the symptoms of anemia. Sickle-cell disease is most common among blacks because the sickle-shaped cells are a protection against malaria, a disease prevalent in parts of Africa. The parasite that causes malaria cannot infect sickle-shaped red blood cells.

Hemolysis is the rupturing of red blood cells. **Hemolytic disease of the newborn,** which is discussed at the end of this chapter (p. 217), is also a type of anemia.

Certain viral illnesses, like influenza, measles, and mumps, cause the white blood cell count to decrease. **Leukopenia** (lu″ko-pe′ne-ah) is a total white blood cell count below 5,000 per cubic millimeter. Other illnesses, like appen-

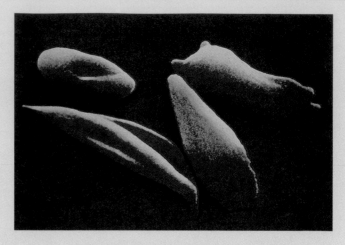

figure 11A Sickle-shaped red blood cells, as seen by scanning electron microscope.

dicitis and bacterial infections, cause the white blood cell count to increase dramatically. **Leukocytosis** (lu″ko-si-to′sis) is a white blood cell count above 10,000 per cubic millimeters.

Illness often causes an increase in a particular type of white blood cell. For this reason, a **differential white blood cell count,** involving the microscopic examination of a blood sample and the counting of each type of white blood cell to a total of 100 cells, may be done as part of the diagnostic procedure. For example, the characteristic finding in the viral disease **mononucleosis** (mon″o-nu″kle-o′sis) is a great number of lymphocytes that are larger than mature lymphocytes and that stain more darkly. This condition takes its name from the fact that lymphocytes are mononuclear.

Leukemia is a form of cancer characterized by uncontrolled production of abnormal white blood cells. These cells accumulate in the bone marrow, lymph nodes, spleen, and liver so that these organs are unable to function properly. **Acute lymphoblastic leukemia** (ALL), which represents over 80% of the acute leukemias in children, also occurs in adults. Chemotherapy is used to destroy abnormal cells and restore normal blood cell production. Intraspinal injection of drugs and craniospinal irradiation are measures that prevent leukemic cells from infiltrating the central nervous system. In general, there is a more favorable prognosis in children between the ages of 2 and 10 years, as compared to either older or younger patients. Prognosis is somewhat better in females because there is a recurrence of leukemia in the testes of 8–16% of males. Remission occurs in 78% of adult patients after chemotherapy, and the median period of remission is 20 months. With chemotherapy, 50–60% of children survive past five years, and of those among this group who do not have a relapse, 85% are considered cured.

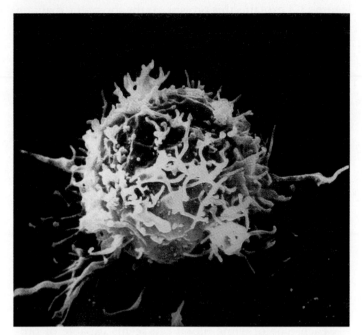

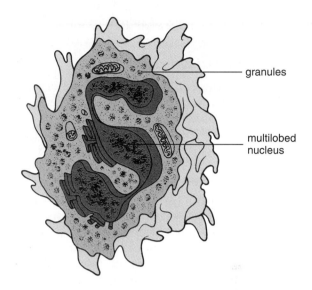

granules

multilobed nucleus

a. Neutrophil

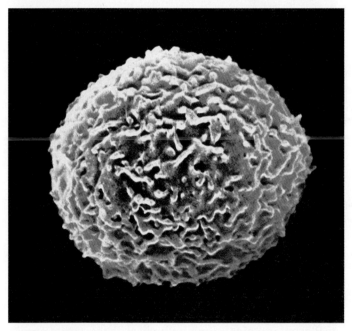

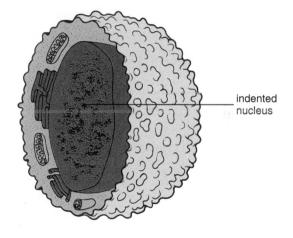

indented nucleus

b. Lymphocyte

Granular Leukocytes

Neutrophils are the most abundant of the white blood cells (see fig. 11.2). They have a multilobed nucleus joined by nuclear threads; therefore, they are also called *polymorphonuclear* (fig. 11.3*a*). Their granules do not significantly take up the stain eosin (a pink to red stain) or a basic stain that is blue to purple. This accounts for the name *neutrophil*.

Neutrophils are the first type of white blood cells to respond to an infection, and they engulf bacteria and cell debris via **phagocytosis.**

Eosinophils have a bilobed nucleus (see fig. 11.2). Their granules take up eosin red stain, which accounts for the name *eosinophil*, meaning red-loving. The function of eosinophils is not completely understood, but they are known to increase in number during a parasitic worm infection or an allergic reaction.

Basophils have a U-shaped or lobed nucleus (see fig. 11.2). The granules take up the basic stain and turn dark blue, which accounts for the name *basophil*. Basophils move to sites of injury and enter the tissues, where they release histamine, a chemical associated with allergic reactions. Histamine dilates blood vessels and causes contraction of smooth muscle in the walls of air passageways.

Agranular Leukocytes

Agranular leukocytes (monocytes and lymphocytes) typically have a spherical or kidney-shaped nucleus (see fig. 11.2). **Monocytes** are the largest of the white blood cells, and after taking up residence in the tissues, they differentiate into even larger macrophages (see fig. 13.4). Macrophages phagocytize pathogens and stimulate other white blood cells to defend the body.

Lymphocytes are of two types, but both have the same appearance (fig. 11.3b). B lymphocytes are responsible for antibody-mediated immunity—that is, they produce antibodies, proteins that combine with antigens, as described on page 216–17. T lymphocytes are responsible for cell-mediated immunity—that is, they directly destroy any cell that bears foreign antigens. B lymphocytes and T lymphocytes are discussed more fully in chapter 13.

> White blood cells are divided into the granular leukocytes and the agranular leukocytes. Each type of white blood cell has a specific role to play in defending the body against disease.

Platelets (Thrombocytes)

Platelets (plāt′lets) (thrombocytes) result from fragmentation of certain large cells, called megakaryocytes, in the red bone marrow (see fig. 11.2 and fig. 11.4). The platelet count in blood is normally 150,000 to 300,000 per cubic millimeter. **Thrombocytopenia** is an insufficient number of platelets (less than 50,000 per cubic millimeter).

Platelets do not have nuclei and are about half the size of red blood cells. They live only about 10 days and are involved in repairing damaged blood vessels and in initiating the process of blood clotting, as discussed on page 215.

> Platelets are fragments of cells that are involved in vessel repair and blood clotting.

Hematopoiesis

The process by which blood cells are formed is called *hematopoiesis.* Blood cells are continuously produced in the red bone marrow of the skull, ribs, and vertebrae, and in the ends of long bones. The bone marrow contains stem cells that constantly divide, producing cells that eventually become the various types of formed elements. Figure 11.4 explains this process. At the top of the diagram is a multipotent stem cell (hemocytoblast) that divides, producing two other types of stem cells. The myeloid stem cell gives rise to the cells that go through a number of stages to become red blood cells, platelets, granular leukocytes, and monocytes. The lymphoid stem cell produces the lymphocytes.

As red blood cells pass through a number of stages, they lose their nuclei and gain hemoglobin. About 2.5 million erythrocytes are produced every second, and an equal number is continuously destroyed by the spleen and liver. Red blood cell production increases when a growth factor called erythropoietin is released by the kidneys (and probably other organs as well). Erythropoietin stimulates stem cells to divide and produce cells that become red blood cells.

Ordinarily, only mature blood cells enter the bloodstream. Blood entering the bone goes into blood sinuses, where the mature blood cells collect and from which they enter the circulatory system.

Each type of white blood cell seems capable of producing specific growth factors that circulate back to the bone marrow and stimulate increased production of white blood cells. The best known growth factor is Gm-CSH (granulocyte-macrophage colony-stimulating hormone).

> Blood cells are formed through the process of hematopoiesis. Stem cells in the red bone marrow continuously divide, giving rise to cells that become the formed elements.

Plasma

Plasma, the liquid portion of blood, is approximately 92% water but contains many different types of molecules, including the plasma proteins (see fig. 11.2). Two types of plasma proteins have special functions. One type, the **gamma globulins** (glob′u-linz), are antibodies that help fight infection. The other is the protein **fibrinogen** (fi-brin′o-jen), which is converted to fibrin threads when blood clotting occurs. Both of these proteins are synthesized in the liver.

Plasma proteins, along with electrolytes, create an osmotic pressure that draws water from the tissues into the blood. This function of the plasma proteins is particularly associated with *albumin,* the smallest and most plentiful of the plasma proteins.

> There are several different plasma proteins, each with specific functions.

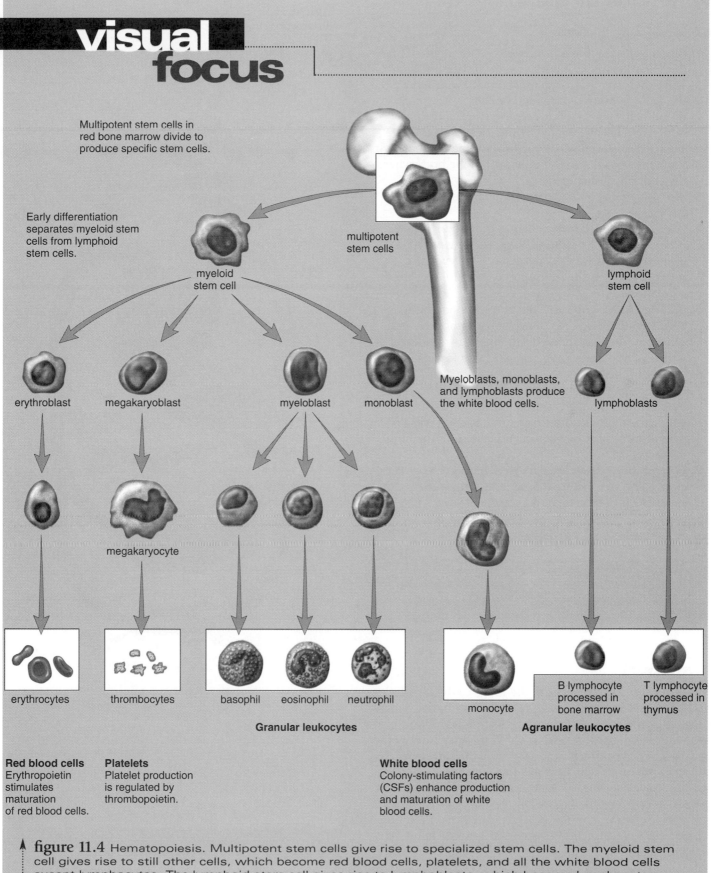

Multipotent stem cells in red bone marrow divide to produce specific stem cells.

Early differentiation separates myeloid stem cells from lymphoid stem cells.

multipotent stem cells

myeloid stem cell

lymphoid stem cell

erythroblast

megakaryoblast

myeloblast

monoblast

Myeloblasts, monoblasts, and lymphoblasts produce the white blood cells.

lymphoblasts

megakaryocyte

erythrocytes

thrombocytes

basophil   eosinophil   neutrophil

Granular leukocytes

monocyte

B lymphocyte processed in bone marrow

T lymphocyte processed in thymus

Agranular leukocytes

Red blood cells
Erythropoietin stimulates maturation of red blood cells.

Platelets
Platelet production is regulated by thrombopoietin.

White blood cells
Colony-stimulating factors (CSFs) enhance production and maturation of white blood cells.

figure 11.4 Hematopoiesis. Multipotent stem cells give rise to specialized stem cells. The myeloid stem cell gives rise to still other cells, which become red blood cells, platelets, and all the white blood cells except lymphocytes. The lymphoid stem cell gives rise to lymphoblasts, which become lymphocytes.

Functions of Blood

Blood has many functions:

1. Blood helps maintain homeostasis. Blood and tissue fluid are the body's internal environment, and as such, their characteristics must remain within certain limits if the body is to remain healthy.
2. Blood helps regulate body temperature. If the body is hot, excess heat generated by muscles is taken to the skin by the blood, where the heat can be dissipated. If the body is cold, heat remains within the body where the vital organs are located.
3. Blood contains buffers that help maintain the pH of the blood.
4. The fluid and electrolyte balance of blood is largely maintained by the kidneys, which excrete or retain electrolytes and water as necessary. However, the blood transports the hormones responsible for this regulation.
5. Blood plays a significant role in infection fighting. Some white blood cells phagocytize pathogens, and others produce antibodies, proteins that combine with and inactivate antigens, which are foreign substances in the blood.
6. Blood keeps the composition of tissue fluid constant by transporting nutrients and oxygen to capillaries, where they diffuse into tissue fluid. It also picks up carbon dioxide at capillaries and takes it to the lungs for excretion.
7. When blood vessels are damaged, blood clots and clotting prevent **hemorrhage** (hem'o-rij)—that is, the escape of blood from blood vessels.

Only the last two functions of blood—transport and clotting—are discussed in detail here.

> Blood has many functions, including helping to maintain homeostasis, regulating body temperature, and fighting infection. Two additional functions are transport and clotting.

Transport

Blood transports oxygen from the lungs to the capillaries and nutrients from the intestine to the capillaries, where they enter tissue fluid. From the tissue fluid, blood also absorbs carbon dioxide and other wastes given off by the cells and carries them away. Carbon dioxide exits the blood at the lungs. The liver produces urea, a waste product that travels by way of the bloodstream to the kidneys, where it is excreted. Figure 11.5 diagrams the major transport functions of blood, indicating how these functions help keep the body's internal environment relatively constant.

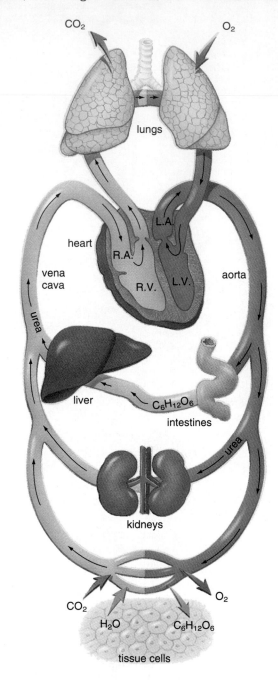

figure 11.5 Diagram of the circulatory system, illustrating the transport function of blood. ($C_6H_{12}O_6$ = glucose; R.A. = right atrium; L.A. = left atrium; R.V. = right ventricle; L.V. = left ventricle.)

> Homeostasis is possible only because blood brings oxygen and nutrients to the cells and removes wastes.

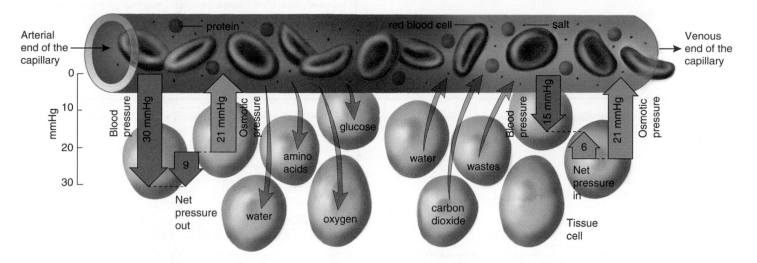

Exchanges between Blood and Tissue Fluid

At the arterial end of a capillary, blood pressure (40 mm Hg) is higher than the osmotic pressure of the blood (25 mm Hg) (fig. 11.6). Osmotic pressure is caused by the presence of salts and particularly by the plasma proteins. Because blood pressure is higher than osmotic pressure, water exits a capillary at the arterial end. This is a **filtration** process because large substances, such as red blood cells and plasma proteins, remain, but small substances, such as water molecules, leave the capillaries. **Tissue fluid,** created by this process, consists of all the components of plasma except the plasma proteins.

Along the length of the capillary, molecular diffusion follows a concentration gradient. The area of greater concentration for oxygen and nutrients is always blood. Therefore, they diffuse out of a capillary. The cells use glucose ($C_6H_{12}O_6$) and oxygen (O_2) in the process of cellular respiration, and they use amino acids for protein synthesis. Following cellular respiration, the cells give off carbon dioxide (CO_2) and water (H_2O). Tissue fluid is always the area of greater concentration for these waste materials; therefore, they diffuse into blood at the capillary.

Because blood pressure is much reduced (10 mm Hg) at the venous end of the capillary, osmotic pressure (25 mm Hg) tends to pull water back into the capillary. Retrieving water by means of osmotic pressure is not completely effective; thus, a portion of the fluid is not picked up at the venous end. This excess tissue fluid enters the lymphatic capillaries and is called lymph. Lymph is returned to the systemic venous blood when the major lymphatic vessels enter the subclavian veins (see fig. 13.1).

> Water, oxygen, and nutrient molecules (for example, glucose and amino acids) exit a capillary near the arterial end. Water and waste molecules (for example, carbon dioxide) enter a capillary near the venous end.

Blood Clotting

When a blood vessel is cut, it immediately constricts to reduce blood flow to the area. Then, the platelets stick to collagenous fibers in the connective tissue layer beneath the endothelium. As more and more platelets congregate, they form a **platelet plug** that can fill a small break.

Another response to blood vessel injury is the formation of a blood clot. At least 12 *clotting factors* in the blood participate in blood clot formation. However, only the roles played by the platelets, **prothrombin,** and fibrinogen are discussed here. Prothrombin and fibrinogen are plasma proteins produced by the liver. Vitamin K is necessary for prothrombin production, and if it is missing from the diet, hemorrhagic disorders develop.

The clotting process (fig. 11.7) consists of a cascade of enzymatic reactions in which each reaction leads to the next. The main events can be summarized as follows: The clotting process is initiated when platelets and damaged tissue release **prothrombin activator,** an enzyme that converts prothrombin to thrombin. This reaction requires the presence of calcium (Ca^{2+}). **Thrombin,** in turn, acts as an enzyme that brings about a change in fibrinogen so that it forms long threads of fibrin. Once formed, **fibrin** threads wind around the platelet plug and provide a framework for the clot. Red blood cells trapped within the fibrin threads make a clot appear red.

figure 11.7 Steps in blood clot formation.

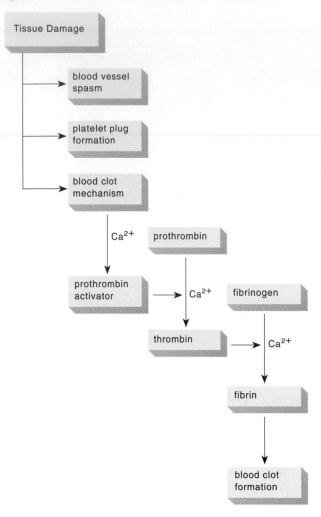

figure 11.8 Blood clots. When blood clots and a solid plug forms, serum is squeezed out. In a blood vessel, the clot helps prevent further blood loss.

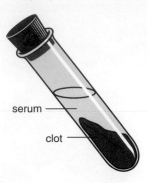

table 11.1	Body Fluids	
	Name	**Composition**
	Blood	Formed elements and plasma
	Plasma	Liquid portion of blood
	Serum	Plasma minus fibrinogen
	Tissue fluid	Plasma minus proteins
	Lymph	Tissue fluid within lymph vessels

In the body, a blood clot that remains stationary in a blood vessel is a **thrombus,** but when it dislodges and is transported in the blood, it is called an **embolus.** If **thromboembolism** is not treated, a heart attack can occur, as is discussed in chapter 12. In time, an enzyme called *plasmin* destroys the fibrin network of a clot and restores the fluidity of plasma.

Blood clotting requires platelets, plasma proteins, and a variety of other materials. A blood clot consists of red blood cells entangled within fibrin threads.

Hemophilia A is the most common of the severe clotting disorders that are inherited. The bleeding problems associated with hemophilia A are the result of a deficiency of factor VIII:C. If the level of factor VIII:C is less than 1% of normal, the affected individual often bleeds in infancy and is afflicted throughout life by hemarthroses (hēm-arth-ro's-ēs) (sing., **hemarthrosis**), or bleeding into joints. This leads to cartilage degeneration in the joints and resorption of underlying bone. Bleeding into muscles leads to necrosis of overlying tissue, as well as pressure-induced nerve damage and muscular atrophy. Gastrointestinal bleeding occurs, and blood also appears in the urine. The most common cause of death is intracranial bleeding accompanied by neurological damage.

If blood is placed in a test tube and allowed to clot, a yellowish fluid rises above the clotted material (fig. 11.8). This fluid, called **serum,** contains all of the components of plasma except fibrinogen and prothrombin. Since a number of different terms have been used to refer to portions of the blood, table 11.1 provides a reference guide.

Blood Groups and Typing

A **transfusion** is the introduction of whole blood or other blood components into a patient's bloodstream. To prevent any complications, the patient's blood type must be known. Two primary systems are used to designate blood type: the ABO system and the Rh system.

ABO System

The body has the ability to recognize foreign substances—that is, substances that do not belong in the body. Such substances, called **antigens,** are often protein molecules that appear on the plasma membrane of, for example, bacteria. The immune system responds to antigens by produc-

table 11.2

Blood Groups

Blood Type	Antigen on Red Blood Cells	Antibody in Plasma	% U.S. African American	% U.S. Caucasian	% U.S. Asian	% Native American
A	A	Anti-B	27	40	28	16
B	B	Anti-A	20	11	27	4
AB	A, B	None	4	4	5	<1
O	None	Anti-A and anti-B	49	45	40	79

ing **antibodies.** When antibodies combine with antigens on bacteria, the bacteria are marked for destruction by the body.

Red blood cells also carry protein molecules on their plasma membranes. Some individuals have A-type protein molecules, some have B-type protein molecules, some have both A and B, and some have neither. If your cells carry only A-type protein molecules, then B-type proteins are antigens to you, and your plasma contains anti-B antibodies. If you were to receive blood that contained red blood cells carrying B-type antigens, **agglutination** (ah-gloo"tĭ-na'shun) would occur; that is, the red blood cells would clump due to an antigen-antibody reaction. Agglutination of red blood cells can cause the blood to stop circulating in small blood vessels, leading to organ damage. Agglutination is followed by the destruction of the red blood cells, or hemolysis, which leads to death.

Table 11.2 lists the different types of blood according to the **ABO system** of typing blood. A person receiving blood from a donor must not have plasma with antibodies that will cause donor cells to agglutinate. Individuals with type O blood are often referred to as *universal donors.* Because their red blood cells lack both the A and B antigens, they can theoretically donate blood to anyone. Similarly, type AB individuals are called *universal recipients.* Because their plasma lacks anti-A and anti-B antibodies, they can theoretically receive blood from any donor. Because of inherent risks, however, only carefully matched blood is actually used for transfusions. The Medical Focus reading on page 218 explains how potential blood donors are screened.

> The ABO blood grouping system consists of four types of blood: A, B, AB, and O. The presence of A and/or B antigens on blood cells determines blood type.

Rh System

Another important antigen in matching blood groups is the **Rh factor,** so named because it was first discovered in the rhesus monkey. People with this particular antigen on their red blood cells are Rh positive (Rh$^+$); those without it are Rh negative (Rh$^-$). Rh-negative individuals do not normally have antibodies to the Rh factor, but their body produces the antibodies after exposure to the Rh factor. When Rh-positive blood is mixed with anti-Rh antibodies, agglutination (clumping) occurs.

During pregnancy, if the mother is Rh negative and the father is Rh positive, the fetus may be Rh positive. The Rh-positive red blood cells of the fetus may leak into the mother's circulatory system, since placental tissues normally break down before and at birth. This causes the mother to produce anti-Rh antibodies. If the mother becomes pregnant with another Rh-positive baby, anti-Rh antibodies may cross the placenta and destroy the fetus's red blood cells. This destruction of red blood cells is called hemolytic disease of the newborn.

The Rh incompatibility problem has been solved by giving Rh-negative women an Rh immune globulin injection midway through the first pregnancy and also no later than 72 hours after giving birth to an Rh-positive child. This injection contains anti-Rh antibodies that attack any of the baby's red blood cells in the mother's blood before these cells can stimulate the mother to produce her own antibodies.

> Each of the four blood types can be positive or negative, depending on whether the Rh factor is present. Hemolytic disease of the newborn is possible when the mother is Rh negative and the father is Rh positive.

Blood Typing

For transfusion purposes, it is important to determine each person's blood type. Figure 11.9 demonstrates a way to use the antibodies derived from plasma to determine blood type. If clumping occurs after a blood sample is exposed to a particular antibody, the person has that type of blood. For example, on laboratory slide 4 in figure 11.9, the blood sample did agglutinate in the presence of anti-A and anti-B antibodies; therefore, the individual has blood type AB. The blood sample did not agglutinate in the presence of

Who Should Not Give Blood

Donors who have:

Ever had hepatitis

Had malaria or have taken drugs to prevent malaria in the last three years

Been treated for syphilis or gonorrhea in the last twelve months

Donors who have AIDS or one of its symptoms:

Unexplained weight loss (10 pounds or more in less than two months)

Night sweats

Blue or purple spots on or under the skin

Long-lasting white spots or unusual sores in the mouth

Lumps in the neck, armpits, or groin for over a month

Diarrhea lasting over a month

Persistent cough and shortness of breath

Fever higher than 37°C for more than ten days

Donors who are at risk for AIDS, that is, if you have:

Taken illegal drugs by needle, even once

Taken clotting factor concentrates for a bleeding disorder, such as hemophilia

Tested positive for any AIDS virus or antibody

Been given money or drugs for sex since 1977

Had a sexual partner within the last twelve months who did any of the above

(For men) had sex *even once* with another man since 1977 or within the last twelve months had sex with a female prostitute

(For women) had sex with a male or female prostitute within the last twelve months or had a male sexual partner who had sex with another man *even once* since 1977

DONORS SHOULD NOT GIVE BLOOD to find out whether they test positive for antibodies to the viruses that cause AIDS. Although the tests for HIV are very good, they are not perfect. HIV antibodies may take weeks to develop after infection with the virus. If donors were infected recently, they may have a negative test result yet be able to infect someone. **It is for this reason that they must not give blood if they are at risk of getting AIDS or other infectious diseases.**

Source: Courtesy of the American Red Cross.

anti-Rh antibody; therefore, the individual is Rh negative. The minus sign after AB signifies that the individual is Rh⁻. On the other hand, laboratory slide 1 shows that the blood sample did not agglutinate in the presence of anti-A and anti-B antibodies; therefore, the individual has blood type O. The blood sample did agglutinate in the presence of anti-Rh antibody; therefore, the individual is Rh positive. The plus sign after the O signifies that the individual is Rh⁺.

> To type blood in the laboratory, blood samples are tested with specific antibodies derived from plasma. If agglutination occurs, red blood cells in the sample have a particular antigen.

Effects of Aging

Anemias, leukemias, and clotting disorders increase in frequency with age. Iron deficiency anemia most frequently results from a poor diet, but pernicious anemia signals that the digestive tract is unable to absorb enough vitamin B_{12}. Leukemia is a form of cancer, which generally increases with age because of both intrinsic (genetic) and extrinsic (environmental) reasons. Thromboembolism, a clotting disorder, may be associated with the progressive development of atherosclerosis in an elderly person. When arteries develop plaque (p. 232), thromboembolism often follows. For many people, atherosclerosis can be controlled by diet and exercise, as discussed in the chapter 12 MedAlert reading, "Preventing Heart Attacks."

figure 11.9 Blood typing. **a.** Antigen is not present, and agglutination does not occur. **b.** Antigen is present, and agglutination occurs. **c.** To type the blood, drops of anti-A antibodies, anti-B antibodies, and anti-Rh antibodies are put on a slide. To each of these, a drop of the person's blood is added. Presence or absence of agglutination tells the blood type.

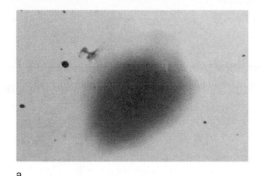

a.

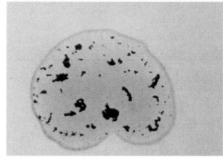

b.

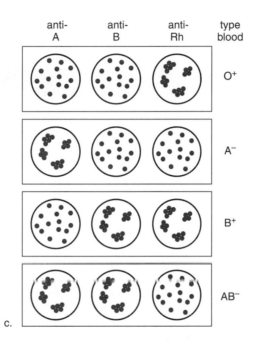

c.

Selected New Terms

Basic Key Terms

agglutination (ah-gloo″tĭ-na′shun), p. 217
deoxyhemoglobin (de-ok″se-he″mo-glo′bin), p. 208
fibrin (fi′brin), p. 215
fibrinogen (fi-brin′o-jen), p. 212
formed element (formed el′ĕ-ment), p. 208
gamma globulin (gam′ah glob′u-lin), p. 212
globin (glo′bin), p. 208
hematocrit (he-mat′o-krit), p. 208
heme (hēm), p. 208
hemoglobin (he″mo-glo′bin), p. 208
lymphocyte (lim′fo-sīt), p. 212
oxyhemoglobin (ok″se-he″mo-glo′bin), p. 208

plasma (plaz′mah), p. 208
platelet (plāt′let), p. 212
platelet plug (plāt′let plug), p. 215
prothrombin (pro-throm′bin), p. 215
prothrombin activator (pro-throm′bin ak′ti-va″tor),
    p. 215
red blood cell (RBC) (red blud sel), p. 208
red blood cell count (red blud sel kownt), p. 208
Rh factor (Rh fak′tor), p. 217
serum (se′rum), p. 216
thrombin (throm′bin), p. 215
white blood cell (WBC) (wīt blud sel), p. 208
white blood cell count (wīt blud sel kownt), p. 208

Clinical Key Terms

acute lymphoblastic leukemia (ah-kūt lim-fo-blas'tik loo-ke'me-ah), p. 210

anemia (ah-ne'me-ah), p. 210

aplastic anemia (a-plas'tik ah-ne'me-ah), p. 210

differential white blood cell count (dif"er-en'shal wīt blud sel kownt), p. 210

embolus (em'bo-lus), p. 216

hemarthrosis (hēm-arth-ro'sis), p. 216

hemolytic anemia (he-mah-lĭ'tik ah-ne'me-ah), p. 210

hemolytic disease of the newborn (he-mah-lĭ'tik dĭ-zez' ov thah nu'born), p. 210

hemophilia A (he-mah-fil'e-ah a), p. 216

hemorrhage (hem'o-rij), p. 214

iron deficiency anemia (i'ern dĭ-fī'shun-se ah-ne'me-ah), p. 210

leukemia (lu-ke'me-ah), p. 210

leukocytosis (lu"ko-si-to'-sis), p. 210

leukopenia (lu"ko-pe'ne-ah), p. 210

mononucleosis (mon"o-nu"kle-o'sis), p. 210

pernicious anemia (per-nĭ'shus ah-ne'me-ah), p. 210

polycythemia (pol"e-si-the'me-ah), p. 210

sickle-cell disease (sĭk'l sel dĭ-zēz'), p. 210

thrombocytopenia (throm"bo-si-to-pe'ne-ah), p. 212

thromboembolism (throm"bo-em'bo-lizm), p. 216

thrombus (throm'bus), p. 216

transfusion (trans-fu'zhun), p. 216

Summary

I. Composition of Blood

Blood is a liquid connective tissue. The liquid portion is termed plasma, and the solid portion consists of the formed elements.

A. Red blood cells do not have nuclei but they do have hemoglobin, which carries oxygen.

B. White blood cells are divided into the granular leukocytes and the agranular leukocytes. Each type of white blood cell has a specific role to play in defending the body against disease.

C. Platelets are fragments of cells that are involved in vessel repair and blood clotting.

D. Blood cells are formed through the process of hematopoiesis. Stem cells in the red bone marrow continuously divide, giving rise to cells that become the formed elements.

E. Plasma proteins have specific functions. Gamma globulins help fight infection. Fibrinogen is necessary to blood clotting. Albumin helps maintain blood volume.

II. Functions of Blood

Blood has many functions, including helping to maintain homeostasis, regulating body temperature, and fighting infection. Two additional functions are transport and clotting.

A. Transport. Homeostasis is possible only because blood brings oxygen and nutrients to the cells and removes wastes. Water, oxygen, and nutrient molecules (for example, glucose and amino acids) exit a capillary near the arterial end. Water and waste molecules (for example, carbon dioxide) enter a capillary near the venous end.

B. Blood clotting. Blood clotting requires many clotting factors and a cascade of reactions. Important contributors always present in the blood are platelets, prothrombin, and fibrinogen. A blood clot consists of red blood cells entangled within fibrin threads.

III. Blood Groups and Typing

A. The ABO blood grouping system consists of four types of blood: A, B, AB, and O. Type O blood has neither the A nor the B antigen on the red blood cells; the other types of blood are designated by the antigens present on the red blood cells.

B. Each of the four blood types can be positive or negative, depending on whether the Rh factor is present. Hemolytic disease of the newborn is possible when the mother is Rh negative and the father is Rh positive.

C. To type blood in the laboratory, blood samples are tested with specific antibodies derived from plasma. If agglutination occurs, red blood cells in the sample have a particular antigen.

Study Questions

1. Name the formed elements, and describe the general structure and function of each. (p. 208)
2. Describe the life cycle of a red blood cell. What factor controls the quantity of red blood cells in the body? (p. 208)
3. State the general function of white blood cells, and categorize the different types. (pp. 208–12)
4. What is the function of stem cells in the red bone marrow? (p. 212)
5. Name three plasma proteins, and list a function for each. (p. 212)
6. Draw a diagram of a capillary, illustrating the exchanges that occur in the tissues. What forces operate to facilitate exchange of molecules across the capillary wall? (p. 215)
7. Name the events that take place after a blood vessel is injured. Which substances are present in the blood at all times, and which appear during the clotting process? (pp. 215–16)
8. Define the terms *blood, plasma, tissue fluid, lymph,* and *serum.* (p. 216)
9. What are the four ABO blood types in humans? For each type, list the antibodies present in the plasma. (p. 217)

Objective Questions

I. Fill in the blanks.
1. The liquid part of blood is called _____ .
2. Red blood cells carry _____ , and white blood cells _____ .
3. Hemoglobin that is carrying oxygen is called _____ .
4. Human red blood cells lack a _____ and only live about _____ days.
5. The most common granular leukocyte is the _____ , a phagocytic white blood cell.
6. B lymphocytes are responsible for _____ immunity, and T lymphocytes are responsible for _____ immunity.

7. At a capillary, _____ , _____ , and _____ leave the arterial end, and _____ and _____ enter the venous end.
8. When a blood clot occurs, fibrinogen has been converted to _____ threads.
9. AB blood has the antigens _____ and _____ on the red blood cells and _____ of these antibodies in the plasma.
10. Hemolytic disease of the newborn can occur when the mother is _____ and the father is _____ .

II. Matching
Match the terms in the key to the descriptions in questions 11–14.
Key:
 a. hematocrit
 b. red blood cell count
 c. white blood cell count
 d. hemoglobin
11. 5,000 to 11,000 per cubic millimeter
12. 4.6 to 6.2 million per cubic millimeter in males
13. Comprises 45% of blood volume
14. 12 to 17 grams per 100 milliliters

Medical Terminology Reinforcement Exercise

Consult Appendix B for help in pronouncing and analyzing the meaning of the terms that follow.

1. hematemesis (hem″ah-tem′ě-sis)
2. erythrocytometry (ě-rith″ro-si-tom′ě-tre)
3. leukocytogenesis (loo″ko-si″to-jen′ě-sis)
4. thrombocytopenia (throm″bo-si″to-pe′ne-ah)
5. hemophobia (he″mo-fo′be-ah)
6. afibrinogenemia (ah-fi″brin-o-jě-ne′me-ah)
7. polycythemia (pol″e-si-the′me-ah)
8. lymphosarcoma (lim″fo-sar-ko′mah)
9. phagocytosis (fag″o-si-to′sis)
10. phlebotomy (flě-bot′o-me)

Website Link

For a listing of the most current Websites related to this chapter, please visit the Mader home page at:

http://www.mhhe.com/maderap

The Circulatory System

Chapter Outline and Learning Objectives

After you have studied this chapter, you should be able to:

Anatomy of the Heart (p. 223)

- Describe the anatomy of the heart, and trace the path of blood through it.
- Name the heart valves, and describe their functions.
- Describe the cardiac cycle and the cardiac conduction system.
- Label and explain a normal electrocardiogram.
- Describe how the heartbeat is regulated.
- Describe the conditions that may cause a heart attack.

Vascular System (p. 234)

- Name the three types of blood vessels, and describe their structure and function.
- Name the two circuits of the circulatory system, and trace the path of blood from the heart to any organ in the body and back to the heart.
- Describe the functions of the fetal circulatory structures.

Features of the Circulatory System (p. 244)

- Define pulse.
- Describe the factors that control blood pressure and blood flow in the arteries, capillaries, and veins.
- Define hypertension, and distinguish between systolic pressure and diastolic pressure.

Effects of Aging (p. 246)

- Anatomical and physiological changes occur in the circulatory system as we age.

Working Together (p. 246)

- The circulatory system works with other systems of the body to maintain homeostasis.

Medical Focus

Congestive Heart Failure (p. 227)
The Electrocardiogram (p. 230)

MedAlert

Preventing Heart Attacks (p. 232)

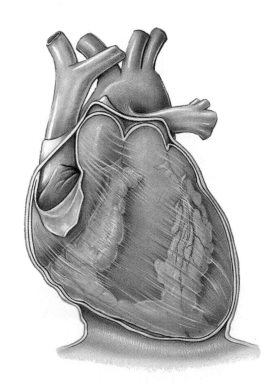

The heart is enclosed by a pericardium whose layers have been wholly or partially dissected away.

figure 12.1 Tissue layers of the heart wall and the pericardium.

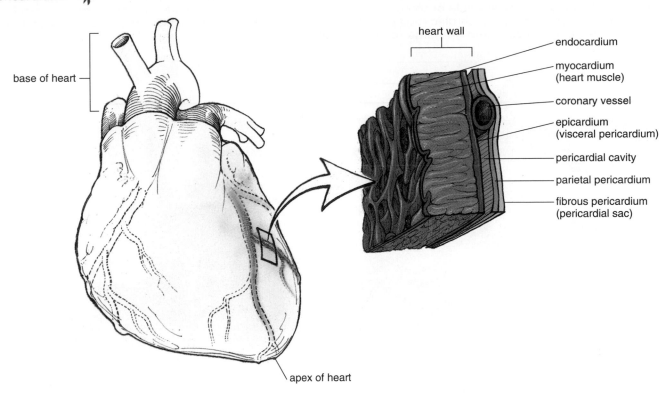

The circulatory system consists of the heart and the blood vessels. The heart pumps the blood, and the blood vessels, which comprise the vascular system, conduct the blood toward and away from the heart. The primary function of the circulatory system is keeping the blood moving in its circular path. The other functions of the circulatory system are the same as those noted for blood in chapter 11.

Anatomy of the Heart

The **heart** is a cone-shaped, muscular organ about the size of a clenched fist. It is located in the thorax between the lungs, anterior to the backbone and posterior to the sternum. Its apex is tilted to the left, and about two-thirds of the heart is located to the left of the body's midline.

Pericardial Membranes

The heart lies within a sac formed by the **pericardial membranes** (fig. 12.1). First is the *fibrous pericardium*, a layer of fibrous connective tissue that adheres to the blood vessels at the heart's base and the sternal wall of the thorax and the diaphragm below. Next comes the *parietal pericardium*, a serous membrane that is separated by a small space, called the *pericardial cavity*, from the *epicardium* (visceral pericardium), another serous membrane. These serous membranes produce a liquid called the *pericardial fluid*, which lubricates them and reduces friction as the heart beats.

The epicardium is a part of the heart wall that also has two other layers. The **myocardium** is the thickest part of the heart wall and is made up of cardiac muscle (see fig. 4.13). When cardiac muscle fibers contract, the heart beats. The inner **endocardium** includes an endothelium formed from simple squamous epithelium that not only lines the heart but also continues into and lines the blood vessels. The endothelium's smooth nature helps prevent blood from clotting unnecessarily.

Chambers of the Heart

The heart has four chambers: The right and left atria (sing. **atrium**) are superior to the right and left **ventricles** (fig. 12.2). The atria are smaller and have thinner walls than the ventricles. Internally, the atria are separated by the **interatrial septum,** and the ventricles are separated by the **interventricular septum** (fig. 12.3). Thus, the heart has a right and left side.

figure 12.2 Surface view of the heart. **a.** The venae cavae bring oxygen-poor blood to the right side of the heart from the body, and the pulmonary arteries take it to the lungs. The pulmonary veins bring oxygen-rich blood from the lungs to the left side of the heart, and the aorta takes it to the body. **b.** The coronary arteries and cardiac veins occur throughout cardiac muscle. The coronary arteries are the first blood vessels to branch off the aorta. They bring oxygen and nutrients to cardiac cells.

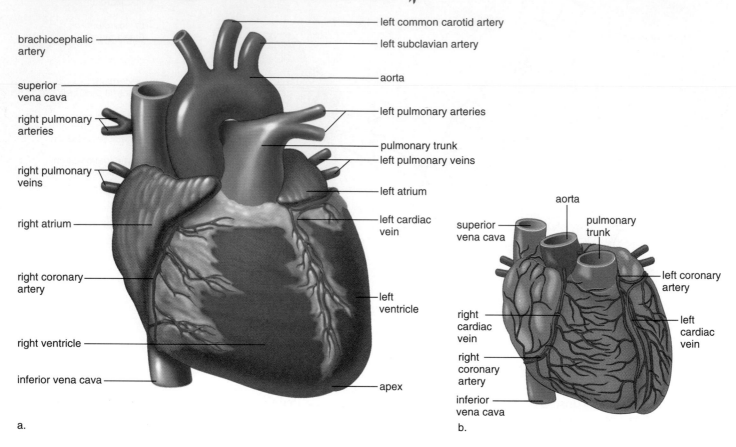

a.

b.

Right Atrium

Three large openings are located in the wall of the *right atrium*. The superior vena cava enters superiorly, and the inferior vena cava enters inferiorly on the posterior side (see fig. 12.3). There is also an opening between the right atrium and the right ventricle that is guarded by a valve appropriately called an **atrioventricular (AV) valve.** This valve, like other **heart valves,** directs the flow of blood and prevents any backflow. This particular AV valve is also known as the **tricuspid valve** because it has three cusps or flaps (fig. 12.4*b*).

Although not shown in figure 12.3, another opening into the right atrium is the coronary sinus, a vein that carries oxygen-poor blood from the heart wall. The coronary sinus opens into the right atrium between the inferior vena cava and the tricuspid valve.

Right Ventricle

The wall of the *right ventricle* contains conical extensions of myocardium called *papillary muscles.* Fibrous cords from the papillary muscles, called the **chordae tendineae,** attach to the cusps of the tricuspid valve. The chordae tendineae support the valve and prevent the cusps from inverting (turning back into the right atrium) when the right ventricle fills with blood and begins to contract.

The pulmonary trunk leaves the right ventricle. This opening has a **semilunar valve** with cusps that resemble half-moons. This valve, called the **pulmonary semilunar valve,** prevents blood from flowing back into the right ventricle (see fig. 12.4).

figure 12.3 Frontal section of heart. **a.** The venae cavae empty into the right atrium, and the pulmonary trunk leaves from the right ventricle. The pulmonary veins enter the left atrium, and the aorta leaves from the left ventricle. **b.** A diagrammatic representation that traces the path of blood through the heart.

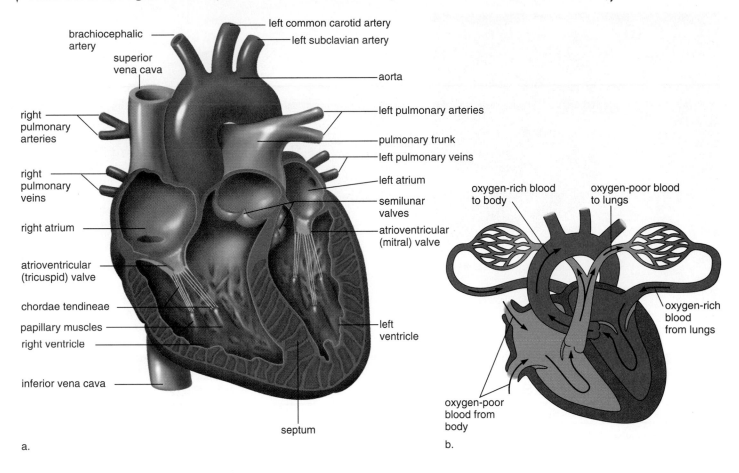

a.

b.

Left Atrium

The *left atrium* is smaller than the right atrium but has thicker walls. Four pulmonary veins, two from each lung, enter the left atrium. The openings do not have valves. An atrioventricular valve between the left atrium and the left ventricle is called the **bicuspid** or mitral **valve** because it has two cusps.

Left Ventricle

The cavity of the *left ventricle* is oval-shaped, while that of the right ventricle is crescent-shaped. It appears to be smaller in size because the walls are thicker.

The papillary muscles in the left ventricle are quite large, and the chordae tendineae are thicker and stronger

than those in the right ventricle. These chordae tendineae and papillary muscles keep the bicuspid valve from inverting into the left atrium when the left ventricle contracts.

The opening by which the aorta leaves the left ventricle is closed by a semilunar valve called the **aortic semilunar valve** (see fig. 12.4). The semilunar cusps of this valve are larger and thicker than those of the pulmonary semilunar valve. Openings just beyond the aortic semilunar valve lead to the coronary arteries, blood vessels that lie on and nourish the heart itself (see fig. 12.2). The heart valves are listed in table 12.1.

The heart has a right and left side and four chambers, consisting of two atria and two ventricles. The heart valves are the tricuspid valve, the pulmonary semilunar valve, the bicuspid valve, and the aortic semilunar valve.

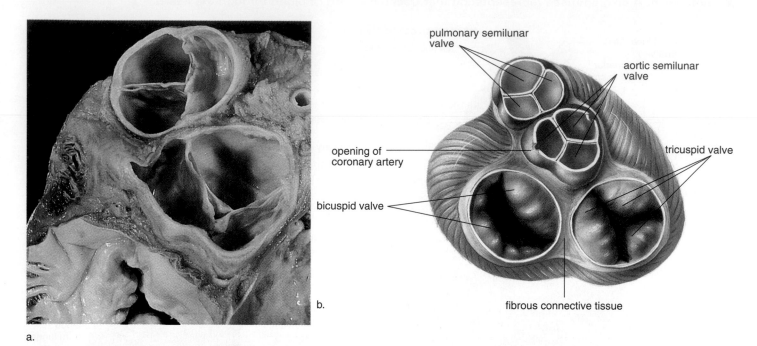

figure 12.4 Valves of the heart. **a.** The two semilunar valves. **b.** Heart valves in a cutaway superior view after removal of the atria. The heart valves are supported by fibrous connective tissue.

pulmonary semilunar valve

aortic semilunar valve

opening of coronary artery

tricuspid valve

bicuspid valve

fibrous connective tissue

a.

b.

Valves of the Heart					
Right Side			**Left Side**		
Valve	**Location**	**Function**	**Valve**	**Location**	**Function**
Tricuspid valve	Right atrioventricular valve	Prevents blood from moving from right ventricle into right atrium during ventricular contraction	Bicuspid (mitral) valve	Left atrioventricular valve	Prevents blood from moving from left ventricle into left atrium
Pulmonary semilunar valve	Entrance to pulmonary trunk	Prevents blood from moving from pulmonary trunk into right ventricle during ventricular relaxation	Aortic semilunar valve	Entrance to aorta	Prevents blood from moving from aorta into left ventricle

Source: David Shier, et al., *Hole's Human Anatomy and Physiology*, 8th ed. Copyright © 1999 The McGraw-Hill Companies, Inc., Dubuque, Iowa.

Congestive Heart Failure

Congestive heart failure occurs when the cardiac output is insufficient to meet the body's needs. The term *congestive* is used because heart failure is accompanied by increased venous volume and pressure. When the left side of the heart fails to pump blood, due perhaps to a heart attack or valve failure, fluid backs up in the lungs and produces pulmonary congestion and edema. The result is shortness of breath and fatigue; if severe, pulmonary edema can be fatal. During the past 20 years, deaths from congestive heart failure have increased by one-third, even though heart attacks are down 25% and strokes are down 40%.

Treatment consists of the three Ds: diuretics (which increase urinary output), digoxin (which increases the heart's contractile force), and dilators (which relax the blood vessels). Surgical repair and replacement are also possible. **Heart transplants** are done, or a piece of muscle is taken from the back, brought into the thorax, and wrapped around the heart.

Double Pump

The heart is a double pump. The right side of the heart sends oxygen-poor blood, which is also high in carbon dioxide, to the lungs, and the left side sends oxygen-rich blood throughout the body. Therefore, the blood actually travels in two circuits: (1) from the heart to the lungs and back to the heart, and (2) from the heart to the body and back to the heart. The right side of the heart is a pump for the first circuit, and the left side is a pump for the second. The left ventricle has the harder job of pumping blood to all parts of the body; therefore, its walls are thicker than those of the right ventricle.

> The right side of the heart pumps oxygen-poor blood to the lungs, and the left side pumps oxygen-rich blood to the tissues.

Path of Blood in the Heart

The path of blood through the heart is traced as follows: Blood enters the right atrium from the **superior** and **inferior venae cavae** (ve'ne ca've), the largest veins in the body. Contraction of the right atrium forces the blood through the tricuspid valve to the right ventricle. The right ventricle pumps the blood through the pulmonary semilu- nar valve, which allows blood to enter the *pulmonary trunk.* The pulmonary trunk divides into the **pulmonary arteries,** which take blood to the lungs.

From the lungs, blood enters the left atrium from the **pulmonary veins.** Contraction of the left atrium forces blood through the bicuspid valve into the left ventricle. The left ventricle then pumps the blood through the aortic semilunar valve into the **aorta** (a-or'tah), the largest artery in the body. The aorta sends blood to all body tissues. Notice that oxygen-poor blood never mixes with oxygen-rich blood and that blood must pass through the lungs before entering the left side of the heart.

Heartbeat

Blood is forced out of the ventricles with each heartbeat. The *stroke volume* is the volume of blood pumped by a ventricle with each beat. The *cardiac output* is the volume of blood pumped by one ventricle per minute. In a resting adult, the cardiac output is usually about 5 liters—approximately the amount of blood in the body. The Medical Focus reading on this page describes congestive heart failure, a condition that occurs when the cardiac output is insufficient to meet the body's needs.

figure 12.5 Stages in the cardiac cycle. **a.** When the atria contract, the ventricles are relaxed and filling with blood. **b.** When the ventricles contract, the atrioventricular valves are closed, the semilunar valves are open, and blood is pumped into the pulmonary trunk and aorta. **c.** When the heart relaxes, both atria and ventricles are filling with blood.

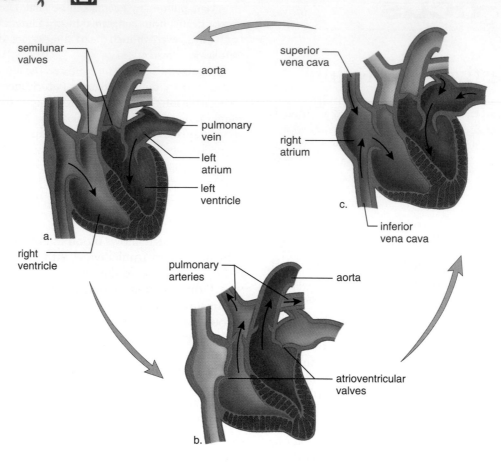

Cardiac Cycle

From the description of the path of blood through the heart, it might seem that the right and left side of the heart beat independently of one another, but actually, they contract together. First, the two atria contract simultaneously; then, the two ventricles contract at the same time. The word **systole** (sis'to-le) refers to contraction of heart muscle, and the word **diastole** (di-as'to-le) refers to relaxation of heart muscle; therefore, atrial systole is followed by ventricular systole.

If the heart contracts, or beats, about 70 times a minute, then each heartbeat lasts about 0.85 second. Each heartbeat, or *cardiac cycle* (fig. 12.5), consists of the following elements:

Time	Atria	Ventricles
1. 0.15 sec.	Systole	Diastole
2. 0.30 sec.	Diastole	Systole
3. 0.40 sec.	Diastole	Diastole

This shows that, while the atria contract, the ventricles relax, and vice versa, and that all chambers rest at the same time for 0.40 second. The short systole of the atria is appropriate since the atria send blood only into the ventricles. The muscular ventricles require a longer systole because they pump blood throughout the whole body. When the word *systole* is used alone, it usually refers to the left ventricular systole.

The heartbeat is divided into three phases: (1) the atria contract during atrial systole, (2) the ventricles contract during ventricular systole, and (3) both the atria and ventricles relax during diastole.

figure 12.6 Conduction system of the heart. The SA node sends out a stimulus, which causes the atria to contract. When this stimulus reaches the AV node, it signals the ventricles to contract. Impulses pass down the two branches of the atrioventricular bundle to the Purkinje fibers and thereafter the ventricles contract.

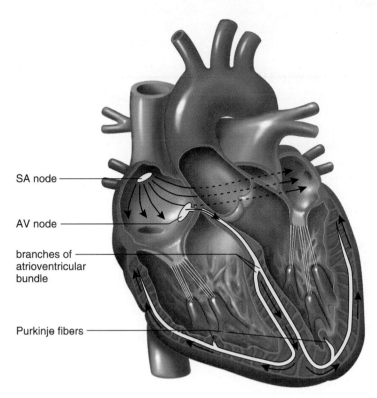

SA node

AV node

branches of atrioventricular bundle

Purkinje fibers

Heart Sounds

A heartbeat produces the familiar "lub dub" sounds. The sounds are due to vibrations caused by pressure changes that occur when the chambers contract and the valves close. The "lub" sound is heard when the ventricles contract and the atrioventricular valves close. This sound lasts longer and has a lower pitch than the "dub" sound, which is heard when the semilunar valves close and the ventricles relax. Heart murmurs, or a slight slush sound after the "lub," are often due to ineffective valves that allow blood to pass back into the atria after the atrioventricular valves have closed.

Rheumatic fever resulting from a streptococcal infection is one cause of a faulty valve, particularly the mitral valve. **Mitral stenosis** is a narrowing of the opening of the bicuspid valve. If operative procedures are unable to open and/or restructure the valve, it may be replaced by an artificial valve.

> The heart sounds are due to the closing of the heart valves.

Cardiac Conduction System

The heart contains specialized cardiac muscle fibers with both muscular and nervous characteristics: They can con-

duct cardiac impulses throughout the myocardium. These fibers are a part of the **cardiac conduction system.**

The heartbeat is intrinsic, meaning that the heart will beat independently of outside nervous stimulation. This ability is due to a specialized mass of cardiac fibers called the **sinoatrial (SA) node.** The SA node is located in the posterior wall of the right atrium, just below the opening for the superior vena cava (fig. 12.6). The SA node is known as the **pacemaker** because it initiates the heartbeat and automatically sends out an excitation impulse every 0.85 second. The atria then contract, and the impulse is sent on to a second node, called the **atrioventricular (AV) node.** This node is located in the wall of the right atrium near the septum just superior to the ventricles. The AV node conducts the impulse to a group of large fibers called the **atrioventricular (AV) bundle** (bundle of His). The AV bundle, which is in the upper part of the interventricular septum, divides into right and left *bundle branches*. These branches give rise to **Purkinje fibers** (pur-kin'je), which cause the ventricles to contract. Contraction begins at the apex and moves toward the base, where the blood vessels are located.

With the contraction of any muscle, including the myocardium, electrolyte changes occur that can be detected by electrical recording devices. A record of the changes that occur during a cardiac cycle is called an

Medical Focus

The Electrocardiogram

a graph that records the electrical activity of the myocardium during a cardiac cycle is called an electrocardiogram or ECG. The change in polarity as the heart's chambers contract and then relax is measured in millivolts by electrodes placed on the skin and wired to a voltmeter (instrument that measures voltage).

An ECG consists of a set of waves: the P wave, a QRS complex, and a T wave (fig. 12A). The P wave represents depolarization of the atria as an impulse started by the SA node travels throughout the atria. The P wave signals that the atria are going to be in systole and that the atrial myocardium is about to contract. The QRS complex represents depolarization of the ventricles following excitation of the Purkinje fibers. It signals that the ventricles are going to be in systole and that the ventricular myocardium is about to contract. The QRS complex shows greater voltage changes because the ventricles have more muscle mass than the atria. The T wave represents repolarization of the ventricles. It signals that the ventricles are going to be in diastole and that the ventricular myocardium is about to relax. Atrial diastole does not show up on an ECG because the voltage changes are masked by the QRS complex.

An ECG records the length of the heartbeat and therefore can be used to detect a rate that is slower or faster than normal. Fewer than 60 heartbeats per minute is called **bradycardia,** and more than 100 heartbeats per minute is called **tachycardia.**

An ECG also detects abnormal heartbeats. The term **arrhythmia** describes a heartbeat displaying an abnormal rhythm. The heart is in **fibrillation** when it beats rapidly, but the contractions are uncoordinated. Fibrillation is common when a person is having a heart attack. The heart can sometimes be defibrillated by briefly applying a strong electrical current to the chest.

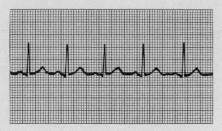

a.

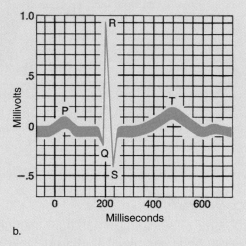

b.

figure 12A Electrocardiogram. **a.** A portion of an electrocardiogram. **b.** An enlarged normal cycle.

Source: David Shier, et al., *Hole's Human Anatomy and Physiology*, 8th ed. Copyright © 1999 The McGraw-Hill Companies, Inc., Dubuque, Iowa.

---

electrocardiogram (ECG). Electrocardiograms are discussed further in the Medical Focus reading on this page.

> The conduction system of the heart includes the SA node, the AV node, the AV bundle, the bundle branches, and the Purkinje fibers. The SA node causes the atria to contract. The AV node and the rest of the conduction system cause the ventricles to contract.

Regulation of the Heartbeat

While, as discussed in the preceding section, the heartbeat is intrinsic, the rate is regulated by the nervous system. A **cardiac control center** in the medulla oblongata of the brain can alter the heartbeat rate by way of the autonomic nervous system (fig. 12.7). Parasympathetic motor impulses conducted by the vagus nerve cause the heartbeat to

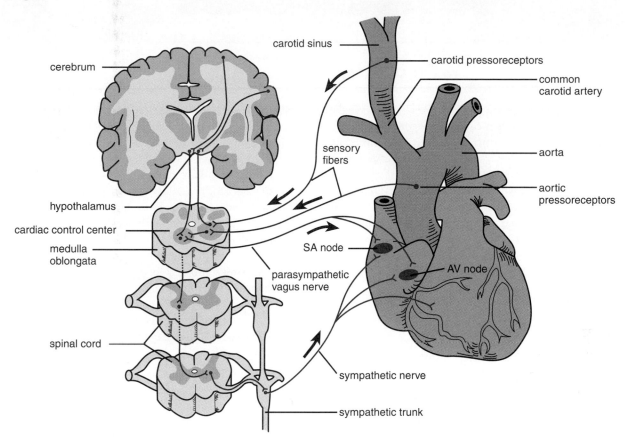

figure 12.7 Regulation of the heart rate. The activities of the SA and AV nodes are regulated by the autonomic nervous system. The brain and spinal cord are shown to the left, and the heart and pressoreceptors are shown to the right.

slow down, and sympathetic motor impulses conducted by sympathetic motor fibers cause the heartbeat to increase.

The cardiac control center receives sensory input from receptors within the circulatory system. For example, *pressoreceptors* (baroreceptors) are present in the aorta just after it leaves the heart and in the carotid arteries, which take blood from the aorta to the brain. If blood pressure falls, as it might if we stand up quickly, the pressoreceptors signal the cardiac control center. Thereafter, sympathetic motor impulses to the heart cause the heartbeat rate to increase. Once blood pressure begins to rise above normal, nerve impulses from the cardiac control center cause the heartbeat rate to decrease. Such reflexes also help control cardiac output and, therefore, blood pressure, as discussed later.

The cardiac control center is under the influence of the cerebrum and the hypothalamus. Therefore, when we feel anxious, the sympathetic motor nerves are activated, and the adrenal medulla releases the hormones norepinephrine and epinephrine. The end result is an increase in heartbeat rate. On the other hand, activities such as yoga and meditation lead to activation of the vagus nerve, which slows the heartbeat rate.

Other factors also influence the heartbeat rate. A cold temperature slows the heartbeat rate, which is why the body temperatures of a person undergoing open-heart surgery is lowered. The correct plasma concentration of electrolytes, such as potassium (K^+) and calcium (Ca^{2+}), is important to a regular heartbeat.

The heartbeat rate is regulated largely by the autonomic nervous system.

MedAlert

Preventing Heart Attacks

*t*he coronary arteries are small blood vessels that serve the needs of cardiac muscle. When coronary arteries become occluded to any degree, coronary heart disease (CHD) is present. Although CHD develops slowly over a period of years, a heart attack (myocardial infarction) can develop quite suddenly. Most heart attacks occur when a blood clot forms in a coronary artery already narrowed by plaque (fig. 12B). Then the portion of the heart deprived of oxygen and nutrients dies, and surrounding tissue may also be damaged.

The following factors increase the risk of CHD:

Male gender or postmenopausal female

Family history of heart attack under age 55

Tobacco usage (for example, smoking cigarettes, chewing tobacco)

Severe obesity (30% or more overweight)

Hypertension (high blood pressure)

Unfavorable blood levels of HDL and LDL cholesterol

Impaired circulation to the brain or the legs

Diabetes mellitus

Hypertension (high blood pressure) is a major factor in the development of cardiovascular disease, and two controllable behaviors contribute to it: smoking cigarettes (including filtered cigarettes) and obesity. While cigarette smoking is a habit to avoid, most of its detrimental side effects can be reversed when the individual stops smoking. Since obese individuals find it very difficult to lose weight, weight control should be a lifelong endeavor.

Investigators have identified several behaviors that may help to reduce the possibility of heart attack and stroke. Exercise is critical. Sedentary individuals have a risk of cardiovascular disease that is about double that of those who are very active. One physician recommends that his patients walk for one hour, three times a week. Stress reduction is also desirable for heart attack and stroke prevention. Daily meditation and yogalike stretching and breathing exercises may help reduce stress.

A diet low in saturated fats and cholesterol retards plaque development and thereby helps reduce the chance of heart attack. Cholesterol is carried in the blood by two types of plasma proteins:

1. LDL (low-density, or "bad," lipoprotein) transports cholesterol to the tissues from the liver.
2. HDL (high-density, or "good," lipoprotein) transports cholesterol out of the tissues to the liver.

A diet low in saturated fat and cholesterol will probably lower the total blood cholesterol level and perhaps the LDL level of some individuals, but most likely will not raise the HDL level. Certain drugs apparently can raise the HDL level, and exercise is also sometimes effective.

Cardiac Disorders

Cardiac disorders are especially associated with **atherosclerosis** (ath″er-o″skle-ro′sis), an accumulation of soft masses of fatty materials, particularly cholesterol, beneath the inner linings of the arteries. Such deposits are called **plaque** (plak), and as they develop, they tend to protrude into the vessel and interfere with blood flow. As discussed in the MedAlert reading on this page, diet can be used to control blood cholesterol level, and therefore plaque, when necessary.

If the coronary artery (see fig. 12.2) is partially occluded (blocked) by the presence of atherosclerosis, the individual may suffer from **ischemic** (is-kem′ik) **heart disease.** Although enough oxygen may normally reach the heart, the individual experiences insufficiency during exercise or stress. At that time, the individual may suffer **angina pectoris** (an-ji′nah pek′to-ris), chest pain that is often accompanied by a radiating pain in the left arm.

Sometimes blood clots in an unbroken blood vessel particularly if plaque is present. A *thrombus* is a stationary blood clot in an unbroken blood vessel, and an *embolus* is a blood clot that is moving along in the bloodstream. **Thromboembolism** is present when a blood clot breaks away from its place of origin and is carried to a new

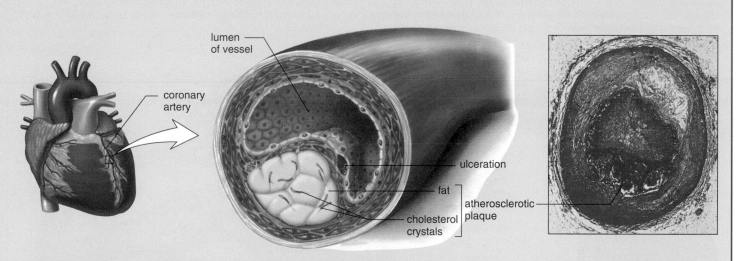

lumen
of vessel

coronary
artery

ulceration

fat

atherosclerotic
plaque

cholesterol
crystals

figure 12B Coronary arteries and plaque. Plaque (in yellow) is an irregular accumulation of cholesterol and other substances. When plaque is present in a coronary artery, a heart attack is more apt to occur because of restricted blood flow.

Some of the cardiovascular risk factors, such as male gender and family history, are inherent in an individual. Other risk factors, however, can be controlled if the individual believes it is worth the effort. The four great admonitions for a healthy life—eating a low-fat, low-cholesterol diet; getting regular exercise; maintaining proper weight; and refraining from smoking—all contribute to keeping the blood cholesterol level low and the blood pressure within a normal range.

Questions

1. In what ways does a low-fat, low-cholesterol diet help prevent CHD?
2. Describe what happens during a myocardial infarction.

location. Thromboembolism leads to heart attacks when the embolus blocks a coronary artery and a portion of the heart dies due to lack of oxygen. Dead tissue is called an infarct, and therefore, the individual who has had a heart attack has had a **myocardial infarction** (mi″o-kar′de-al in-fark′shun).

Myocardial infarction is often preceded by atherosclerosis, angina pectoris, and thromboembolism.

Two surgical procedures are associated with **occluded coronary arteries.** In **thrombolytic therapy,** a plastic tube is threaded into an artery of an arm or leg and is guided through a major blood vessel toward the heart. Once the

tube reaches a blockage, a balloon attached to the end of the tube can be inflated to break up the clot, a procedure called **balloon angioplasty.** In some cases, a small metal-mesh cylinder called a vascular stent is inserted into a blood vessel during balloon angioplasty. The stent functions to hold the vessel open and decreases the risk of future occlusion. Alternately, streptokinase may be injected to dissolve the clot. In a **coronary bypass operation,** a portion of a blood vessel from another part of the body, such as a large vein in the leg, is sutured from the aorta to the coronary artery, past the point of obstruction. This procedure allows blood to flow normally again from the aorta to the heart.

figure 12.8 Blood vessels. The walls of arteries and veins have three layers. The inner layer is composed largely of endothelium with a basement membrane that has elastic fibers; the middle layer is smooth muscle tissue; the outer layer is connective tissue (largely collagen fibers). **a.** Arteries have a thicker wall than veins because they have a larger middle layer than veins. **b.** Capillary walls are one-cell-thick endothelium. **c.** Veins are larger in diameter than arteries, so that collectively veins have a larger holding capacity than arteries. **d.** Scanning electron micrograph of an artery and vein.

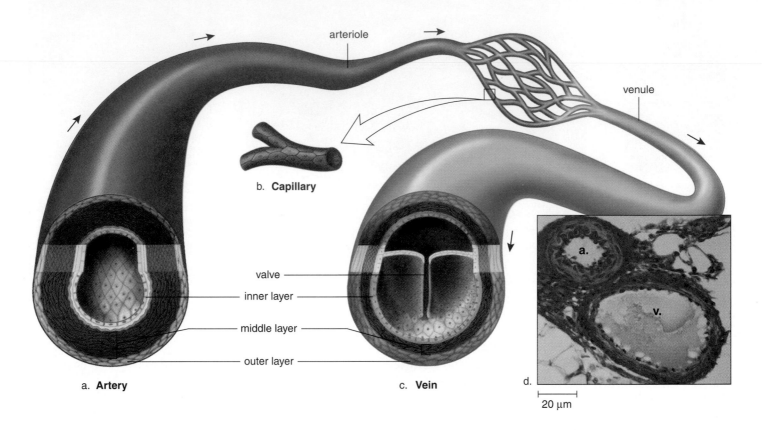

Vascular System

Blood Vessels

The blood vessels comprise the vascular system. Blood vessels are of three types: arteries, capillaries, and veins (fig. 12.8).

Arteries and Arterioles

Arteries (fig. 12.8*a*) transport blood away from the heart. They have thick walls composed of an inner endothelium layer (tunica interna), an outer connective tissue layer (tunica externa), and also a thick middle layer (tunica media) of elastic fibers and smooth muscle. The elastic fibers enable an artery to expand and accommodate the sudden in- crease in blood volume that results after each heartbeat. Arterial walls are sometimes so thick that they are supplied with blood vessels.

Arterioles are small arteries just visible to the naked eye. The middle layer of these vessels has some elastic tissue but is composed mostly of smooth muscle whose fibers encircle the arteriole. Contraction of smooth muscle cells is under involuntary control by the autonomic nervous system. If the muscle fibers contract, the lumen (cavity) of the arteriole decreases; if the fibers relax, the lumen of the arteriole enlarges. Whether arterioles are constricted or dilated affects blood pressure. The greater the number of vessels dilated, the lower the resistance to blood flow, and hence, the lower the blood pressure, and vice versa.

figure 12.9 Anatomy of a capillary bed. Capillary beds form a maze of vessels that lie between an arteriole and a venule. Blood can move directly between the arteriole and the venule by way of a shunt. When precapillary sphincter muscles are closed, blood flows through the shunt. When precapillary sphincter muscles are open, the capillary bed is open, and blood flows through the capillaries.

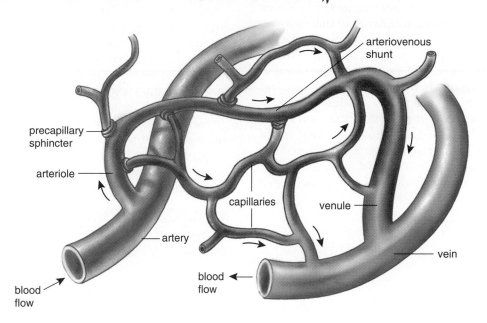

Arteriosclerosis The plaques associated with atherosclerosis may also restrict nutrition of the smooth muscle tissue and elastic fibers comprising the artery wall, causing these tissues to deteriorate and lesions to form. This is accompanied by the deposition of calcium salts and the formation of nonelastic scar tissue, resulting in increased rigidity of the vessel wall. This process of hardening of the arteries, or **arteriosclerosis,** not only contributes to hypertension but also increases the risk of a heart attack or stroke.

A stroke, also called a **cerebrovascular accident (CVA),** occurs when a portion of the brain is deprived of oxygen. Two reasons for strokes are hypertension, which can cause a cerebral artery to burst, or a blood clot, which can prevent blood flow to the brain.

Capillaries

Arterioles branch into **capillaries** (fig. 12.8*b*), which are extremely narrow, microscopic blood vessels with a wall composed of only one layer of endothelial cells. *Capillary beds* (a network of many capillaries) are present in all regions of the body; consequently, a cut in any body tissue draws blood. Capillaries are an important part of the circulatory system because nutrient and waste molecules are exchanged only across their thin walls. Oxygen and glucose diffuse out of capillaries into the tissue fluid that surrounds cells, and carbon dioxide and other wastes diffuse into the capillaries (see fig. 11.6). Since capillaries serve the needs of the cells, the heart and other vessels of the circulatory system can be considered a means by which blood is conducted to and from the capillaries.

Not all capillary beds (fig. 12.9) are open or in use at the same time. For instance, after a meal, the capillary beds of the digestive tract are usually open, and during muscular exercise, the capillary beds of the skeletal muscles are open.

Most capillary beds have a shunt that allows blood to move directly from arteriole to a venule (a small vessel leading to a vein) when the capillary bed is closed. Sphincter muscles, called *precapillary sphincters,* encircle the entrance to each capillary. When the capillary bed is closed, the capillary sphincters are constricted, preventing blood from entering the capillaries; when the capillary bed is open, the capillary sphincters are relaxed. As would be expected, the larger the number of capillary beds open, the lower the blood pressure.

Veins and Venules

Veins and smaller vessels called **venules** carry blood from the capillary beds to the heart. First, the venules drain the blood from the capillaries and then join together to form a vein. The wall of a vein is much thinner than that of an artery because the middle layer of muscle and elastic fibers is thinner (see fig. 12.8c). Within some veins, especially in the major veins of the arms and legs, **valves** allow blood to flow only toward the heart when they are open and prevent the backward flow of blood when they are closed.

At any given time, more than half of the total blood volume is found in the veins and venules. If blood is lost due to, for example, hemorrhaging, sympathetic nervous stimulation causes the veins to constrict, providing more blood to the rest of the body. In this way, the veins act as a blood reservoir.

> Arteries and arterioles carry blood away from the heart, veins and venules carry blood to the heart, and capillaries join arterioles to venules.

Varicose Veins and Phlebitis   **Varicose veins** are abnormal and irregular dilations in superficial (near the surface) veins, particularly those in the lower legs. Varicose veins in the rectum, however, are commonly called piles, or more properly, **hemorrhoids.** Varicose veins develop when the valves of the veins become weak and ineffective due to backward pressure of the blood. The problem can be aggravated when venous blood flow is obstructed by crossing the legs or by sitting in a chair so that its edge presses against the back of the knees.

Phlebitis (flĭ-bi′tus), or inflammation of a vein, is a more serious condition, particularly when a deep vein is involved. When blood in a large, unbroken vein clots, thromboembolism can occur. In this instance the embolus, which is a blood clot moving along in the bloodstream, may eventually come to rest in a pulmonary arteriole, blocking circulation through the lungs. This condition, termed **pulmonary embolism,** can result in death.

Path of Circulation

The vascular system, which is diagrammatically represented in figure 12.10, can be divided into two circuits: the **pulmonary circulation,** which circulates blood through the lungs, and the **systemic circulation,** which serves the needs of the body's tissues.

Pulmonary Circulation

The path of blood through the lungs can be traced as follows: Blood from all regions of the body first collects in the right atrium and then passes into the right ventricle, which pumps it into the pulmonary trunk. The pulmonary trunk divides into the pulmonary arteries, which divide into the arterioles of the lungs. The arterioles then take blood to the pulmonary capillaries, where carbon dioxide and oxygen are exchanged. The blood then enters the pulmonary venules and flows through the pulmonary veins back to the left atrium. Since the blood in the pulmonary arteries is oxygen-poor but the blood in the pulmonary veins is oxygen-rich, it is not correct to say that all arteries carry blood that is high in oxygen and that all veins carry blood that is low in oxygen. In fact, it is just the reverse in the pulmonary system.

> The pulmonary arteries transport oxygen-poor blood to the lungs, and the pulmonary veins return oxygen-rich blood to the heart.

Systemic Circulation

The systemic circulation includes all of the other arteries and veins of the body. The largest artery in the systemic circuit is the aorta, and the largest veins are the superior and inferior venae cavae. The superior vena cava collects blood from the head, chest, and arms, and the inferior vena cava collects blood from the lower body regions. Both venae cavae enter the right atrium. The aorta and venae cavae are the major pathways for blood in the systemic system.

The path of systemic blood to any organ in the body begins in the left ventricle, which pumps blood into the aorta. Branches from the aorta go to the major body regions and organs. Tracing the path of blood to any organ in the body requires only mentioning the aorta, the proper branch of the aorta, the organ, and the returning vein to the vena cava. In many instances, the artery and vein that serve the same organ have the same name. For example, the path of blood to the kidneys is: left ventricle; aorta; renal artery; arterioles, capillaries, venules; renal vein; inferior vena cava; right atrium. In the systemic circuit, unlike the pulmonary system, arteries contain oxygen-rich blood and appear bright red, while veins contain oxygen-poor blood and appear purplish.

> The systemic circulation transports blood from the left ventricle of the heart to the arteries, arterioles, and capillaries, and then from the capillaries to the venules and veins to the right atrium of the heart. It serves the body proper.

figure 12.10 Blood vessels in the pulmonary and systemic circulations. Except for the pulmonary system, the veins (blue-colored) carry oxygen-poor blood, and the arteries (red-colored) carry oxygen-rich blood. Arrows indicate the direction of blood flow. Lymph vessels collect excess tissue fluid and return it to the subclavian veins in the shoulders.

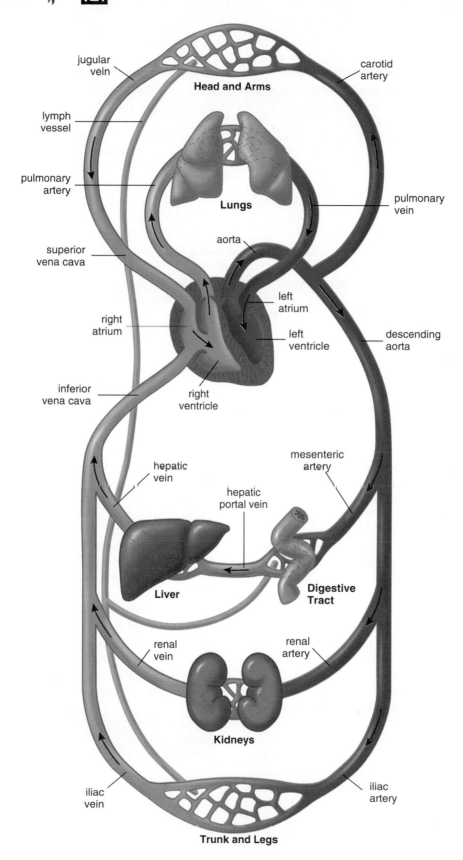

table 12.2

Aorta and Its Principal Branches

Portion of Aorta	Major Branch	Regions Supplied
Ascending aorta	Coronary arteries	Heart
Aortic arch	Brachiocephalic	
	Right common carotid	Right side of head
	Right subclavian	Right arm
	Left common carotid	Left side of head
	Left subclavian	Left arm
Descending aorta		
Thoracic aorta	Intercostal	Thoracic wall
Abdominal aorta	Celiac artery	Stomach, spleen, and liver
	Superior mesenteric	Small and large intestine (ascending and transverse colons)
	Renal artery	Kidney
	Gonadal artery	Ovary or testis
	Inferior mesenteric	Lower digestive system (transverse and descending colons, and rectum)
	Common iliac	Pelvic organs and legs

table 12.3

Principal Veins That Join the Venae Cavae

Vein	Region Drained	Vena Cava
Right and left brachiocephalic veins	Head, neck, and upper extremities	Form superior vena cava
Right and left common iliac veins	Lower extremities	Form inferior vena cava
Right testicular vein (or right ovarian vein)	Gonad	Enters inferior vena cava
Right and left renal veins	Kidneys	Enters inferior vena cava
Right and left hepatic veins	Liver, digestive tract, and spleen	Enters inferior vena cava

The Major Systemic Arteries   By examining the path of the aorta (fig. 12.11) after it leaves the heart, one can see why it is divided into the *ascending aorta,* the *aortic arch,* and the *descending aorta.* The coronary arteries, which supply blood to the heart, branch off of the ascending aorta.

Three major arteries branch off the aortic arch: the **brachiocephalic** artery, the **left common carotid** artery, and the **left subclavian** artery. The brachiocephalic artery divides into the **right common carotid** and the **right subclavian** arteries. These blood vessels serve the head (right and left common carotids) and arms (right and left subclavians).

The descending aorta is divided into the *thoracic aorta,* which branches off to the organs within the thoracic cavity, and the *abdominal aorta,* which branches off to the organs in the abdominal cavity. The major branches of the aorta are listed in table 12.2, and most of them are shown in figure 12.11.

The descending aorta ends when it divides into the **common iliac** arteries that branch into the **internal iliac** artery and the **external iliac** artery. The internal iliac artery serves the pelvic organs, and the external iliac artery serves the legs.

The Major Systemic Veins   Figure 12.12 shows the major veins of the body. The **external** and **internal jugular** veins drain blood from the brain, head, and neck. The external jugular veins enter the **subclavian** veins that, along with the internal jugular veins, enter the **brachiocephalic** veins. These vessels merge, giving rise to the superior vena cava.

In the abdominal cavity, as discussed in more detail later, the **hepatic portal vein** receives blood from the abdominal viscera and enters the liver. Emerging from the liver, the **hepatic** veins enter the inferior vena cava.

In the pelvic region, veins from the various organs enter the **internal iliac** veins, while the veins from the legs enter the **external iliac** veins. The internal and external iliac veins become the **common iliac** veins that merge, forming the inferior vena cava. Table 12.3 lists the principal veins that enter the venae cavae.

All the arteries in the systemic circulation can be traced from the aorta. All the veins in the systemic circulation can be traced to the venae cavae.

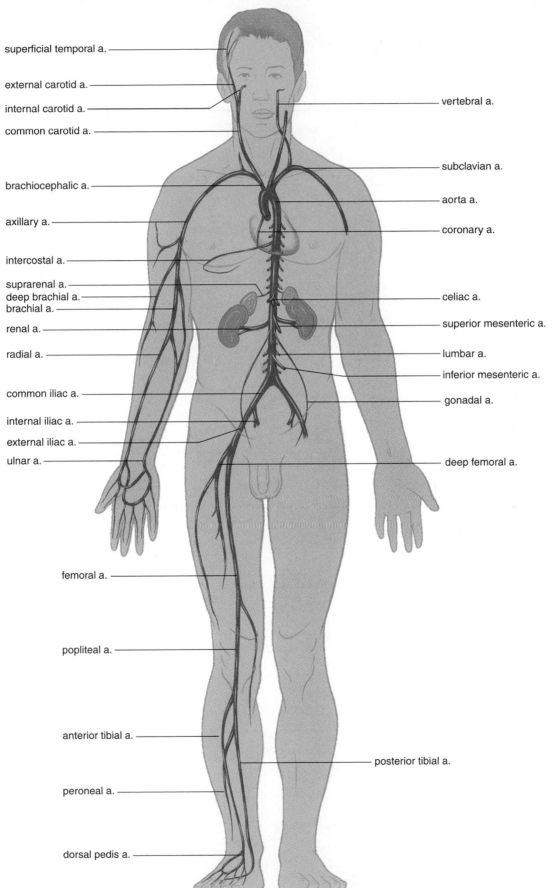

superficial temporal a.

external carotid a.

internal carotid a.

common carotid a.

vertebral a.

subclavian a.

brachiocephalic a.

aorta a.

axillary a.

coronary a.

intercostal a.

suprarenal a.

deep brachial a.

brachial a.

celiac a.

superior mesenteric a.

renal a.

radial a.

lumbar a.

inferior mesenteric a.

common iliac a.

internal iliac a.

gonadal a.

external iliac a.

ulnar a.

deep femoral a.

femoral a.

popliteal a.

anterior tibial a.

posterior tibial a.

peroneal a.

dorsal pedis a.

figure 12.12 Major veins of the human body in anterior view. (v. = vein.)

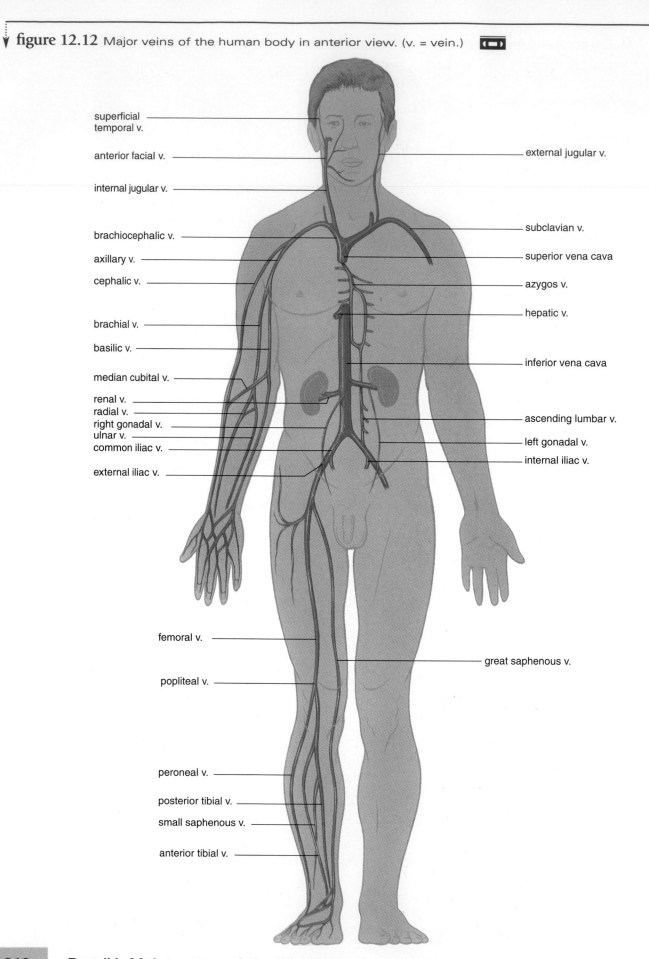

superficial temporal v.

anterior facial v.

internal jugular v.

brachiocephalic v.

axillary v.

cephalic v.

brachial v.

basilic v.

median cubital v.

renal v.

radial v.

right gonadal v.

ulnar v.

common iliac v.

external iliac v.

external jugular v.

subclavian v.

superior vena cava

azygos v.

hepatic v.

inferior vena cava

ascending lumbar v.

left gonadal v.

internal iliac v.

femoral v.

popliteal v.

peroneal v.

posterior tibial v.

small saphenous v.

anterior tibial v.

great saphenous v.

figure 12.13 Arterial blood supply to the brain, including the circle of Willis. (a. = artery.)

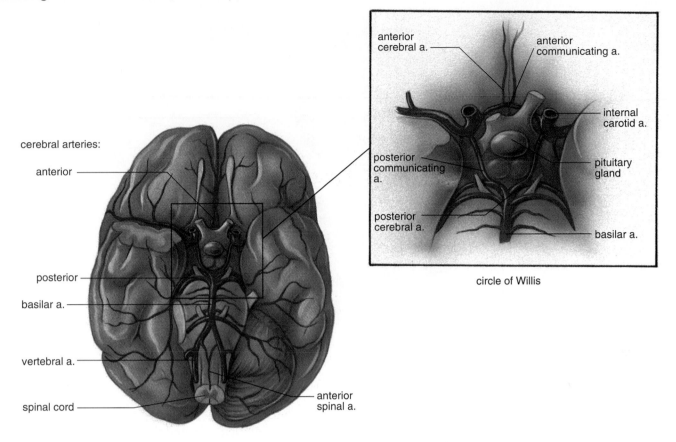

cerebral arteries:

anterior

posterior

basilar a.

vertebral a.

spinal cord

anterior spinal a.

anterior cerebral a.

anterior communicating a.

internal carotid a.

posterior communicating a.

pituitary gland

posterior cerebral a.

basilar a.

circle of Willis

Vital Systemic Circulatory Routes

Blood Supply to the Brain　The brain is supplied with oxygenated blood in arteries (vertebral and internal carotids) that give off branches. These branches join to form the **circle of Willis**, a circle in the region of the pituitary gland (fig. 12.13). The value of having the blood vessels join in this way is that, if one becomes blocked, the brain still can receive blood via three other routes.

> Circulation to the brain includes the circle of Willis, which protects the brain from reduced blood supply.

Blood Supply to the Heart　The **coronary arteries** (see fig. 12.2), which are a part of the systemic circulation, are extremely important because they serve the myocardium. (The heart is not nourished by the blood in its chambers.) The right and left coronary arteries arise from the aorta, just beyond the aortic semilunar valve. They lie on the exterior surface of the heart, where they branch off in various directions into smaller arteries and arterioles. The coronary capillary beds join to form venules, which converge into the *cardiac veins.* The cardiac veins follow the path of the coronary arteries and finally empty into the *coronary sinus,* an enlarged vein on the posterior surface of the heart. The coronary sinus enters the right atrium. Although the coronary arteries receive blood under high pressure, they have a very small diameter and can become *occluded,* or blocked, as discussed in the MedAlert reading "Preventing Heart Attacks" on page 232.

> Circulation to the myocardium is dependent upon the proper functioning of the coronary arteries.

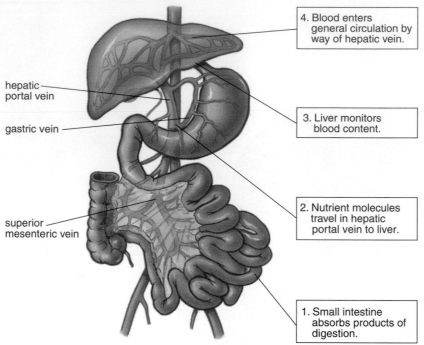

hepatic portal vein

gastric vein

superior mesenteric vein

4. Blood enters general circulation by way of hepatic vein.

3. Liver monitors blood content.

2. Nutrient molecules travel in hepatic portal vein to liver.

1. Small intestine absorbs products of digestion.

Blood Supply to the Liver   The **hepatic portal system** (fig. 12.14) carries blood from the stomach, intestines, and other organs to the liver. A portal system is one that begins and ends in capillaries; thus, there are two sets of capillaries between an artery and a final vein. The superior mesenteric artery brings blood to the small intestine, where the first set of capillaries occurs. Various veins join to form the hepatic portal vein, which takes blood to the liver, where the second set of capillaries occurs. Then, the hepatic veins leave the liver to enter the inferior vena cava.

The hepatic portal system carries blood from the stomach and intestines to the liver.

Fetal Circulation

As figure 12.15 shows, the fetus has four circulatory features that are not present in adult circulation:

1. **Foramen ovale,** or *oval window,* an opening between the two atria. This window is covered by a flap of tissue that acts as a valve.
2. **Ductus arteriosus,** or *arterial duct,* a connection between the pulmonary artery and the aorta.
3. **Umbilical arteries** and **vein,** vessels that travel to and from the placenta, leaving waste and receiving nutrients.

4. **Ductus venosus,** or *venous duct,* a connection between the umbilical vein and the inferior vena cava.

All of these features can be related to the fact that the fetus does not use its lungs for gas exchange, since it receives oxygen and nutrients from the mother's blood at the placenta.

The path of blood in the fetus can be traced, beginning from the right atrium (fig. 12.15). Most of the blood that enters the right atrium passes directly into the left atrium by way of the foramen ovale because the blood pressure in the right atrium is somewhat greater than that in the left atrium. The rest of the fetal blood entering the right atrium passes into the right ventricle and out through the pulmonary trunk, but because of the ductus arteriosus, most blood then passes into the aorta. Notice that by whatever route blood takes, most of it reaches the aorta instead of the lungs.

Blood within the aorta travels to the various branches, including the iliac arteries, which connect to the umbilical arteries leading to the placenta. Exchange between maternal and fetal blood takes place at the placenta. Blood in the umbilical arteries is oxygen-poor, but blood in the umbilical vein, which travels from the placenta, is oxygen-rich. The umbilical vein enters the ductus venosus, which passes directly through the liver. The ductus venosus then joins with the inferior vena cava, a vessel that contains oxygen-poor blood. The vena cava returns this mixture to the right atrium.

figure 12.15 Fetal circulation. Arrows indicate the direction of blood flow.

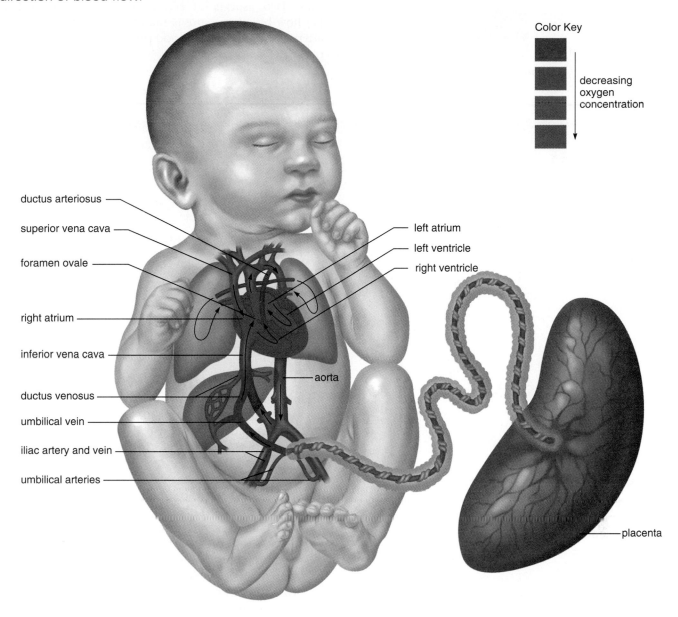

Color Key

decreasing oxygen concentration

ductus arteriosus

superior vena cava

foramen ovale

right atrium

inferior vena cava

ductus venosus

umbilical vein

iliac artery and vein

umbilical arteries

left atrium

left ventricle

right ventricle

aorta

placenta

The most common of all cardiac defects in the newborn is the persistence of the foramen ovale. With the tying of the umbilical cord and the expansion of the lungs, blood enters the lungs in quantity. Return of this blood to the left side of the heart usually causes a small valve located on the left side of the interatrial septum to close the foramen ovale. Incomplete closure occurs in nearly one out of four individuals, but even so, blood rarely passes from the right atrium to the left atrium because either the opening is small or it closes when the atria contract. In a small number of cases, the passage of oxygen-poor blood from the right side to the left side of the heart is sufficient to cause **cyanosis,** a bluish cast to the skin. This condition can now be corrected by open-heart surgery.

The ductus arteriosus closes because endothelial cells divide and block the duct. Remains of the arterial duct and parts of the umbilical arteries and vein are later transformed into connective tissue.

Fetal circulation includes four unique features: (1) the foramen ovale, (2) the ductus arteriosus, (3) the umbilical arteries and vein, and (4) the ductus venosus. These features are necessary because the fetus does not use its lungs for gas exchange.

figure 12.16 Sites where the pulse can be taken.

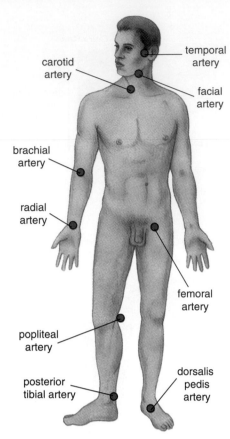

carotid
artery

temporal
artery

facial
artery

brachial
artery

radial
artery

femoral
artery

popliteal
artery

posterior
tibial artery

dorsalis
pedis
artery

Features of the Circulatory System

When the left ventricle contracts, blood is sent out into the aorta under pressure.

Pulse

The surge of blood entering the arteries causes their elastic walls to swell, but then they almost immediately recoil. This alternate expanding and recoiling of an arterial wall can be felt as a **pulse** in any artery that runs close to the surface. The pulse is often checked by placing several fingers on the radial artery, which lies near the outer border of the palm side of the wrist near the thumb. The pulse can also be obtained at other locations (fig. 12.16). The pulse rate indicates the heartbeat rate and also gives information about the strength and rhythm of the heartbeat.

The pulse rate indicates the heartbeat rate.

Blood Pressure

Blood pressure is the force of blood against a blood vessel wall. Two aspects of blood pressure are considered: (1) how blood pressure is maintained in the arteries and arterioles, and (2) how blood pressure varies in other parts of the circulatory system.

Maintaining Blood Pressure

The two factors that affect blood pressure are cardiac output (p. 227) and peripheral resistance. The following factors can cause cardiac output and peripheral resistance to rise:

↑Cardiac output	↑Peripheral resistance
Heart rate increase	Arterial constriction
Blood volume increase	

The pressoreceptors mentioned earlier (see fig. 12.7) regulate the heartbeat rate and also blood pressure. For example, when a person stands up quickly and blood pressure falls, the pressoreceptors signal a cardiac control center and a vasomotor control center in the medulla oblongata. Impulses conducted along sympathetic nerve fibers then cause heartbeat rate to increase and the arterioles to constrict to a greater degree. The end result is a rise in blood pressure.

Certain hormones also cause blood pressure to rise. Epinephrine and norepinephrine increase the heart rate, as mentioned in chapter 10. The renin-angiotensin-aldosterone mechanism (also discussed in chapter 10) constricts arterioles and leads to an increase in blood volume when sodium and water are reabsorbed. ADH (antidiuretic hormone) released by the posterior pituitary also causes a rise in blood volume and therefore blood pressure (see fig. 10.11).

What factors lead to a reduction in blood pressure? If blood pressure rises above normal, the pressoreceptors signal the cardiac control center and vasomotor control center in the medulla oblongata. Subsequently, the heart rate decreases and the arterioles dilate. In addition, recall that atrial natriuretic factor opposes the actions of aldosterone and ADH by decreasing blood pressure (see fig. 10.11). Therefore, blood pressure is under the control of both the nervous system and the endocrine system.

Blood pressure is dependent on cardiac output and peripheral resistance. These are regulated by the activation of autonomic nerve impulses and by the release of certain hormones.

figure 12.17 Blood pressure changes throughout the systemic circulation. Blood pressure decreases with distance from the left ventricle.

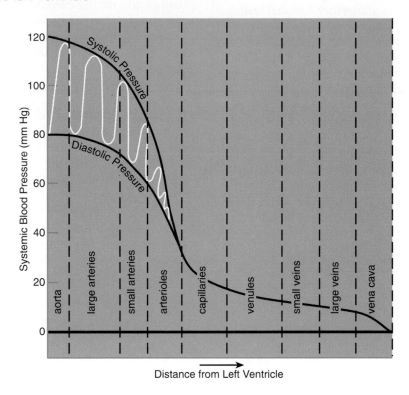

Blood Flow in Arteries, Arterioles, and Capillaries

Blood pressure accounts for the movement of blood in the arteries and arterioles. Blood pressure decreases with distance from the left ventricle of the heart. Blood pressure is higher in the arteries than in the arterioles, and there is a sharp drop in pressure when blood reaches the capillaries (fig. 12.17). This decrease in pressure may be correlated with the increase in the total cross-sectional area of the vessels; there are more arterioles than arteries, and many more capillaries than arterioles. Also, blood moves much slower in the capillaries than it does in the aorta, to allow time for the exchange of molecules between the blood and the tissues.

Blood Flow in Veins and Venules

Movement of blood in the veins is not due to blood pressure, but to skeletal muscle contraction. When skeletal muscles contract, they press against the weak walls of the veins, causing the blood to move past a valve (see fig.

12.8c). Once past the valve, blood is prevented from flowing backward. Blood flow gradually increases in the venous system, due to a progressive reduction in the cross-sectional area, as small venules join to form veins. Because blood pressure in the veins is low, the return of venous blood to the heart depends on several adaptations. As already mentioned, the squeezing action produced by skeletal muscle contraction and the presence of one-way valves move blood in veins. These adaptations, along with pressure differences between the thoracic and abdominal cavities, help return venous blood to the heart. As noted in chapter 14, when a person inhales, the diaphragm contracts, creating a partial vacuum in the thoracic cavity. Because pressure is higher in the abdominal cavity, the pressure difference causes blood to move from the abdominal to the thoracic veins.

Blood pressure steadily decreases with distance from the heart's left ventricle. Blood pressure causes the flow of blood in the arteries and arterioles. Skeletal muscle contraction causes the flow of blood in the venules and veins, and valves prevent backflow of blood.

Hypertension

Blood pressure is usually measured in the brachial artery with a sphygmomanometer, an instrument that records changes in terms of millimeters of mercury. Normal resting blood pressure for a young adult is 120/80. The higher number is the **systolic pressure,** the pressure recorded in an artery when the left ventricle contracts, and the lower number is the **diastolic pressure,** the pressure recorded in an artery when the left ventricle relaxes. About 20% of all Americans are estimated to have **hypertension** (hi"per-ten'shun), or high blood pressure. Reasons for the development of hypertension are varied, as discussed in the MedAlert reading on page 232. One possible reason is associated with the kidney's secretion of renin (chapter 10). The renin-angiotensin-aldosterone system leads to absorption of sodium and high blood pressure. The same effect can be brought about directly by excess salt in the diet.

> Blood pressure is a measure of the systolic and diastolic pressures. Hypertension is elevated blood pressure.

Effects of Aging

The heart generally grows larger with age, primarily because of fat deposition in the epicardium and myocardium. In many middle-aged people, the heart is covered by a layer of fat, and the number of collagenous fibers in the endocardium increases. With age, the valves, particularly the aortic semilunar valve, become thicker and more rigid.

As a person ages, the myocardium loses some of its contractile power and some of its ability to relax. The resting heart rate decreases throughout life, and the maximum possible rate during exercise also decreases. With age, the contractions become less forceful; the heart loses about 1% of its reserve pumping capacity each year after age 30.

In the elderly, arterial walls tend to thicken with plaque and become inelastic, signaling that atherosclerosis and arteriosclerosis are present. The chances of coronary thrombosis and heart attack increase with age. Increased blood pressure was once believed to be inevitable with age, but now hypertension is known to be the result of other conditions, such as kidney disease and atherosclerosis. The MedAlert reading for this chapter describes how diet and exercise in particular can help prevent atherosclerosis.

The occurrence of varicose veins increases with age, particularly in people who are required to stand for long periods. Thromboembolism as a result of varicose veins can lead to death if a blood clot settles in a major branch of a pulmonary artery. (This disorder is called pulmonary embolism.)

Working Together

The accompanying illustration shows how the circulatory system works with other organ systems of the body to maintain homeostasis. All the organ systems of the body are interrelated.

> **BodyWorks CD-ROM** 💿
> The module accompanying chapter 12 is Cardiovascular System.

working together   Circulatory System

Integumentary System

Blood vessels deliver nutrients and oxygen to skin, carry away wastes; blood clots if skin is broken.

Skin prevents water loss; helps regulate body temperature; protects blood vessels.

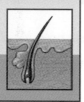

Skeletal System

Blood vessels deliver nutrients and oxygen to bones; carry away wastes.

Rib cage protects heart; red bone marrow produces blood cells; bones store Ca^{2+} for blood clotting.

Muscular System

Blood vessels deliver nutrients and oxygen to muscles; carry away wastes.

Muscle contraction keeps blood moving in heart and blood vessels.

Nervous System

Blood vessels deliver nutrients and oxygen to neurons; carry away wastes.

Brain controls nerves that regulate the heart and dilation of blood vessels.

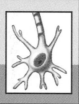

Endocrine System

Blood vessels transport hormones from glands; blood services glands; heart produces atrial natriuretic hormone.

Epinephrine increases blood pressure; ADH, aldosterone, and atrial natriuretic hormone factors help regulate blood volume; growth factors control blood cell formation.

How the Circulatory System works with other body systems

Lymphatic System/Immunity

Blood vessels transport leukocytes and antibodies; blood services lymphoid organs and is source of tissue fluid that becomes lymph.

Lymphoid organs produce and store formed elements; lymphatic vessels transport leukocytes and return tissue fluid to blood vessels; spleen serves as blood reservoir, filters blood.

Respiratory System

Blood vessels transport gases to and from lungs; blood services respiratory organs.

Gas exchange in lungs rids body of carbon dioxide, helping to regulate the pH of blood; breathing aids venous return.

Digestive System

Blood vessels transport nutrients from digestive tract to body; blood services digestive organs.

Digestive tract provides nutrients for plasma protein formation and blood cell formation; liver detoxifies blood, makes plasma proteins, destroys old red blood cells.

Urinary System

Blood vessels deliver wastes to be excreted; blood pressure aids kidney function; blood services urinary organs.

Kidneys filter blood and excrete wastes; maintain blood volume, pressure, and pH; produce renin and erythropoietin.

Reproductive System

Blood vessels transport sex hormones; vasodilation causes genitals to become erect; blood services reproductive organs.

Sex hormones influence cardiovascular health; sexual activities stimulate cardiovascular system.

Selected New Terms

Basic Key Terms

aorta (a-or'tah), p. 227

arteriole (ar-te're-ol), p. 234

artery (ar'ter-e), p. 234

atrioventricular (AV) node (a"tre-o-ven-trik'u-lar nōd), p. 229

atrioventricular valve (a"tre-o-ven-trik'u-lar valv), p. 224

atrium (a'tre-um), p. 223

bicuspid valve (bi-kus'pid valv), p. 225

capillary (kap'ĭ-lar"e), p. 235

circle of Willis (ser'kl ov wil'is), p. 241

coronary artery (kor'ŏ-na-re ar'ter-e), p. 241

diastole (di-as'to-le), p. 228

endocardium (en"do-kar'de-um), p. 223

heart (hart), p. 223

heart valve (valv), p. 224

hepatic portal system (hĕ-pat'ik por'tal sis'tem), p. 242

inferior vena cava (in-fēr'e-or ve'nah ka'vah), p. 227

interatrial septum (in"ter-a'tre-al sep'tum), p. 223

interventricular septum (in"ter-ven-trik'u-ler sep'-tum), p. 223

myocardium (mi"ō-kar'de-um), p. 223

pacemaker (pās'māk-er), p. 229

pericardial membrane (per"ĭ-kar'de-al mem'brān), p. 223

pulmonary circulation (pul'mo-ner"e ser"ku-la'shun), p. 236

pulse (puls), p. 244

Purkinje fiber (pur-kin'je fi'ber), p. 229

semilunar valve (sem"e-lu'nar valv), p. 224

sinoatrial (SA) node (si"no-a'tre-al nōd), p. 229

superior vena cava (su-pēr'e-or ve'nah ka'vah), p. 227

systemic circulation (sis-tem'ik ser"ku-la'shun), p. 236

systole (sis'to-le), p. 228

tricuspid valve (tri-kus'pid valv), p. 224

vein (vān), p. 236

ventricle (ven'trĭ-kl), p. 223

venule (ven' ŭl), p. 236

Clinical Key Terms

angina pectoris (an-ji'nah pek'to-ris), p. 232

arrhythmia (ah-rith'me-ah), p. 230

arteriosclerosis (ar-te"re-o-sklĕ-ro'sis), p. 235

atherosclerosis (ath"er-o"sklĕ-ro'sis), p. 232

bradycardia (brăd"e-kar'de-ah), p. 230

cerebrovascular accident (CVA), (ser"e-bro-vas'ku-lar ak'si-dent), p. 235

congestive heart failure (kon-jes'tiv hart fāl'yer), p. 227

coronary bypass operation (kor'ŏ-na-re bi'pas op-er-a'shun), p. 233

cyanosis (si"ah-no'sis), p. 243

electrocardiogram (e-lek"tro-kar'de-o-gram"), p. 230

fibrillation (fi"brĭ-la'shun), p. 230

heart transplant (hart tranz'plant), p. 227

hemorrhoid (hem'roid), p. 236

hypertension (hi"per-ten'shun), p. 246

ischemic heart disease (is-kem'ik hart dĭ-zĕz'), p. 232

mitral stenosis (mi'tral sten-o'sis), p. 229

myocardial infarction (mi"o-kar'de-al in-fark'shun), p. 233

occluded coronary arteries (ŏ-klood'ed kor'ŏ-na-re ar'ter-ēz), p. 233

phlebitis (flĭ-bi'tus), p. 236

plaque (plak), p. 232

pulmonary embolism (pul'mo-ner"e em'bo-lizm), p. 236

tachycardia (ta"ke-kar'de-ah), p. 230

thromboembolism (throm"bo-em'bol-lizm), p. 232

thrombolytic therapy (throm"bo-lĭ'tik ther'ah-pe), p. 233

varicose vein (var'ĭ-kos vān), p. 236

Summary

I. Anatomy of the Heart
A. The heart has a right and left side and four chambers, consisting of two atria and two ventricles. The heart valves are the tricuspid valve, the pulmonary semilunar valve, the bicuspid valve, and the aortic semilunar valve.

B. The right side of the heart pumps blood to the lungs, and the left side pumps blood to the tissues.

C. The heartbeat is divided into three phases: (1) the atria contract, (2) the ventricles contract, and (3) both the atria and ventricles rest. To put it another way: When the atria are in systole, the ventricles are in diastole, and vice versa; finally, all chambers are in diastole.

D. The heart sounds are due to the closing of the heart valves.

E. The conduction system of the heart includes the SA node, the AV node, the AV bundle, the bundle branches, and the Purkinje fibers. The SA node causes the atria to contract. The AV node and the rest of the conductive system cause the ventricles to contract.

F. The heartbeat rate is regulated largely by the autonomic nervous system.

G. Myocardial infarction is often preceded by atherosclerosis, angina pectoris, and thromboembolism.

II. Vascular System
A. Arteries and arterioles carry blood away from the heart, veins and venules carry blood to the heart, and capillaries join arterioles to venules.

B. The pulmonary arteries transport oxygen-poor blood to the lungs, and the pulmonary veins return oxygen-rich blood to the heart.

C. The systemic circulation transports blood from the left ventricle of the heart to the arteries, arterioles, and capillaries, and then from the capillaries to the venules and veins to the right atrium of the heart. It serves the body proper.

D. All the arteries in the systemic circulation can eventually be traced from the aorta.

E. All the veins in the systemic circulation can be traced to the venae cavae.

F. Circulation to the brain includes the circle of Willis, which protects the brain from reduced blood supply.

G. Circulation to the heart is dependent upon the proper functioning of the coronary arteries.

H. The hepatic portal system carries blood from the stomach and intestines to the liver.

I. Fetal circulation includes four unique features: (1) the foramen ovale, (2) the ductus arteriosus, (3) the umbilical arteries and vein, and (4) the ductus venosus. These features are necessary because the fetus does not use its lungs for gas exchanges.

III. Features of the Circulatory System
A. The pulse rate indicates the heartbeat rate.

B. Blood pressure is dependent on cardiac output and peripheral resistance. These are regulated by the activation of autonomic nerve impulses and by the release of certain hormones.

C. Blood pressure steadily decreases with distance from the heart's left ventricle. Blood pressure causes the flow of blood in the arteries and arterioles. Skeletal muscle contraction causes the flow of blood in the venules and veins, and valves prevent backflow of blood.

D. Blood pressure is a measure of the systolic and diastolic pressures. Hypertension is elevated blood pressure.

Study Questions

1. Describe the structure of the heart, including its chambers and valves. (pp. 223–25)

2. Trace the path of blood in the pulmonary circuit, as it travels from and returns to the heart. (p. 227)

3. Describe the cardiac cycle (using the terms *systole* and *diastole*), and explain the heart sounds. (pp. 228–29)

4. Describe the circulatory disorders that can lead to a heart attack. (pp. 232–33)

5. What types of blood vessels are in the body? Discuss their structure and function. (pp. 234–36)

6. Trace the path of blood from the mesenteric arteries to the aorta, indicating which of the vessels are in the systemic circulation and which are in the pulmonary circulation. (pp. 236–38)

7. What is blood pressure, and what does 120/80 mean? (pp. 244, 245)

8. In which type of vessel is blood pressure highest? In which type of vessel is it lowest? In which type of vessel is blood velocity lowest and why? Why is this beneficial? What factors assist venous return of the blood? (p. 245)

Fill in the blanks.

1. When the left ventricle contracts, blood enters the _____ .
2. The right side of the heart pumps blood to the _____ .
3. The _____ node is known as the pacemaker.
4. Arteries are blood vessels that take blood _____ the heart.
5. The blood vessels that serve the heart are the _____ arteries and veins.
6. The major blood vessels taking blood to the arms are the _____ arteries and veins. Those taking blood to the legs are the _____ arteries and veins.
7. Blood vessels to the brain end in a circular path known as the _____ .
8. The human body contains a hepatic portal system that takes blood from the _____ to the _____ .
9. The force of blood against the walls of a vessel is termed _____ .
10. Blood moves in arteries due to _____ and in veins due to _____ .
11. The blood pressure recorded when the left ventricle contracts is called the _____ pressure, and the pressure recorded when the left ventricle relaxes is called the _____ pressure.
12. The two factors that affect blood pressure are _____ and _____ .
13. In the fetus, the opening between the two atria is called the _____ , and the connection between the pulmonary artery and the aorta is called the _____ .
14. The valve between the left atrium and left ventricle is the _____ or mitral valve.

Medical Terminology Reinforcement Exercise

Consult Appendix B for help in pronouncing and analyzing the meaning of the terms that follow.

1. cryocardioplegia (kri-o-kar″de-o-ple′je-ah)
2. echocardiography (ek″o-kar″de-og′rah-fe)
3. percutaneous transluminal coronary angioplasty (per″ku-ta′ne-us trans″loo′mĭ-nal kor′ŏ-na-re an′je-o-plas″te)
4. vasoconstriction (vas″o-kon-strik′shun)
5. valvuloplasty (val′vu-lo-plas″te)
6. arteriosclerosis (ar-te″re-o-sklĕ-ro′sis)
7. tachycardia (tak″e-kar′de-ah)
8. antihypertensive (an″tĭ-hi″per-ten′siv)
9. arrhythmia (ah-rith′me-ah)
10. thromboendarterectomy (throm″bo-end″ar-ter-ek′to-me)

Website Link

For a listing of the most current Websites related to this chapter, please visit the Mader home page at:

http://www.mhhe.com/maderap

chapter 13

The Lymphatic System and Immunity

Chapter Outline and Learning Objectives

After you have studied this chapter, you should be able to:

Lymphatic System (p. 252)
- Describe the structure and functions of the lymphatic system.
- Describe the structure and function of lymph nodes.
- Describe the structures and functions of the thymus, spleen, and red bone marrow.

Immunity (p. 256)
- Describe the body's nonspecific defense mechanisms.
- Contrast antibody-mediated immunity with cell-mediated immunity.

Immunotherapy (p. 262)
- Describe how to provide an individual with active and passive immunity.
- Give examples of immunotherapeutic drugs.

Immunological Side Effects and Illnesses (p. 266)
- Give examples of how the immune system overdefends and underdefends the body.

Effects of Aging (p. 267)
- Anatomical and physiological changes occur in the lymphatic system as we age.

Working Together (p. 267)
- The lymphatic system works with other systems of the body to maintain homeostasis.

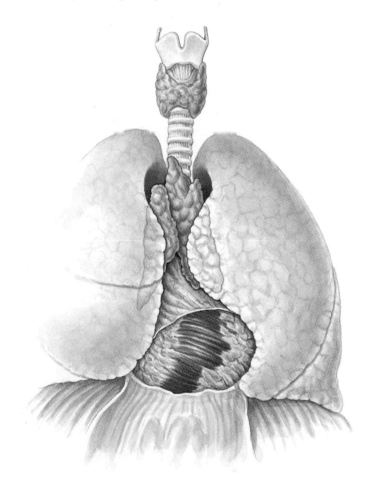

The thymus, which is found between the lungs, is essential to the development of certain lymphocytes, called T lymphocytes.

figure 13.1 Lymphatic system. **a.** Distribution of lymphatic and lymphoid organs. **b.** Thymus gland and micrograph of a lobule. **c.** Lymph node and micrograph of lymph nodule.

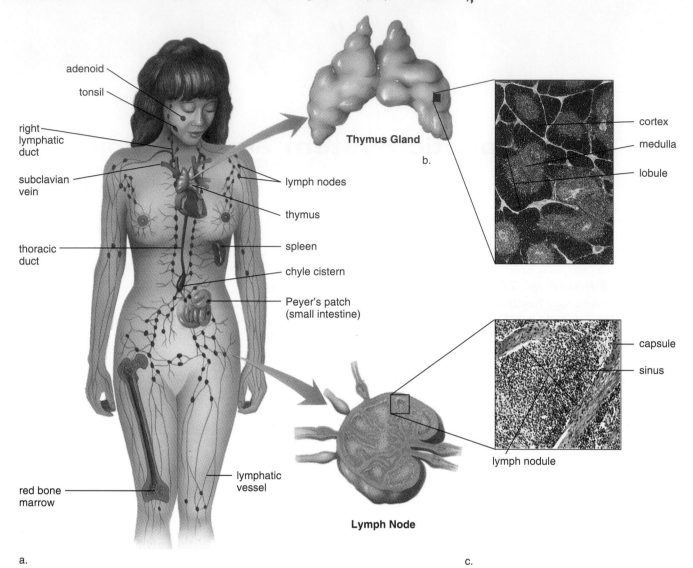

Thymus Gland

b.

cortex

medulla

lobule

adenoid

tonsil

right lymphatic duct

subclavian vein

thoracic duct

lymph nodes

thymus

spleen

chyle cistern

Peyer's patch (small intestine)

lymphatic vessel

red bone marrow

capsule

sinus

lymph nodule

Lymph Node

a.

c.

Lymphatic System

The **lymphatic** (lim-fat′ik) **system** consists of lymphatic vessels and the lymphoid organs (fig. 13.1). This system, which is closely associated with the cardiovascular system, has three main functions:

1. Lymphatic vessels take up excess tissue fluid and return it to the bloodstream.
2. Lymphatic capillaries absorb fats at the intestinal villi and transport them to the bloodstream (p. 303).
3. The lymphatic system helps to defend the body against disease.

The lymphatic system: (1) collects excess tissue fluid, (2) absorbs fat molecules in the intestines, and (3) plays a major role in the body's defense against disease.

Lymphatic Vessels

Lymphatic vessels contain lymph. **Lymph** (limf), which is excess tissue fluid, consists mostly of water and some plasma proteins that have leaked out of capillaries. Lymphatic vessels are quite extensive; every region of the body is supplied richly with lymphatic capillaries. The structure of the larger lymphatic vessels is similar to that of cardio-

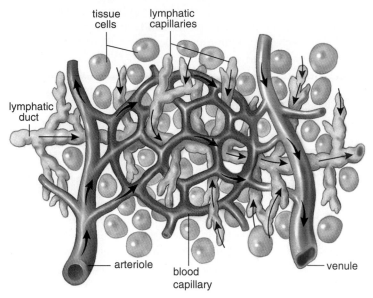

figure 13.2 Location and structure of lymphatic capillaries. Arrows indicate the flow of lymph in lymphatic vessels and blood in blood vessels. ✗

tissue cells

lymphatic capillaries

lymphatic duct

arteriole

blood capillary

venule

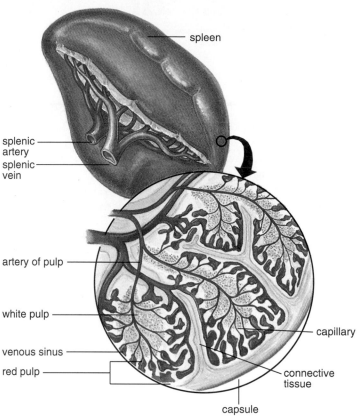

figure 13.3 Structure of the spleen. ✗

spleen

splenic artery

splenic vein

artery of pulp

white pulp

capillary

venous sinus

red pulp

connective tissue

capsule

vascular veins, including the presence of *valves*. Also, the movement of lymph within these vessels is dependent upon skeletal muscle contraction. When the muscles contract, lymph is squeezed past a valve that closes, preventing lymph from flowing backward.

The lymphatic system is a one-way system. The system begins with lymphatic capillaries that lie near blood capillaries. The lymphatic capillaries take up fluid that exited from, and was not reabsorbed by, the blood capillaries (fig. 13.2). Once tissue fluid enters the lymphatic vessels, it is called lymph. The lymphatic capillaries join to form lymphatic vessels that merge in the thoracic cavity before entering one of two ducts: the thoracic duct or the right lymphatic duct.

The *thoracic duct* is much larger than the right lymphatic duct. It serves the lower extremities, abdomen, left arm, and the left side of the head and neck. In the thorax, the left thoracic duct enters the left subclavian vein. The *right lymphatic duct* serves only the right arm and the right side of the head and the neck; it enters the right subclavian vein.

The lymphatic system is a one-way system. Lymph flows from a lymphatic capillary to ever-larger lymphatic vessels, and finally, to a lymphatic duct, which enters a subclavian vein.

Lymphoid Organs

The lymphoid organs are so-called because they contain lymphocytes. The lymphoid organs of special interest are the spleen, lymph nodes, thymus, and red bone marrow (see fig. 13.1).

Spleen

The **spleen** is located in the upper left abdominal cavity, just beneath the diaphragm. Its construction is similar to that of a lymph node: The outer connective tissue divides the organ into sinus-containing lobules (fig. 13.3). In the spleen, however, the sinuses are filled with blood instead of lymph. Since the blood vessels of the spleen can expand, this organ serves as a blood reservoir and makes blood available in times of low pressure or when the body needs extra oxygen in the blood.

The lobules of the spleen contain red pulp and white pulp. Red pulp contains red blood cells, lymphocytes, and macrophages. White pulp forms nodules that contain only lymphocytes and macrophages. Both types of pulp help to purify the blood that passes through the spleen. If the spleen ruptures due to injury, it can be removed. Although the spleen's functions are duplicated by other organs, an individual whose spleen has been removed is slightly more susceptible to certain bacterial infections because of a loss of B lymphocytes, discussed later in the chapter.

The spleen is divided into sinus-containing lobules in which blood is cleansed by lymphocytes and macrophages.

Lymph Nodes and Illnesses

*t*he internal structure of a lymph node is designed to filter out any foreign material from the lymph. An infection that causes swelling and tenderness of nearby lymph nodes is called **lymphadenitis** (lim-fad″ĕ-ni′tis). If the infection is not contained, then **lymphangitis** (lim″fan-ji′tis), an infection of the lymphatic vessels, may result. Red streaks can be seen through the skin, indicating that the infection may spread to the bloodstream.

Failure of the lymphatic vessels to remove tissue fluid results in an accumulation of tissue fluid and is called **edema** (ĕ-de′mah). A dramatic example of edema occurs when a parasitic roundworm clogs the lymphatic vessels, resulting in tremendous swelling of the arm, leg, or external genitals, a condition called **elephantiasis** (ĕ″lah-fun-ti′ah-sis). Edema can also be due to a low osmotic pressure of the blood, as when plasma proteins are excreted by the kidneys. In this case, more tissue fluid than usual forms, and lymphatic vessels are unable to absorb it. **Pulmonary edema** is a life-threatening condition associated with congestive heart failure. Due to a weak heart, blood backs up in the pulmonary circulation, causing an increase in blood pressure, which leads to excess tissue fluid. The walls of the air sacs in the lungs may rupture, and the patient may suffocate.

When surgery is used in the diagnosis and/or treatment of cancer, regional lymph nodes are usually removed for examination. The presence or absence of tumor cells in the nodes can be used to determine how far the disease has spread and aid in the decision concerning additional treatment, such as radiation or chemotherapy. Cancer of lymphoid tissue itself is called **lymphoma.** In **Hodgkin's disease,** billions of lymphoma cells create swollen lymph nodes in the neck. The lymphoma cells can migrate and grow in the spleen, liver, and bone marrow. Prognosis is good, however, if Hodgkin's disease is diagnosed early.

Lymph Nodes

At certain points along lymphatic vessels are small (about 1–25 mm), ovoid or round structures called **lymph nodes.** A lymph node has a fibrous connective tissue capsule (see fig. 13.1). Connective tissue also divides a node into nodules, each of which contains a sinus (open space) filled with many lymphocytes and macrophages. As lymph passes through the sinuses, it is purified of infectious organisms and any other debris.

While nodules usually occur within lymph nodes, they can also occur singly or in groups. The **tonsils,** which are partly encapsulated lymph nodules, are located in the pharynx. The pharyngeal tonsils, or adenoids, are situated in the nasopharynx. The palatine tonsils are located in the posterior lateral wall of the oropharynx. The lingual tonsils are at the base of the tongue in the oropharynx (see fig. 14.2). Other nodules called *Peyer's patches* are located within the intestinal wall.

The lymph nodes occur in groups in certain regions of the body. For example, the inguinal nodes are in the groin, and the axillary nodes are in the armpits (see fig. 13.1). The Medical Focus reading on this page examines illnesses associated with the lymph nodes.

Lymph nodes are divided into sinus-containing nodules in which the lymph is cleansed of infectious organisms and debris.

Thymus

The **thymus** is located along the trachea atop the heart and posterior to the sternum in the upper thoracic cavity. This gland varies in size, but it is larger in children than in adults and may eventually almost disappear. The thymus is divided into lobules by connective tissue. T lymphocytes (or T cells), discussed later in the chapter, mature in these lobules. Those in the interior (medulla) are more mature than those in the exterior (cortex) of a lobule. Mature T lymphocytes have survived a critical test: If any show the ability to react with "self" cells, they die. If they have the potential to attack a foreign cell, they leave the thymus.

The thymus secretes *thymosin,* a hormone that is believed to be an inducing factor; that is, it causes pre-T cells to become T (for thymus) cells. Thymosin may also have other functions in immunity.

The thymus is divided into lobules, where lymphocytes are produced.

figure 13.4 The five types of white blood cells, which differ according to structure and function. The frequency of each type of cell is given as a percentage of the total.

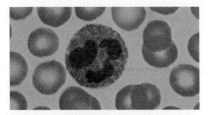

Neutrophil
40–70%
Phagocytizes
primarily bacteria

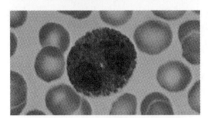

Eosinophil
1–4%
Phagocytizes and destroys
antigen-antibody
complexes

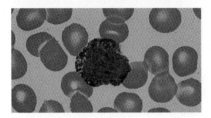

Basophil
0–1%
Releases histamine
when stimulated

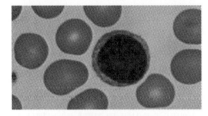

Lymphocyte
20–45%
B type produces antibodies
in blood and lymph;
T type kills virus-
containing cells.

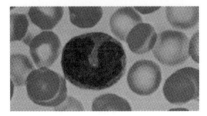

Monocyte
4–8%
Becomes macrophage—
phagocytizes bacteria and
viruses

Red Bone Marrow

Red bone marrow is the site of origination for all blood cells, including the white blood cells that function in immunity (fig. 13.4). In the adult, red bone marrow is present only in the bones of the skull, sternum, ribs, clavicle, pelvis, and spinal column, and in the ends of the femur and humerus.

Bone Marrow Biopsy and Transplant

*b*one marrow biopsies are performed to study bone marrow samples for abnormalities or for possible donation. The procedure can be done in the doctor's office. With the patient lying on his or her stomach or side, a large needle is positioned perpendicular to the pelvis and pushed into the bone, using a screwing motion. When the needle is deep enough in the bone to be anchored, a syringe is attached in order to remove a sample of bone marrow.

Bone marrow transplants are sometimes performed in an effort to cure leukemia or other types of blood disorders. First, the recipient's diseased bone marrow is destroyed by powerful drugs and radiation therapy. The donor's marrow, which has been treated as necessary, is then injected into the recipient's bloodstream. The injected stem cells are expected to migrate to the recipient's marrow, perform hematopoiesis, and produce new formed elements. As with any other transplant, bone marrow transplants require careful matching of donor and recipient tissue and administration of drugs to suppress the immune system to avoid transplant rejection.

Red bone marrow consists of a network of connective tissue fibers, called reticular fibers, produced by reticular cells. These cells, along with the cells that develop into blood cells, are packed around thin-walled venous sinuses. Differentiated blood cells (see fig. 11.4) enter the bloodstream at these sinuses. The Medical Focus reading on this page explains bone marrow biopsies and transplants.

Red bone marrow produces white blood cells necessary for the development of immunity.

Immunity

Immunity is the body's ability to defend itself against infectious organisms, foreign cells, and even body cells that have gone awry, such as cancer cells. Immunity includes nonspecific and specific defenses.

Nonspecific Defenses

Three nonspecific defenses are useful against all types of pathogens: barriers to entry, the inflammatory reaction, and protective proteins.

Barriers to Entry

The skin and the mucous membrane lining the respiratory and digestive tracts are mechanical barriers to entry by **pathogens,** which are disease-causing agents like bacteria and viruses. Secretions of the oil glands in the skin contain chemicals that weaken or kill bacteria. The respiratory tract is lined with ciliated cells that sweep mucus and trapped particles into the throat, where they can be expectorated or swallowed. In addition, the stomach has an acidic pH that inhibits the growth of many types of bacteria. A mix of bacteria that normally reside in the intestine and other organs, such as the vagina, prevent pathogens from colonizing vulnerable tissues.

Inflammatory Reaction

When the skin is broken due to a minor injury, a series of events occurs that is known as the **inflammatory reaction** because of the reddening and swelling at the site of the injury. Figure 13.5 illustrates the participants in the inflammatory reaction. One participant, the mast cells, may be derived from basophils and are found lodged within connective tissues.

Following an injury, capillaries and several tissue cells are apt to rupture and to release **histamine,** a molecule that causes a capillary to dilate and to become more permeable. The enlarged capillaries cause the skin to redden, and their increased permeability allows proteins and fluids to escape, resulting in swelling. The increased fluid pressure in the injured region stimulates free nerve endings, resulting in pain.

Any break in the skin allows pathogens to enter the body. Neutrophils and monocytes are amoeboid and can squeeze through capillary walls to enter tissue fluid, where they carry on **phagocytosis** (fag"o-si-to'sis), also called cell eating (see fig. 3.9). When a neutrophil phagocytizes a bacterium, an intracellular vacuole is formed. The engulfed bacterium is destroyed by hydrolytic enzymes when the vacuole combines with one of the neutrophil's granules.

Monocytes differentiate into **macrophages** (mak'ro-faj"ez)—large, phagocytic cells that can devour dozens of invaders and still survive. Some organs, like the liver, kidney, spleen, and brain, have resident macrophages that routinely act as scavengers, devouring old blood cells, bits of dead tissue, and other debris. Macrophages are also capable of bringing about an explosive increase in the number of leukocytes by liberating a growth factor that stimulates the production and release of white blood cells, usually neutrophils from bone marrow.

As an infection is being overcome, some neutrophils die. These neutrophils, along with dead tissue, cells, bacteria, and living white blood cells, form **pus,** a thick, yellowish fluid. The presence of pus indicates that the body is trying to overcome the infection.

> The inflammatory reaction is a "call to arms": It marshals phagocytic white blood cells to the site of bacterial invasion.

Protective Proteins

The **complement system,** often simply called complement, consists of a number of plasma proteins designated by the letter C and a number or letter. Once a complement protein is activated, it, in turn, activates another protein in a set series of reactions. Only a small amount of activated protein is needed because of a cascade response: Each protein in the series is capable of activating many proteins next in line.

Complement is activated when pathogens enter the body. One series of reactions is complete when complement proteins form holes in bacterial cell walls and membranes. These holes allow fluids and salt to enter the bacterial cell until it bursts (fig. 13.6).

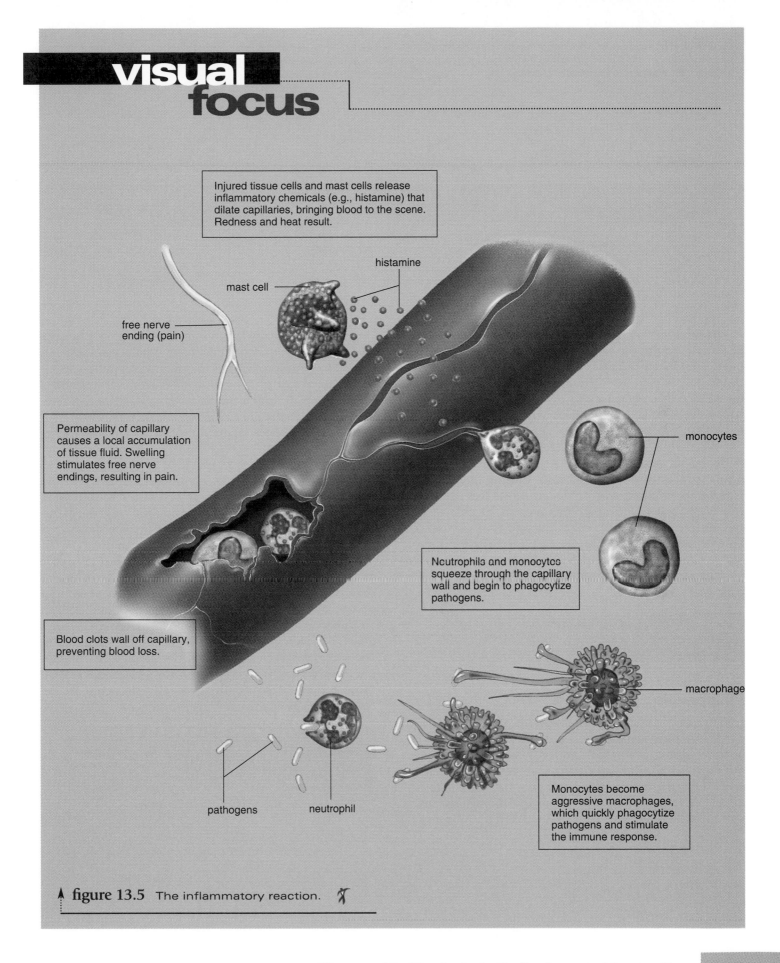

Injured tissue cells and mast cells release inflammatory chemicals (e.g., histamine) that dilate capillaries, bringing blood to the scene. Redness and heat result.

histamine

mast cell

free nerve ending (pain)

Permeability of capillary causes a local accumulation of tissue fluid. Swelling stimulates free nerve endings, resulting in pain.

monocytes

Neutrophils and monocytes squeeze through the capillary wall and begin to phagocytize pathogens.

Blood clots wall off capillary, preventing blood loss.

macrophage

pathogens   neutrophil

Monocytes become aggressive macrophages, which quickly phagocytize pathogens and stimulate the immune response.

figure 13.5 The inflammatory reaction.

figure 13.6 Action of the complement system against a bacterium. When complement system in plasma is activated, complement proteins form holes in bacterial cell walls and membranes, allowing fluids and salts to enter until the cell eventually bursts.

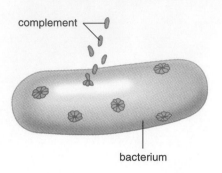
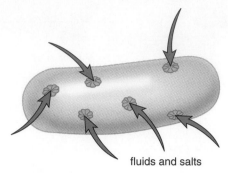
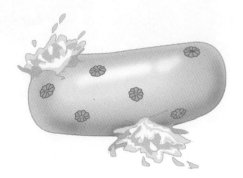

Complement proteins form holes in the bacterial cell wall and membrane.

Holes allow fluids and salts to enter the bacterium.

Bacterium expands until it bursts.

Complement also activates chemicals that attract phagocytes to the site and induce inflammation. Complement completes certain immune responses, which accounts for its name. For example, some proteins bind to the surface of pathogens already coated with antibodies, which ensures that the pathogens will be phagocytized (see fig. 13.5).

When viruses infect a tissue cell, the infected cell produces and secretes interferon. **Interferon** binds to receptors on the surface of noninfected cells, causing them to prepare for possible attack by producing substances that interfere with viral replication. Interferon is specific to the species; therefore, only human interferon can be used in humans. Formerly, collecting enough interferon for clinical and research purposes was difficult, but interferon is now a product of **biotechnology** and is produced by genetically engineered bacteria.

> Nonspecific defenses against pathogens include barriers to entry, the inflammatory reaction, and protective proteins (the complement system).

Specific Defenses

Sometimes, the body relies on a specific defense against a particular **antigen** (usually a protein molecule that the body recognizes as foreign). Antigens occur on pathogens, but they can also be part of a foreign cell or a cancerous cell. Ordinarily, a person does not become immune to his or her body's normal cells; therefore, the immune system is able to distinguish "self" from "nonself" (fig. 13.7).

Immunity usually lasts a long time. For example, once a person has had measles, he or she usually cannot be infected by the measles virus a second time. Immunity is primarily the result of the action of **B lymphocytes** and **T lymphocytes.** B, which stands for bone marrow, lymphocytes mature in the bone marrow, and T, which stands for thymus, lymphocytes mature in the thymus gland. B lymphocytes, also called *B cells*, become plasma cells that produce **antibodies,** proteins capable of combining with and inactivating antigens. These antibodies are secreted into the blood, lymph, and mucus. In contrast, T lymphocytes, also called *T cells*, do not produce antibodies. Instead, certain T cells directly attack cells bearing antigens they recognize. Other T cells regulate the immune response.

Lymphocytes can recognize antigens because they have receptor molecules on their surfaces. The shape of the receptors on any particular lymphocyte is appropriate to a portion of one specific antigen. The receptor and antigen are said to fit together like a lock and key. During a person's lifetime, the body is estimated to encounter 1 million different antigens; therefore, it needs 1 million different lymphocytes with specific receptors for protection against those antigens. Despite this great diversity, none of the lymphocytes ordinarily attacks the body's own cells.

> Two types of lymphocytes provide specific immunity. B cells produce and secrete antibodies that combine with antigens. Certain T cells directly attack antigen-bearing cells, and others regulate the immune response.

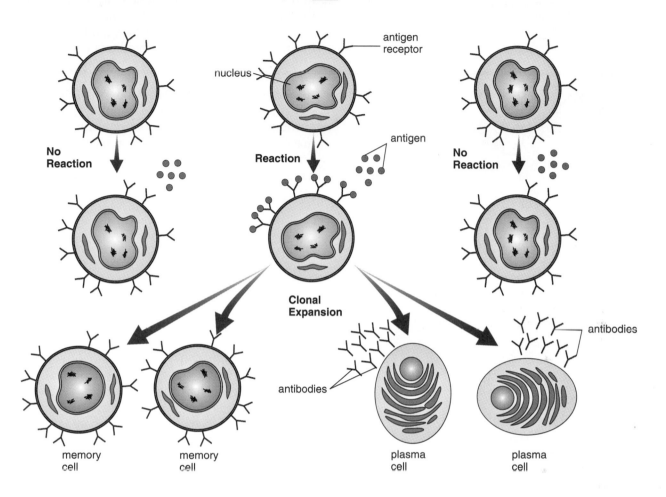

figure 13.7 Clonal selection theory as it applies to B cells. An antigen activates the appropriate B cell, which undergoes clonal expansion if stimulated by a helper T cell. During the process, many plasma cells, which produce specific antibodies against this antigen, are produced. Memory B cells, which retain the ability to secrete these antibodies, are produced also.

Antibody-Mediated Immunity

The receptor on a B cell is called a membrane-bound antibody because antibodies are the secreted form of the B-cell receptor. A B cell is activated when it encounters a bacterial cell or a toxin bearing an appropriate antigen; that is, the B cell has the potential to divide many times and produce many **plasma cells** that will secrete antibodies against this antigen (fig. 13.7). All of the plasma cells derived from one parent B lymphocyte are called clones, and a clone produces one type of antibody. A B cell does not clone until its antigen is present. The *clonal selection theory* states that the antigen selects the B cell that will produce a clone of plasma cells.

Once antibody production is sufficient, the antigen disappears from the system, and the development of plasma cells ceases. However, some members of a clone do not participate in antibody production; instead, they remain in the bloodstream as **memory B cells**, which are capable of producing the antibody specific to a particular antigen for some time, even as long as a lifetime in some instances. As long as these cells are present, the individual is said to be actively immune—that is, future antibody production is possible because the memory B cells can form into more plasma cells if the same antigen invades the system again.

Defense by B cells is called **antibody-mediated immunity,** or humoral immunity, because B cells produce antibodies that are present in the bloodstream.

> B cells are responsible for antibody-mediated immunity. Each one produces a specific type of antibody to counteract a particular infection.

figure 13.8 Structure of the most common antibody (IgG). **a.** An IgG antibody contains two heavy (long) amino acid chains and two light (short) amino acid chains, there are two variable regions, where a particular antigen is capable of binding with the antibody. **b.** Computer model of an antibody.

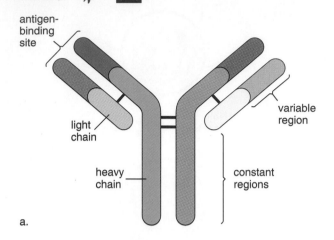

a.

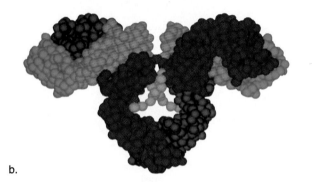

b.

Each antibody contains variable regions that bind to an antigen in a lock-and-key manner and a constant region (fig. 13.8). Antibodies can be classified according to their constant regions. Table 13.1 lists the different classes of antibodies and their specific functions. Most antibodies in the blood belong to the class IgG (immunoglobulin G).

The antigen-antibody reaction can take several forms, but frequently, the reaction produces complexes of antigens combined with antibodies. When viruses and bacterial toxins have combined with specific antibodies, they cannot attach to target cells. An antigen-antibody complex, sometimes called the immune complex, marks the antigen for destruction by other forces. For example, the complex may be engulfed by neutrophils or macrophages, or it may activate complement. In this case, complement makes pathogens more susceptible to phagocytosis.

An antibody combines with its antigen in a lock-and-key manner. The antibody-antigen reaction can lead to complexes that contain several molecules of antibody and antigen. These complexes are then destroyed by phagocytic cells or complement.

Cell-Mediated Immunity

T cells are responsible for **cell-mediated immunity.** Their receptors "see" an antigen held in a cleft formed by a **major histocompatibility (MHC) protein**[1] that lies on the T-cell surface. The importance of MHC proteins was first recognized when researchers discovered that they contribute to specificity of tissues and make it difficult to transplant tissue from one person to another. In other words, the donor and the recipient must be histo (tissue) compatible (the same or nearly so) for a transplant to be successful.

table 13.1

Antibodies

Classes	Presence	Function
IgG	Main antibody type in circulation	Attacks pathogens and bacterial toxins; enhances phagocytosis
IgA	Main antibody type in secretions, such as saliva and milk	Attacks pathogens and bacterial toxins
IgE	Antibody type found as membrane-bound receptor on basophils in blood and on mast cells in tissues	Responsible for allergic reactions
IgM	Antibody type found in circulation; largest antibody	Activates complement, clumps cells
IgD	Antibody type found as a membrane-bound receptor	Functions unknown

[1]Often called human leukocyte associated protein in humans.

figure 13.9 T-cell activation. A macrophage phagocytizes a microbe and digests it in a vesicle. An antigen from the microbe is combined with an MHC protein, and the complex is presented to the T cell. Then the T cell is ready to find and destroy other cells that bear the same antigen. 🏃 ▭

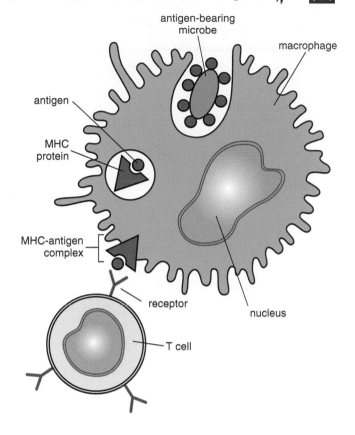

figure 13.10 Cell-mediated immunity. Scanning electron micrograph of killer T cells attacking a cancer cell. 🏃 ▭

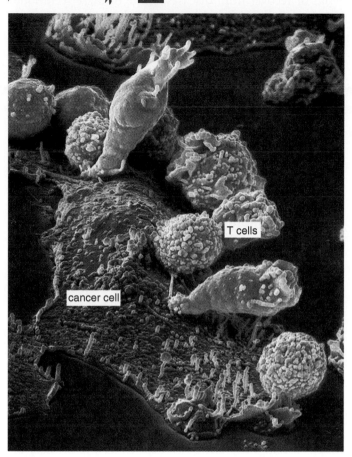

T cells are activated by macrophages that "present" a complex consisting of the antigen within an MHC protein to them (fig. 13.9). Langerhans cells also present antigens to T cells. Langerhans cells occur in the skin and are responsible for initiating a defense against bacteria that have penetrated the outermost layers of epidermis and against superficial skin cancers.

After being activated, some T cells, called *helper T cells,* stimulate B cells. They also release lymphokines, "messenger" proteins that stimulate the immune system. The AIDS virus attacks helper T cells, which accounts for the fact that a person with AIDS has an impaired immune system.

Other types of T cells divide and become "killer cells." *Killer cells* exhibit cytotoxicity, meaning that direct contact between a killer cell and a target cell (bearing a specific antigen) causes death of the target cell (fig. 13.9). Killer T cells specialize in providing resistance against foreign antigens. They attack any tissue cell that bears a foreign antigen, including virus-infected cells, the cells of a transplanted organ or donated skin graft, or cancer cells. Cancer cells are abnormal cells that most likely display altered antigens. As long as T cells are capable of recognizing newly developed cancer cells, cancer cannot grow or spread (fig. 13.10). When a T cell attacks a cell, it releases chemicals that perforate the cell's membrane, causing the cell to burst before shriveling up and dying.

T cells are responsible for cell-mediated immunity. They directly attack cells that bear antigens. For example, they protect the body against cancer but also attack transplanted organs.

Immunotherapy

The immune system can be manipulated to help people avoid or recover from diseases. Some of these techniques have been utilized for a long time, and others are relatively new.

Active Immunity

Active immunity, which provides long-lasting protection against a disease-causing organism, develops after an individual is infected with a virus or bacterium. In many instances today, however, it is not necessary to suffer an illness to become immune because it is possible to be artificially immunized against a disease. One recommended immunization schedule for children is given in the Medical Focus reading on the next page.

Immunization involves the use of **vaccines,** substances that contain an antigen to which the immune system responds. Traditionally, vaccines have been the pathogens themselves, which have been treated so that they are no longer able to cause disease. New methods of producing vaccines, however, are being developed through biotechnology. For example, as discussed in the Medical Focus reading on the next page, biotechnology is responsible for a new hepatitis B vaccine. Vaccines for other serious illnesses, such as malaria, Lyme disease, and AIDS, are also being developed. The MedAlert reading on page 264 discusses the AIDS vaccine.

After a vaccine is given, it is possible to determine the amount of antibody present in a sample of blood serum, which is called the **antibody titer** (ti'ter). A primary response follows the first exposure to an antigen. For a period of several days, no antibodies are present; then, there is a slow rise in the titer, followed by a gradual decline (fig. 13.11).

A secondary response may occur after a second exposure. If so, the titer rises rapidly to a level much greater than before. The second exposure is often called the *booster* because it "boosts" the antibody titer to a higher level. The antibody titer then may be high enough to prevent disease symptoms even if the individual is exposed to the disease. If so, the individual is immune to that particular disease.

A good secondary response can be related to the number of plasma cells and memory cells in the blood serum. Although it is customary to test the serum, the memory cells in lymph nodes and the spleen are probably more important than those that are blood-borne. Upon second exposure, these cells are already present, and antibodies can be produced rapidly.

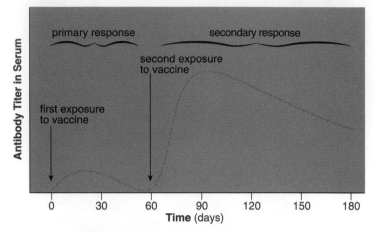

figure 13.11 Immunization responses. The primary response after the first exposure to a vaccine is minimal, but the secondary response after the second exposure may show a dramatic rise in the amount of antibody present in blood serum.

Active (long-lived) immunity can be induced by vaccines when a person is well and in no immediate danger of contracting an infectious disease. Active immunity is dependent upon the presence of memory cells in the body.

Passive Immunity

Passive immunity occurs when an individual is given antibodies (immunoglobulins) to combat a disease. Since these antibodies are not produced by the individual's B cells, passive immunity is short-lived. For example, newborn infants possess passive immunity because antibodies have crossed the placenta from the mother's blood. These antibodies soon disappear, however, so that within a few months, infants become more susceptible to infections. Breast-feeding prolongs the passive immunity an infant receives from the mother because of antibodies in the mother's milk.

Even though passive immunity does not last, it is occasionally used to prevent illness in a patient who has been unexpectedly exposed to an infectious disease. Usually, the person receives an injection of an antibody-containing serum that may have been taken from donors who have recovered from the illness. In the past, horses were immunized, and serum was taken from them to provide the antibodies needed to combat diphtheria, botulism, and

Immunization: The Great Protector

*i*mmunization protects children and adults from diseases. The success of immunization is witnessed by the fact that the smallpox vaccination is no longer required because the disease has been eradicated. However, parents today often fail to get their children immunized because they do not realize the importance of immunizations or cannot bear the expense. Newspaper accounts of an outbreak of measles at a U.S. college or hospital, therefore, are not uncommon because many adults were not immunized as children.

Figure 13A shows a recommended immunization schedule for children, and the United States is now committed to the goal of immunizing all children against the common types of childhood diseases listed. Diphtheria, whooping cough, and Haemophilus influenza infection are all life-threatening respiratory diseases. Tetanus is characterized by muscular rigidity, including a locked jaw. These extremely serious infections are all caused by bacteria; the rest of the diseases listed are caused by viruses. Polio is a type of paralysis; measles and rubella, sometimes called German measles, are characterized by skin rashes; and mumps is characterized by enlarged parotid and other salivary glands.

Adults, rather than children, are more likely to contract a disease through sexual contact. Although hepatitis B virus (HVB) is a blood-borne pathogen, it is spread in the United States mainly by sexual contact and intravenous drug use. Health-care workers who are exposed to blood or blood products are also at risk. Maternal-neonatal transmission is also a possibility. Recovery from an initial bout of hepatitis (inflammation of the liver) can lead to chronic hepatitis and then cancer of the liver. Fortunately, biotechnology (see chapter 19) has now produced a hepatitis B vaccine. Other vaccines for sexually transmitted diseases that soon may be available through biotechnology are AIDS (p. 264), herpes, and chlamydia vaccines.

Even though bacterial infections (for example, tetanus) can be cured by antibiotic therapy, it is better to be immunized. Some patients are allergic to antibiotics, and the reaction can be fatal. In addition, antibiotics not only kill off disease-causing bacteria, they also reduce the number of beneficial bacteria in the intestinal tract and elsewhere. These beneficial bacteria may have checked the spread of pathogens that now are free to multiply and to invade the body. This is why antibiotic therapy is often followed by a secondary infection, such as a vaginal yeast infection in women. Antibiotic therapy also leads to resistant bacterial strains that are difficult to cure, even with antibiotics. Resistant strains of bacteria now cause gonorrhea, a disease that has no vaccine.

Therefore, everyone should avail themselves of appropriate vaccinations. Preventing a disease by becoming actively immune to it is preferable to becoming ill and needing antibiotic therapy to be cured.

Vaccine	Age (Months)	Age (Years)
HepB* (hepatitis B)	Birth, 2, 4, 6, 12–15	11–12
DTP† (diphtheria, tetanus, whooping cough)	2, 4, 6, 15–18	4–6
Td‡ (adult tetanus)		11–12, 14–16
OPV§ (oral polio vaccine)	2, 4, 6, 12–15	4–6
Hib§ (Haemophilus influenza, type b)	2, 4, 6, 12–15	
MMR¶ (measles, mumps, rubella)	12–15 and 1 month later	4–6, 11–12

* Three doses will be required for kindergarten entry.
† Five doses recommended for school entry.
‡ First Td needed 10 years after last DTP.
§ Doses 3 and 4 should be given according to manufacturer's guidelines.
¶ A second dose given at least 1 month after first dose required for kindergarten entry
Source: Iowa Department of Public Health, July 1998.

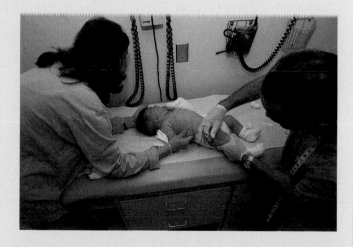

figure 13A Suggested immunization schedule for infants and young children.

AIDS Epidemic

*a*cquired immunodeficiency syndrome (AIDS) is caused by a group of related retroviruses known as HIV (human immunodeficiency viruses). In the United States, AIDS is usually caused by HIV-1, which enters a host by attaching itself to a plasma protein called a CD4 receptor. HIV-1 infects helper T cells, the type of lymphocyte which stimulates B cells to produce antibodies and cytotoxic T cells to destroy virus-infected cells. Macrophages, which present antigens to helper T cells and thereby stimulate them, are also under attack.

HIV is a retrovirus, meaning that its genetic material consists of RNA instead of DNA. Once inside the host cell, HIV uses a special enzyme called reverse transcriptase to make a DNA copy (called cDNA) of its genetic material. Now cDNA integrates into a host chromosome, where it directs the production of more viral RNA. Each strand of viral RNA brings about synthesis of an outer protein coat called a capsid. The viral enzyme protease is necessary to the formation of capsids. Capsids assemble with RNA strands to form viruses which bud from the host cell.

Transmission of AIDS

Infection spreads when infected cells in body secretions, such as semen, and in blood are passed to another individual. To date, as many as 50 million people worldwide may have contracted HIV and almost 14 million have died. A new infection is believed to occur every 15 seconds, the majority in heterosexuals. HIV infections are not distributed equally throughout the world. Most infected people live in Africa (66%) where the infection first began, but new infections are now occurring at the fastest rate in Southeast Asia and the Indian subcontinent.

HIV is transmitted by sexual contact with an infected person, including vaginal or rectal intercourse and oral/genital contact. Also, needle-sharing among intravenous drug users is high-risk behavior. Babies born to HIV-infected women may become infected before or during birth, or through breast-feeding after birth.

HIV first spread through the homosexual community, and male-to-male sexual contact still accounts for the largest percentage of new AIDS cases in the United States. But the largest increases of HIV infections are occurring through heterosexual contact or by intravenous drug use. Now, women account for 20% of all newly diagnosed cases of AIDS. The rise in the incidence of AIDS among women of reproductive age is paralleled by a rise in the incidence of AIDS in children younger than 13.

Phases of an HIV Infection

The Centers for Disease Control and Prevention recognize three stages of an HIV-1 infection called Category A, B, and C. During a Category A stage, the helper T lymphocyte count is 500 per mm^3 or greater. For a period of time after the initial infection with HIV, people don't usually have any symptoms at all. A few (1–2%) do have mononucleosis-like symptoms that may include fever, chills, aches, swollen lymph nodes, and an itchy rash. These symptoms disappear, however, and there are no other symptoms for quite some time. Although there are no symptoms, the person is highly infectious. Although there is a large number of viruses in the plasma, the HIV blood test is not yet positive because it tests for the presence of antibodies and not for the presence of HIV itself. This means that HIV can still be transmitted before the HIV blood test is positive.

Several months to several years after a nontreated infection, the individual will probably progress to category B in which the helper T lymphocyte count is 200 to 499 per mm^3. During this stage there will be swollen lymph nodes in the neck, armpits, or groin that persist for three months or more. Other symptoms that indicate category B are severe fatigue not related to exercise or drug use; unexplained persistent or recurrent fevers, often with night sweats; persistent

tetanus. Occasionally, a patient who received these antibodies developed a disease called serum sickness, indicating that the serum contained proteins that the individual's immune system recognized as foreign.

A new method of producing antibodies has been developed. Lymphocytes are removed from the body and exposed in vitro (in laboratory glassware) to a particular antigen. The stimulated lymphocytes are fused with a can-

cer cell so that they continuously divide, producing a clone of cells capable of making only one type of antibody. These antibodies, called **monoclonal** (mon"o-klōn'al) **antibodies,** do not cause serum sickness.

> Passive immunity is needed when an individual is in immediate danger of succumbing to an infectious disease. Passive immunity is short-lived because the antibodies are administered to and not made by the individual.

cough not associated with smoking, a cold, or the flu; and persistent diarrhea.

When the individual develops non-life-threatening but recurrent infections, it is a signal that the disease is progressing. One possible infection is thrush, a fungal infection that is identified by the presence of white spots and ulcers on the tongue and inside the mouth. The fungus may also spread to the vagina, resulting in a chronic infection there. Another frequent infection is herpes simplex, with painful and persistent sores on the skin surrounding the anus, the genital area, and/or the mouth.

Previously, the majority of infected persons proceeded to category C, in which the helper T lymphocyte count is below 200 per mm³ and the lymph nodes degenerate. The patient, who is now suffering from AIDS, characterized by severe weight loss and weakness due to persistent diarrhea and coughing, will most likely contract an opportunistic infection. An **opportunistic infection** is one that only has the opportunity to occur because the immune system is severely weakened. Persons with AIDS die from one or more opportunistic diseases such as *Pneumocystis carinii* pneumonia, *Mycobacterium tuberculosis*, Toxoplasmic encephalitis, Kaposi's sarcoma or invasive cervical cancer. This condition has been added to the list because the incidence of AIDS has now increased in women.

Treatment for AIDS

Therapy usually consists of combining two drugs that inhibit reverse transcriptase with another that inhibits protease, an enzyme needed for formation of a viral capsid. This multidrug therapy, when taken according to the manner prescribed, seems to usually prevent mutation of the virus to a resistant strain. The sooner drug therapy begins after infection, the better the chances that the immune system will not be destroyed by HIV. And medication must be continued indefinitely. Unfortunately, an HIV strain resistant to all known drugs has been reported—persons who become infected with this strain have no drug therapy available to them.

The likelihood of transmission from mother to child at birth can be lessened if the mother takes an inhibitor of reverse transcriptase called AZT and the child is delivered by cesarean section.

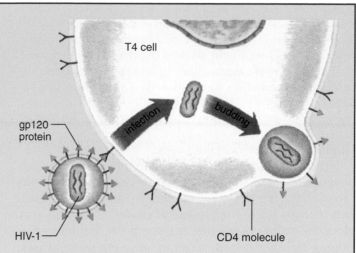

figure 13B Future vaccine for AIDS. HIV-1, the most common cause of AIDS in the United States, has an envelope protein called gp120, which allows it to attach to CD4 molecules that project from a T4 cell. Infection of the T4 cell follows, and HIV eventually buds from the infected T4 cell. If the immune system can be trained by the use of a vaccine to attack and destroy all cells that bear gp120, a person could not be infected with HIV-1.

Many investigators are working on a vaccine for AIDS. Some are trying to develop a vaccine in the traditional way. Others are working on subunit vaccines that utilize just a single HIV protein as the vaccine (fig. 13B) For example, a protein from the viral envelope has been produced by genetic engineering of bacteria. This protein is undergoing clinical trial as a vaccine in Thailand. Also, investigators have found that viral cDNA for this protein can act as a vaccine because it causes cells to produce and display the protein at their plasma membrane.

Questions

1. What is the difference between an HIV blood test and a T4 cell count? What can the results of these tests confirm?
2. Why is it common for AIDS patients to die as a result of infections that are normally nonfatal?

Lymphokines

Lymphokines (also called *cytokines*) are the immunostimulatory proteins released by T cells. They are being investigated as possible adjunct therapy for cancer and AIDS because they stimulate white blood cell formation and/or function. Both interferon and various other types of lymphokines called interleukins have been used as immunotherapeutic drugs, particularly to potentiate the ability of the individual's T cells (and possibly B cells) to fight cancer.

Interferon, discussed previously on page 258, is a substance produced by leukocytes, fibroblasts, and probably most cells in response to a viral infection. When produced by T cells, interferon is called a lymphokine. Interferon is still being investigated as a possible cancer drug, but so far, it has proven effective only in certain patients, and the exact reasons as yet cannot be discerned.

Theoretically, cancer cells that carry an altered protein on their cell surface should be attacked and destroyed by killer T cells. The presence of a developing cancer may indicate that killer T cells have not been activated. In that case, lymphokines might awaken the immune system and lead to the cancer's destruction. In one technique being investigated, researchers first withdraw T cells from the patient and activate the cells by culturing them in the presence of an interleukin. The cells are then reinjected into the patient, who is given doses of interleukin to maintain the killer activity of the T cells.

Those who are actively engaged in interleukin research believe that interleukins soon will be used as adjuncts for vaccines, for the treatment of chronic infectious diseases, and perhaps for the treatment of cancer. Chemicals that block the actions of interleukins may prove helpful in preventing skin and organ rejection, autoimmune diseases, and allergies.

The interleukins and other lymphokines show some promise of potentiating the individual's own immune system.

Immunological Side Effects and Illnesses

The immune system protects the body from disease because it can differentiate "self" from "nonself." Sometimes, however, the immune system is underprotective, allowing cancer to develop, or is overprotective, preventing an individual from receiving certain types of blood.

Allergies

Allergies are caused by an overactive immune system that forms antibodies to substances that are usually not recognized as foreign substances. Unfortunately, allergies usually are accompanied by coldlike symptoms, or even at times, by severe systemic reactions, such as *anaphylactic shock*, a sudden drop in blood pressure and respiratory difficulties that can lead to death.

Of the five varieties of antibodies (see table 13.1), IgE antibodies cause allergies. IgE antibodies stick on the membranes of basophils in the blood and of *mast cells*, which are found in the tissues. When an *allergen*, an antigen that provokes an allergic reaction, attaches to the IgE antibodies on mast cells, the mast cells release histamine and other substances that cause mucus secretion and airway constriction, resulting in characteristic allergy symptoms (fig. 13.12). On occasion, basophils and other white blood cells release these chemicals into the bloodstream. The resulting increased capillary permeability can lead to fluid loss and shock.

Allergy shots sometimes prevent the onset of allergic symptoms. Injections of the allergen cause the body to ac-

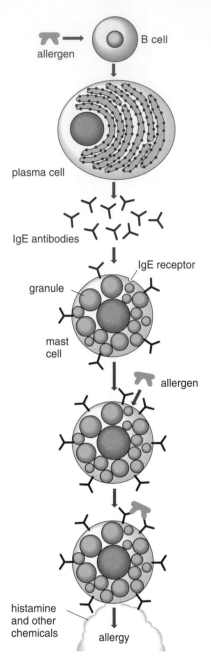

figure 13.12 Allergic response. IgE receptor combines with an allergen, and histamine is released.

cumulate large quantities of IgG antibodies, which combine with environmental allergens before they have a chance to reach the IgE antibodies located on the membranes of mast cells.

Tissue Rejection

Certain organs, such as the skin, heart, and kidneys, could be transplanted easily from one person to another if the body did not attempt to *reject* them. Rejection occurs because killer T cells bring about disintegration of foreign tissue in the body.

Organ rejection can be controlled in two ways: (1) by careful selection of the organ to be transplanted and (2) by **immunosuppression**—inactivation of the immune system. It is best if the transplanted organ has the same type of MHC proteins as those of the recipient because killer T cells treat foreign MHC proteins as antigens.

The immunosuppressive drug cyclosporine has been in use for years. A new drug, tacrolimus (formerly known as FK-506), shows some promise, especially in liver transplant patients. Both drugs, which act by inhibiting the production of interleukin-2, are known to adversely affect the kidneys as a toxic side effect.

AutoImmune Diseases

Certain human illnesses occur because killer T cells attack an individual's own tissues. Some bacteria are known to produce toxic products called superantigens. A superantigen can bind to an MHC protein on a macrophage without first being processed by the macrophage. Subsequently, killer T cells that interact with the affected macrophage learn to recognize and attack the body's own cells. This chain of events is hypothesized to cause at least some autoimmune diseases.

In the autoimmune disease **myasthenia gravis** (mi"as-the'ne-ah grah'vis), neuromuscular junctions do not obey nervous stimuli, resulting in muscular weakness. In **multiple sclerosis (MS)**, the myelin sheath of nerve fibers is attacked, which results in various neuromuscular disorders. A person with the autoimmune disease **systemic lupus erythematosus (SLE)** has various symptoms, including fever and malaise, and at some time during the course of the illness, signs of kidney failure. In the autoimmune disease **rheumatoid arthritis,** the joints are affected. Type I diabetes and heart damage following rheumatic fever are also suspected to be autoimmune illnesses. As yet, there are no cures for autoimmune diseases, but research is progressing.

Immune Deficiency

Immune deficiency represents a breakdown in some aspect of the immune system's ability to protect the body against disease. AIDS (p. 264) is an example of an acquired immune deficiency. As a result of a weakened immune system, AIDS patients show a greater susceptibility to a variety of diseases, including opportunistic infections, such as pneumocystis pneumonia. They also have a higher risk of cancer. One type of cancer, Kaposi's sarcoma (a malignant form of skin cancer), is common among AIDS patients, although it is rare in the general population.

Immune deficiency may also be congenital (that is, inherited). Infrequently, a child may be born with an impaired B- or T-cell system caused by a defect in lymphocyte development. In **severe combined immunodeficiency disease (SCID)**, both antibody and cell-mediated immunity are lacking or inadequate. Without treatment, even common infections can be fatal. The use of bone marrow transplants (p. 255) containing healthy stem cells has met with some success in the treatment of affected children.

Effects of Aging

The thymus gland degenerates with age. Having reached its maximum size in early childhood, it begins to shrink after puberty and has virtually disappeared by old age. As the gland decreases in size, so does the number of T cells. The T cells remaining do not function as well; therefore, the chance of cancer increases with age.

Among the elderly, the B cells sometimes fail to form clones. When they do form clones, the antibodies released may not function well. Therefore, infections are more common among the elderly. Also, the antibodies are more likely to attack the body's own tissues, increasing the incidence of autoimmune diseases.

Working Together

The accompanying illustration shows how the lymphatic system works with other organ systems of the body to maintain homeostasis. All the organ systems of the body are interrelated.

Integumentary System

Lymphatic vessels pick up excess tissue fluid; immune system protects against skin infections.

Skin serves as a barrier to pathogen invasion; Langerhans' cells phagocytize pathogens; protects lymphatic vessels.

Skeletal System

Lymphatic vessels pick up excess tissue fluid; immune system protects against infections.

Red bone marrow produces leukocytes involved in immunity.

Muscular System

Lymphatic vessels pick up excess tissue fluid; immune system protects against infections.

Skeletal muscle contraction moves lymph; physical exercise enhances immunity.

Nervous System

Lymphatic vessels pick up excess tissue fluid; immune system protects against infections of nerves.

Microglial cells engulf and destroy pathogens.

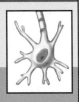

Endocrine System

Lymphatic vessels pick up excess tissue fluid; immune system protects against infections.

Thymus is necessary to maturity of T lymphocytes.

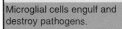
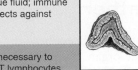

How the Lymphatic System works with other body systems

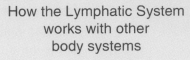

Circulatory System

Lymphoid organs produce and store formed elements; lymphatic vessels transport leukocytes and return tissue fluid to blood vessels; spleen serves as blood reservoir, filters blood. Blood vessels transport leukocytes and antibodies; blood services lymphoid organs and is source of tissue fluid that becomes lymph.

Respiratory System

Lymphatic vessels pick up excess tissue fluid; immune system protects against respiratory tract and lung infections.

Tonsils and adenoids occur along respiratory tract; breathing aids lymph flow; lungs carry out gas exchange.

Digestive System

Lacteals absorb fats; Peyer's patches prevent invasion of pathogens; appendix contains lymphoid tissue.

Digestive tract provides nutrients for lymphoid organs; stomach acidity prevents pathogen invasion of body.

Urinary System

Lymphatic system picks up excess tissue fluid, helping to maintain blood pressure for kidneys to function; immune system protects against infections.

Kidneys control volume of body fluids, including lymph.

Reproductive System

Immune system does not attack sperm or fetus, even though they are foreign to the body.

Sex hormones influence immune functioning; acidity of vagina helps prevent pathogen invasion of body; milk passes antibodies to newborn.

Selected New Terms

Basic Key Terms

active immunity (ak′tiv ĭ-myu′nĭ-te), p. 262

antibodies (an″tĭ-bod″ez), p. 258

antibody-mediated immunity (an″tĭ-bod″e-me″de-āt′ed ĭ-myu′nĭ-te), p. 259

antigen (an′tĭ-jen), p. 258

B lymphocytes (B lim′fo-sītz), p. 258

cell-mediated immunity (sel me″de-āt′ed ĭ-myu′nĭ-te), p. 260

complement system (kom′plĕ-ment sis′tem), p. 256

histamine (his′tah-min), p. 256

immunity (ĭ-mu′nĭ-te), p. 256

inflammatory reaction (in-flam′ah-to″re re-ak′shun), p. 256

lymph (limf), p. 252

lymphatic system (lim-fat′ik sis′tem), p. 252

lymphokines (lim′fo-kīnz), p. 265

macrophages (mak′ro-fāj″ez), p. 256

memory B cells (mem′o-re B selz), p. 259

monoclonal antibodies (mon″o-klōn′al an″tĭ-bod″ez), p. 264

passive immunity (pă′siv ĭ-myu′nĭ-te), p. 262

phagocytosis (fag″o-si-to′sis), p. 256

plasma cells (plaz′mah selz), p. 259

pus (pus), p. 256

red bone marrow (red bōn mar′o), p. 255

T lymphocytes (T lim′fo-sītz), p. 258

Clinical Key Terms

AIDS (acquired immunodeficiency syndrome) (ah-kwīr′d ĭ-myu″no-dĭ-fĭ′shun-se sin′drōm), p. 264

allergies (al′er-jēz), p. 266

antibody titer (an″tĭ-bod″e ti′ter), p. 262

edema (ĕ-de′mah), p. 254

elephantiasis (ĕ″lah-fun-ti′ah-sis), p. 254

Hodgkin's disease (hoj′kinz dĭ-zēz), p. 254

immunization (ĭ-myu-nĭ-za′shun), p. 262

immunosuppression (im″u-no-sŭ-pres″shun), p. 267

interferon (in″ter-fēr′on), p. 258

lymphadenitis (lim-fad″ĕ-ni′tis), p. 254

lymphangitis (lim″fan-jī′tis), p. 254

lymphoma (lim-fo′mah), p. 254

multiple sclerosis (MS) (mul′tĭ-pl skler-o′sis), p. 267

myasthenia gravis (mi″as-the′ne-ah grah′vis), p. 267

opportunistic infection (op″per-tu-nis′tik in-fek′shun), p. 265

pulmonary edema (pul′mo-ner″e ĕ-de′mah), p. 254

rheumatoid arthritis (ru′mah-toid ar-thri′tis), p. 267

severe combined immunodeficiency disease (SCID) (s′ĕ-vēr′ kum-bīnd′ ĭ-myu″no-dĭ-fĭ′shun-se di-zēz), p. 267

systemic lupus erythematosus (sis-tem′ik lu′pus er-ĭ-the-mah-to′sus), p. 267

vaccines (vak′sēnz), p. 262

Summary

I. Lymphatic System

A. The lymphatic system: (1) collects excess tissue fluids, (2) absorbs fat molecules in the intestines, and (3) plays a major role in the body's defense against disease.

B. The lymphatic system is a one-way system. Lymph flows from a lymphatic capillary to ever-larger lymphatic vessels, and finally, to a lymphatic duct, which enters a subclavian vein.

C. Lymph nodes are divided into sinus-containing nodules in which the lymph is cleansed of infectious organisms and debris.

D. The thymus is divided into lobules, where T lymphocytes are produced.

E. The spleen is divided into sinus-containing lobules in which blood is cleansed by lymphocytes and macrophages.

F. Red bone marrow produces white blood cells necessary for the development of immunity.

II. Immunity

A. Nonspecific defenses. Nonspecific defenses against pathogens include barriers to entry, the inflammatory reaction, and protective proteins (the complement system). The inflammatory reaction is a "call to arms": It marshals phagocytic white blood cells to the site of bacterial invasion.

B. Specific defense. Two types of lymphocytes provide specific immunity. B cells produce and secrete antibodies that combine with antigens. Certain T cells directly attack antigen-bearing cells, and others regulate the immune response.

1. Antibody-mediated immunity. B cells are responsible for antibody-mediated immunity. Each one produces a specific type of antibody to counter a particular infection. An antibody combines with its antigen in a lock-and-key manner. The antibody-antigen reaction can lead to complexes that contain several molecules of antibody and antigen. These complexes are then destroyed by phagocytic cells or complement.

2. Cell-mediated immunity. T cells are responsible for cell-mediated immunity. They directly attack cells that bear antigens. For example, they protect the body from cancer but also attack transplanted organs.

III. **Immunotherapy**
   A. Active (long-lived) immunity can be induced by vaccines when a person is well and in no immediate danger of contracting an infectious disease. Active immunity is dependent upon the presence of memory cells in the body.

B. Passive immunity is needed when an individual is in immediate danger of succumbing to an infectious disease. Passive immunity is short-lived because the antibodies are administered to and not made by the individual.
   C. The interleukins and other lymphokines show some promise of potentiating the individual's own immune system.

IV. **Immunological Side Effects and Illnesses**
   A. Allergic symptoms are caused by the release of histamine and other substances from mast cells.
   B. When a transplanted organ is

rejected, the immune system attacks cells that bear different MHC proteins from those of the individual.
   C. Autoimmune diseases seem to be preceded by an infection that results in an attack by killer T cells on the body's own organs.
   D. Immune deficiency is caused by a failure of some aspect of the body's immune system.

Study Questions

1. What is the lymphatic system, and what are its three primary functions? (p. 252)
2. Describe the structures and functions of the lymph nodes, thymus, spleen, and red bone marrow. (pp. 253–55)
3. What are the body's nonspecific defense mechanisms? (pp. 256–58)
4. Describe the inflammatory reaction, and give a role for each type of cell and chemical that participates in it. (pp. 256–57)

5. For which type of immunity are B cells responsible? What is the clonal selection theory? (p. 259)
6. Describe the structure of an IgG antibody, including the terms *variable regions* and *constant regions*. (p. 260)
7. Name two types of T cells, and state their functions. (pp. 260–61)

8. How is active immunity achieved? How is passive immunity achieved? (pp. 262–64)
9. Discuss allergies, tissue rejection, autoimmune diseases, and immune deficiency as they relate to the immune system. (pp. 266–67)
10. Name two immune-deficiency diseases. (p. 267)

Objective Questions

Fill in the blanks.

1. Lymphatic vessels contain _____ , which close, preventing lymph from flowing backward.
2. _____ and _____ are two types of white blood cells produced and stored in lymphoid organs.
3. Lymph nodes cleanse the _____ , while the spleen cleanses the _____ .
4. _____ and _____ are phagocytic white blood cells.
5. T lymphocytes have matured in the _____ .

6. A stimulated B cell divides and differentiates into antibody-secreting _____ cells and also into _____ cells that are ready to produce the same type of antibody at a later time.
7. B cells are responsible for _____-mediated immunity.
8. Killer T cells are responsible for _____-mediated immunity.
9. Immunization with _____ brings about active immunity.
10. Allergic reactions are associated with the release of _____ from mast cells.

11. Whereas _____ immunity occurs when an individual is given antibodies to combat a disease, _____ immunity occurs when an individual develops the ability to produce antibodies against a specific antigen.
12. Barriers to entry, protective proteins, and the inflammatory reaction are all examples of _____ , defenses.
13. Proteins that function to form holes in bacterial cell walls comprise the _____ system.

Medical Terminology Reinforcement Exercise

Consult Appendix B for help in pronouncing and analyzing the meaning of the terms that follow.

1. metastasis (mĕ-tas'tah-sis)
2. allergist (al'er-jist)
3. immunosuppressant (im"u-no-sū-pres'ant)
4. immunotherapy (ĭ-mu"no-ther'ah-pe)
5. macrophage (mak'ro-faj)
6. splenorrhagia (sple"no-ra'je-ah)
7. thymusectomy (thi"mus-ek'to-me)
8. lymphadenopathy (lim-fad"ĕ-nop'ah-the)
9. lymphangiography (lim-fan"-je-og'rah-fe)
10. lymphedema (lim"fe-de'mah)

Website Link

For a listing of the most current Websites related to this chapter, please visit the Mader home page at:

http://www.mhhe.com/maderap

c h a p t e r 14

The Respiratory System

Chapter Outline and Learning Objectives

After you have studied this chapter, you should be able to:

Respiratory Organs (p. 273)
- Describe the organization of the respiratory system and the process of respiration.
- Describe the structures and functions of the respiratory system organs.
- Describe the structure and function of the respiratory membrane.

Mechanisms of Breathing (p. 278)
- Describe the mechanisms by which breathing occurs, including explanations of inspiration and expiration.
- Tell where the respiratory center is located, and explain how it controls normal breathing rate.
- Describe vital capacity and its relationship to other measurements of breathing capacities.

Respiration Mechanisms and Transport of Gases (p. 282)
- Describe the process of gas exchange in the lungs and the tissues.
- Explain how oxygen and carbon dioxide are transported in the blood.

Respiratory Infections and Lung Disorders (p. 285)
- Name and describe the various infections of the respiratory tract.
- Describe the effects of smoking on the respiratory tract and on overall health.

Effects of Aging (p. 289)
- Anatomical and physiological changes occur in the respiratory system as we age.

Working Together (p. 289)
- The respiratory system works with other systems of the body to maintain homeostasis.

Visual Focus
Inspiration Versus Expiration (p. 280)
Lower Respiratory Tract Disorders (p. 288)

Medical Focus
Modified Breathing Rates (p. 278)
Heimlich Maneuver (p. 282)

MedAlert
The Most Often Asked Questions about Smoking, Tobacco, Health, and . . . the Answers (p. 290)

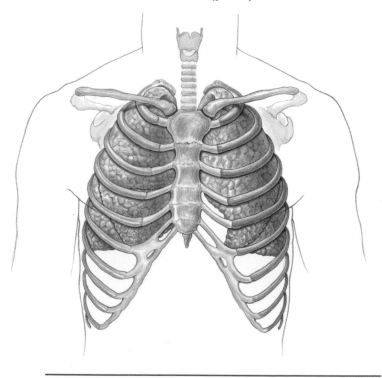

The lungs in the thoracic cavity are protected by the rib cage.

figure 14.1 Respiratory organs. **a.** Inset shows an enlargement of a section of lung. Gas exchange occurs in alveolar sacs, which are surrounded by a capillary network. Notice that the pulmonary arteriole carries oxygen-poor blood (colored blue) and that the pulmonary venule carries oxygen-rich blood (colored red). **b.** Path of air.

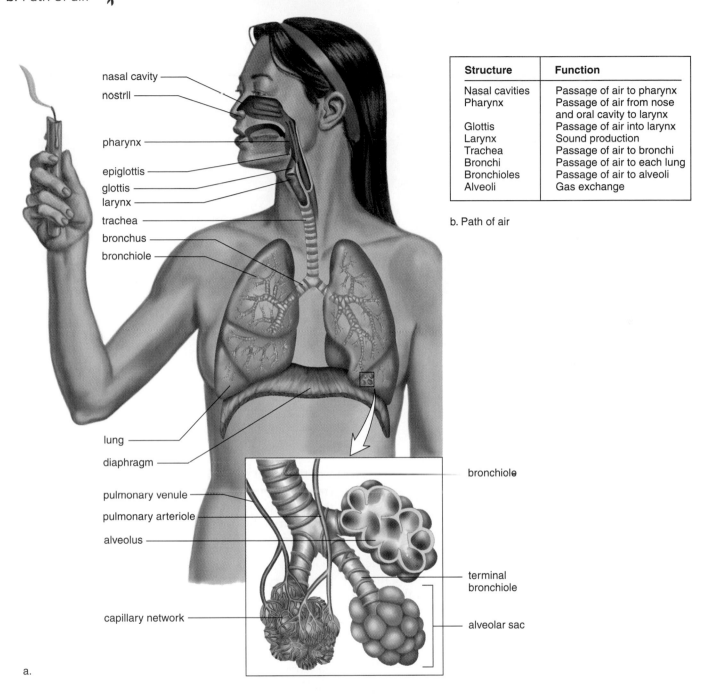

Structure	Function
Nasal cavities	Passage of air to pharynx
Pharynx	Passage of air from nose and oral cavity to larynx
Glottis	Passage of air into larynx
Larynx	Sound production
Trachea	Passage of air to bronchi
Bronchi	Passage of air to each lung
Bronchioles	Passage of air to alveoli
Alveoli	Gas exchange

b. Path of air

nasal cavity
nostril
pharynx
epiglottis
glottis
larynx
trachea
bronchus
bronchiole
lung
diaphragm
pulmonary venule
pulmonary arteriole
alveolus
capillary network
bronchiole
terminal bronchiole
alveolar sac

a.

The process of **respiration** includes: (1) moving air in and out of the lungs, (2) exchanging gases between air and blood, (3) exchanging gases between blood and body cells, and (4) the use of oxygen and production of carbon dioxide by the cells.

Respiratory Organs

The respiratory organs (fig. 14.1) are the nose (nasal cavities), pharynx, larynx, trachea, bronchi, and lungs. This listing also reflects the sequence of organs through which air passes into the body.

figure 14.2 Structure of the nasal cavity, pharynx, and larynx in sagittal section.

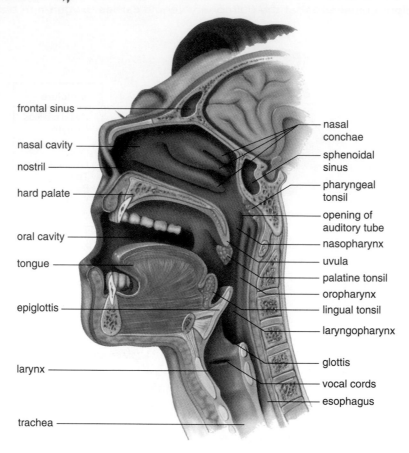

- frontal sinus
- nasal cavity
- nostril
- hard palate
- oral cavity
- tongue
- epiglottis
- larynx
- trachea
- nasal conchae
- sphenoidal sinus
- pharyngeal tonsil
- opening of auditory tube
- nasopharynx
- uvula
- palatine tonsil
- oropharynx
- lingual tonsil
- laryngopharynx
- glottis
- vocal cords
- esophagus

Nose

The **nose** contains two nostrils, openings that enter the *nasal cavities*. The nasal cavities are narrow canals separated from one another by a median septum. The surface area of the cavity walls is increased by bone projections called conchae (kong'ke) (sing., **concha** [kong'kah]) (fig. 14.2). The conchae are covered by a ciliated mucous membrane that filters, warms, and humidifies the air as it passes through the nose. Olfactory cells are located in the upper medial portion of each nasal cavity, in the lining of the conchae.

The *lacrimal* (tear) *glands* in the upper lateral corner of the orbit drain into the nasal cavities by way of tear ducts. This is why crying produces a runny nose.

The **paranasal sinuses** are air-filled spaces within the skull. Each is named for the bone where found. The paranasal sinuses open into the nasal cavities and are also lined with a mucous membrane. Inflammation of this membrane is called sinusitis. If the passageways to the sinuses are blocked, a partial vacuum is created, causing a sinus headache.

The nasal cavities are separated from the oral cavity (mouth) by the hard and soft palates, and they empty into the nasopharynx. When we swallow, the uvula, a posterior extension of the soft palate, moves back and blocks the nasopharynx so that food does not ordinarily enter the nose.

Pharynx

The **pharynx** (far'inks), commonly called the throat, connects the nasal and oral cavities to the larynx. The pharynx has three parts: (1) The nasal cavities open into the *nasopharynx*, (2) the oral cavity (mouth) opens into the *oropharynx*, and (3) the *laryngopharynx* opens into the larynx (fig. 14.2).

The auditory (eustachian) tubes lead from the nasopharynx to the middle ears. The middle ears contain air, and the auditory tubes keep the pressure in the middle ears equal to that in the nasopharynx. The pharyngeal tonsils, or adenoids, are situated in this part of the pharynx.

The oropharynx is the middle portion of the pharynx. The palatine tonsils, located in the posterior lateral wall,

figure 14.3 Larynx and vocal cords. **a.** Anterior view of the larynx. The larynx contains various cartilages; the larger thyroid cartilage forms the anterior surface of the larynx. **b.** Posterior view of the larynx. **c.** Photograph of the glottis and vocal cords, viewed from above. The false vocal cords are nonfunctional, and only the true vocal cords are involved in voice production.

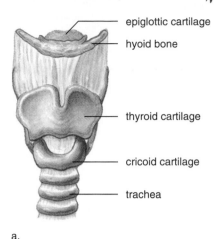

epiglottic cartilage

hyoid bone

thyroid cartilage

cricoid cartilage

trachea

a.

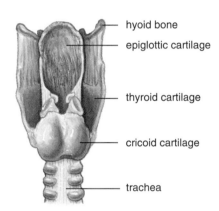

hyoid bone

epiglottic cartilage

thyroid cartilage

cricoid cartilage

trachea

b.

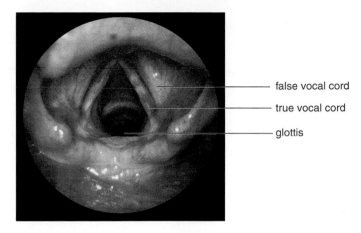

false vocal cord

true vocal cord

glottis

c.

can be observed if the tongue is depressed. There are also lingual tonsils at the base of the tongue. The tonsils are part of the immune system and help protect us from disease. Sometimes, they become diseased and are surgically removed.

The laryngopharynx is a passageway for both air and food. Air that enters from either the nose or oral cavity is sent on to the larynx. Food enters the esophagus, a tube that leads to the stomach.

> The nasal cavities, which filter, warm, and humidify incoming air, open into the pharynx. Food and air passages join in the pharynx, which conducts air to the larynx and food to the esophagus.

Larynx

The **larynx** (lar′inks), or voice box, can be envisioned as a triangular box whose apex, the *thyroid cartilage* (Adam's apple), is located at the front of the neck (fig. 14.3). At the top of the larynx is a variable-sized opening called the **glottis.** When food is being swallowed, a flap of tissue called the **epiglottis** covers the glottis so that no food passes into the larynx. If, by chance, food or some other substance does gain entrance to the larynx, reflex coughing usually occurs, expelling the substance.

The larynx is called the voice box because the vocal cords are inside the larynx. The *vocal cords* are mucous membrane folds supported by elastic ligaments stretched across the glottis (fig. 14.3c). When air passes through the glottis, the vocal cords vibrate, producing sound. At the time of puberty, the growth of the larynx and the vocal cords is much more rapid and accentuated in the male than in the female, causing the male to have a more prominent Adam's apple and a deeper voice. The voice "breaks" in the young male due to his inability to control the longer vocal cords. These changes cause the male voice to have a lower pitch.

The high or low pitch of the voice is regulated when speaking and singing by changing the tension on the vocal cords. The greater the tension, as when the glottis becomes more narrow, the higher the pitch. When the glottis is wider, the pitch is lower (fig. 14.3c). The loudness, or intensity, of the voice depends upon the amplitude of the vibrations (that is, the degree to which vocal cords vibrate).

> The vocal cords are stretched across the glottis inside the larynx, often called the voice box. When swallowing, the glottis is covered by the epiglottis.

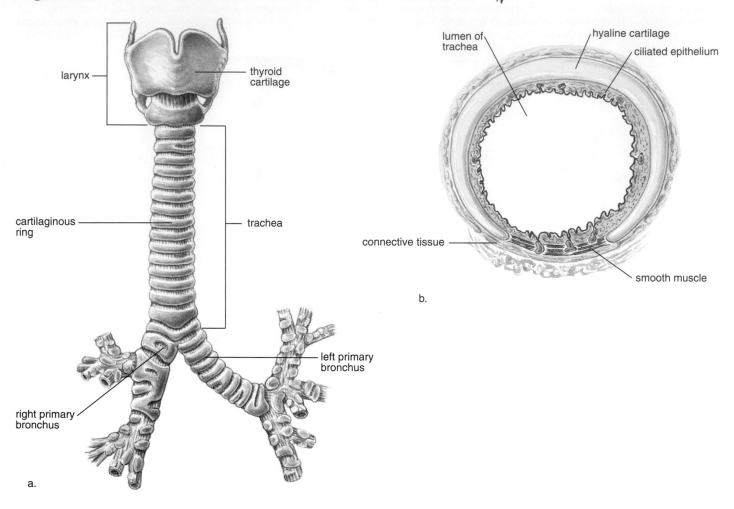

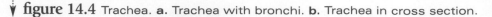

Trachea

The **trachea** (tra′ke-ah), commonly called the windpipe, extends from the larynx into the thoracic cavity, where it divides into two primary bronchi. The wall of the trachea contains C-shaped hyaline cartilage rings that prevent it from collapsing and hold it open for the passage of air (fig. 14.4). A ciliated mucous membrane (see fig. 4.4) with many goblet cells lines the trachea. The cilia beat upward, carrying mucus, dust, and occasional bits of food that went the wrong way into the pharynx, where the accumulation may be swallowed or expectorated.

If the trachea is blocked because of illness or accidental swallowing of a foreign object, an operation called a **tracheostomy** can be performed. During this procedure, a tube inserted via an incision in the trachea acts as an artificial air intake and exhaust duct.

Bronchi and Bronchioles

The trachea divides into two primary bronchi (sing., **bronchus** [brong′kus]) that enter the right and left lungs at a *hilus*, a recessed region through which nerves, ducts, and blood vessels pass into an organ. The primary bronchi are constructed similarly to the trachea. The right primary bronchus is shorter and wider than the left. The primary bronchi divide into secondary bronchi, which, in turn, branch into tertiary bronchi. The tertiary bronchi branch into a number of smaller passages called the **bronchioles** (brong′ke-ōlz) (fig. 14.5).

The wall of a bronchus contains hyaline cartilage wedges that keep the bronchus open as it enters a lung. The bronchiole wall contains bundles of smooth muscle, whose contraction constricts these airways. Each terminal bronchiole gives off branches to an alveolar sac enclosing many air sacs, called alveoli (sing., **alveolus** [al-ve′o-lus]), which make up the lungs.

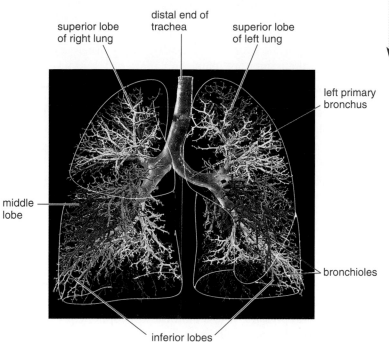

figure 14.5 Bronchial tree. The branches of each bronchiole are similarly colored.

superior lobe of right lung

distal end of trachea

superior lobe of left lung

left primary bronchus

middle lobe

bronchioles

inferior lobes

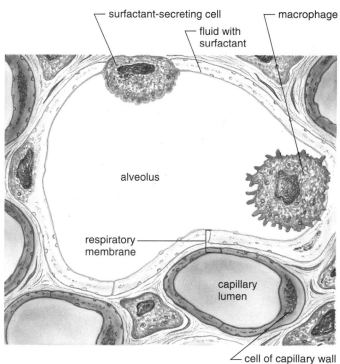

figure 14.6 The respiratory membrane. Gases must pass through the walls of the alveolus and the capillary for gas exchange to occur. The surfactant-secreting cells produce an agent that lowers surface tension and prevents collapse of the alveoli. Macrophages phagocytize any debris in the alveolus.

surfactant-secreting cell

macrophage

fluid with surfactant

alveolus

respiratory membrane

capillary lumen

cell of capillary wall

As discussed later, **lung cancer** usually begins in a bronchus. Callusing and a loss of cilia due to irritants lead to tumor formation.

Asthma occurs when airflow is obstructed due to widespread narrowing of the airways. Severe episodes are associated with bronchospasms caused by contractions of the smooth muscle in the walls of the bronchioles. A typical asthma attack begins with breathlessness and difficulty in forcing air out of the lungs. Usually, there is wheezing, but with severe attacks, wheezing may not be audible. An attack, which may last several hours, is followed by prolonged coughing with mucus production.

Lungs

Within the lungs, the alveoli are surrounded by a network of capillaries (see fig. 14.1). Both the alveolar wall and the capillary wall consist of a layer of simple squamous epithelium and a basement membrane. These two walls together are called the **respiratory membrane** because this is where gas exchange occurs (fig. 14.6).

The alveolar wall contains surfactant-secreting cells. **Surfactant** lowers surface tension and prevents alveoli from collapsing. Some newborn babies do not produce

surfactant, and the result is infant **respiratory distress syndrome**. In the absence of surfactant, a hyaline membrane composed of cellular debris and protein fibers forms a lining within the alveoli. This interferes with gas exchange and leads to an inadequate oxygen supply. Since surfactant is not produced until about the eighth month of fetal development, premature babies are more likely to suffer from this disorder.

The lungs contain approximately 300 million alveoli, with a total cross-sectional area of 50 to 70 square meters. This is about 40 times the surface area of the skin. Because of their many air spaces, the lungs are very light; normally, a piece of lung tissue dropped in a glass of water will float.

Externally, the lungs are large, cone-shaped organs separated from one another by the heart and other structures of the mediastinum. Each lung has a narrow and rounded apex that extends above the level of the clavicle. The base of

Chapter 14 The Respiratory System   277

each lung is broad and concave to fit upon the convex surface of the diaphragm. The other lung surfaces follow the contours of the ribs and of the organs present in the thoracic cavity.

The left lung, which is somewhat smaller than the right, has a superior lobe and an inferior lobe. The right lung has a superior lobe, a middle lobe, and an inferior lobe. Each lobe contains a specific number of bronchopulmonary segments, each of which receives a tertiary bronchus.

> Air moves from the nasal cavities to the pharynx, to the trachea, to the bronchi, to the bronchioles, and finally, to the approximately 300 million alveoli in the lungs. Gas exchange occurs at the alveoli.

Mechanisms of Breathing

To understand **ventilation**, the manner in which air is drawn into and expelled out of the lungs, it is necessary to remember first that when a person breathes, there is a continuous column of air from the pharynx to the alveoli of the lungs; that is, the air passages are open.

Second, it should be noted that the lungs lie within the sealed-off thoracic cavity. The ribs, which are hinged to the vertebral column at the back and to the sternum at the front, along with the muscles that lie between them, compose the top and sides of the thoracic cavity. The **diaphragm**, a dome-shaped, horizontal muscle, forms the thoracic cavity floor. The lungs themselves are enclosed by the **pleural membranes.** One of these, the *parietal membrane*, adheres closely to the walls of the thoracic cavity and diaphragm, while the other, the *visceral membrane*, is fused to the lungs. The two pleural layers lie close to one another, being separated only by a thin film of fluid. Normally, the intrapleural pressure is less than the atmospheric pressure. This reduced pressure is important because when, by design or accident, air enters the intrapleural space, the lungs collapse, making inspiration impossible. (Presence of air in the intrapleural space is called a *pneumothorax.*)

> The lungs are completely enclosed and, by way of the pleural membranes, adhere to the walls of the thoracic cavity.

Inspiration

The **respiratory center,** located in the medulla oblongata, consists of a group of neurons that exhibit an automatic rhythmic discharge that triggers inspiration. Carbon dioxide (CO_2) and hydrogen ions (H^+) are the primary stimuli that cause changes in the activity of this center. The respiratory center is not affected by low oxygen (O_2) levels. Chemoreceptors in the **carotid bodies** (located in the carotid arteries) and in the **aortic bodies** (located in the aorta) respond primarily to hydrogen ion concentration but also to the level of carbon dioxide and oxygen in blood. These bodies communicate with the respiratory center. When levels of carbon dioxide and hydrogen rise, the rate and depth of breathing increase. (Do not confuse the carotid and aortic bodies with the carotid and aortic sinuses, which monitor blood pressure.) The Medical Focus reading on this page describes the modified breathing rates that result when disease affects the respiratory center.

figure 14.7 Nervous control of breathing. During inspiration, the respiratory center stimulates the intercostal (rib) muscles to contract via the intercostal nerves, and the diaphragm to contract via the phrenic nerve. Should the lung volume increase above 1.5 liters, stretch receptors send inhibitory nerve impulses to the respiratory center via the vagus nerve. In any case, expiration occurs due to a lack of stimulation from the respiratory center to the diaphragm and intercostal muscles.

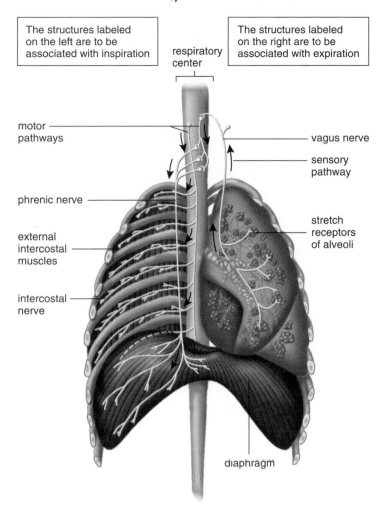

The structures labeled on the left are to be associated with inspiration

respiratory center

The structures labeled on the right are to be associated with expiration

motor pathways

vagus nerve

sensory pathway

phrenic nerve

external intercostal muscles

stretch receptors of alveoli

intercostal nerve

diaphragm

The **respiratory center** sends out nerve impulses by way of nerves to the diaphragm and the rib cage (fig. 14.7). In its relaxed state, the diaphragm is dome-shaped, but upon stimulation, it contracts and lowers. Also, the external intercostal muscles contract, and the rib cage moves upward and outward. Both of these contractions increase the size of the thoracic cavity, which expands the lungs. When the lungs expand, air pressure within the enlarged alveoli lowers and is immediately rebalanced by air rushing in through the nose or the mouth.

Inspiration is the active phase of breathing (fig. 14.8). During this time, the diaphragm and the intercostal (rib) muscles contract, intrapleural pressure decreases even more, the lungs expand, and air rushes in. Note that the air comes in because the lungs already have opened up; air does not force the lungs open. This is why humans are sometimes said to breathe by *negative pressure*. The creation of a partial vacuum causes air to enter the lungs.

Stimulated by nerve impulses, the rib cage lifts up and out, and the diaphragm lowers to expand the thoracic cavity and lungs, allowing inspiration to occur.

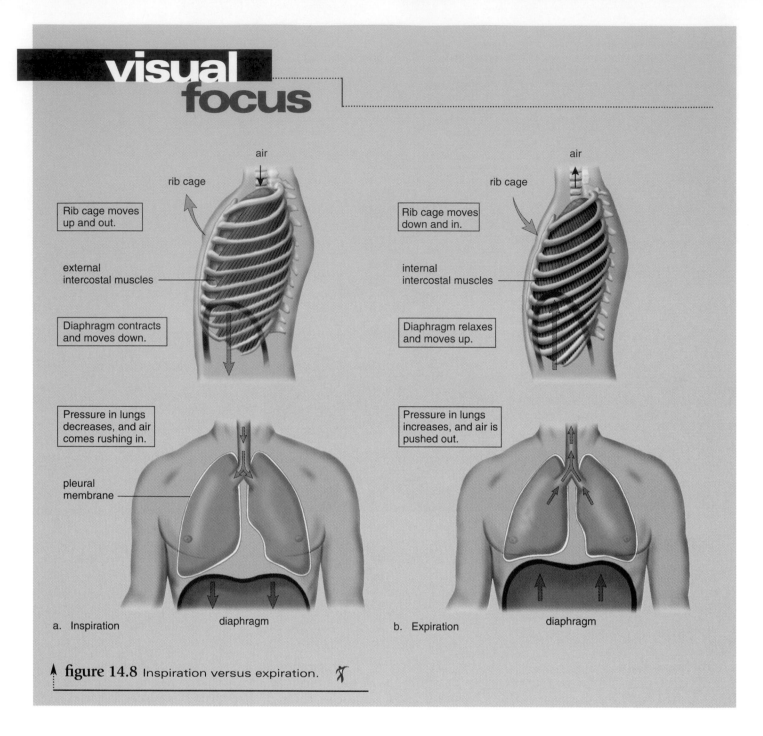

a. Inspiration

b. Expiration

figure 14.8 Inspiration versus expiration.

Expiration

When the respiratory center stops sending signals to the diaphragm and the rib cage, the diaphragm relaxes and resumes its dome shape. The abdominal organs press up against the diaphragm, and the rib cage moves down and inward (fig. 14.8). Now the elastic lungs recoil, and air is pushed out.

The respiratory center acts rhythmically to bring about breathing at a normal rate and volume. If by chance we inhale more deeply, the lungs are expanded, and the alveoli stretch. This stimulates stretch receptors in the alveolar walls, and they initiate inhibitory nerve impulses that travel from the inflated lungs to the respiratory center. This causes the respiratory center to stop sending out nerve impulses.

While inspiration is the active phase of breathing, **expiration** is normally passive—that is, the diaphragm and external intercostal muscles are relaxed during expiration. In deeper and more rapid breathing, expiration can also be active. Contraction of internal intercostal muscles

figure 14.9 Vital capacity. **a.** This individual is using a spirometer, which measures the maximum amount of air that can be inhaled and exhaled. When she inspires, a pen moves up, and when she expires, a pen moves down. **b.** The resulting pattern, such as the one shown here, is called a spirograph.

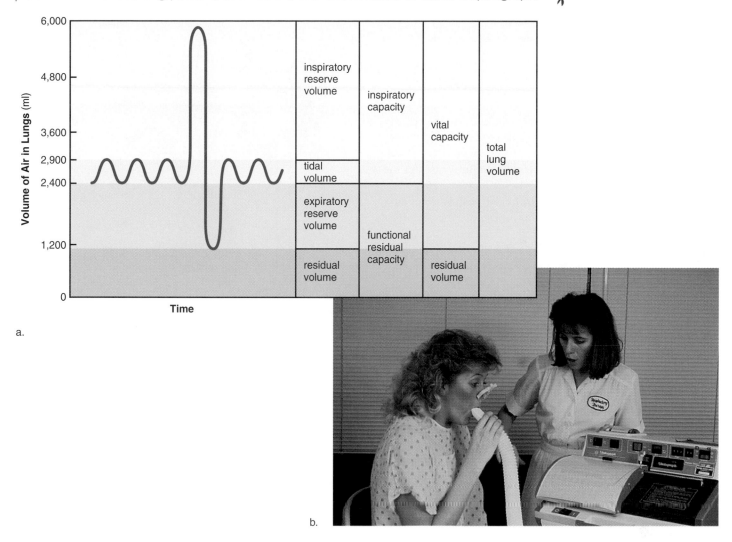

a.

b.

can force the rib cage to move downward and inward. Also, when the abdominal wall muscles are contracted, increased pressure helps to expel air. Air can also be moved into and out of the lungs for reasons not associated with respiration (for example, coughing, sneezing, yawning).

> When nervous stimulation ceases, the rib cage lowers and the diaphragm rises, allowing the lungs to recoil and expiration to occur.

Lung Capacities

The amount of air moved in and out with each breath is called the **tidal volume (TV).** Normally, the tidal volume is about 500 milliliters (ml), but a person can increase the amount inhaled and exhaled by deep breathing.

The total volume of air that can be moved in and out of the lungs during a single breath is called the **vital capac-**ity (VC) (fig. 14.9). A person can increase inspiration to as much as 3,100 ml. The increase beyond tidal volume is called the **inspiratory reserve volume (IRV).** Similarly, a person can increase expiration beyond tidal volume by contracting the thoracic muscles. This volume of air is called the **expiratory reserve volume (ERV)** and measures approximately 1,400 ml of air. Vital capacity is the sum of the tidal volume plus the inspiratory reserve and expiratory reserve volumes.

Note that in figure 14.9, even after very deep breathing, some air (about 1,000 ml) remains in the lungs. This volume of air is called the **residual volume (RV)** and is no longer useful for gas exchange purposes. In some lung diseases, such as emphysema and asthma, the residual volume builds up because the individual has difficulty emptying the lungs. Thus, the lungs tend to be filled with useless air and, as can be seen from examining figure 14.9, vital capacity is reduced.

Heimlich Maneuver

*a*n average of eight Americans choke to death each day on food lodged in the larynx. A simple process termed the abdominal thrust, or **Heimlich** (Hīm'lik) **maneuver** (fig. 14A), named after American surgeon Henry Heimlich, can save the life of a person who is choking: If the victim is standing or sitting, (1) stand behind the victim or the victim's chair, and wrap your arms around the person's waist; (2) grasp your fist with your other hand, and place the fist against the victim's abdomen, slightly above the navel and below the rib cage; (3) press your fist into the abdomen with a quick, upward thrust; (4) repeat several times if necessary.

The Heimlich maneuver may also be used on a drowning victim who is placed in a facedown position, face turned to one side. The procedure forces water and debris out of the victim's lungs.

To perform the Heimlich maneuver on a choking infant, place the infant face up on a firm surface, or hold the child in your lap. Use the pads of the index and middle fingers instead of your fist. Place the fingers on the infant's abdomen, below the rib cage and slightly above the navel. Gently press into the abdomen with a quick upward thrust. This usually dislodges the object, although you may need to repeat the process several times.

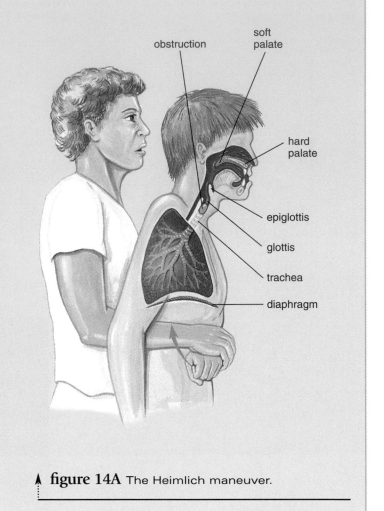

figure 14A The Heimlich maneuver.

Dead Space

Some of the inspired air never reaches the lungs; instead, it fills the conducting airways. These passages are not used for gas exchange and, therefore, are said to contain *dead space*. Breathing more slowly and deeply ensures that a greater percentage of the tidal volume actually reaches the lungs.

If a person breathes through a tube, there is an increase in the amount of dead space and in the amount of air that never reaches the lungs. Any device that increases the amount of dead space beyond maximal inhalation capacity means death to the individual because the air inhaled never reaches the alveoli.

Any blockage of air passages may require extreme measures, such as the Heimlich maneuver described in the Medical Focus reading on this page.

The air used for gas exchange excludes both the residual volume in the lungs and that in the dead space of the respiratory tract.

Respiration Mechanisms and Transport of Gases

External respiration is the exchange of gases between alveolar air and blood across the respiratory membrane (see fig. 14.6) in the lungs. **Internal respiration** is the exchange of gases between blood and tissue cells. For these two events to occur, gases must be transported from the lungs to the tissues and back to the lungs again.

Partial Pressures of Gases

During external and internal respiration, gases diffuse down their concentration gradients. During diffusion, molecules move from the area of higher concentration to the area of lower concentration. With reference to respiration, gases move from an area of higher pressure to an area of lower pressure. Further, the pressure of a gas determines the rate at which it will diffuse from one area to another.

figure 14.10 Partial pressures. Differences in partial pressures cause gas exchange between the air of the alveolus and the blood of the capillary. Similarly, differences in partial pressures cause gas exchange in the tissues.

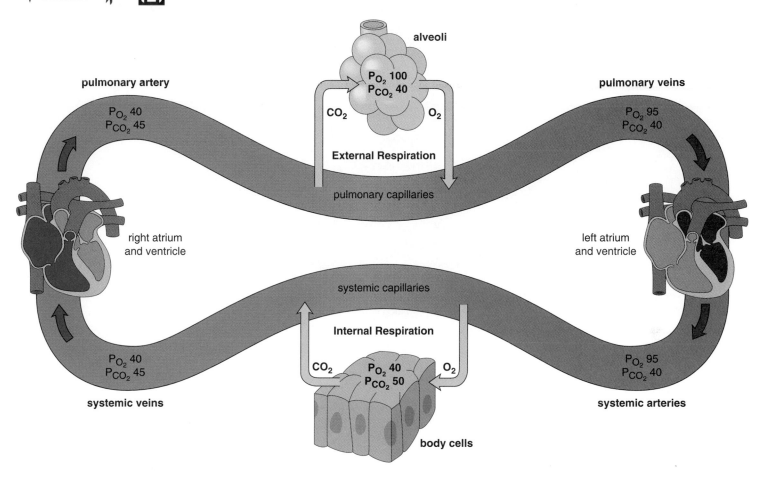

When there is a mixture of gases, as in the atmosphere, each gas contributes to the total weight or pressure; therefore, each gas has a partial pressure, symbolized by the letter P. For example, we know that atmospheric pressure is 760 mm Hg (millimeters mercury) and that air is 21% oxygen. This means that the partial pressure of oxygen (P_{O_2}) in the atmosphere is $0.21 \times 760 = 160$ mm Hg. Similarly, we know that air is 0.04% carbon dioxide. This means that the partial pressure of carbon dioxide (P_{CO_2}) is $0.0004 \times 760 = 0.3$ mm Hg.

Figure 14.10 gives the partial pressure of gases in the lungs and the tissues. In the lungs, since P_{CO_2} is higher in venous blood than in the alveolus, carbon dioxide diffuses out of the blood; since the P_{O_2} is lower in venous blood than in the alveolus, oxygen diffuses into the blood.

In the tissues, since P_{CO_2} is higher in the tissues than in arterial blood, carbon dioxide diffuses into the blood; since P_{O_2} is higher in arterial blood than in the tissues, oxygen diffuses out of the blood (fig. 14.10).

The differences in the partial pressures of carbon dioxide and oxygen in the lungs and the tissues account for gas exchange. In the lungs, carbon dioxide diffuses out of the blood, and oxygen diffuses into the blood. This is external respiration. In the tissues, oxygen diffuses out of the blood, and carbon dioxide diffuses into the blood. This is internal respiration.

Transport of Gases

Some oxygen is dissolved in plasma, but most is carried by hemoglobin, the respiratory pigment, in red blood cells. At the P_{O_2} of alveolar air, hemoglobin is about 100% saturated with oxygen. The heme portion of hemoglobin contains iron, which combines with oxygen; still, the equation for **oxyhemoglobin** formation is:

$$Hb + O_2 \rightarrow HbO_2$$

figure 14.11 Internal respiration. When blood reaches the tissues, oxyhemoglobin (HbO_2) becomes deoxyhemoglobin (Hb) as oxygen (O_2) leaves the blood and enters the tissues. Carbon dioxide (CO_2) enters the plasma and then red blood cells. Some carbon dioxide combines with hemoglobin; forming carbaminohemoglobin ($HbCO_2$); the rest is converted to bicarbonate ions (HCO_3^-), which are carried in the plasma.

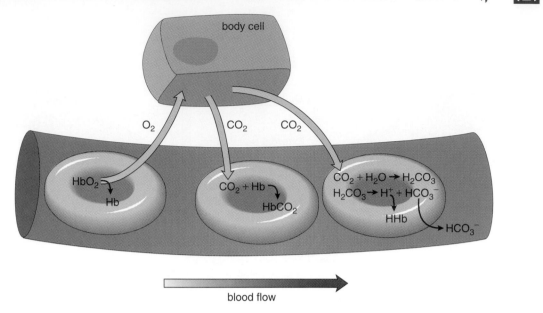

Oxyhemoglobin, which forms in the lungs, is bright red, and this explains the color of arterial blood.

Oxygen is bound only loosely to heme and, at the P_{O_2} in the tissues, hemoglobin readily gives oxygen up. Tissues that are actively using oxygen, such as contracting muscles, have a lower-than-normal P_{O_2}, and this causes more oxygen to be released. Also, hemoglobin more readily releases oxygen in the warmer and more acidic conditions created by the release of carbon dioxide by actively metabolizing tissues. Hemoglobin that is not carrying oxygen is called deoxyhemoglobin (dHb). Deoxyhemoglobin is dark purple, which explains the color of venous blood.

Red blood cells that have given up oxygen in the tissues are now ready to take part in the transport of carbon dioxide (fig. 14.11). At the P_{CO_2} of the tissues, carbon dioxide readily enters the blood, and some dissolves in plasma. Another portion combines with the amino groups of hemoglobin to form **carbaminohemoglobin:**

$$Hb + CO_2 \rightarrow HbCO_2$$

However, most of the carbon dioxide is transported as **bicarbonate ions** (HCO_3^-). This ion appears after carbon dioxide has combined with water to form carbonic acid. Carbonic acid dissociates (breaks down) to a hydrogen ion and a bicarbonate ion:

$$H_2O + CO_2 \rightarrow H_2CO_3 \rightarrow H^+ + HCO_3^-$$

An enzyme in red blood cells, called carbonic anhydrase, speeds up this reaction. The released hydrogen ions, which could drastically change the pH of the blood, are absorbed by the globin portion of hemoglobin, which therefore acts as a buffer. The buffering ability of hemoglobin plays a vital role in maintaining blood pH. The bicarbonate ions diffuse out of the red blood cells to be carried in the plasma. Electrolyte balance is maintained when chloride ions (Cl^-) from the plasma enter the red blood cells. This is called the chloride shift.

Once venous blood reaches the lungs, the lower P_{CO_2} of alveolar air prompts the reaction just described to occur in reverse:

$$HCO_3^- + H^+ \rightarrow H_2CO_3 \rightarrow H_2O + CO_2$$

Under the influence of carbonic anhydrase, bicarbonate ions join with hydrogen ions released by hemoglobin, and the resulting carbonic acid splits into carbon dioxide and water. The carbon dioxide diffuses out of the blood into the alveoli for expiration. Now hemoglobin is ready again to transport oxygen (fig. 14.12).

Some oxygen is dissolved in plasma, but most forms a loose association with hemoglobin. Therefore, oxygen is transported in the blood within red blood cells. Carbon dioxide is transported: (1) dissolved in plasma, (2) in carbaminohemoglobin, and (3) mostly as bicarbonate ions.

figure 14.12 External respiration. When blood reaches the lungs, bicarbonate ions (HCO_3^-) are converted to carbon dioxide (CO_2), which diffuses out of the blood into the alveolus. Carbaminohemoglobin ($HbCO_2$) also releases carbon dioxide, which diffuses into the alveolus. Deoxyhemoglobin (Hb) takes up oxygen and becomes oxyhemoglobin (HbO_2).

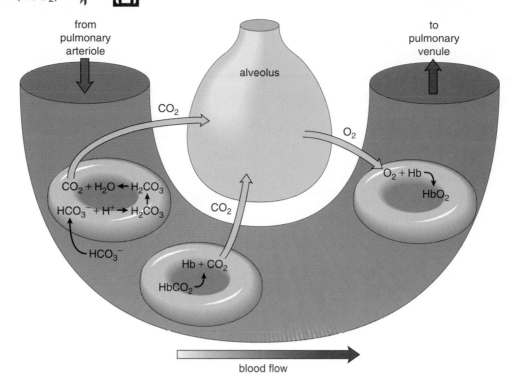

Respiratory Infections and Lung Disorders

Pathogens frequently spread from one individual to another by way of the respiratory tract. Droplets from a single sneeze may be loaded with billions of bacteria or viruses. The mucous membranes are protected by mucus production and by the constant beating of the cilia, but if the number of infective agents is large and/or the individual's resistance is reduced, an upper respiratory infection can result.

Upper Respiratory Tract Infections

The upper respiratory tract consists of the nose, the pharynx, and the larynx. Upper respiratory infections (URI) can spread from the nasal cavities to the sinuses, to the middle ears, and to the larynx. Viral infections sometimes lead to secondary bacterial infections. What we call "strep throat" is a primary bacterial infection caused by *Streptococcus pyogenes* that can lead to a generalized upper respiratory infection and even a systemic (affecting the body as a whole) infection. While antibiotics have no effect on viral infections, they are successfully used for most bacterial infections, including strep throat.

Common Cold

A cold is a viral infection that usually begins as a scratchy sore throat, followed by a watery mucous discharge from the nasal cavities. Fevers are rare, and symptoms are usually mild, requiring little or no medication. Although colds have a short duration, the immunity they provide is also brief. Since there are estimated to be over 150 cold-causing viruses, acquiring immunity to all of them is very difficult.

Cold vaccines are not in wide use, and because colds are viral in nature, antibiotics are not helpful. Since viruses reproduce inside cells, it is difficult to develop drugs that will kill the virus without affecting the cell itself.

Influenza

Influenza (flu) is a viral infection of the respiratory tract that is usually accompanied by aches and pains in the joints and fever. Influenza usually lasts longer than a cold. Immunity is possible, but only the vaccine developed for the particular flu virus prevalent that season can usually be successful in protecting the individual. Since influenza viruses constantly mutate, immunity does not build up, and a new viral illness rapidly spreads from person to person and from place to place. Pandemics, in which a newly

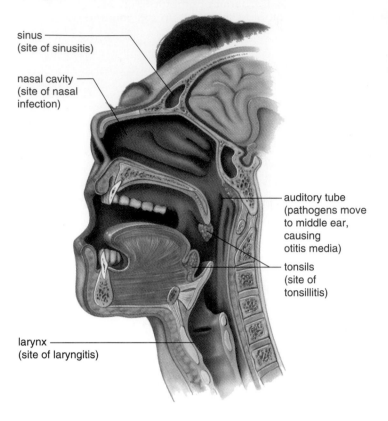

figure 14.13 Upper respiratory infections.

sinus
(site of sinusitis)

nasal cavity
(site of nasal
infection)

auditory tube
(pathogens move
to middle ear,
causing
otitis media)

tonsils
(site of
tonsillitis)

larynx
(site of laryngitis)

mutated influenza virus spreads throughout the world, occur regularly, about every 10 years.

Sinusitis

Sinusitis is an infection of the sinuses, cavities within the facial skeleton that drain into the nasal cavities (fig. 14.13). Only about 1–3% of upper respiratory infections are accompanied by sinusitis. Sinusitis develops when nasal congestion blocks the tiny openings leading to the sinuses. Symptoms include postnasal discharge as well as facial pain that worsens when the patient bends forward. Pain and tenderness usually occur over the lower forehead or over the cheeks. If the latter, toothache is also a complaint. Successful treatment depends on restoring proper drainage of the sinuses. Even a hot shower and sleeping upright can be helpful. Otherwise, spray decongestants are preferred over oral antihistamines, which thicken rather than liquefy the material trapped in the sinuses.

Otitis Media

Otitis media is a bacterial infection of the middle ear. The middle ear is not a part of the respiratory tract, but this in-

fection is considered here because it is a complication often seen in children who have a nasal infection. Infection can spread by way of the **auditory tube** that leads from the nasopharynx to the middle ear (fig. 14.13). Pain is the primary symptom of a middle ear infection. A sense of fullness, hearing loss, vertigo (dizziness), and fever may also be present. Antibiotics almost always bring about a full recovery, and a recurrence is most likely due to a new infection. Drainage tubes (called tympanostomy tubes) are sometimes placed in the eardrum of children with multiple recurrences to help prevent buildup of fluid in the middle ear and the possibility of hearing loss. Normally, the tubes slough out with time.

Tonsillitis

Tonsillitis occurs when tonsils become inflamed and enlarged. Tonsils are masses of lymphatic tissue that occur in the pharynx (fig. 14.13). The tonsils in the dorsal wall of the nasopharynx are often called adenoids. The tonsils remove many of the pathogens that enter the pharynx; therefore, they are a first line of defense against invasion of the body. If tonsillitis occurs frequently and enlargement makes breathing difficult, the tonsils can be removed surgically in a **tonsillectomy.** Fewer tonsillectomies are performed today than in the past because it is now known that tonsils serve an important function in defending the body against infection.

Laryngitis

Laryngitis is an infection of the larynx with an accompanying hoarseness leading to the inability to talk in an audible voice (fig. 14.13). Usually laryngitis disappears with treatment of the upper respiratory infection. Persistent hoarseness without the presence of an upper respiratory infection is one of the warning signs of cancer and therefore should be looked into by a physician.

> Upper respiratory infections due to viruses, such as colds and influenza, are not treatable by antibiotics, but bacterial infections, such as otitis media and strep throat, usually respond to antibiotic therapy.

Lower Respiratory Tract Infections and Disorders

Acute bronchitis, pneumonia, and tuberculosis, three infections of the lungs, formerly caused a large percentage of deaths in the United States but are now controlled by antibiotics. Other illnesses, such as emphysema and lung cancer, are not due to infections but, in most instances, to cigarette smoking.

Bronchitis

Bronchitis can be acute or chronic. Acute bronchitis is usually caused by a secondary bacterial infection of the bronchi, resulting in a heavy mucus discharge with persistent coughing. Acute bronchitis usually responds to antibiotic therapy.

Chronic bronchitis is not necessarily due to infection, but rather to a constant irritation of the lining of the bronchi. The irritation causes the bronchi lining to become inflamed and undergo degenerative changes, with the loss of cilia preventing normal cleansing action (fig. 14.14). The affected individual coughs frequently and is more susceptible to upper respiratory infections. Chronic bronchitis most often affects cigarette smokers.

Asthma

Asthma is a disease of the bronchi and bronchioles that is marked by wheezing, breathlessness, and sometimes cough and expectoration of mucus. The airways are unusually sensitive to specific irritants, which can include a wide range of allergens such as pollen, animal dander, dust, cigarette smoke, and industrial fumes. Even cold air, however, can be an irritant. When exposed to the irritant, the smooth muscle in the bronchioles undergoes spasms. It now appears that chemical mediators given off by immune cells in the bronchioles result in the spasms. Most asthma patients have some degree of bronchial inflammation that reduces the diameter of the airways and contributes to the seriousness of an attack. Asthma is not curable but is treatable. There are inhalers that control the inflammation and hopefully prevent an attack, but there are also inhalers that stop the muscle spasms should an attack occur.

Pneumonia

Most forms of **pneumonia** are caused by bacteria or viruses that infect the lungs. The demise of AIDS patients is usually due to a particularly rare form of pneumonia caused by the protozoan *Pneumocystis carinii*. Sometimes, pneumonia is localized in specific lobes of the lungs, which become inoperative as they fill with fluid (fig. 14.14). The more lobes involved, the more serious the infection.

Pulmonary Tuberculosis

Tuberculosis is caused by the tubercle bacillus. Exposure to tuberculosis is detected by the use of a skin test in which a highly diluted extract of the bacilli is injected into the patient's skin. A person who has never been in contact with the bacilli will show no reaction, but one who has developed immunity to the organism will show an area of inflammation that peaks in about 48 hours.

If bacilli invade the lung tissue, tissue cells build a protective capsule around the bacilli, isolating them from the rest of the body. This tiny capsule is called a *tubercle* (fig. 14.14). If body resistance is high, the imprisoned organisms may die, but if resistance is low, the organisms may eventually be liberated. If a chest X ray detects the presence of tubercles, the individual is put on appropriate drug therapy to ensure the localization of the disease and the eventual destruction of any live bacterial organisms.

Tuberculosis killed about 100,000 people a year in the United States before the middle of the 20th century, when antibiotic therapy brought it largely under control. In recent years, however, the incidence of tuberculosis has risen, particularly among AIDS patients, the homeless, and the rural poor. Worse, the new strains are resistant to the usual antibiotic therapy. Therefore, some physicians would like to again quarantine patients in sanitoriums.

Emphysema

Emphysema refers to the destruction of lung tissue, with accompanying ballooning or inflation of the lungs due to trapped air. The problems of emphysema stem from the destruction and collapse of the bronchioles. When this occurs, the alveoli are cut off from renewed oxygen supply, and the air within them is trapped. The trapped air often causes alveolar walls to rupture (fig. 14.14), and a loss of elasticity also makes breathing difficult. The victim is breathless and may have a cough. Since the surface area for gas exchange is reduced, not enough oxygen reaches the heart and the brain. Even so, the heart works furiously to force more blood through the lungs, which can lead to a heart condition. Lack of oxygen to the brain can make the person feel depressed, sluggish, and irritable. As mentioned in the MedAlert reading on page 290, chronic bronchitis and emphysema, two conditions most often caused by smoking, together are called **chronic obstructive pulmonary disease (COPD).**

Pulmonary Fibrosis

Inhaling particles such as silica (sand), coal dust, and asbestos (fig. 14.14) can lead to **pulmonary fibrosis,** in which fibrous connective tissue accumulates in the lungs. Breathing capacity can be seriously impaired, and cancer can develop. In the past, asbestos was widely used as a fireproofing and insulating agent, resulting in unwarranted exposure.

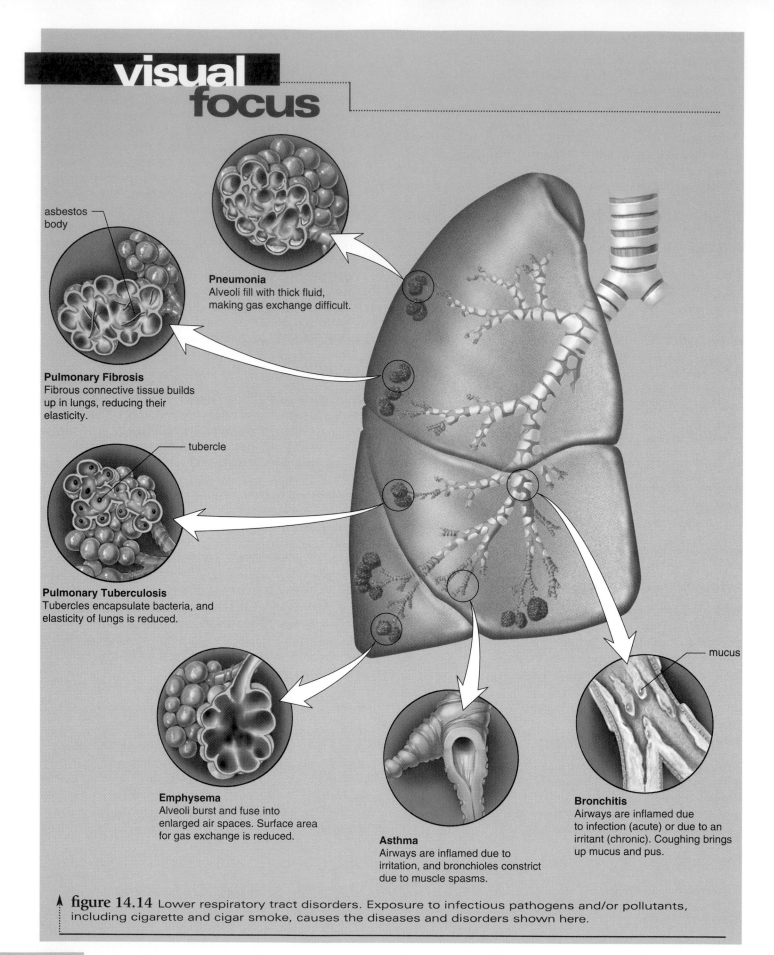

asbestos body

Pneumonia
Alveoli fill with thick fluid, making gas exchange difficult.

Pulmonary Fibrosis
Fibrous connective tissue builds up in lungs, reducing their elasticity.

tubercle

Pulmonary Tuberculosis
Tubercles encapsulate bacteria, and elasticity of lungs is reduced.

Emphysema
Alveoli burst and fuse into enlarged air spaces. Surface area for gas exchange is reduced.

Asthma
Airways are inflamed due to irritation, and bronchioles constrict due to muscle spasms.

mucus

Bronchitis
Airways are inflamed due to infection (acute) or due to an irritant (chronic). Coughing brings up mucus and pus.

figure 14.14 Lower respiratory tract disorders. Exposure to infectious pathogens and/or pollutants, including cigarette and cigar smoke, causes the diseases and disorders shown here.

Lung Cancer

For many years, lung cancer was more prevalent in men than in women, but recently, it has surpassed breast cancer as a cause of death in women. This can be linked to an increase in the number of women who smoke today. Autopsies on smokers have revealed the progressive steps by which the most common form of lung cancer develops. The first event appears to be thickening and callusing of the cells lining the bronchi. (Callusing occurs whenever cells are exposed to irritants.) Then, a loss of cilia makes it impossible to prevent dust and dirt from settling in the lungs. Following this, cells with atypical nuclei appear in the callused lining. A tumor consisting of disordered cells with atypical nuclei is considered to be cancer *in situ* (at one location). A final step occurs when some of these cells break loose and penetrate other tissues, a process called metastasis. Now the cancer has spread. The tumor may grow until the bronchus is blocked, cutting off the air supply to that lung. The entire lung then collapses, the secretions trapped in the lung spaces become infected, and pneumonia or a lung abscess (localized area of pus) results. The only treatment that offers a possibility of cure is to remove a lobe or the lung completely before secondary growths have had time to form. This operation is called **pneumonectomy.**

The incidence of lung cancer is over 20 times higher in individuals who smoke than in those who do not. In addition, current research indicates that passive smoking—simply breathing in air filled with cigarette smoke—can also cause lung cancer and other illnesses associated with smoking. The MedAlert reading on page 290 notes the various illnesses that are apt to occur when a person smokes. If a person stops smoking, and if the body tissues are not already cancerous, they usually return to normal over time.

Acute bronchitis, pneumonia, and tuberculosis are caused by bacteria. Chronic bronchitis, emphysema, and lung cancer occur more frequently in smokers than in nonsmokers.

Effects of Aging

Respiratory fitness decreases with age. Maximum breathing capacities decline, while the likelihood of fatigue increases. Inspiration and expiration are not as effective in older persons. With age, weakened intercostal muscles and increased inelasticity of the rib cage combine to reduce the inspiratory reserve volume, while the lungs' inability to recoil reduces the expiratory reserve volume. More residual air is found in the lungs of older people.

With age, gas exchange in the lungs is not as efficient, not only due to changes in the lungs but also due to changes in the blood capillaries. The walls of the alveoli and capillaries thicken, and the gases cannot diffuse as rapidly as they once did.

In the elderly, the ciliated cells of the trachea are reduced in number, and those remaining are not as effective as they once were. Respiratory diseases, such as those discussed in the chapter, are more prevalent in older people than in the general public. Pneumonia and other respiratory infections are among the leading causes of death in older persons.

Working Together

The accompanying illustration shows how the respiratory system works with other organ systems of the body to maintain homeostasis. All the organ systems of the body are interrelated.

BodyWorks CD-ROM
The module accompanying chapter 14 is Respiratory System.

The Most Often Asked Questions about Smoking, Tobacco, and Health, and . . . the Answers

Is There a Safe Way to Smoke?
No. All cigarettes can cause damage, and smoking even a small amount is dangerous. Cigarettes are perhaps the only legal product whose advertised and intended use—that is, smoking them—will hurt the body. Some people try to make smoking safer by smoking fewer cigarettes, but most smokers find this difficult. Some people think that switching from high tar/nicotine cigarettes to those with low tar/nicotine content makes smoking safer, but this does not always happen. . . The health benefits from switching would be insignificant compared with the benefits of quitting altogether. . . .

Does Smoking Cause Cancer?
Yes, it causes lung cancer (fig. 14B). Tobacco use is responsible for about 30% of all cancer deaths in the United States. Cigarette smoking causes about 87% of lung cancer deaths. Besides lung cancer, cigarette smoking is also a major cause of cancers of the mouth, larynx (voice box), and esophagus (swallowing tube). In addition, smoking increases the risk of cancer of the bladder, kidney, pancreas, stomach, and uterine cervix.

What Are the Chances of Being Cured of Lung Cancer?
Very low; the five-year survival rate is only 13%. Most forms of the disease start without producing any warning signs, so that it is rarely detected in the early stages, when it is more likely to be cured. The past 15 years have brought little significant progress in the earlier diagnosis or treatment of lung cancer. Fortunately, lung cancer is a largely preventable disease. That is, by not smoking it can be prevented. . . .

Do Cigarettes Cause Other Lung Diseases?
Yes. Cigarette smoking causes other lung diseases that can be just as dangerous as lung cancer. It leads to chronic bronchitis—a disease where the airways produce excess mucus, which forces the smoker to cough frequently. Cigarette smoking is also the major cause of emphysema—a disease that slowly destroys a person's ability to breathe. . . . Chronic obstructive pulmonary disease (COPD), which includes chronic bronchitis and emphysema, kills about 81,000 people each year; cigarette smoking is responsible for 65,000 of these deaths.

Why Do Smokers Have "Smoker's Cough"?
Cigarette smoke contains chemicals that irritate the air passages and lungs. When a smoker inhales these substances, the body tries to protect itself by coughing. The well-known "early

morning" cough of smokers happens for a different reason. Normally, cilia (tiny, hairlike formations that line the airways) beat outward and "sweep" harmful material out of the lungs. Cigarette smoke, however, decreases this sweeping action, so some of the poisons in the smoke remain in the lungs. . . . Smokers are more likely to get pneumonia because damaged or destroyed cilia cannot protect the lungs from bacteria and viruses that float in the air. . . .

If You Smoke But Don't Inhale, Is There Any Danger?
Yes. Wherever smoke touches living cells, it does harm. So, even if smokers don't inhale—including pipe and cigar smokers—they are at an increased risk for lip, mouth, and tongue cancer. Because it is virtually impossible to avoid inhaling tobacco smoke totally, these smokers also have an increased chance of getting lung cancer. Lung cancer is much more likely to occur in a person who has always smoked cigars or pipes than in a person who has never smoked at all.

Does Cigarette Smoking Affect the Heart?
Yes. Smoking cigarettes increases the risk of heart disease, which is America's number one killer. About 150,000 Americans die each year from heart attacks and other forms of heart disease caused by smoking. Smoking, high blood pressure, high blood cholesterol, and lack of exercise are all risk factors for heart disease. Smoking alone doubles the risk of heart disease. When a person smokes and has other risk factors, his or her chance of getting heart disease increases dramatically. For example, if smoking is combined with high blood pressure or high cholesterol, then the risk goes up four times. Put all three together—smoking, high blood pressure, and high cholesterol—and the risk goes up eight times. Smokers who have already had one heart attack are also more likely than nonsmokers to have another attack.

Is There Any Risk for Pregnant Women and Their Babies?
Pregnant women who smoke endanger the health and lives of their unborn babies. When a pregnant woman smokes, she really is smoking for two because the nicotine, carbon monoxide, and other dangerous chemicals in smoke enter the mother's bloodstream and then pass into the baby's body. Women who smoke during pregnancy risk having a miscarriage, or a premature or stillborn baby. Their babies are also more likely to be underweight, by an average of $1/2$ pound. . . .

Does Smoking Cause Any Special Health Problems for Women?

Yes. Nonsmoking women who use oral contraceptives ("the Pill") double their chances of having a heart attack. However, when women use the Pill and smoke, they are 10 times more likely to suffer a heart attack than nonsmoking women who don't take the Pill. Women who smoke and use the Pill have an increased risk of stroke and blood clots in the legs as well. Women who smoke also run the risk of having trouble getting pregnant; the more they smoke, the more likely it is that they will have difficulty. Some studies show that female smokers, especially the elderly, are at a higher risk for osteoporosis (a disease that weakens the bones and makes them more likely to break) than nonsmoking women. In addition, women who smoke increase their chances of getting cancer of the uterine cervix.

What Are Some of the Short-term Effects of Smoking Cigarettes?

Almost immediately, smoking can make it hard to breathe. Within a short time, it can also worsen asthma and allergies. Nicotine reaches the brain only 7 seconds after taking a puff (faster than it takes heroin to reach the brain), where it produces a variety of effects. . . .

Are There Any Other Risks to the Smoker?

Yes. There are many other risks. As we already mentioned briefly, smoking cigarettes causes stroke, which is the third leading cause of death in America. Smoking causes lung cancer, but if a person smokes and is exposed to radon or asbestos, the risk increases even more. Smokers are also more likely to have and die from stomach ulcers than nonsmokers. In addition, cigarettes can interact with medication the smoker is taking in unwanted ways—like preventing the drug from doing what it is supposed to do. . . .

What Are the Dangers of Passive Smoking?

Passive smoking causes lung cancer in healthy nonsmokers. Children whose parents smoke are more likely to suffer from pneumonia or bronchitis in the first 2 years of life than children who come from smoke-free households. Nonsmokers who are married to smokers have a 30% greater risk for developing lung cancer than nonsmokers married to other nonsmokers. . . .

Are Chewing Tobacco and Snuff Safe Alternatives to Cigarette Smoking?

No, they are not. Many people who use chewing tobacco or snuff believe it can't harm them because there is no smoke. Wrong. Smokeless tobacco contains nicotine, the same addicting drug found in cigarettes. Snuff dippers also take in an average of over 10 times more cancer-causing substances (called nitrosamines) than cigarette smokers. In fact, some brands of smokeless tobacco contain as much as 20,000 times the legal limit of nitrosamines that are permitted in certain foods and consumer products (such as bacon, beer, and baby bottle nipples). While not inhaled through the lungs, the juice from smokeless tobacco is absorbed through the lining of the mouth. There, it can cause sores and white patches, which often lead to cancer of the mouth. . . .

Questions

1. Why are smokers less able to prevent bacteria and viruses that float in the air from entering their lungs?
2. Smoking causes arterioles to constrict. Relate this to coronary heart disease and low-birth-weight babies in smokers.
3. Smoking brings impurities into the lungs. Relate this to the increased likelihood of cancer and emphysema in smokers.

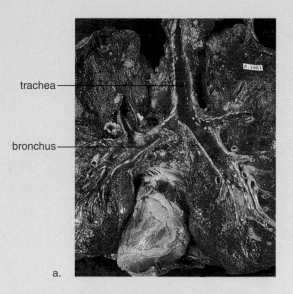

trachea
bronchus

a.

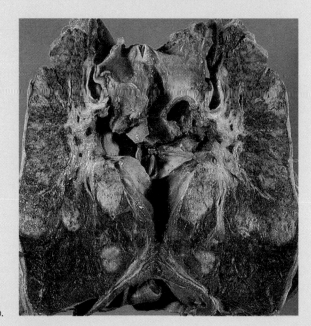

b.

Figure 14B Normal lung versus cancerous lung. **a.** Normal lung with heart in place. Notice the healthy red color. **b.** Lung of a heavy smoker cut in half. Notice how black the lung is except where cancerous tumors have formed.

working together   Respiratory System

Integumentary System

Gas exchange in lungs provides oxygen to skin and rids body of carbon dioxide from skin.

Skin helps protect respiratory organs and helps regulate body temperature.

Skeletal System

Gas exchange in lungs provides oxygen and rids body of carbon dioxide.

Rib cage protects lungs and assists breathing; bones provide attachment sites for muscles involved in breathing.

Muscular System

Lungs provide oxygen for contracting muscles and rid the body of carbon dioxide from contracting muscles.

Muscle contraction assists breathing; physical exercise increases respiratory capacity.

Nervous System

Lungs provide oxygen for neurons and rid the body of carbon dioxide produced by neurons.

Respiratory centers in brain regulate breathing rate.

Endocrine System

Gas exchange in lungs provides oxygen and rids body of carbon dioxide.

Epinephrine promotes ventilation by dilating bronchioles; growth factors control production of red blood cells that carry oxygen.

How the Respiratory System works with other body systems

Circulatory System

Gas exchange in lungs rids body of carbon dioxide, helping to regulate the pH of blood; breathing aids venous return.

Blood vessels transport gases to and from lungs; blood services respiratory organs.

Lymphatic System/Immunity

Tonsils and adenoids occur along respiratory tract; breathing aids lymph flow; lungs carry out gas exchange.

Lymphatic vessels pick up excess tissue fluid; immune system protects against respiratory tract and lung infections.

Digestive System

Gas exchange in lungs provides oxygen to the digestive tract and excretes carbon dioxide from the digestive tract.

Breathing is possible through the mouth because digestive tract and respiratory tract share the pharynx.

Urinary System

Lungs excrete carbon dioxide, provide oxygen, and convert angiotensin I to angiotensin II, leading to kidney regulation.

Kidneys compensate for water lost through respiratory tract; work with lungs to maintain blood pH.

Reproductive System

Gas exchange increases during sexual activity.

Sexual activity increases breathing; pregnancy causes breathing rate and vital capacity to increase.

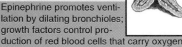
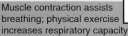
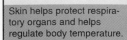

Basic Key Terms

alveolus (al-ve′o-lus), p. 276

aortic body (a-or′tik bod′e), p. 278

bicarbonate ion (bi′kar′bo-nāt i′on), p. 284

bronchiole (brong′ke-ōl), p. 276

bronchus (brong′kus), p. 276

carbaminohemoglobin (kar-bah-me″no-he″mo-glo′bin), p. 284

carotid body (kar-rah′tid bod′e), p. 278

concha (kong′kah), p. 274

diaphragm (di′ah-fram), p. 278

epiglottis (e″pi-glot′is), p. 275

expiration (eks″pĭ-ra′shun), p. 280

expiratory reserve volume (ERV) (ik-spi′rah-tor-e re-zerv′ vol′yūm), p. 281

external respiration (eks-ter′nal res″pĭ-ra′shun), p. 282

glottis (glot′is), p. 275

inspiration (in″spĭ-ra′shun), p. 279

inspiratory reserve volume (IRV) (in-spi′rah-tor-e re-zerv′ vol′yūm), p. 281

internal respiration (in-ter′nal res″pĭ-ra′shun), p. 282

larynx (lar′inks), p. 275

nose (nōz), p. 274

oxyhemoglobin (ok″se-he″mo-glo′bin), p. 283

paranasal sinus (par-ah-na′zul si′nus), p. 274

pharynx (far′inks), p. 274

pleural membrane (ploor′al mem′brān), p. 278

residual volume (RV) (rĭ-zĭ′jah-wahl vol′yūm), p. 281

respiration (res″pĭ-ra′shun), p. 273

surfactant (sur-fak′tant), p. 277

tidal volume (TV) (tīd′al vol′ūm), p. 281

trachea (tra′ke-ah), p. 276

ventilation (ven″tĭ-la′shun), p. 278

vital capacity (VC) (vi′tal kah-pas′i-te), p. 281

Clinical Key Terms

apnea (ap′ne-ah), p. 278

asthma (az′mah), p. 287

bronchitis (bron-ki′tis), p. 287

Cheyne-Stokes respiration (shān-stōks res″pĭ-ra′shun), p. 278

chronic obstructive pulmonary disease (kron′ik ob-struk′tiv pul′mo-ner″e dĭ-zēz′), p. 287

emphysema (em″fĭ-se′mah), p. 287

Heimlich maneuver (Hīm′lik mah-nu′ver), p. 282

hyperpnea (hi″per-ne′ah), p. 278

influenza (in″flu-en′zah), p. 285

laryngitis (lar-in-ji′tis), p. 286

lung cancer (lung′kan-ser), p. 277

otitis media (o-ti′tis me′-de-ah), p. 286

pneumonectomy (nu-mah-nek′tah-me), p. 289

pneumonia (nu-mo′nyah), p. 287

pulmonary fibrosis (pul′mo-ner″e fi-bro′sis), p. 287

respiratory distress syndrome (rĭ-spi′rah-tor-e dis-tres′ sin′drōm), p. 277

sinusitis (sīn-u-si′tis), p. 286

tonsillitis (ton-sil-li′tis), p. 286

tracheostomy (tra″ke-os′tah-me), p. 276

tuberculosis (tu″ber-ku-lo′sis), p. 287

Summary

I. Respiratory Organs

A. The nasal cavities, which filter, warm, and humidify incoming air, open into the pharynx.

B. Food and air passages join in the pharynx, which conducts air to the larynx and food to the esophagus.

C. The larynx is the voice box. It contains the vocal cords at the sides of the glottis, an opening covered by the epiglottis when food is being swallowed.

D. Air moves from the trachea (windpipe) and the two bronchi, held open by cartilage, into the lungs. The bronchi branch and rebranch into bronchioles within the lungs. Bronchioles lead to air sacs called alveoli.

E. Air moves from the nasal cavities to the pharynx, to the trachea, to the bronchi, to the bronchioles, and finally, to the approximately 300 million alveoli in the lungs. Gas exchange occurs at the alveoli.

II. Mechanisms of Breathing

The lungs are completely enclosed and, by way of the pleural membranes, adhere to the walls of the thoracic cavity.

A. Inspiration. Stimulated by nerve impulses, the rib cage lifts up and out, and the diaphragm lowers to expand the thoracic cavity and lungs, allowing inspiration to occur.

B. Expiration. When nervous stimulation ceases, the rib cage lowers and the diaphragm rises, allowing the lungs to recoil and expiration to occur.

C. Lung capacities. Tidal volume is the amount of air inhaled and exhaled with each breath. Vital capacity is the total volume of air that can be moved in and out of the lungs during a single breath. Some air remains in the lungs after expiration. This is called the residual volume. Passages within the airways are called dead space because no gas exchange takes place in the airways.

III. **Respiration Mechanisms and Transport of Gases**

A. Partial pressures of gases. The differences in the partial pressures of carbon dioxide and oxygen in the lungs and the tissues account for gas exchange. In the lungs, carbon dioxide diffuses out of the blood, and oxygen diffuses into the blood. This is external respiration. In the tissues, oxygen diffuses out of the blood, and carbon dioxide diffuses into the blood. This is internal respiration.

B. Transport of gases. Some oxygen is dissolved in plasma, but most forms a loose association with hemoglobin. Therefore, oxygen is transported in the blood within red blood cells. Carbon dioxide is transported: (1) dissolved in plasma, (2) in carbaminohemoglobin, and (3) mostly as bicarbonate ions.

IV. **Respiratory Infections and Lung Disorders**

A. Upper respiratory tract infections. Upper respiratory infections due to viruses, such as colds and influenza, are not treatable by antibiotics, but bacterial infections, such as otitis media and strep throat, usually respond to antibiotic therapy.

B. Lower respiratory tract infections and disorders. Acute bronchitis, pneumonia, and tuberculosis are caused by bacteria. Chronic bronchitis, emphysema, and lung cancer occur more frequently in smokers than in nonsmokers.

Study Questions

1. What are the four parts of respiration? In which of these is oxygen actually used up and carbon dioxide produced? (p. 273)

2. List the parts of the respiratory tract. What are the special functions of the nasal cavities, larynx, and alveoli? (pp. 273–78)

3. What are the steps in inspiration and expiration? How is the breathing rate controlled? (pp. 278–81)

4. Why is it impossible to breathe through a very long tube? (p. 282)

5. What physical process is believed to explain gas exchange? (pp. 282–83)

6. Explain gas exchange for both internal and external respiration. (pp. 282–83)

7. How is carbon dioxide carried in the blood? How is oxygen carried? (pp. 283–84)

8. Name and describe some infections of the respiratory tract. (pp. 285–89)

9. What are emphysema and pulmonary fibrosis, and how do they affect a person's health? (p. 287)

10. What is the most common cause of lung cancer? Why is the incidence of lung cancer rising among women? (pp. 289–91)

Objective Questions

Fill in the blanks.

1. In tracing the path of air, the _____ immediately follows the pharynx.
2. The lungs contain air sacs called _____ .
3. The breathing rate is primarily regulated by the amount of _____ in the blood.
4. Air enters the lungs after they have _____ .
5. Gas exchange is dependent on the physical process of _____ .
6. During external respiration, oxygen _____ the blood.
7. During internal respiration, carbon dioxide _____ the blood.
8. Carbon dioxide is carried in the blood as _____ ions.
9. The most likely cause of emphysema and chronic bronchitis is _____ .
10. Most cases of lung cancer actually begin in the _____ .
11. The amount of air moved in and out of the respiratory system with each normal breath is called the _____ .
12. The total amount of air that can be moved in and out of the lungs during a single breath is called the _____ .
13. The _____ closes the opening into the larynx during swallowing.
14. The respiratory membrane consists of the walls of the _____ and _____ .

Medical Terminology Reinforcement Exercise

Consult Appendix B for help in pronouncing and analyzing the meaning of the terms that follow.

1. eupnea (ūp-ne'ah)
2. nasopharyngitis (na"zo-far"in-ji'tis)
3. tracheostomy (tra"ke-os'to-me)
4. pneumonomelanosis (nu-mo"no-mel"ah-no'sis)
5. pleuropericarditis (ploor"o-per"ĭ-kar"di'tis)
6. bronchoscopy (brong-kos'ko-pe)
7. dyspnea (disp'ne-ah)
8. laryngospasm (lah-rin'go-spazm)
9. hemothorax (he"mo-tho'raks)
10. otorhinolaryngology (o"to-ri"no-lar"in-gol'o-je)
11. apnea (ap'ne-ah)
12. hypoxemia (hi"pok-se'me-ah)

Website Link

For a listing of the most current Websites related to this chapter, please visit the Mader home page at:

http://www.mhhe.com/maderap

chapter 15

The Digestive System

Chapter Outline and Learning Objectives

After you have studied this chapter, you should be able to:

Digestive System (p. 297)
- Trace the path of food in the digestive tract, and describe the general structure and function of each organ mentioned.
- Describe peristalsis, and state its function.
- Describe the wall of the small intestine, and relate its anatomy to nutrient absorption.
- Name the hormones produced by the digestive tract that help control digestive secretions.
- Name the accessory organs of digestion, and describe their contributions to the digestive process.

Mechanical and Chemical Digestion (p. 310)
- Name two structures which are involved in mechanical digestion.
- Name and state the functions of digestive enzymes for carbohydrates, proteins, and fats.

Nutrition (p. 310)
- State the functions of glucose, fats, and amino acids in the body.
- Define the terms *essential fatty acid*, *essential amino acid*, and *vitamin*.
- Describe the functions of major vitamins and minerals in the body.

Effects of Aging (p. 316)
- Anatomical and physiological changes occur in the digestive system as we age.

Working Together (p. 316)
- The digestive system works with other systems of the body to maintain homeostasis.

Medical Focus
Viral Hepatitis (p. 306)
Human Teeth (p. 311)
Antioxidants (p. 313)

MedAlert
Constipation (p. 308)

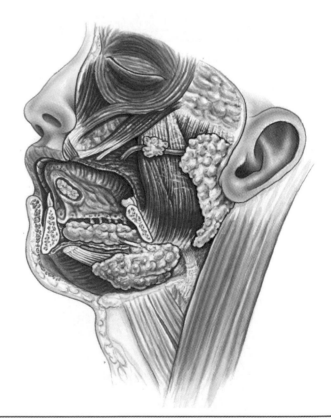

The salivary glands (yellow) are located near the ear and beneath the tongue. Their secretions aid the digestion of starch in the mouth.

figure 15.1 Major organs of the digestive tract.

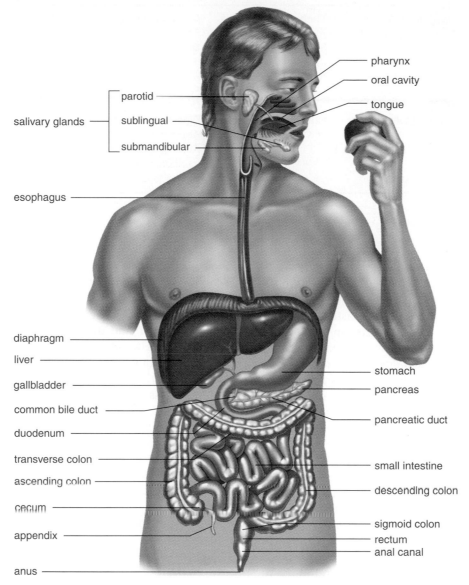

Digestion takes place within a tube called the **alimentary canal,** which begins at the mouth and ends at the anus. Digestive enzymes are secreted into the tube by glands located in the tube lining or nearby. While the term *digestion*, strictly speaking, means the breakdown of food by enzymatic action, we will expand the term to include both the physical and chemical processes that reduce food to small, soluble molecules. The digestive system contributes to homeostasis by providing the body with the nutrients needed to sustain the life of cells.

Digestive System

The digestive system includes the organs listed in table 15.1 and illustrated in figure 15.1. The functions of the digestive system are to:

1. Ingest the food
2. Break it down into small molecules that can cross plasma membranes
3. Absorb these nutrient molecules
4. Eliminate nondigestible wastes

table 15.1

Path of Food		
Organ	**Special Features**	**Functions**
Mouth	Teeth, tongue	Chewing of food
	Tongue	Formation of bolus
	Salivary glands	Digestion of starch
Pharynx		Swallowing
Esophagus		Passageway
Stomach		Food storage; acidity kills bacteria
	Gastric glands	Digestion of protein
Small intestine	Intestinal glands	Digestion of all foods
	Villi	Absorption of nutrients
Large intestine		Absorption of water; storage of nondigestible remains
Anus		Defecation

Mouth

The **mouth** receives the food in its *oral cavity.* The taste of food results from the stimulation of chemoreceptors in taste buds on the tongue and in the nasal cavities. Therefore, smell is a very important component of taste perception.

The roof of the mouth has two parts: An anterior **hard palate** separates the oral cavity from the nasal passages, and a posterior **soft palate** separates the oral cavity from the nasopharynx. The hard palate contains several skull bones—namely, portions of the maxillae and palatine bones. The soft palate is a muscular structure and ends in the *uvula,* a cone-shaped process.

The tongue is made up of skeletal muscle covered by a mucous membrane. Intrinsic muscles that have their origin and insertion in the tongue itself change the shape of the tongue. Extrinsic muscles insert into the tongue but have their origin outside the tongue, such as on a skull bone. The extrinsic muscles move the tongue about and account for its maneuverability. A fold of mucous membrane (lingual frenulum) on the underside of the tongue attaches the tongue to the floor of the mouth. If the frenulum is too short, the individual cannot speak clearly and is said to be tongue-tied. The floor of the oral cavity and underside of the tongue are richly supplied with blood vessels, and soluble medications will enter the circulation directly if placed beneath the tongue.

There are three pairs of **salivary glands** called the parotid, sublingual, and submandibular glands. The *parotid glands* are located in front of and below the ears. Each parotid gland has a duct that opens on the inner surface of the cheek, just at the location of the second upper molar. Mumps, an acute viral infection of the parotid salivary glands, causes them to swell, making it difficult to open the mouth. The *sublingual glands* lie beneath the tongue proper, and their ducts open into the floor of the oral cavity. The *submandibular glands* lie in the posterior floor of the oral cavity beneath the base of the tongue. The ducts from the submandibular glands open on either side of the lingual frenulum.

The salivary glands produce about 1 liter of saliva a day, which enters the mouth by way of their ducts. Saliva contains mucus and a digestive enzyme called **salivary amylase.** Mucus binds and lubricates the food so it is easier to swallow, and salivary amylase begins the process of digesting the food by breaking down starch. Saliva also contains chemicals that protect against bacterial infection.

The teeth, which are described in the Medical Focus reading on page 311, carry out **mastication,** or chewing of food. The tongue assists in mastication by moving the food between the teeth. Mastication breaks down the food into small portions, making the work of the digestive enzymes more efficient. The tongue forms the chewed food into a small mass called a **bolus** in preparation for swallowing.

The salivary glands produce saliva, which enters the mouth, where teeth masticate the food and the tongue forms a bolus prior to swallowing.

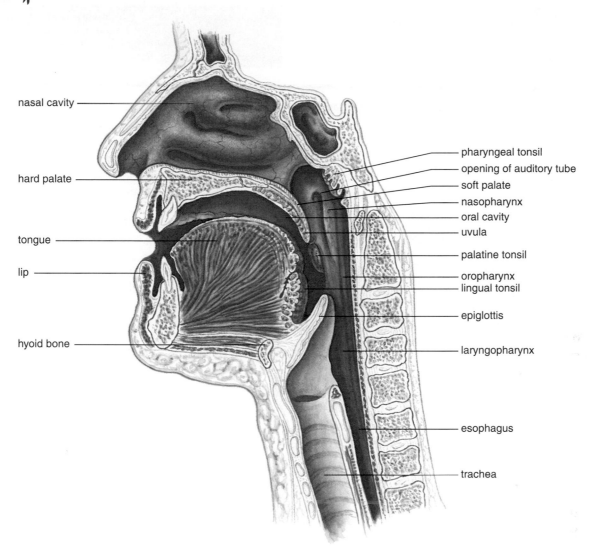

nasal cavity

hard palate

tongue

lip

hyoid bone

pharyngeal tonsil

opening of auditory tube

soft palate

nasopharynx

oral cavity

uvula

palatine tonsil

oropharynx

lingual tonsil

epiglottis

laryngopharynx

esophagus

trachea

Pharynx

Swallowing, a reflex action that moves food into the esophagus, occurs in the **pharynx** (far'inks) (fig. 15.2), a region that opens into the nose, mouth, and larynx. During swallowing, food normally enters only the esophagus, a long, muscular tube that extends to the stomach, because the nasal and laryngeal passages are blocked. The nasopharyngeal openings are covered when the soft palate moves back. The opening to the larynx at the top of the trachea, called the **glottis**, is covered when the trachea moves up under a flap of tissue, called the epiglottis. Therefore, breathing does not occur when swallowing. This process is easy to observe in the up-and-down movement of the Adam's apple, the ventral cartilage of the larynx, when a person eats.

The air and food passages merge in the pharynx. When a person swallows, the air passage is usually blocked off, and food must enter the esophagus.

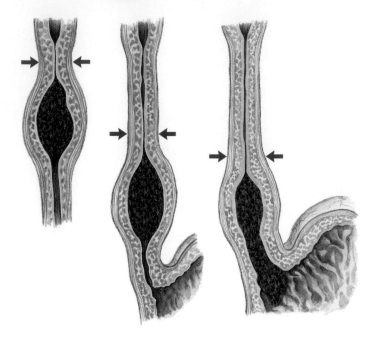

figure 15.3 Peristalsis in the digestive tract. The three drawings show how a peristaltic wave moves through a single section of the esophagus over time.

Esophagus

The **esophagus** is a tube with a pharyngoesophageal sphincter near the pharynx and a gastroesophageal sphincter near the stomach. **Sphincters** are muscles that encircle tubes and act as valves; tubes close when sphincters contract, and they open when sphincters relax.

Swallowing pushes the bolus past the pharyngoesophageal sphincter into the esophagus, which conducts it through the thoracic cavity to the stomach. Rhythmical contraction of the digestive tract wall, called **peristalsis** (fig. 15.3), begins in the esophagus and pushes food along the entire tract. Occasionally, peristalsis begins even though there is no food in the esophagus. This produces the sensation of a lump in the throat.

The esophagus stretches in three smooth curves from the back of the pharynx to just below the diaphragm, where it meets the stomach at an angle. When food reaches the stomach, the gastroesophageal sphincter relaxes, allowing the bolus to pass through. Normally, this sphincter prevents the acidic contents of the stomach from entering the esophagus. However, **heartburn,** which feels like a burning pain rising up into the throat, occurs when some of the contents of the stomach escape into the esophagus. When

vomiting occurs, contraction of the abdominal muscles and diaphragm propels the contents of the stomach upward through the esophagus.

> The esophagus takes food to the stomach. Peristalsis, a rhythmical contraction of the digestive tract wall, pushes food through the esophagus, as well as the rest of the digestive tract.

Stomach

The **stomach** (fig. 15.4) is a thick-walled, J-shaped organ that lies in the upper left quadrant of the abdominal cavity beneath the diaphragm. It is continuous with the esophagus superiorly and the duodenum of the small intestine inferiorly. The stomach stores food and starts the digestion of protein. The wall of the stomach has three layers of muscle and contains deep folds called **rugae** (roo'je), which flatten out as the stomach fills. The muscular wall of the stomach churns, mixing its contents.

The mucosal lining of the stomach contains millions of microscopic digestive glands, called **gastric glands** (the word *gastric* always refers to the stomach). The gastric glands produce gastric juice, which contains a digestive enzyme, called **pepsin,** and hydrochloric acid (HCl). Pepsin digests protein. The acidity of the stomach is beneficial in that it kills most bacteria present in food. Although HCl does not digest food, it breaks down the connective tissue of meat and activates gastric enzymes present in gastric juice. It also stimulates the stomach to produce a hormone called *gastrin.* Gastrin goes into the bloodstream, and when it circulates back to the stomach, the gastric glands continue to secrete (table 15.2).

Normally, the wall of the stomach is protected by a thick layer of mucus, but if, by chance, HCl penetrates this mucus, an ulcer can form. An **ulcer** is an open sore in the stomach wall caused by the gradual disintegration of a mucous membrane (fig. 15.4). The most frequent cause of an ulcer has been shown to be an infection caused by the bacterium *Helicobacter pylori.*

Normally, the stomach empties in 2–6 hours. When food leaves the stomach, it is a pasty, semisolid, acidic mixture called **chyme** (kīm). Chyme leaves the stomach and enters the small intestine by way of the *pyloric sphincter.* This sphincter repeatedly opens and closes, allowing chyme to enter the small intestine in small amounts only. This assures that digestion in the small intestine will proceed at a slow and thorough rate.

> The stomach expands and stores food. While food is in the stomach, the stomach churns, mixing food with the acidic gastric juices.

figure 15.4 Stomach anatomy. **a.** The stomach has a thick wall with deep folds (rugae) that can expand to allow more food to be stored. **b.** The mucosa of the stomach has gastric pits. The pits are openings for the gastric glands, which secrete mucus, hydrochloric acid (HCl), and the digestive enzyme pepsin. **c.** View of a bleeding ulcer by using an endoscope (a tubular instrument bearing a tiny lens and a light source) that can be inserted into the abdominal cavity.

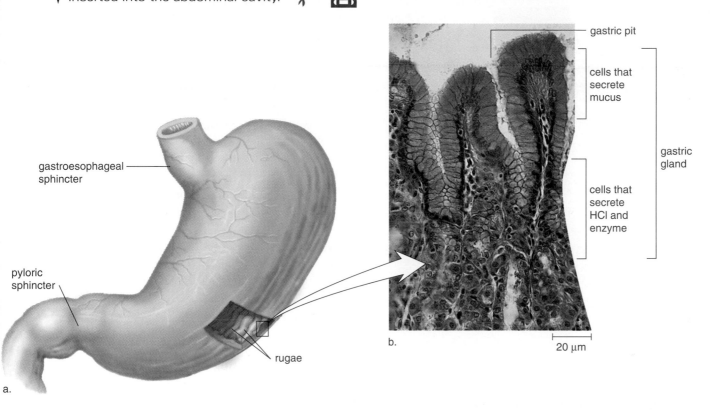

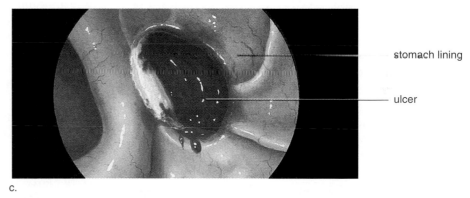

table 15.2	Hormones of the Digestive Tract		
	Hormone	**Source and Secretion**	**Function**
	Gastrin	Stomach wall, in response to the presence of food rich in protein	Stimulates gastric glands to increase their secretory activity
	Secretin	Intestinal wall, in response to food rich in fat	Stimulates pancreas to secrete sodium bicarbonate
	Cholecystokinin (CCK)	Intestinal wall, in response to chyme entering the small intestine	Stimulates pancreas to secrete digestive enzymes; stimulates gallbladder to contract and release bile

figure 15.5 Regions of the small intestine. The duodenum is attached to the stomach. The jejunum leads to the ileum, which is attached to the large intestine.

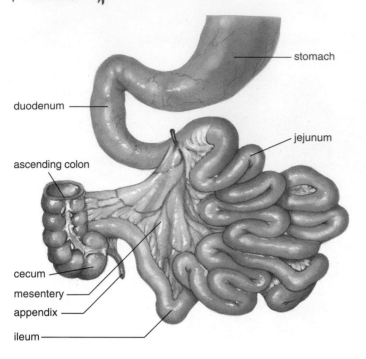

Small Intestine

The **small intestine** (fig. 15.5) is named for its small diameter (compared to that of the large intestine), but perhaps it should be called the long intestine. The small intestine averages about 3 meters (10 feet) in length, compared to the large intestine, which is about 1.5 meters (5 feet) in length. The small intestine is found in the central and lower portion of the abdominal cavity, where it is supported by a fan-shaped mesentery. It receives secretions from the liver and pancreas, chemically and mechanically breaks down chyme, absorbs nutrient molecules, and transports undigested material to the large intestine.

The small intestine has these regions:

Duodenum (du"o-de'num) The first 25 centimeters (10 inches) contain distinctive glands that secrete mucus and also receive the pancreatic secretions and the bile from the liver through a common duct. Folds and villi (fig. 15.6) are more numerous at the end than at the beginning.

Jejunum (jĕ-joo'num) The next 1 meter (3 feet) contains folds and villi, more at the beginning than at the end.

Ileum (il'e-um) The last 2 meters (6–7 feet) contain fewer folds and villi than the jejunum. The ileum wall contains Peyer's patches, aggregates of lymph nodules mentioned in chapter 13.

The small intestine has the following layers, as does the rest of the digestive tract (see fig. 15.6):

Serosa The serosa is a very thin, outermost layer of squamous epithelium supported by connective tissue. The serosa secretes a serous fluid that keeps the outer surface of the intestines moist so that the organs of the abdominopelvic cavity slide against one another.

Muscularis Two layers of smooth muscle comprise this section: The inner, circular layer of cells encircles the intestines; the outer, longitudinal layer lies in the same direction as the intestines.

Submucosa This broad band of loose connective tissue contains blood vessels and nerves.

Mucosa This layer of columnar epithelium, supported by connective tissue and smooth muscle, lines the central cavity and contains glandular epithelial cells that secrete mucus. Digestive glands, if present, are in this layer.

The wall of the small intestine has circular folds with fingerlike projections, called **villi** (vil'i) (sing., *villus*). The epithelial cells of each villus have extensions called microvilli (fig. 15.6). A large number of villi with their microvilli increase the small intestine's surface area for nutrient absorption and give the intestinal wall a soft, velvety appearance.

Each villus contains blood vessels and a small lymph vessel called a **lacteal.** The nutrient molecules that result from digestion cross the walls of the villi until all have been absorbed. Sugar and amino acids cross the epithelial cells to enter the blood; fats enter the lacteals.

As already mentioned, the plasma membranes of epithelial cells making up the villi contain microvilli, thousands of microscopic projections. In electron micrographs, the microvilli give the cells a fuzzy border, collectively called a *brush border*. The microvilli bear the intestinal digestive enzymes, which are therefore called the *brush-border enzymes.* These enzymes finish the digestion of chyme to small molecules that can be absorbed.

The intestinal wall also secretes the hormones *secretin* and *cholecystokinin (CCK)* (see table 15.2). These hormones travel in the bloodstream and stimulate the pancreas to secrete its juices. CCK also stimulates the gallbladder to contract and release bile.

figure 15.6 Wall of the small intestine. **a.** The intestinal wall has folds that **(b)** bear fingerlike projections called villi. **c.** The products of digestion are absorbed by villi, which contain blood vessels and a lacteal. **d.** Each villus has many microscopic extensions called microvilli.

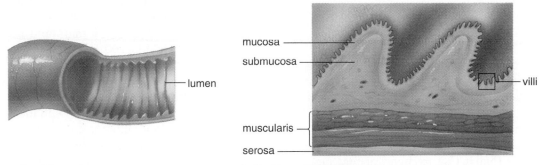

a. **Small intestine**

lumen

b. **Section of intestinal wall**

mucosa
submucosa
villi
muscularis
serosa

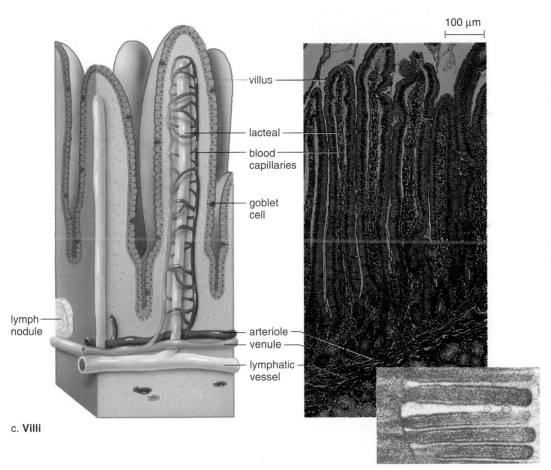

100 µm

villus

lacteal

blood capillaries

goblet cell

lymph nodule

arteriole
venule
lymphatic vessel

c. **Villi**

d. **Electron micrograph of microvilli**

Accessory Organs

The salivary glands, the pancreas, liver, and gallbladder are accessory organs of digestion. Ducts transport pancreatic juices from the pancreas and bile from the liver and gallbladder to the duodenum. The small intestine helps regulate the release of these juices by secreting hormones (see table 15.2).

Pancreas

The **pancreas** lies deep in the peritoneal cavity, resting on the posterior abdominal wall. It is an elongated and somewhat flattened organ that has both an endocrine function (see chapter 10) and an exocrine function. The exocrine function of the pancreas is discussed here. Most of the pancreatic cells produce pancreatic juice, which contains sodium bicarbonate ($NaHCO_3$), a chemical that neutralizes the acidic pH of the chyme, and digestive enzymes for carbohydrates, proteins, and fats. **Pancreatic amylase** digests starch, **trypsin** and chymotrypsin digest protein, and **lipase** digests fat. In other words, the pancreas secretes enzymes for the digestion of all major types of food. These enzymes travel by way of the pancreatic duct to the duodenum of the small intestine. Any blockage that prevents pancreatic juice from entering the duodenum may result in a serious condition called **acute pancreatitis**. In this disorder, pancreatic juice backs up into the pancreatic duct, causing digestion of the pancreatic tissue itself.

> The pancreas produces pancreatic juice, which contains bicarbonate and digestive enzymes for carbohydrates, proteins, and fats.

Liver

The **liver,** which is the largest gland in the body, lies mainly in the right upper quadrant of the peritoneal cavity, under the diaphragm. There are two main lobes, the right lobe and the smaller left lobe, which crosses the midline and lies above the stomach (fig. 15.7a). These two lobes are separated by the falciform ligament, which secures the liver to the anterior abdominal wall and the diaphragm. The liver contains approximately 100,000 lobules that serve as the structural and functional units of the liver (fig. 15.7b). Triads consisting of the following structures are located between the lobules: (1) a branch of the hepatic artery, which brings oxygenated blood to the liver; (2) a branch of the hepatic portal vein, which brings nutrients from the intestines; and (3) a bile duct, which takes bile away from the liver. The central veins within the lobules enter the hepatic vein. Figure 12.14 shows that the liver lies between the hepatic portal vein and the hepatic vein.

In some ways, the liver acts as the gatekeeper to the blood. As blood from the intestines passes through the liver, the liver removes poisonous substances and works to keep the contents of the blood constant. It also removes and stores iron and the fat-soluble vitamins A, D, E, and K. The liver makes the plasma proteins from amino acids, and these have important functions within the blood itself.

The liver under the influence of the pancreatic hormones insulin and glucagon maintains the blood glucose level at about 100 mg/100 ml ± 0.1%, even though a person eats intermittently. Any excess glucose in the hepatic portal vein is removed and stored by the liver as glycogen. Between eating periods, glycogen is broken down to glucose, which enters the hepatic vein; in this way, the glucose content of the blood remains constant.

If, by chance, the supply of glycogen or glucose is depleted, the liver converts amino acids to glucose molecules. In this process, ammonia is given off and is converted to **urea,** the common nitrogenous waste product of humans. After its formation in the liver, urea is transported to the kidneys for excretion.

The liver produces **bile.** Bile is a yellowish green fluid because it contains the pigments bilirubin and biliverdin, which result from the breakdown of hemoglobin—the pigment found in red blood cells. Bile also contains bile salts that emulsify fats once bile reaches the duodenum of the small intestine. When fats are emulsified, they break down into droplets that can be acted upon by a digestive enzyme called lipase from the pancreas. Emulsification is a process that can be witnessed by adding oil to water in a test tube. The oil has no tendency to mix with the water, but if a liquid detergent is added and the contents of the tube are shaken, the oil breaks down and disperses into the water.

The following are significant functions of the liver:

1. Storage of glucose as glycogen after eating and the breakdown of glycogen to glucose between meals to maintain a constant blood glucose level.
2. Production of urea from the breakdown of amino acids.
3. Destruction of old red blood cells and conversion of hemoglobin to the breakdown products in bile (bilirubin and biliverdin).
4. Production of bile, which is stored in the gallbladder before entering the small intestine, where it emulsifies fats.
5. Production of the plasma proteins.
6. Detoxification of the blood by removing poisonous substances and metabolizing them.
7. In addition, the liver is involved in the storage of iron and the fat-soluble vitamins A, D, E, and K.

figure 15.7 Macroscopic and microscopic anatomy of the liver. **a.** The liver has two lobes viewed anteriorly (left) and inferiorly (right). **b.** Cross section of a hepatic lobule, illustrating microscopic structure.

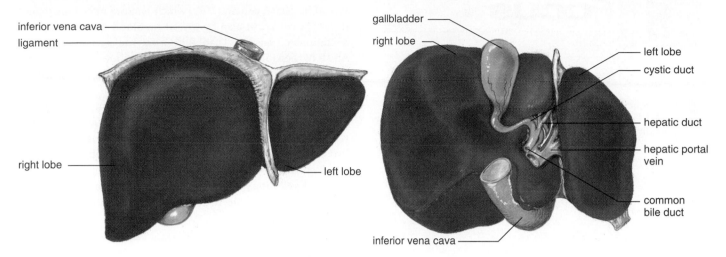

inferior vena cava

ligament

right lobe

left lobe

gallbladder

right lobe

left lobe

cystic duct

hepatic duct

hepatic portal vein

common bile duct

inferior vena cava

a.

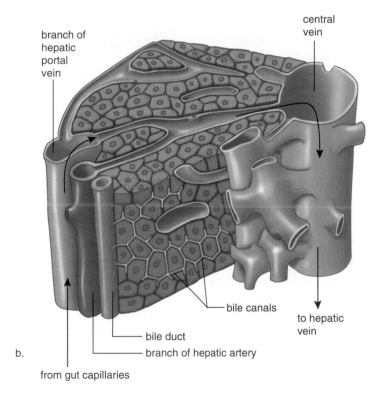

branch of hepatic portal vein

central vein

bile canals

to hepatic vein

bile duct

branch of hepatic artery

from gut capillaries

b.

Viral Hepatitis

$\mathcal{V}$iral hepatitis causes flulike symptoms, including fatigue, fever, headache, nausea, vomiting, muscle aches, and dull pain in the upper right quadrant of the abdomen. Jaundice, a yellowish cast to the skin, is also seen. Hepatitis A, due to HAV (hepatitis A virus), is usually acquired from sewage-contaminated drinking water. Restaurant employees are admonished to scrub their hands after using the washroom as a way to prevent the spread of this disease. A vaccine is available and currently is given to individuals at risk.

Hepatitis B, caused by HBV (hepatitis B virus), is usually spread by sexual contact. HBV, which is more infectious than the AIDS virus, can also be spread by blood transfusions or contaminated needles. Thankfully, a vaccine is now available. An HDV (hepatitis D virus) infection sometimes accompanies an HBV infection, and if so, liver function is expected to deteriorate more rapidly.

Hepatitis C, called the post-transfusion form of hepatitis because it is usually acquired by contact with infected blood, is also of great concern. Infection can lead to chronic hepatitis, liver cancer, and death. A vaccine is not yet available.

Hepatitis E, caused by HEV (hepatitis E virus), is usually seen in developing countries. Only imported cases, in travelers or visitors to endemic regions, have been reported in the United States.

Serious Liver Disorders   Jaundice, hepatitis, and cirrhosis are three life-threatening diseases that affect the entire liver and hinder its ability to repair itself. A person with **jaundice** has a yellowish tint to the whites of the eyes and lightly pigmented skin. Bilirubin has been deposited in the skin, due to an abnormally large amount in the blood. In *hemolytic jaundice,* red blood cells are broken down in abnormally large amounts; in *obstructive jaundice,* a bile duct is blocked, or the liver cells are damaged. Obstructive jaundice often occurs when gallstones block the common bile duct.

Jaundice can also result from **hepatitis,** inflammation of the liver, as discussed in the Medical Focus reading on this page.

Cirrhosis (sah-ro′sis) is a chronic disease of the liver in which the organ first becomes fatty. Liver tissue is then replaced by inactive fibrous scar tissue. In alcoholics, who often develop cirrhosis of the liver, the condition most likely is caused by the excessive amounts of alcohol (a toxin) the liver is forced to break down.

The liver has amazing regenerative powers and can recover if the rate of regeneration exceeds the rate of damage. During liver failure, however, there may not be enough time for the liver to heal itself. Liver transplantation is usually the preferred treatment, but artificial livers have been developed and tried in a few cases. One type consists of a cartridge containing cloned liver cells. As the patient's blood passes through cellulose acetate tubing, it is serviced in the same manner as with a normal liver. In the meantime, the patient's liver has a chance to recover.

The liver is a very critical organ, with numerous important functions, and any malfunction is a matter of considerable concern. The liver receives blood from the small intestine by way of the hepatic portal vein.

Gallbladder

The **gallbladder** is a pear-shaped, muscular sac attached to the ventral surface of the liver (see fig. 15.7). The liver produces bile, which enters the many bile ducts associated with hepatic lobules. These bile ducts join to form the common bile duct that enters the duodenum. Any excess bile backs up through the cystic duct into the gallbladder, where it is stored.

Bile, which contains bile salts, bile pigments, cholesterol, and electrolytes, becomes concentrated in the gallbladder as water is reabsorbed. Normally cholesterol stays in solution, but sometimes it may come out of solution and form crystals. This may happen if the liver secretes too much cholesterol and/or the bile becomes too concentrated in the gallbladder. The crystals can become larger and larger until they form gallstones. As mentioned, if the gallstones leave the gallbladder and block the common bile duct, obstructive jaundice may occur. Gallbladder disease can be very painful and accompanied by nausea and vomiting. In such cases, the gallbladder can be surgically removed. Today, it is possible to use a laser that results in only four tiny cuts, each a quarter to half inch long.

The liver produces bile, which is stored in the gallbladder. Digestive enzymes and bile enter the duodenum via ducts.

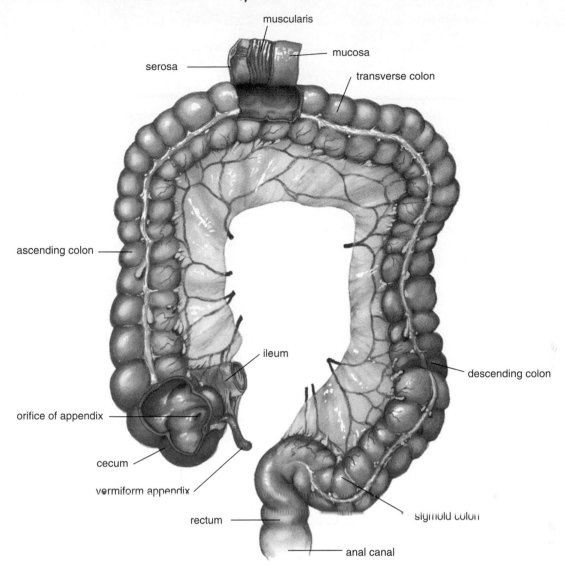

figure 15.8 The large intestine. Notice the colon has four regions: the ascending colon, the transverse colon, the descending colon, and the sigmoid colon.

muscularis

serosa

mucosa

transverse colon

ascending colon

ileum

descending colon

orifice of appendix

cecum

vermiform appendix

rectum

sigmoid colon

anal canal

Large Intestine

The **large intestine** (fig. 15.8), which includes the cecum, **colon,** rectum, and anal canal, is larger in diameter than the small intestine (6.5 centimeters compared to 2.5 centimeters). It begins in the lower right quadrant of the peritoneal cavity. The **cecum** (se'kum), which lies inferior to this point, has a small projection called the **vermiform appendix** (*vermiform* means wormlike). Superior to the cecum, the large intestine is termed the **ascending colon.** At the level of the liver, the large intestine bends sharply and becomes the **transverse colon.** At the left abdominal wall, the large intestine bends again to become the **descending** colon. In the pelvic region, the large intestine turns medially to form an S-shaped bend known as the **sigmoid colon.** The last 20 centimeters of the large intestine, the **rectum,** ends in the anal canal, which opens at the **anus.**

The large intestine absorbs water and electrolytes. It also prepares and stores nondigestible material **(feces)** for defecation at the anus. In addition to nondigestible remains, feces also contain bile pigments, which give them color, and large quantities of bacteria, particularly *Escherichia coli.*

The *E. coli* live off any substances that were not digested earlier. When they break down this material, they

Constipation

*t*he colon of the large intestine has four regions: the ascending colon, the transverse colon, the descending colon, and the sigmoid colon (fig. 15A). Water is removed from the nondigestible remains that enter the ascending colon from the small intestine. At this point, bacteria begin their action; they use cellulose as an energy source as they produce fatty acids and vitamins that can also be used by their host. They also release hydrogen gas and sulfur-containing compounds that contribute to human flatulence (gas). Feces, which consist of nondigested remains, bacteria, and sloughed-off intestinal cells, begin to form in the transverse colon. From there, they are propelled down the descending colon toward the rectum by periodic, firm contractions. When sufficient feces are in the rectum (130–200 grams), a defecatory urge is felt. The defecation reflex contracts the rectal muscles and relaxes the internal anal sphincter, a ring of muscle that closes off the rectum. Then, feces move toward the anus. A pushing motion, along with relaxation of the external anal sphincter, propels feces from the body. Since these activities are under voluntary control, it is possible to control defecation.

Defecation normally occurs from three times a week to three times a day; therefore, some variation in occurrence is nothing to worry about. However, if the frequency of defecation declines and if defecation becomes difficult, then constipation is present. If constipation is a continuing problem, a physician can help record the movement of materials through the large intestine via several tests. The patient swallows about 20 small markers that will show up on an X ray. At intervals during the following week, X rays are taken, and the number and locations of the markers are noted. If muscle contraction of the intestinal wall is insufficient, the markers move slowly along their course. Injured nerves, certain drugs, and prolonged overuse of stimulatory laxatives can bring about this difficulty. On the other hand, markers may move normally at first and then slow down considerably in the descending colon and rectum. Habitual disregard of the defecatory urge may have caused this problem, or a cancerous polyp might be obstructing normal movement. If the former is the case, it is possible to retrain the rectum to work properly. Sitting on the toilet about 20 minutes each morning can encourage a return of the reflexes that have disappeared, but straining is not recommended.

Temporary constipation due to traveling, pregnancy, or medication can sometimes be relieved by increasing dietary fiber, drinking plenty of water, and getting moderate amounts of exercise. The use of oral laxatives (agents that aid emptying of the intestine) is a last resort. Bulk-forming laxatives, such as those that contain bran, psyllium, and methyl cellulose, are considered best because they promote the defecation reflex. Laxatives that contain osmotic agents, such as carbohydrates or salts (lactulose, milk of magnesia, or Epsom salts), cause

---

emit odorous molecules that cause the characteristic fecal odor. Some of the vitamins (vitamin K and some B complex vitamins), amino acids, and other growth factors produced by these bacteria are absorbed by the intestinal lining. In this way, *E. coli* and other bacteria perform a service for the human body.

> The large intestine does not digest food. It absorbs water and some electrolytes and stores nondigestible material prior to defecation.

Diarrhea and Constipation

Two common everyday complaints associated with the large intestine are diarrhea and constipation. In **diarrhea,** too little water has been absorbed; in **constipation,** too much water has been absorbed.

The major causes of diarrhea are infection of the lower tract and nervous stimulation. In the case of infection, such as food poisoning caused by eating contaminated food, the intestinal wall becomes irritated, and peristalsis increases. Lack of water absorption is a protective measure, and the diarrhea that results serves to rid the body of the infectious organisms. When a person is under stress, the nervous system sometimes stimulates the intestinal wall, resulting in diarrhea. Loss of water due to diarrhea may lead to dehydration, a serious condition in which the body tissues lose their normal water content.

When a person is constipated, the feces are dry and hard. The MedAlert reading on this page discusses the causes of constipation and how it can be prevented. Chronic constipation is associated with the development of hemorrhoids (see chapter 12).

> Diarrhea occurs when too little water has been absorbed in the large intestine; constipation occurs when too much water has been absorbed.

water to move into rather than out of the colon. Stool softeners (mineral oil or those that contain docusate) should be used sparingly. Mineral oil reduces the absorption of fat-soluble vitamins, and docusate can cause liver damage. Laxatives that contain chemical stimulants (such as phenolphthalein in Ex-Lax and Feen-A-Mint) can damage the defecation reflex and lead to a dependence on their use. Aside from laxatives, rectal suppositories are sometimes helpful in providing lubrication and stimulating the defecation reflex. Enemas introduce water into the colon and, therefore, also help stimulate defecation.

Questions

1. Why is it correct to say that defecation has both involuntary and voluntary components?
2. Which of these components is sometimes damaged by overuse of laxatives?

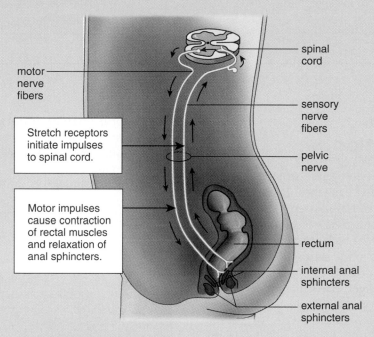

motor nerve fibers

spinal cord

sensory nerve fibers

Stretch receptors initiate impulses to spinal cord.

pelvic nerve

Motor impulses cause contraction of rectal muscles and relaxation of anal sphincters.

rectum

internal anal sphincters

external anal sphincters

figure 15A Defecation reflex. The accumulation of feces in the rectum causes it to stretch, which initiates a reflex action resulting in rectal contraction.

Disorders of the Large Intestine

The appendix is a fingerlike projection from the cecum of the large intestine. Unfortunately, the appendix can become infected, resulting in **appendicitis,** a very painful condition in which the fluid content of the appendix can increase to the point that it bursts. The appendix should be removed before it bursts to avoid a generalized infection of the peritoneal membrane of the abdominal cavity.

Diverticulosis (di″ver-tĭ-kyu'lo′sis) is characterized by the presence of diverticula, or saclike pouches, of the colon. Ordinarily, these pouches cause no problems. But about 15% of people with diverticulosis develop an inflammation known as diverticulitis. The symptoms of diverticulitis are similar to those of appendicitis—cramps or steady pain with local tenderness. Fever, loss of appetite, nausea, and vomiting may also occur. Today, high-fiber diets are recommended to prevent the development of these conditions and of cancer of the colon.

The colon is subject to the development of **polyps,** small growths that generally appear on epithelial tissue, such as the epithelial tissue that lines the digestive tract. Whether polyps are benign or cancerous, they can be removed individually, along with a portion of the colon, if necessary. If the last portion of the rectum and the anal canal must be removed, then the intestine is sometimes attached to the abdominal wall through a procedure known as a **colostomy** (kah-los'tah-me), and the digestive remains are collected in a plastic bag fastened around the opening. Recently, the use of metal staples has permitted surgeons to join the colon to a piece of rectum that formerly was considered too short.

table 15.3

Comparison of Enzymes

Enzyme	Source	Optimum pH*	Type of Food Digested	Product
Salivary amylase	Salivary glands	Neutral	Starch	Maltose
Pepsin	Stomach	Acidic	Protein	Peptides
Pancreatic amylase	Pancreas	Basic	Starch	Maltose
Lipase	Pancreas	Basic	Fat	Glycerol, fatty acids
Trypsin, chymotrypsin	Pancreas	Basic	Protein	Peptides
Nucleases	Pancreas	Basic	RNA, DNA	Nucleotides
Peptidases	Intestine	Basic	Peptides	Amino acids
Maltase, sucrase, lactase	Intestine	Basic	Maltose, sucrose, lactose	Glucose and other monosaccharides

*pH is discussed on page 22.

Mechanical and Chemical Digestion

Mechanical digestion breaks food up into small pieces that are more accessible to digestive enzymes. It includes the chewing of food by teeth which is discussed in the Medical Focus on the next page. The churning of food in the stomach is another good example of mechanical digestion.

Chemical digestion refers to the action of the digestive enzymes. As mentioned previously, the various digestive juices contain enzymes that digest particular types of food (table 15.3). Each of these enzymes will now be considered in regard to the digestion of a ham sandwich, which contains starch (a carbohydrate in the bread), protein (in the ham), and fat (butter on the bread).

Salivary amylase, an enzyme present in saliva, acts on starch. The end product of this reaction—maltose—however, is not small enough to cross plasma membranes to any degree; therefore, further digestion is required before absorption is possible.

In the stomach, the enzyme pepsin is present in gastric juices, which act on protein. The end product of this reaction—peptides—is, again, too large to cross plasma membranes; therefore, further digestion is required before absorption is possible.

Pancreatic juice, which enters the small intestine at the duodenum, contains digestive enzymes for all major types of food. Pancreatic amylase, like salivary amylase, acts on starch. The enzymes trypsin and chymotrypsin, like pepsin, act on protein. And the enzyme lipase acts on fat droplets after fat has been emulsified by bile salts. The end products of the lipase reaction—glycerol and fatty acids—are small enough to enter the cells of the intestinal villi. In the villi, glycerol and fatty acids rejoin to form fat, which enters the lacteals (see fig. 15.6), a branch of the lymphatic system.

The brush border of the small intestines contains enzymes that complete the digestion of both starch and protein, forming small molecules that can cross the cells of the intestinal villi. Glucose is the end product of starch breakdown, and amino acids are the end product of protein breakdown. These end products enter the blood capillaries of the villi. These blood capillaries join to form venules and veins, which empty into the hepatic portal vein, a vessel that goes to the liver.

> Digestive enzymes present in digestive juices break down food into the following nutrient molecules: glucose, amino acids, fatty acids, and glycerol. Glucose and amino acids are absorbed into the blood capillaries of the intestinal villi. Fatty acids and glycerol are also absorbed but then rejoin to produce fat, which enters the lacteals.

Nutrition

The body requires many different types of organic molecules and various types of inorganic ions and molecules in the diet each day. Laypersons tend to think of the term *organic* as meaning that the food was grown without the use of pesticides, but it actually refers to molecules associated with living things. Our bodies primarily contain carbohydrates, fats, proteins, and nucleic acids, the organic molecules discussed in chapter 2. Inorganic molecules are associated with nonliving things; for example, the ocean contains salt (sodium chloride). We all know, however, that sodium chloride plays a significant role in the body.

Nutrition involves an interaction between food and the living organism, and a nutrient is a substance that the body uses to maintain health. A balanced diet contains all the essential nutrients and includes a variety of foods, proportioned as shown in figure 15.9.

Human Teeth

during the first 2 years of life, the 20 deciduous, or baby, teeth appear. Eventually, the deciduous teeth are replaced by the adult teeth. Normally, adults have 32 teeth (fig. 15B). One-half of each jaw has teeth of four different types: (1) two chisel-shaped incisors for biting, (2) one pointed canine (cuspid) for tearing, (3) two fairly flat premolars (bicuspids) for grinding, and (4) three more flattened molars for crushing. The last molars, called the wisdom teeth, may fail to come in, or if they do, they may grow in crooked and are useless. Frequently, wisdom teeth are extracted.

Each tooth (fig. 15C) has a crown and a root. The crown has a layer of enamel, an extremely hard outer covering of calcium compounds; dentin, a thick layer of bonelike material; and an inner pulp, which contains the nerves and blood vessels. Dentin and pulp are also in the root. **Caries,** tooth decay commonly called cavities, occur when bacteria within the mouth break down sugar and give off acids that corrode the teeth. Fluoride treatments, particularly in children, can make the enamel stronger and more resistant to decay.

Gum disease is more likely as we age. One example of gum disease is inflammation of the gums, called **gingivitis,** that may spread to the periodontal membrane (fig. 15B) that lines the tooth socket. When this occurs, the individual develops **periodontitis,** characterized by a loss of bone and loosening of the teeth, which may have to be pulled. Daily brushing and flossing of teeth, along with stimulation of the gums, help prevent these conditions.

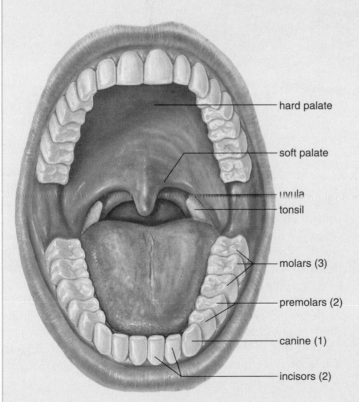

- hard palate
- soft palate
- uvula
- tonsil
- molars (3)
- premolars (2)
- canine (1)
- incisors (2)

figure 15B The mouth, showing the adult teeth.

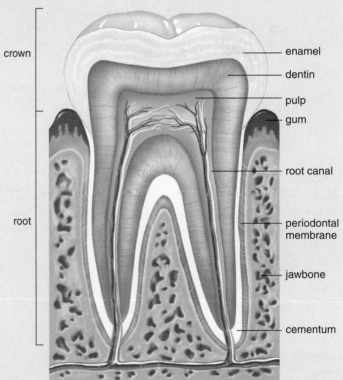

crown

- enamel
- dentin
- pulp
- gum
- root canal

root

- periodontal membrane
- jawbone
- cementum

figure 15C Longitudinal section of a molar.

figure 15.9 Food guide pyramid: A guide to daily food choices. The U.S. Department of Agriculture uses a pyramid to show the ideal diet because it emphasizes the importance of including grains, fruits, and vegetables in the diet. Meats and dairy products are needed in limited amounts; fats, oils, and sweets should be used sparingly.

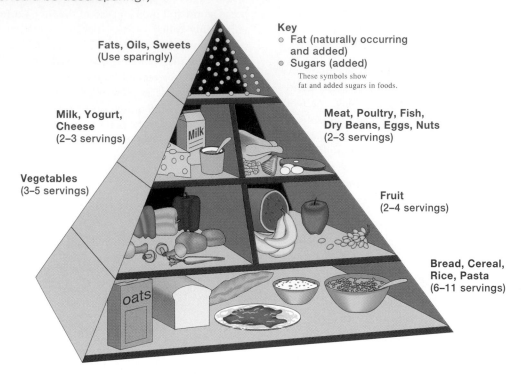

Key
○ Fat (naturally occurring and added)
● Sugars (added)
These symbols show fat and added sugars in foods.

Fats, Oils, Sweets
(Use sparingly)

Milk, Yogurt, Cheese
(2–3 servings)

Meat, Poultry, Fish, Dry Beans, Eggs, Nuts
(2–3 servings)

Vegetables
(3–5 servings)

Fruit
(2–4 servings)

Bread, Cereal, Rice, Pasta
(6–11 servings)

Following digestion, nutrients enter the blood in the circulatory system, which distributes them to the tissues, where they are utilized by the body's cells. Mitochondria use glucose to produce a constant supply of ATP for the cell. In other words, glucose is the body's immediate energy source. Since the brain's only source of energy is glucose, it needs a constant supply.

The liver is able to chemically alter ingested fats to suit the body's needs, with the exception of linoleic acid, a fatty acid that it is unable to produce. Since linoleic acid is required for construction of plasma membranes, it is considered an **essential fatty acid.** Essential molecules must be present in food because the body is unable to manufacture them.

If glucose is not available, fats can be metabolized into their components, which are then used as an energy source. Therefore, fats are said to be a long-term energy source. When adipose tissue cells store fats, the body increases in weight. Cells have the capability of converting excess sugar molecules into fats for storage, which accounts for the fact that carbohydrates can also contribute to weight gain.

Amino acids from protein digestion are used by the cells to construct their own proteins, including the enzymes that carry out metabolism. Protein formation requires 20 different types of amino acids. Of these, nine are required in the diet because the body is unable to produce them. These are termed the **essential amino acids.** The body produces the other amino acids by simply transforming one type into another type. Some protein sources, such as meat, are complete in the sense that they provide all the different types of amino acids. Vegetables supply the body with amino acids, but they are incomplete sources because at least one of the essential amino acids is absent. A combination of certain vegetables, however, can provide all of the essential amino acids.

Vitamins

Vitamins are *vital* to life because they play essential roles in cellular metabolism. Since the body is unable to produce them, vitamins must be present in the diet. Vitamins are organic molecules, but they differ radically from

Antioxidants

Over the past 20 years, numerous statistical studies have been done to determine whether a diet rich in fruits and vegetables protects against cancer. The vitamins C and E and beta-carotene, which is converted to vitamin A in the body, are especially abundant in fruits and vegetables and seem to have a special function in cells.

Cellular metabolism generates **free radicals,** unstable molecules that can attack and damage other molecules, such as DNA, carbohydrates (for example, collagen), proteins (for example, an enzyme), and lipids, that are found in plasma membranes. The damage to these cellular molecules may lead to disorders, perhaps even cancer. In addition, plaque formation in arteries may begin when arterial linings are injured by damaged cholesterol molecules.

The most common free radical in cells is oxygen in the unstable form O_3^-. Vitamins C, E, and A are believed to defend the body against free radicals, and therefore, they are termed **antioxidants.** To receive adequate amounts of these vitamins, you should eat five servings of fruits and vegetables daily. Any one of the following is considered "one serving":

- 1 cup of raw leafy greens, such as lettuce or spinach
- 1/2 cup of raw or cooked vegetables, such as broccoli, cauliflower, peas, green beans, and so on
- one average carrot or one medium potato
- one medium apple, orange, banana, or similar-sized fruit
- 1/2 cup of grapes or cut fruit, such as diced pineapple
- 1/4 cup of dried fruit, such as raisins
- 3/4 cup of pure fruit or vegetable juice

Dietary supplements may provide a potential safeguard against cancer and cardiovascular disease, but taking supplements instead of improving intake of fruits and vegetables is not the solution. Fruits and vegetables provide hundreds of beneficial compounds that cannot be obtained from a vitamin pill. They enhance each other's absorption or action and perform independent biological functions.

---

carbohydrates, fats, and proteins. They are much smaller in size and are not broken down to be used as building blocks or as a source of energy. Instead, the body protects them and provides many of them with protein carriers that transport them in the blood to the cells. In the cells, vitamins become helpers in metabolic processes that break down or synthesize other organic molecules. Because vitamins can be used over and over again, they are required in very small amounts only.

Vitamins fall into two groups: fat-soluble vitamins (vitamins A, D, E, and K) and water-soluble vitamins (the B complex vitamins and vitamin C) (table 15.4). Most of the water-soluble vitamins are coenzymes, or enzyme helpers, that help speed up specific reactions. The functions of the fat-soluble vitamins, some of which have been previously discussed, are more specialized. Vitamin A, as noted in chapter 9, is used to synthesize the visual pigments. Vitamin D is needed to produce a hormone that regulates calcium and phosphorus metabolism (see chapter 5). Vitamin E, as discussed in the Medical Focus reading on this page, is an antioxidant. Vitamin K is required to form *prothrombin,* a substance necessary for normal blood clotting (see chapter 11).

Minerals

In contrast to vitamins, **minerals** are inorganic elements (table 15.5). An element, you will recall, is one of the basic substances of matter that cannot be broken down further into simpler substances. Minerals sometimes occur as a single atom, in contrast to vitamins, which contain many atoms, and carbohydrates, such as starch, which contain thousands of atoms. Minerals cannot lose their identity, no matter how they are handled. Because they are indestructible, no special precautions are needed to preserve them when cooking.

Minerals are divided into macronutrients, which are needed in gram amounts per day, and micronutrients (trace elements), which are needed in only microgram amounts per day. The macronutrients sodium, magnesium, phosphorus, chlorine, potassium, and calcium serve as constituents of cells and body fluids, and as structural components of tissues. The micronutrients have very specific functions, as noted in table 15.5. As research continues, more elements will be added to the list of those considered essential.

table 15.4

Vitamins: Their Role in the Body and Food Sources

Vitamins	Role in Body	Good Food Sources
Fat-Soluble Vitamins		
Vitamin A	Assists in the formation and maintenance of healthy skin, hair, and mucous membranes; aids in the ability to see in dim light (night vision); is essential for proper bone growth, tooth development, and reproduction	Deep yellow/orange and dark green vegetables and fruits (carrots, broccoli, spinach, cantaloupe, sweet potatoes); cheese, milk, and fortified margarine
Vitamin D	Aids in the formation and maintenance of bones and teeth; assists in the absorption and use of calcium and phosphorus	Milk fortified with vitamin D; tuna, salmon, or cod liver oil; also made in the skin when exposed to sunlight
Vitamin E	Protects vitamin A and essential fatty acids from oxidation; prevents plasma membrane damage	Vegetable oils and margarine; nuts; wheat germ and whole-grain breads and cereals; green, leafy vegetables
Vitamin K	Aids in synthesis of substances needed for clotting of blood; helps maintain normal bone metabolism	Green, leafy vegetables, cabbage, and cauliflower; also made by bacteria in intestines of humans, except for newborns
Water-Soluble Vitamins		
Vitamin C	Is important in forming collagen, a protein that gives structure to bones, cartilage, muscle, and vascular tissue; helps maintain capillaries, bones, and teeth; aids in absorption of iron; helps protect other vitamins from oxidation	Citrus fruits, berries, melons, dark green vegetables, tomatoes, green peppers, cabbage, potatoes
B Complex Vitamins		
Thiamin	Helps in release of energy from carbohydrates; promotes normal functioning of nervous system	Whole-grain products, dried beans and peas, sunflower seeds, nuts
Riboflavin	Helps body transform carbohydrates, proteins, and fats into energy	Nuts, yogurt, milk, whole-grain products, cheese, poultry, leafy, green vegetables
Niacin	Helps body transform carbohydrates, proteins, and fats into energy	Nuts, poultry, fish, whole-grain products, dried fruit, leafy greens, beans; can be formed in the body from tryptophan, an essential amino acid found in protein
Vitamin B_6	Aids in the use of fats and amino acids; aids in the formation of protein	Sunflower seeds, beans, poultry, nuts, bananas, dried fruit, leafy green vegetables
Folic acid	Aids in the formation of hemoglobin in red blood cells; aids in the formation of genetic material	Nuts, beans, whole-grain products, fruit juices, dark green, leafy vegetables
Pantothenic acid	Aids in the formation of hormones and certain nerve-regulating substances; helps in the metabolism of carbohydrates, proteins, and fats	Nuts, beans, seeds, poultry, dried fruit, milk, dark green, leafy vegetables
Biotin	Aids in the formation of fatty acids; helps in the release of energy from carbohydrates	Occurs widely in foods, especially eggs; made by bacteria in the human intestine
Vitamin B_{12}	Aids in the formation of red blood cells and genetic material; helps in the functioning of the nervous system	Milk, yogurt, cheese, fish, poultry, eggs; not found in plant foods unless fortified (such as in some breakfast cereals)

Source: From David C. Nieman, et al., *Nutrition,* Revised 1st ed. Copyright ©1992 Wm. C. Brown Communications, Inc., Dubuque, Iowa. Reprinted by permission of C. V. Mosby, St. Louis, MO.

table 15.5

Minerals: Their Role in the Body and Food Sources

Minerals	Role in Body	Good Food Sources
Macronutrients		
Calcium	Is used for building bones and teeth and for maintaining bone strength; also is involved in muscle contraction, blood clotting, and maintenance of plasma membranes	All dairy products; dark green, leafy vegetables; beans, nuts, sunflower seeds, dried fruit, molasses, canned fish
Phosphorus	Is used to build bones and teeth; to release energy from carbohydrates, proteins, and fats; and to form genetic material, plasma membranes, and many enzymes	Beans, sunflower seeds, milk, cheese, nuts, poultry, fish, lean meats
Magnesium	Is used to build bones, to produce proteins, to release energy from muscle carbohydrate stores (glycogen), and to regulate body temperature	Sunflower and pumpkin seeds, nuts, whole-grain products, beans, dark green vegetables, dried fruit, lean meats
Sodium	Regulates body-fluid volume and blood acidity; aids in transmission of nerve impulses	Most of the sodium in the American diet is added to food as salt (sodium chloride) in cooking, at the table, or in commercial processing; animal products contain some natural sodium
Chloride	Is a component of gastric juice and aids in acid-base balance	Table salt, seafood, milk, eggs, meats
Potassium	Assists in muscle contraction, the maintenance of fluid and electrolyte balance in the cells, and the transmission of nerve impulses; also aids in the release of energy from carbohydrates, proteins, and fats	Widely distributed in foods, especially fruits and vegetables, beans, nuts, seeds, and lean meats
Micronutrients (Trace Elements)		
Iron	Is involved in the formation of hemoglobin in the red blood cells of the blood and myoglobin in muscles; also is a part of several enzymes and proteins	Molasses, seeds, whole-grain products, fortified breakfast cereals, nuts, dried fruits, beans, poultry, fish, lean meats
Zinc	Is involved in the formation of protein (growth of all tissues), in wound healing, and in prevention of anemia; is a component of many enzymes	Whole-grain products, seeds, nuts, poultry, fish, beans, lean meats
Iodine	Is an integral component of thyroid hormones	Table salt (fortified), dairy products, shellfish, and fish
Fluoride	Is involved in maintenance of bone and tooth structure	Fluoridated drinking water is the best source; also found in tea, fish, wheat germ, kale, cottage cheese, soybeans, almonds, onions, milk
Copper	Is vital to enzyme systems and in manufacturing red blood cells; is needed for utilization of iron	Nuts, oysters, seeds, crab, wheat germ, dried fruit, whole grains, legumes
Selenium	Functions in association with vitamin E; may assist in protecting tissues and plasma membranes from oxidative damage; may also aid in preventing cancer	Nuts, whole grains, lean pork, cottage cheese, milk, molasses, squash
Chromium	Is required for maintaining normal glucose metabolism; may assist insulin function	Nuts, prunes, vegetable oils, green peas, corn, whole grains, orange juice, dark green vegetables, legumes
Manganese	Is needed for normal bone structure, reproduction, and the normal functioning of the central nervous system; is a component of many enzyme systems	Whole grains, nuts, seeds, pineapple, berries, legumes, dark green vegetables, tea
Molybdenum	Is a component of enzymes; may help prevent dental caries	Tomatoes, wheat germ, lean pork, legumes, whole grains, strawberries, winter squash, milk, dark green vegetables, carrots

Source: From David C. Nieman, et al., *Nutrition,* Revised 1st ed. Copyright ©1992 Wm. C. Brown Communications, Inc., Dubuque, Iowa. Reprinted by permission of C. V. Mosby, St. Louis, MO.

Effects of Aging

The incidence of gastrointestinal disorders increases with age. Periodontitis, which is common in elderly people, leads to the loss of teeth and the need for false teeth.

The esophagus, which rarely causes any difficulties in younger people, is more prone to disorders in the elderly. The portion of the esophagus normally found inferior to the diaphragm can protrude into the thoracic cavity, causing an esophageal hiatal hernia. In some cases, the lower esophageal sphincter opens inappropriately and allows chyme to regurgitate into the esophagus, causing heartburn. Or in some older persons, chest pain may occur when this sphincter fails to open and a bolus cannot enter the stomach. Eventually, the esophagus may develop a diverticulum that allows food to collect abnormally.

Peristalsis generally slows within the digestive tract as the muscular wall loses tone. Peptic ulcers increase in frequency with age. Failure of older people to consume sufficient dietary fiber can result in diverticulosis and constipation. Constipation and hemorrhoids are frequent complaints among the elderly, as is fecal incontinence.

The liver shrinks with age and receives a smaller blood supply than in younger years. Notably, it needs more time to metabolize drugs and alcohol. With age, gallbladder difficulties occur; there is an increased incidence of gallstones and cancer of the gallbladder. In fact, cancer of the various organs of the gastrointestinal tract is seen more often among the elderly. For example, most cases of pancreatic cancer occur in people over the age of 60.

Working Together

The accompanying illustration shows how the digestive system works with other organ systems of the body to maintain homeostasis. All the organ systems of the body are interrelated.

BodyWorks CD-ROM
The module accompanying chapter 15 is Digestive System.

working together   Digestive System

Integumentary System

Digestive tract provides nutrients needed by skin.

Skin helps to protect digestive organs; helps to provide vitamin D for Ca^{2+} absorption.

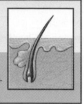

Skeletal System

Digestive tract provides Ca^{2+} and other nutrients for bone growth and repair.

Bones provide support and protection; hyoid bone assists swallowing.

Muscular System

Digestive tract provides glucose for muscle activity; liver metabolizes lactic acid following anaerobic muscle activity.

Smooth muscle contraction accounts for peristalsis; skeletal muscles support and help protect abdominal organs.

Nervous System

Digestive tract provides nutrients for growth, maintenance, and repair of neurons and neuroglial cells.

Brain controls nerves, which innervate smooth muscle and permit tract movements.

Endocrine System

Stomach and small intestine produce hormones.

Hormones help control secretion of digestive glands and accessory organs; insulin and glucagon regulate glucose storage in liver.

How the Digestive System works with other body systems

Circulatory System

Digestive tract provides nutrients for plasma protein formation and blood cell formation; liver detoxifies blood, makes plasma proteins, destroys old red blood cells.

Blood vessels transport nutrients from digestive tract to body; blood services digestive organs.

Lymphatic System/Immunity

Digestive tract provides nutrients for lymphoid organs; stomach acidity prevents pathogen invasion of body.

Lacteals absorb fats; Peyer's patches prevent invasion of pathogens; appendix contains lymphoid tissue.

Respiratory System

Breathing is possible through the mouth because digestive tract and respiratory tract share the pharynx.

Gas exchange in lungs provides oxygen to digestive tract and excretes carbon dioxide from digestive tract.

Urinary System

Liver synthesizes urea; digestive tract excretes bile pigments from liver and provides nutrients.

Kidneys convert vitamin D to active form needed for Ca^{2+} absorption; compensate for any water loss by digestive tract.

Reproductive System

Digestive tract provides nutrients for growth and repair of organs and for development of fetus.

Pregnancy crowds digestive organs and promotes heartburn and constipation.

Basic Key Terms

ascending colon (ah-send'ing ko'lon), p. 307
bile (bīl), p. 304
bolus (bo'lus), p. 298
cecum (se'kum), p. 307
chyme (kīm), p. 300
descending colon (de-send'ing ko'lon), p. 307
duodenum (du"o-de'num), p. 302
esophagus (ĕ-sof'ah-gus), p. 300
essential amino acids (ĕ-sen'shal ah-me'no as'idz), p. 312
essential fatty acids (ĕ-sen'shal fat'e as'idz), p. 312
gallbladder (gawl'blad-der), p. 306
gastric glands (gas'trik glandz), p. 300
glottis (glot'is), p. 299
hard palate (hard pal'at), p. 298
ileum (il'e-um), p. 302
jejunum (jĕ-joo'num), p. 302
large intestine (larj in-tes'tin), p. 307
liver (liv'er), p. 304
mastication (mas"ti-ka"shun), p. 298
mineral (min'er-al), p. 313
mouth (mowth), p. 298
pancreas (pan'kre-as), p. 304
peristalsis (per"ĭ-stal'sis), p. 300
pharynx (far'inks), p. 299
rectum (rek'tum), p. 307
rugae (roo'je), p. 300
salivary gland (sal'ĭ-ver-e gland), p. 298

sigmoid colon (sig'moid ko'lon), p. 307
small intestine (smawl in-tes'tin), p. 302
soft palate (soft pal'at), p. 298
sphincter (sfingk'ter), p. 300
stomach (stum'ak), p. 300
transverse colon (trans-vers' ko'lon), p. 307
urea (u-re'ah), p. 304
vermiform appendix (ver'mĭ-form ah-pen'diks), p. 307
villi (vil'i), p. 302
vitamin (vi'tah-min), p. 312

Clinical Key Terms

acute pancreatitis (ah-kūt' pan"kre-ah-ti'tis), p. 304
appendicitis (ah"pen-dĭ-si'tis), p. 309
caries (kar'ēz), p. 311
cirrhosis (sah-ro'sis), p. 306
colostomy (kah-los'tah-me), p. 309
constipation (kon-stĭ-pa'shun), p. 308
diarrhea (di-ah-re'ah), p. 308
diverticulosis (di"ver-tĭ-kyu-lo'sis), p. 309
gingivitis (jin-jah-vi'tis), p. 311
heartburn (hart'bern), p. 300
hepatitis (hĕ-pah-ti'tis), p. 306
jaundice (jon'dis), p. 306
periodontitis (per"e-o-don-ti'tis), p. 311
polyp (pah'lip), p. 309
ulcer (ul'ser), p. 300
vomiting (vă'mit-ing), p. 300

Summary

I. **Digestive System**
The functions of the digestive system are to ingest the food, break it down into small molecules that can cross plasma membranes, absorb these nutrient molecules, and eliminate nondigestible wastes.
   A. The salivary glands produce saliva that enters the mouth, where teeth masticate the food and the tongue forms a bolus prior to swallowing.
   B. The air and food passages merge in the pharynx. When a person swallows, the air passage is usually blocked off, and food must enter the esophagus.
   C. The esophagus takes food to the stomach. Peristalsis, a rhythmical contraction of the digestive tract wall, pushes food through the esophagus, as well as the rest of the digestive tract.
   D. The stomach expands and stores food. While food is in the stomach, the stomach churns, mixing food with the acidic gastric juices.
   E. The small intestine is 3 meters long and has three divisions (duodenum, ileum, and jejunum). Its walls have fingerlike projections called villi, in which nutrient molecules are absorbed into the circulatory and lymphatic systems.
   F. The pancreas, liver, and gallbladder are three accessory organs of digestion.
      1. The pancreas produces pancreatic juice, which contains bicarbonate and digestive enzymes for carbohydrates, proteins, and fats. Digestive enzymes and bile enter the duodenum via ducts.
      2. The liver is a critical organ, with numerous important functions, and any malfunction is a matter of

considerable concern. The liver receives blood from the small intestine by way of the hepatic portal vein.

   3. The bile produced by the liver is stored in the gallbladder.
 G. The large intestine consists of the cecum, colons (ascending, transverse, descending, and sigmoid), and the rectum, which ends at the anus. The large intestine does not digest food. It absorbs water and some electrolytes, and stores nondigestible material prior to defecation. Diarrhea occurs when too little water has been absorbed in the large intestine; constipation occurs when too much water has been absorbed.

II. **Mechanical and Chemical Digestion**
 A. Mechanical digestion includes chewing of food by teeth and churning of food by the stomach.
 B. During chemical digestion, enzymes present in digestive juices break down food into the following nutrient molecules: glucose, amino acids, fatty acids, and glycerol. Glucose and amino acids are absorbed into the blood capillaries of the intestinal villi. Fatty acids and glycerol are also absorbed but then rejoin to produce fat, which enters the lacteals.

III. **Nutrition**
 A. Glucose is an immediate energy source for the body, while fats are a long-term energy source. Amino acids build proteins. Essential amino acids cannot be produced by the body and must be in the diet.
 B. Vitamins are organic molecules that play essential roles in cellular metabolism. They are necessary components of the diet, and each has specific functions in the body.
 C. Minerals are inorganic elements. Macronutrients are minerals needed in gram amounts per day. Micronutrients (trace elements) are minerals needed in only microgram amounts per day.

Study Questions

1. List the parts of the digestive tract, anatomically describe them, and state the contribution of each to the digestive process. (pp. 298–303, 307–08)
2. What is peristalsis, and when does it begin? (p. 300)
3. What are gastrin, secretin, and CCK? Where are they secreted? What are their functions? (p. 301)
4. Describe the wall of the small intestine, and explain its role in nutrient absorption. (p. 302)
5. List the accessory organs of digestion, and describe their roles in food digestion. (p. 304)
6. List seven functions of the liver. How does the liver maintain a constant blood glucose level? (p. 304)
7. What is the common intestinal bacterium? What do these bacteria do for the body? (pp. 307–08)
8. Give a step-by-step explanation of the digestion of starch, protein, and fat. (p. 310)

Objective Questions

Fill in the blanks.

1. In the mouth, salivary _____ digests starch.
2. When we swallow, the _____ covers the opening to the larynx.
3. The _____ takes food to the stomach, where _____ is primarily digested.
4. The gastric juices are _____ and, therefore, they usually destroy any bacteria in the food.
5. The large intestine has a colon with four _____ and a(n) _____ , which leads to the _____ .
6. The pancreas transports digestive juices to the _____ , the first part of the small intestine.
7. After a meal, the liver stores glucose as _____ .
8. The gallbladder stores _____ , a substance that _____ fat.
9. Pancreatic juice contains _____ and _____ for digesting protein, _____ for digesting fat, and _____ for digesting starch.
10. The products of digestion are absorbed into the cells of the _____ , fingerlike projections of the intestinal wall.

Medical Terminology Reinforcement Exercise

Consult Appendix B for help in pronouncing and analyzing the meaning of the terms that follow.

1. stomatoglossitis (sto″mah-to-glos-si′tis)
2. glossopharyngeal (glos″o-fah-rin′je-al)
3. esophagectasia (ĕ-sof″ah-jek-ta′se-ah)
4. gastroenteritis (gas″tro-en-ter-i′ tis)
5. colostomy (ko-los′to-me)
6. sublingual (sub-ling′gwal)
7. gingivoperiodontitis (jin″jĭ-vo-per″e-o-don-ti′tis)
8. dentalgia (den-tal′je-ah)
9. pyloromyotomy (pi-lo″ro-mi-ot′o-me)
10. cholangiogram (ko-lan′je-o-gram)
11. cholecystolithotripsy (ko″le-sis″to-lith′o-trip″se)
12. proctosigmoidoscopy (prok″to-sig″moi-dos′ko-pe)

Website Link

For a listing of the most current Websites related to this chapter, please visit the Mader home page at:

http://www.mhhe.com/maderap

c h a p t e r 16

The Urinary System

Chapter Outline and Learning Objectives

After you have studied this chapter, you should be able to:

Name the organs of excretion, and tell what wastes they excrete.

Urinary System (p. 322)
- List and discuss the functions of the urinary system.
- Describe the macroscopic and microscopic anatomy of the kidney.
- State the parts of a kidney nephron, and relate them to the gross anatomy of the kidney.
- Name and describe the structure and function of each organ in the urinary system.
- Trace the path of urine, and describe how urination is controlled.

Urine Formation (p. 325)
- State, in general, the characteristics of normal urine.
- Describe the three steps in urine formation, and relate them to parts of a nephron.

Regulatory Functions of the Kidney (p. 329)
- Describe how the kidneys help maintain the fluid, electrolyte, and acid-base balance of blood.
- Name and explain how three hormones, aldosterone, antidiuretic hormone, and atrial natriuretic hormone work together to maintain blood volume and pressure.

Effects of Aging (p. 333)
- Anatomical and physiological changes occur in the urinary system as we age.

Working Together (p. 333)
- The urinary system works with other systems of the body to maintain homeostasis.

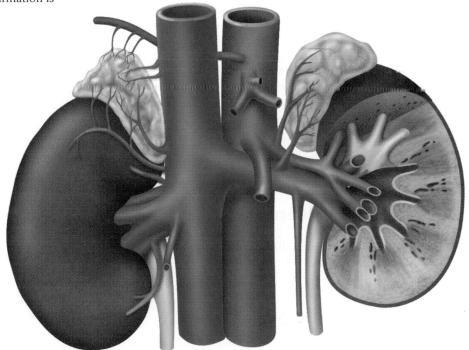

The kidneys are served by major blood vessels of the body.

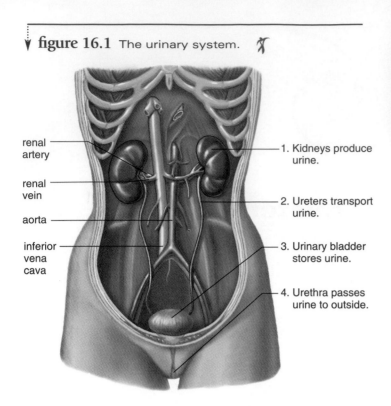

figure 16.1 The urinary system.

renal artery

renal vein

aorta

inferior vena cava

1. Kidneys produce urine.
2. Ureters transport urine.
3. Urinary bladder stores urine.
4. Urethra passes urine to outside.

Urinary System	
Organ	Function
Kidneys	Produce urine
Ureters	Transport urine
Bladder	Stores urine
Urethra	Eliminates urine

table 16.1

The **kidneys,** which excrete nitrogenous wastes, are the primary excretory organs, but other organs also function in excretion: The skin excretes water and salts in sweat; the lungs excrete carbon dioxide as a gas; and the liver excretes bile pigments that come from the breakdown of hemoglobin by way of the intestine.

The terms *defecation* and *excretion* do not refer to the same process. *Defecation* refers to the elimination of feces from the body. *Excretion,* on the other hand, refers to the elimination of metabolic waste products of metabolism. For example, undigested food and bacteria, which make up feces, have never been a part of the functioning of the body, while electrolytes in the urine are excretory substances because they were once metabolites in the body.

Urinary System

The urinary system includes the kidneys and associated structures, which are illustrated in figure 16.1 and listed in table 16.1.

Functions of the Urinary System

The primary functions of the urinary system are carried out by the kidneys:

1. The kidneys excrete nitrogenous wastes, such as urea, uric acid, creatinine, and ammonium.

2. The kidneys maintain blood volume by regulating the amount of water excreted.
3. The kidneys monitor blood composition by regulating electrolyte excretion. Sodium (Na^+), is the most significant, but potassium (K^+), bicarbonate (HCO_3^-), and calcium (Ca^{2+}), among others, are also important.
4. The kidneys monitor blood pH chiefly by regulating the excretion of certain ions, such as hydrogen (H^+).
5. The kidneys secrete the enzyme renin, which helps maintain blood pressure.
6. The kidneys secrete the growth factor erythropoietin, which stimulates red blood cell production.

Urinary Organs

The kidneys produce urine, and the other urinary organs play a role in eliminating urine from the body.

Kidneys

The kidneys are paired organs located near the small of the back in the lumbar region on either side of the vertebral column. They lie in depressions against the deep muscles of the back behind the peritoneum, where they receive some protection from the lower rib cage. Each kidney is usually held in place by connective tissue, called renal fascia. Masses of adipose tissue adhere to each kidney. A sharp blow to the back can dislodge a kidney, which is then called a **floating kidney.**

Each kidney is a bean-shaped, reddish-brown organ about the size of a fist. It is covered by a tough capsule of fibrous connective tissue, called the renal capsule. A depression (the hilum) on the concave side is where the renal blood vessels and the ureters exit (fig. 16.2).

When a kidney is sectioned longitudinally, it is possible to detect three regions: (1) an outer granulated layer called the **renal cortex;** (2) a radially striated, or lined, layer called the **renal medulla;** and (3) an inner space, or cavity, called the **renal pelvis,** which is continuous with the ureter (fig. 16.2*a*).

Microscopically, each kidney is composed of over 1 million **nephrons** (fig. 16.2*b* and *c*). A nephron makes urine, which is transported through a collecting duct to the renal pelvis. In the medulla, groups of nephrons and

figure 16.2 Anatomy of the kidneys. **a.** Section of a kidney showing intact ureter and blood vessels. **b.** Longitudinal section without the blood supply. Now it is easier to see the renal pelvis which connects with a ureter. **c.** Enlargement showing the placement of nephrons in relation to the renal medulla and renal cortex. **d.** Diagram of a single nephron and its blood supply.

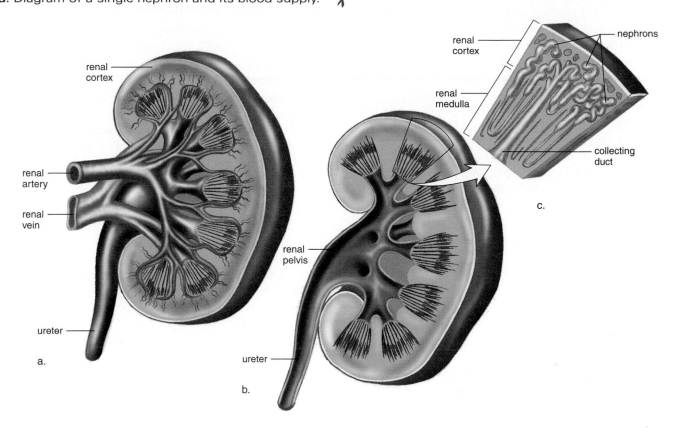

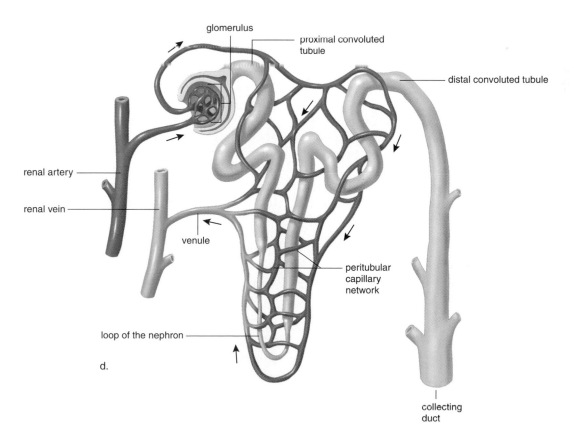

collecting ducts are massed in cone shapes called *renal pyramids.* The collecting ducts in each renal pyramid empty into a *calyx* of the renal pelvis (fig. 16.2a).

Nephrons   A nephron has several parts (fig. 16.2c). The end of the nephron is pushed in on itself to form a cuplike structure called a **glomerular** (glo-mer'u-ler) **capsule** (Bowman's capsule). Within this cup is a cluster of capillaries called the **glomerulus.** The **proximal** (meaning near the glomerular capsule) **convoluted tubule** is next. At the end of the proximal convoluted tubule, there is a U-turn called the loop of the nephron (loop of Henle). This leads to the **distal** (meaning far from the glomerular capsule) **convoluted tubule,** which enters a **collecting duct.** Each nephron has its own blood supply, and a **peritubular capillary network** surrounds the tubules and the loop of the nephron.

> Macroscopically, the kidneys are divided into the renal cortex, renal medulla, and renal pelvis. Microscopically, they are made up of over 1 million nephrons.

Ureters

The **ureters** are tubes about 25 centimeters long that convey urine from the kidneys to the bladder. Each descends from the renal pelvis, behind the parietal peritoneum, to enter the bladder posteriorly on the inferior surface.

The wall of a ureter has three layers. The inner layer is a mucosa (mucous membrane), the middle layer consists of smooth muscle, and the outer layer is a fibrous coat of connective tissue. Peristaltic contractions cause urine to enter the bladder even if a person is lying down. Urine enters the bladder in spurts that occur at the rate of one to five per minute.

Urinary Bladder

The **urinary bladder** stores urine until it can conveniently be expelled from the body. The bladder holds up to 600 milliliters (approximately 1 pint) of urine and becomes overdistended at 750 ml. (You get the urge to void at approximately 250 milliliters, and you become uncomfortable at approximately 500 milliliters. When the bladder becomes overdistended, you may lose the urge to void.) The bladder is located in the pelvic cavity, below the parietal peritoneum and posterior to the symphysis pubis. In the male, the bladder lies anterior to the rectum, seminal vesicles, and vas deferens. In the female, it is anterior to the uterus and upper vagina (fig. 16.3b).

The ureters enter the bladder, and the urethra exits the bladder. A sphincter called the internal urethral sphincter is found at the urethra's exit location.

The wall of the urinary bladder has four layers. A mucosa (mucous membrane) and a submucosa (connective tissue and elastic fibers) are followed by a coarse bundle of smooth muscle that makes up the detrusor muscle. Finally, the bladder's upper surface has a serous membrane that consists of parietal peritoneum. The rest of the bladder has an outer fibrous connective tissue coat.

Urethra

The **urethra** is a thin-walled tube that extends from the urinary bladder to an external opening called the external urethral orifice. The urethra has an inner mucosa (mucous membrane), and the wall consists of smooth muscular and connective tissues. As already mentioned, the internal urethral sphincter occurs where the urethra leaves the bladder, and an external urethral sphincter is located where the urethra exits the pelvic cavity.

The urethra differs in length in females and males. In females, the urethra lies ventral to the vagina and is only about 4 centimeters (1.6 inches) long. The short length of the female urethra and its proximity to the vaginal and anal openings help explain why females are prone to urinary tract infections. In males, the urethra averages 20 centimeters (8 inches) in length when the penis is relaxed. As the male urethra leaves the bladder, it is encircled by the prostate gland.

The genital (reproductive) and urinary systems in females are completely separate, but in males, the two systems share the urethra. During urination in males, the urethra carries urine; during sexual orgasm, it transports semen. This double function of the urethra does not alter the path of urine, and it is important to realize that urine is found only in those structures listed in table 16.1.

Urination

Urination, also called **micturition** (mik"tu-rish'un), occurs in the following manner: When the urinary bladder fills with urine, stretch receptors in the bladder wall transmit impulses to the central nervous system (CNS). Thereafter, parasympathetic impulses leave the sacral portion of the spinal cord and go to the internal urethral sphincter at the base of the bladder. Relaxation of this sphincter follows. However, urination does not take place until the cerebrum sends somatic impulses to an external sphincter located in the urethra. When this sphincter relaxes, micturition takes place. A very painful condition arises when small kidney stones **(renal calculi)** form and are passed through the urinary tract. Large kidney stones may have to be removed surgically, or, as part of a new treatment, smashed with ultrasonic waves.

> Urination (micturition) does not occur until parasympathetic and somatic impulses lead to the relaxation of, respectively, the internal and external urethral sphincters.

figure 16.3 Urinary bladder. **a.** As the bladder fills with urine, sensory impulses go to the spinal cord and then the brain. When urination occurs, motor nerve impulses cause the bladder to contract and internal and external sphincters to open. **b.** Location of the bladder in the female and male.

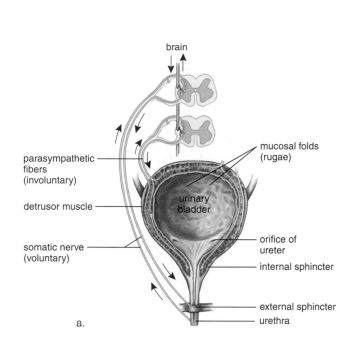

a.

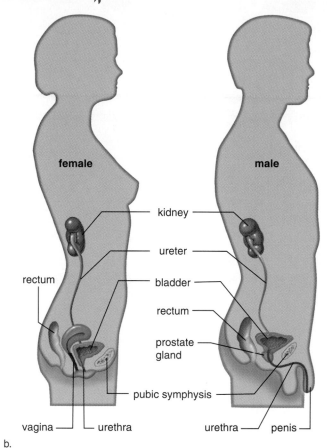

b.

Urine Formation

Table 16.2 lists the parts of a nephron, their locations within the kidney, and their contributions to urine formation. Electrolytes are excreted in urine and their proper concentration in the blood is vital in maintaining the proper osmotic pressure (potential for osmosis to occur) and salt balance, which is discussed later in the chapter.

Steps in Urine Formation

Urine formation requires the following steps (fig. 16.4):

1. *Glomerular filtration* occurs at the glomerular capsule. During glomerular filtration, water, nutrient molecules, and waste molecules move from the blood in the glomerulus to the fluid (filtrate) inside the glomerular capsule. The blood has been *filtered* because blood cells and large molecules, like most proteins, remain within the blood, while small

table 16.2

Nophron		
Name of Part	**Location in Kidney**	**Function**
Glomerular capsule	Cortex	Glomerular filtration
Proximal convoluted tubule	Cortex	Tubular reabsorption
Loop of the nephron	Medulla	Extrusion of sodium and reabsorption of water
Distal convoluted tubule	Cortex	Tubular secretion
Collecting duct	Medulla	Reabsorption of water; excretion of urine

Glomerular Filtration

Water, salts, nutrient molecules, and waste molecules move from the glomerulus to the inside of the glomerular capsule. These small molecules are called the glomerular filtrate.

Tubular Reabsorption

Nutrient and salt molecules are actively reabsorbed from the proximal convoluted tubule into the peritubular capillary network, and water flows passively.

Tubular Secretion

Certain molecules are actively secreted from the peritubular capillary network into the distal convoluted tubule.

proximal convoluted tubule

glomerular capsule

efferent arteriole

distal convoluted tubule

H_2O
glucose

drugs
creatinine

H^+

uric acid
urea
glucose
H_2O
amino
acids
salts

amino
acids

salts

glomerulus

afferent arteriole

artery

venule

vein

loop of the nephron

collecting duct

peritubular
capillary
network

H_2O
salts
urea
uric acid
NH_4^+
creatinine

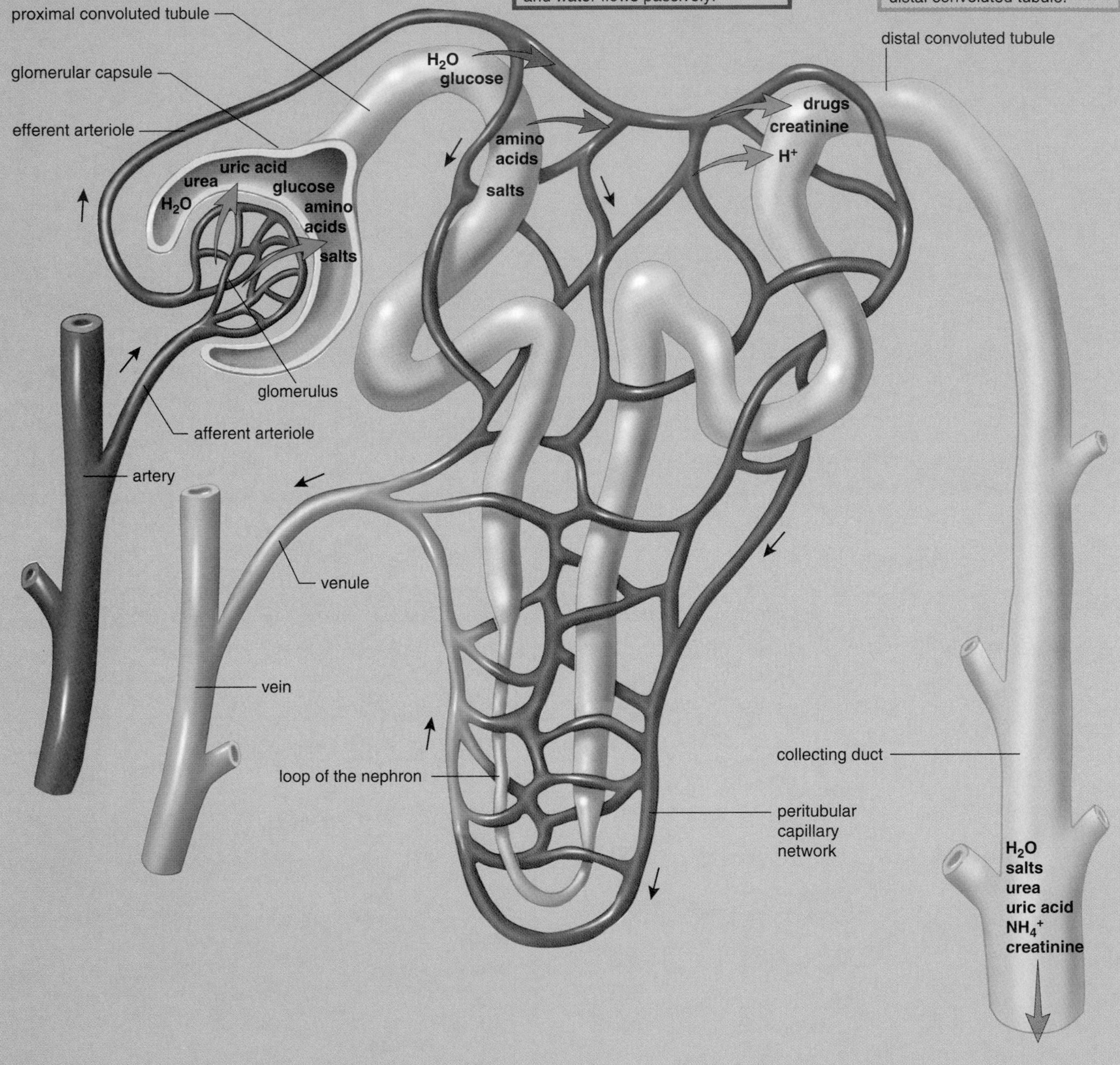

figure 16.4 Steps in urine formation. The three main steps in urine formation are color-coded to arrows that show the movement of molecules into or out of the nephron at specific locations. In the end, urine is composed of the substances within the collecting duct (see gray arrow).

Medical Focus

Illnesses Detected by Urinalysis

*U*rinalysis, or examination of the urine, indicates if any abnormal substances are present in the urine. The presence of glucose in the urine usually indicates that the individual has diabetes mellitus, a condition in which either the liver fails to store glucose as glycogen or the cells fail to take up glucose. In both cases, the blood glucose level is abnormally high. This makes the filtrate level of glucose high, and because the proximal convoluted tubule absorbs only an amount of glucose appropriate to the normal blood glucose level, glucose appears in the urine.

The presence of albumin and/or blood cells in the urine indicates that the glomerulus is more permeable than usual, as occurs in renal disease. When plasma proteins are excreted in the urine, the blood's osmotic pressure is reduced, and capillaries fail to take up water. Tissue fluid accumulates, and edema, particularly in the abdomen, occurs. As blood volume, and therefore blood pressure, decreases, the kidneys absorb more salt and water, but this, in the end, serves only to increase the edema. The best treatment is to cure the underlying cause of the edema.

Insufficient urine suggests kidney failure, which leads to **uremia,** or a very high blood urea nitrogen level (BUN). Death from kidney failure, however, is not due to the buildup of nitrogenous wastes; rather, it is due to an imbalance of electrolytes. Studies have shown that if urea is high but can be stabilized at normal levels, the patient usually recovers from the symptoms of uremia. An electrolyte imbalance, however, particularly the accumulation of potassium in the blood, interferes with the heartbeat and leads to heart failure.

---

molecules, such as glucose and urea, leave the blood to enter the tubule.

2. *Tubular reabsorption* occurs primarily at the proximal convoluted tubule. During tubular reabsorption, nutrient and salt (NaCl) molecules are actively reabsorbed from the proximal convoluted tubule into the peritubular capillary, and water follows passively.

3. *Tubular secretion* occurs primarily at the distal convoluted tubule. During tubular secretion, large waste molecules, such as creatinine, are actively secreted into the distal convoluted tubule. This step in urine formation is minor in comparison to the first two steps.

Concentrated Urine

Humans excrete a urine that contains only waste molecules dissolved in a minimum amount of water. This concentrated urine results because water is reabsorbed not only at the proximal convoluted tubule but also along the entire length of the nephron, particularly at the loop of the nephron and the collecting duct.

Urine has a light yellow or amber color due to the pigment urochrome, a breakdown product of hemoglobin, that is formed in the liver. At least 95% of urine volume is water; the remaining 5% is composed of organic nitrogenous wastes and excess electrolytes (table 16.3). The chief nitrogenous waste in humans is *urea*, formed in the liver as a part of amino acid metabolism. *Uric acid* occurs when

table 16.3	**Composition of Urine**	
	Water	95%
	Solids	5%
	Organic nitrogenous wastes (per 1,500 ml of urine)	
	Urea	30 g
	Creatinine	1–2 g
	Ammonia	1–2 g
	Uric acid	1 g
	Electrolytes	25 g
	Positive	*Negative*
	Sodium Na$^+$	Chlorides Cl$^-$
	Potassium K$^+$	Sulfates SO$_4$$^{2-}$
	Magnesium Mg^{2+}	Phosphates PO$_4$$^{3-}$
	Calcium Ca^{2+}	

nucleotides (see chapter 12) are broken down in cells. If uric acid is present in excess, it will precipitate out of the plasma. Uric acid crystals sometimes collect in the joints, producing a painful ailment called **gout.** *Creatinine* is an end product of muscle metabolism. Ordinarily, glucose and albumin are absent from urine, as discussed in the Medical Focus reading on this page.

Wastes, nutrients, and water are all filtered into a nephron, but nutrients and water are reabsorbed so that humans excrete a concentrated solution of wastes.

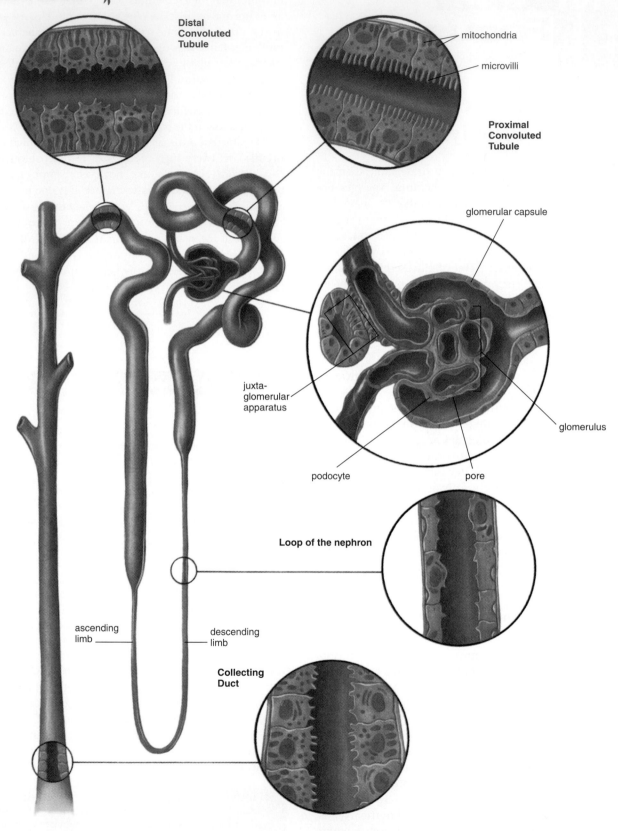

figure 16.5 Microscopic structure of a nephron. Blowups show the tissue composition of various parts as described in the text.

Distal Convoluted Tubule

mitochondria

microvilli

Proximal Convoluted Tubule

glomerular capsule

juxta-glomerular apparatus

glomerulus

podocyte

pore

Loop of the nephron

ascending limb

descending limb

Collecting Duct

Special Features of the Nephron

The cells along the length of the nephron (fig. 16.5) are specialized to carry on their respective functions.

The *juxtaglomerular apparatus* occurs at a region of contact between the afferent arteriole and the distal convoluted tubule. Cells in this region are involved in regulating sodium (Na^+) reabsorption from the distal convoluted tubules and collecting ducts and in maintaining blood volume, as is discussed later in the chapter.

The inner layer of the glomerular capsule is made up of *podocytes* that have long cytoplasmic processes. The podocytes cling to the capillary walls of the glomerulus and leave pores through which filtration can take place.

The cells lining the proximal convoluted tubule have numerous microvilli, about 1 micron in length, that increase the surface area for reabsorption. In addition, the cells contain numerous mitochondria, which produce the energy necessary for active transport. Glucose is an example of a molecule that ordinarily is reabsorbed completely because the supply of carrier molecules for it is plentiful. However, every substance has a maximum rate of transport, and after all its carriers are in use, any excess in the filtrate will appear in the urine. For example, if the blood glucose concentration is higher than normal, as in patients with diabetes mellitus, more glucose molecules than normal will move into the glomerular capsule during filtration. Not all of this glucose will be reabsorbed, and glucose will appear in the urine.

The loop of the nephron, which is lined with squamous epithelium, is made up of a descending and ascending limb. The ascending limb extrudes sodium so that the tissues of the medulla become hypertonic to the fluid in the descending limb and the collecting duct. These features allow the loop of the nephron to perform its function of concentrating the urine.

The cells of the distal convoluted tubule also have numerous mitochondria, but they lack microvilli. This is consistent with their role in actively moving molecules from the blood into the tubule.

> Each region of the nephron is anatomically suited to its task in urine formation.

Regulatory Functions of the Kidney

The kidney is involved in the fluid, electrolyte, and acid-base balance of the blood.

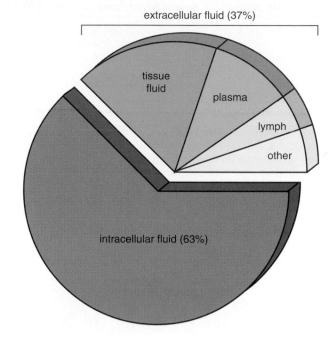

figure 16.6 Location of fluids in the body.

Fluid and Electrolyte Balance

The average adult male is about 60% water by weight. The average adult female is only about 50% water by weight because females generally have more subcutaneous adipose tissue, which contains less water. About two-thirds of this water is inside the cells (called intracellular fluid), and the rest is largely distributed in the plasma, tissue fluid, and lymph (called extracellular fluid). Water is also present in such fluids as cerebrospinal fluid and synovial fluid; in figure 16.6, these fluids are referred to as "other" fluids.

Maintenance of the water content in all body fluids requires that the body is in fluid balance: The total water intake should equal the total water loss. Table 16.4 shows how water enters the body; for example, metabolic water is created by chemical reactions in cells, food contains water, and a person drinks water. The osmolarity (ability to bring about osmosis) of the blood is constantly monitored within the hypothalamus, which determines whether or not a person is thirsty and takes a drink of water. Water is lost from the body in a variety of ways—for example, from sweating, feces formation, evaporation from the lungs, and urine formation.

table 16.4

Fluid Balance

Water Input	Average ml/day and % of Total	Water Output	Average ml/day and % of Total
Liquids	1,000; 40%	Urine	1,300; 52%
Food	1,000; 40%	Sweat	650; 26%
Metabolic water	500; 20%	Exhaled air	450; 18%
		Feces;	100; 4%
	Total 2,500; 100%		Total 2,500; 100%

figure 16.7 The renin-angiotensin-aldosterone system. The organs named in the boxes above the blood vessel release substances into the bloodstream to bring about the secretion of aldosterone, which causes reabsorption of sodium ions and a subsequent rise in blood pressure.

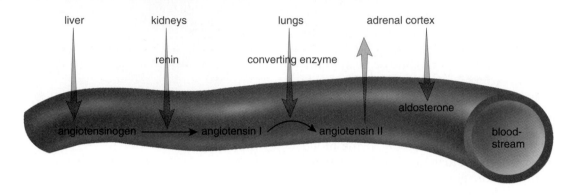

Hormonal Control

Aldosterone, antidiuretic hormone (ADH), and atrial natriuretic hormone (ANH) work together to control blood volume and pressure.

Aldosterone **Aldosterone,** a hormone secreted by the adrenal cortex, primarily maintains sodium (Na$^+$) and potassium (K$^+$) balance. The adrenal cortex is the outer portion of the adrenal glands, which lie atop the kidneys.

Blood volume is constantly monitored by the afferent arteriole cells within the juxtaglomerular apparatus (see fig. 16.5). When blood volume and therefore blood pressure is not sufficient to promote glomerular filtration, afferent arteriole cells secrete the enzyme renin. *Renin* changes angiotensinogen (a large plasma protein produced by the liver) into angiotensin I. Later, angiotensin I is converted to angiotensin II in the lungs by angiotensin-converting enzyme. Angiotensin II, a powerful vasoconstrictor, stimulates the adrenal cortex to release aldosterone (fig. 16.7). When aldosterone is released, sodium (Na$^+$) is reabsorbed into the blood at the distal convoluted tubules and collecting ducts of the nephron. The increase in Na$^+$ in blood causes water to be reabsorbed, leading to an increase in blood volume and blood pressure.

The renin-angiotensin-aldosterone system seems to be always active in some people who have hypertension. **Diuretics** are drugs that have been developed to counteract hypertension. They inhibit the reabsorption of Na$^+$ so that less water is reabsorbed from the nephron.

Antidiuretic Hormone **Antidiuretic hormone (ADH)** is released by the posterior lobe of the pituitary when the solutes in blood become more concentrated, due to a lack of water intake once Na$^+$ has been reabsorbed. To understand the action of this hormone, consider its name. *Diuresis* means increased amount of urine, and *antidiuresis* means decreased amount of urine. When ADH is present, more water is reabsorbed from the distal convoluted tubules and collecting ducts, and the amount of urine decreases. In practical terms, if an individual does not drink much water on a certain day, the posterior lobe of the pituitary releases ADH, causing more water to be reabsorbed and the blood volume to be maintained at a normal level, resulting in less urine formation. On the other hand, if an individual drinks a large amount of water and does not perspire much, the posterior lobe of the pituitary does not release ADH, causing more water to be excreted and the blood volume to be maintained at a normal level, resulting in a greater amount of urine formation.

Drinking alcohol causes diuresis because alcohol inhibits ADH secretion. The dehydration that follows is believed to contribute to the symptoms of a hangover.

Atrial Natriuretic Hormone The actions of aldosterone and ADH are opposed by **atrial natriuretic hormone (ANH)**. This hormone is released by cardiac cells when the atria of the heart are stretched, due to increased blood volume. ANH inhibits renin secretion by the juxtaglomerular apparatus and aldosterone secretion by the adrenal cortex. Its effect, therefore, is to cause the excretion of Na^+—that is, natriuresis. When Na^+ is excreted, so is water, and therefore, blood volume and blood pressure decrease.

> Blood volume and blood pressure are raised when aldosterone and ADH are secreted. The actions of aldosterone and ADH are opposed by ANH, and in this way, normal blood volume and pressure are maintained.

Electrolytes

The osmolarity of body fluids, including plasma, is dependent upon the concentration of substances—particularly electrolytes—within the fluids. **Electrolytes** are compounds and molecules that are able to ionize and, thus, carry an electrical current. The most common electrolytes in the plasma are sodium (Na^+), potassium (K^+), and bicarbonate (HCO_3^-). Na^+ and K^+ are termed *cations* because they are positively charged, and HCO_3^- is termed an *anion* because it is negatively charged. The kidneys control blood composition by regulating electrolyte excretion.

Sodium The movement of Na^+ across an axon membrane, you will recall, is necessary to the formation of a nerve impulse and muscle contraction. The concentration of Na^+ in the blood is also the best indicator of the blood's osmolarity.

Potassium The movement of K^+ across an axon membrane is also necessary to the formation of a nerve impulse and muscle contraction. Abnormally low K^+ concentrations in the blood, as might occur if diuretics are abused, can lead to cardiac arrest.

Bicarbonate Ion HCO_3^- is the form in which carbon dioxide is carried in the blood. The bicarbonate ion has a very important function in that it helps maintain the pH of the blood, as is discussed in the paragraphs that follow.

Other Ions The plasma contains many other ions. For example, calcium ions (Ca^{2+}) and phosphate ions (HPO_4^{2-}) are important to bone formation and cellular metabolism. Their absorption from the intestine and excretion by the kidneys is regulated by hormones, which was discussed in chapter 10.

> The kidneys monitor blood composition by regulating excretion of sodium, potassium, bicarbonate, and other ions.

Acid-Base Balance

The hydrogen concentration [H^+] of body fluids is important because proteins such as cellular enzymes function properly only when the pH is maintained at about 7.4. As you may recall from the discussion of pH in chapter 2 (p. 22), acids decrease the pH of solutions, and bases increase the pH of solutions.

The pH of the blood stays near 7.4 because the blood is buffered. A **buffer** is a chemical or combination of chemicals that can take up excess hydrogen (H^+) or excess hydroxide (OH^-). One of the most important buffers in the blood is carbonic acid (H_2CO_3) and the bicarbonate ion (HCO_3^-):

$$H_2CO_3 \rightleftharpoons H^+ + HCO_3^-$$

If the pH of the blood rises (less acidity), carbonic acid dissociates to release H^+. If the pH of the blood decreases (more acidity), the bicarbonate ion combines with H^+ to give carbonic acid. Proteins also help buffer the blood because they are charged in such a way that they can combine with either H^+ or OH^-.

The kidneys are the final adjusters of pH, contributing to homeostasis by maintaining the blood pH level within a narrow range. The entire nephron takes part in this process. The excretion of hydrogen ions (H^+) and ammonium (NH_4^+), together with the reabsorption of sodium (Na^+) and bicarbonate ions (HCO_3^-), is adjusted to keep the pH within normal bounds. If blood is acidic, hydrogen ions are excreted in combination with ammonium, while sodium and bicarbonate ions are reabsorbed. This restores the pH because $NaHCO_3$ is a base. If blood is basic, fewer hydrogen ions are excreted, and fewer sodium and bicarbonate ions are reabsorbed.

> The kidneys contribute to homeostasis by making adjustments in the excretion of hydrogen ions and ammonium, and in the reabsorption of sodium and bicarbonate ions, to maintain the blood pH level within a narrow range.

Prostate Enlargement

*t*he prostate gland, which is part of the male reproductive system, surrounds the urethra at the point where the urethra leaves the urinary bladder (fig. 16A). The prostate gland produces and adds a fluid to semen as semen passes through the urethra within the penis. At about age 50, the prostate gland often begins to enlarge, growing from a walnut to a lime and even to a lemon size. This condition is called **benign prostatic hyperplasia (BPH).** As it enlarges, the prostate squeezes the urethra, causing urine to back up first into the bladder, then into the ureters, and finally, perhaps, into the kidneys.

The treatment for BPH may involve the administration of prescription drugs designed to shrink prostate tissue, or the direct destruction of prostate tissue, for example, by the application of microwaves to a specific portion of the prostate. In many cases, a physician may decide that prostate tissue should be removed surgically. In most of these cases, rather than performing abdominal surgery, which requires an incision, the physician will gain access to the prostate via the urethra. This operation is called transurethral resection of the prostate (TURP). Recently, a startling finding revealed that the death rate during the 5 years following TURP is much higher than that following abdominal surgery.

Prostate enlargement is due to a prostate enzyme (5a-reductase) that acts on the male sex hormone testosterone, converting it into a substance that promotes prostate growth. The drug nafarelin prevents the release of a brain hormone that leads to testosterone production. When it is administered, approximately half of the patients report relief of urinary symptoms even after drug treatment is halted. However, the patients experience impotency (the inability to achieve an erection) and other side effects, such as hot flashes.

Two new drugs have also recently been tried on patients. One of these—finasteride—inhibits the action of 5a-reductase and gradually shrinks the prostate. Again, however, impotency is a side effect, and only 31% of the patients experienced improvement in the rate of urine flow. Another drug—terazosin—which is on the market for hypertension because it relaxes arterial walls, also relaxes muscle tissue in the prostate. Improved urine flow was experienced by 70% of the patients taking this drug. However, the drug has no effect on the prostate's overall size. Researchers speculate that combined drug therapy may, in the end, bring about the desired results: improved urine flow and a shrinkage of the prostate without troublesome side effects.

Questions

Why would you expect:

1. the drugs nafarelin and finasteride to shrink the prostate, but not the drug terazosin?
2. impotency to be a side effect of prostate drug therapy?

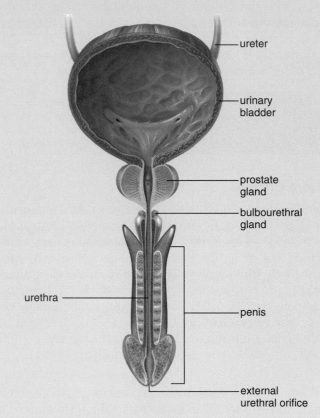

figure 16A Longitudinal section of a male urethra leaving the bladder. Note the position of the prostate gland, which can enlarge to obstruct urine flow.

Effects of Aging

Urinary disorders are significant causes of illness and death among the elderly. Total renal function in an elderly individual may be only 50% of that of the young adult. With increasing age, the kidneys decrease in size and have significantly fewer nephrons. However, vascular changes may play a more significant role in declining renal efficiency than renal tissue loss. Microscopic examination shows many degenerate glomeruli through which blood no longer flows and many other glomeruli that are completely destroyed.

Kidney stones occur more frequently with age, possibly as a result of improper diet, inadequate fluid intake, and kidney infections. Infections of the urethra, bladder, ureters, and kidneys increase in frequency among the elderly. Enlargement of the prostate occurs in males, and as is discussed in the MedAlert reading on page 332, this can lead to urine retention and kidney disease. Cancer of the prostate and bladder are the most common cancers of the urogenital system.

The involuntary loss of urine, which is called **incontinence,** increases with age. The bladder of an elderly person has a capacity of less than half that of a young adult and often contains residual urine. Therefore, urination is more urgent and frequent.

Working Together

The accompanying illustration shows how the urinary system works with other organ systems of the body to maintain homeostasis. All the organ systems of the body are interrelated.

BodyWorks CD-ROM
The module accompanying chapter 16 is Genitourinary System.

working together   Urinary System

Integumentary System

Kidneys compensate for water loss due to sweating; activate vitamin D precursor made by skin.

Skin helps regulate water loss; sweat glands carry on some excretion.

How the Urinary System works with other body systems

Circulatory System

Kidneys filter blood and excrete wastes; maintain blood volume, pressure, and pH; produce renin and erythropoietin.
Blood vessels deliver waste to be excreted; blood pressure aids kidney function; heart produces atrial natriuretic hormone.

Skeletal System

Kidneys provide active vitamin D for Ca^{2+} absorption and help maintain blood level of Ca^{2+}, needed for bone growth and repair.

Bones provide support and protection.

Lymphatic System/Immunity

Kidneys control volume of body fluids, including lymph.

Lymphatic system picks up excess tissue fluid, helping to maintain blood pressure for kidneys to function; immune system protects against infections.

Muscular System

Kidneys maintain blood levels of Na^+, K^+, and Ca^{2+}, which are needed for muscle innervation, and eliminate creatinine, a muscle waste.

Smooth muscular contraction assists voiding of urine; skeletal muscles support and help protect urinary organs.

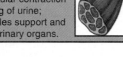

Respiratory System

Kidneys compensate for water lost through respiratory tract; work with lungs to maintain blood pH.

Lungs excrete carbon dioxide, provide oxygen, and convert angiotensin I to angiotensin II, leading to kidney regulation.

Nervous System

Kidneys maintain blood levels of Na^+, K^+, and Ca^{2+}, which are needed for nerve conduction.

Brain controls nerves, which innervate muscles that permit urination.

Digestive System

Kidneys convert vitamin D to active form needed for Ca^{2+} absorption; compensate for any water loss by digestive tract.

Liver synthesizes urea; digestive tract excretes bile pigments from liver and provides nutrients.

Endocrine System

Kidneys keep blood values within normal limits so that transport of hormones continues.

ADH and aldosterone, and atrial natriuretic hormone regulate reabsorption of Na^+ by kidneys.

Reproductive System

Semen is discharged through the urethra in males; kidneys excrete wastes and maintain electrolyte levels for mother and child.

Penis in males contains the urethra and performs urination; prostate enlargement hinders urination.

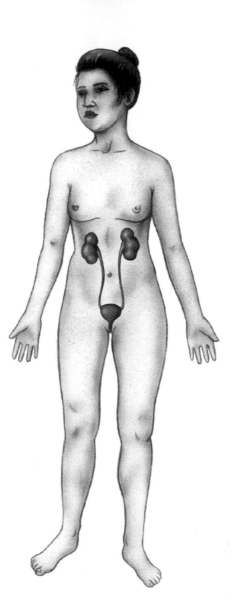

Basic Key Terms

aldosterone (al-dos′ter-ōn), p. 330

antidiuretic hormone (ADH) (an″tĭ-di″u-ret′ik hor′mōn), p. 330

atrial natriuretic hormone (a′tre-al na″tre-u-ret′ik hor′mon), p. 331

buffer (buf′er), p. 331

collecting duct (kah-lekt′ing dukt), p. 324

distal convoluted tubule (dis′tal kon′vo-lūt-ed tu′bul), p. 324

electrolytes (e-lek′tro-lītz), p. 331

glomerular capsule (glo-mer′u-ler kap′sul), p. 324

glomerulus (glo-mer′u-lus), p. 324

kidney (kid′ne), p. 322

micturition (mik″tu-rish′un), p. 324

nephron (nef′ron), p. 322

peritubular capillary network (per″ĭ-tu′bu-lar kap′ĭ-lar″e net′werk), p. 324

proximal convoluted tubule (prok′sĭ-mal kon′vo-lūt-ed tu′bul), p. 324

renal cortex (re′nul kor′teks), p. 322

renal medulla (re′nul mĕ-dul′ah), p. 322

renal pelvis (re′nul pel′vis), p. 322

ureters (u-re′terz), p. 324

urethra (u-re′thrah), p. 324

urinary bladder (u′ri-ner″e blad′der), p. 324

Clinical Key Terms

benign prostatic hyperplasia (BPH) (bĭ-nīn′ prah-sta′tik hi″per-pla′ze-ah), p. 332

diuretics (di-u-ret′iks), p. 330

floating kidney (flōt′ing kid′ne), p. 322

gout (gout′), p. 327

incontinence (in-con′tin-ents), p. 333

uremia (u-re′me-ah), p. 327

urinalysis (yer-ĭ-nal′ĭ-sis), p. 327

Summary

Various organs excrete metabolic wastes, but the kidneys, which excrete nitrogenous wastes, are the primary excretory organs.

I. Urinary System

A. The kidneys perform urinary system functions: They excrete nitrogenous wastes, maintain blood volume, regulate blood composition, monitor blood pH, secrete an enzyme that helps maintain blood pressure, and secrete a growth factor that stimulates red blood cell production.

B. Macroscopically, the kidneys are divided into the renal cortex, renal medulla, and renal pelvis. Microscopically, they are made up of over 1 million nephrons.

C. The organs of the urinary system consist of the kidneys, ureters, bladder, and urethra. The kidneys produce urine, the ureters transport urine, the bladder stores urine, and the urethra eliminates urine.

D. Urination (micturition) does not occur until parasympathetic and somatic impulses lead to the relaxation of, respectively, the internal and external urethral sphincters.

II. Urine Formation

A. The steps in urine formation are glomerular filtration, tubular reabsorption, and tubular secretion.

B. Wastes, nutrients, and water are all filtered into a nephron, but nutrients and water are reabsorbed so that humans excrete a concentrated solution of wastes.

C. Each region of the nephron is anatomically suited to its task in urine formation.

III. Regulatory Functions of the Kidney

A. Blood volume and blood pressure are raised when aldosterone and antidiuretic hormone (ADH) are secreted. The actions of aldosterone and ADH are opposed by atrial natriuretic hormone (ANH), and in this way, normal blood volume and pressure are maintained.

B. The kidneys monitor blood composition by regulating excretion of sodium, potassium, bicarbonate, and other ions.

C. The kidneys contribute to homeostasis by making adjustments in the excretion of hydrogen ions and ammonia, and in the reabsorption of sodium and bicarbonate ions, to maintain the blood pH level within a narrow range.

Study Questions

1. Name several excretory organs and the substances they excrete. (p. 322)
2. List and explain six urinary system functions carried out by the kidneys. (p. 322)
3. Describe the macroscopic anatomy of the kidney. (pp. 322–23)
4. Name the parts of the nephron, and describe the anatomy of each part. (pp. 322–24)
5. Name the urinary organs, and describe each organ's role in eliminating urine from the body. (pp. 322–24)
6. Describe how urine is made by explaining what happens at each part of the nephron. (pp. 325–27)
7. Explain the following terms: glomerular filtration, tubular reabsorption, and tubular excretion. (pp. 325–26)
8. What is the composition of urine? (p. 327)
9. Name three nitrogenous end products, and explain how each is formed in the body. (p. 327)
10. Describe how hormones regulate the fluid balance in the body. (pp. 330–31)
11. How does the nephron regulate the pH of the blood? (p. 331)

Objective Questions

Fill in the blanks.

1. The lungs are organs of excretion because they rid the body of

    _____ .
2. The capillary tuft inside the glomerular capsule is called the _____ .
3. Urine leaves the bladder in the

    _____ .
4. _____ is a substance that is found in the filtrate, is reabsorbed, and is in urine.
5. Tubular secretion takes place at the _____ , a portion of the nephron.
6. The primary nitrogenous end product of humans is _____ .
7. _____ is a substance that is found in filtrate, is not reabsorbed, and is concentrated in urine.
8. In addition to excreting nitrogenous wastes, the kidneys adjust the _____ , _____ and _____ balance of the blood.
9. Reabsorption of water from the collecting duct is regulated by the hormones _____ and _____ .
10. A _____ is a chemical that can combine with either hydrogen ions or hydroxide ions, depending on the pH of the solution.
11. The accumulation of uric acid crystals in a joint cavity produces a condition called _____ .
12. Urine is carried from the kidneys to the urinary bladder by a pair of organs called _____ .
13. The outer granulated layer of the kidney is the renal _____ whereas the inner striated layer is the renal

    _____ .

Medical Terminology Reinforcement Exercise

Consult Appendix B for help in pronouncing and analyzing the meaning of the terms that follow.

1. hematuria (hem″ah-tu-re′ah)
2. oliguria (ol″ĭ-gu′re-ah)
3. polyuria (pol″e-u′re-ah)
4. extracorporeal shock wave lithotripsy (ESWL) (eks″trah-kor-po′re-al lith″o-trip′se)
5. antidiuretic (an″tĭ-di″u-ret′ik)
6. urethratresia (u-re″thrah-tre′ze-ah)
7. cystopyelonephritis (sis″to-pi″e-lo-ne-fri′tis)
8. nocturia (nok-tu′re-ah)
9. glomerulonephritis (glo-mer″u-lo-ne-fri′tis)
10. ureterovesicostomy (u-re″ter-o-ves″i-kos′to-me)

Website Link

For a listing of the most current Websites related to this chapter, please visit the Mader home page at:

http://www.mhhe.com/maderap

chapter 17

The Reproductive System

Chapter Outline and Learning Objectives

After you have studied this chapter, you should be able to:

■ State the functions of the reproductive system.

Male Reproductive System (p. 338)

■ Describe the macroscopic and microscopic anatomy of the testes.
■ State the path of sperm, from their site of production to the site of fertilization.
■ Name the glands and describe the secretions that contribute to the composition of semen.
■ Describe the anatomy of the penis and the events preceding and during ejaculation.
■ Discuss hormonal regulation in the male.
■ Name at least six actions of testosterone, including both primary and secondary sexual characteristics.

Female Reproductive System (p. 343)

■ Describe the macroscopic and microscopic anatomy of the ovaries.
■ Label a diagram of the external female genitals.
■ Contrast male orgasm with female orgasm.
■ Describe the ovarian and uterine cycles.
■ Discuss hormonal regulation in the female, including feedback control.
■ Name at least six actions of estrogen and progesterone, including mention of both primary and secondary sexual characteristics.

Control of Reproduction and Sexually Transmitted Diseases (p. 351)

■ List several means of birth control, and describe their effectiveness.
■ Describe the symptoms of AIDS, genital herpes, genital warts, gonorrhea, chlamydia, and syphilis.

Effects of Aging (p. 357)

■ Anatomical and physiological changes occur in the reproductive system as we age.

Working Together (p. 357)

■ The reproductive system works with other systems of the body to maintain homeostasis.

Visual Focus

Anatomy of Ovary and Follicle (p. 344)
Regulation of Estrogen and Progesterone Secretion by Negative Feedback (p. 348)

Medical Focus

Ovarian Cancer (p. 346)
Alternative Methods of Reproduction (p. 353)

MedAlert

Endometriosis (p. 352)

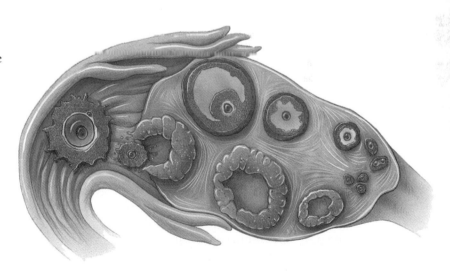

A drawing of an ovary in sagittal section shows the maturation of an egg that soon may be fertilized by a sperm.

Part V

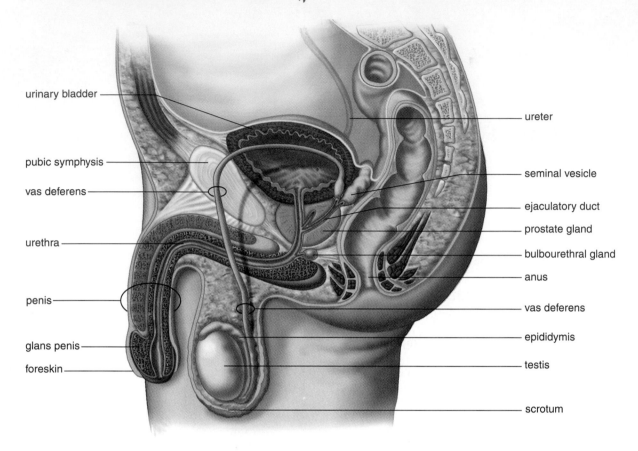

figure 17.1 The male reproductive system in sagittal section. The testes produce sperm. The seminal vesicles, the prostate gland, and the bulbourethral glands provide a fluid medium. Notice that the penis in this drawing is not circumcised—the foreskin is present.

urinary bladder

pubic symphysis

vas deferens

urethra

penis

glans penis

foreskin

ureter

seminal vesicle

ejaculatory duct

prostate gland

bulbourethral gland

anus

vas deferens

epididymis

testis

scrotum

The reproductive system does not begin to fully function until **puberty,** which is usually between the ages of 11 and 13 in girls and 14 and 16 in boys. Following puberty, the individual is capable of producing offspring. The reproductive system has the following functions:

1. Males produce sperm within testes, and females produce eggs within ovaries.
2. Males nurture and transport the sperm until they exit the penis, and females transport the eggs in uterine tubes to the uterus.
3. Sexual intercourse occurs when the male penis penetrates the female vagina.
4. Pregnancy is the period of time during which the offspring develops within the female body.
5. The testes and ovaries produce the sex hormones that have a profound effect on the body, including the masculinization and feminization of the skin, skeleton, voice, muscles, and brain.

Collectively, the sperm and eggs are called the **gametes,** while the testes and ovaries are the **gonads.** The gonads produce the sex hormones that are necessary for their own growth and maintenance and for the growth and mainte-

nance of the accessory reproductive organs. The accessory reproductive organs include the ducts through which the sperm and egg pass after leaving the gonads, the **external genitals** (that is, the external sexual organs), and other structures mentioned in this chapter.

Male Reproductive System

Figure 17.1 shows the male reproductive system, and table 17.1 lists the anatomical parts of this system.

Testes

The **testes** (sing., *testis*) lie outside the abdominal cavity of the male, within the **scrotum.** The testes begin to develop inside the abdominal cavity but descend into the scrotal sacs during the last two months of fetal development. If the testes do not descend, a condition called **cryptorchidism** (krip-tor'kah-dizm) occurs. If the male is not treated or operated on to place the testes in the scrotum, *sterility* (the inability to produce offspring) usually results. This is because

Male Reproductive System

table 17.1

Organ	Function
Testis	Produces sperm and sex hormones
Epididymis	Stores sperm as they mature
Vas deferens	Conducts and stores sperm
Seminal vesicle	Contributes to seminal fluid
Prostate gland	Contributes to seminal fluid
Urethra	Conducts sperm
Bulbourethral (Cowper's) gland	Contributes to seminal fluid
Penis	Serves as organ of copulation

the internal temperature of the body (37°C) is too high to produce viable sperm; the temperature in the scrotum is about 34°C. Wearing tight clothing can increase scrotal temperature and reduce sperm production. When the body is cold, the testes are normally held closer to the body to maintain an optimum temperature.

Seminiferous Tubules

Fibrous connective tissue forms the wall of each testis and divides the testis into lobules (fig. 17.2). Each lobule contains one to three tightly coiled **seminiferous** (se"mĭ-nif'er-us) **tubules** with a combined length of approximately 1.5 meters (5 feet). A cross section through a tubule when viewed with a microscope shows that each tubule is packed with cells undergoing **spermatogenesis** (production of sperm) (fig. 17.2c). Sperm cells are derived from undifferentiated germ cells called spermatogonia (sing., *spermatogonium*) that lie just inside the outer wall of a tubule and divide mitotically, producing new spermatogonia. Newly formed spermatogonia move away from the outer wall, increase in size, and become primary spermatocytes that undergo **meiosis**, which reduces the chromosome number. Secondary spermatocytes divide to produce four spermatids, also with 23 chromosomes. Spermatids then differentiate into **spermatozoa** (sperm). Also present in the tubules are the *sustentacular* (Sertoli) *cells*, which support, nourish, and regulate the spermatogenic cells.

Sperm

The mature sperm (fig. 17.2d) has three distinct parts: a tail, a middle piece, and a head. The *tail* is a flagellum, the *middle piece* contains energy-producing mitochondria, and the *head* contains the 23 chromosomes within a nucleus. Adhering to the nucleus is a specialized structure called the **acrosome**, which contains enzymes that facilitate penetration of the oocyte (egg). The human egg is surrounded by several layers of cells and a thick membrane; the acrosomal enzymes help a sperm digest its way into an egg.

Interstitial Cells

The male sex hormones, the androgens, are secreted by cells that lie between the seminiferous tubules. Therefore, they are called **interstitial cells** (fig. 17.2b). The most important of the androgens is testosterone, whose functions are discussed later in the chapter.

> The testes contain the interstitial cells and seminiferous tubules, in which spermatogenesis occurs. Sperm have a single flagellum and an acrosome-capped head in which 23 chromosomes reside in a nucleus.

Genital Tract

Sperm are produced in the testes, but they mature and are stored in the **epididymides** (ep"ĭ-did'ĭ-mah-dēz) (fig. 17.2a). Each epididymis is a tightly coiled tubule 5–6 meters (about 17 feet) in length, located just outside each testis. Each epididymis joins with a **vas (ductus) deferens,** which ascends through the *inguinal canal* and enters the abdomen, where it curves around the bladder and empties into the urethra (see fig. 17.1). Sperm are also stored in the first part of a vas deferens.

Spermatic Cords

The testes are suspended in the scrotum by the *spermatic cords,* each of which consists of fibrous connective tissue and muscle fibers that enclose the vas deferens, the blood vessels, and the nerves. The region of the inguinal canal, where the spermatic cord passes into the abdomen, remains a weak point in the abdominal wall. As such, it is frequently the site of hernias. A **hernia** is an opening or separation of some part of the abdominal wall through which a portion of an internal organ, usually the intestine, protrudes.

Seminal Fluid

At the time of ejaculation, sperm leave the penis in **seminal fluid (semen).** Three types of glands add secretions to seminal fluid: the seminal vesicles, the prostate gland, and the bulbourethral (Cowper's) glands. The **seminal vesicles** (see fig. 17.1) are located at the base of the bladder, and each has a duct that joins with a vas deferens. The **prostate gland** is a single, doughnut-shaped gland that surrounds the upper portion of the urethra, just below the bladder. In older men, the prostate can enlarge and press on the urethra, making urination painful and difficult. This condition is discussed in the MedAlert reading in chapter 16. The second most common cancer in men is **prostate cancer,** and detection requires vigilance. **Bulbourethral glands** are pea-sized organs that lie posterior to the prostate, on either side of the urethra.

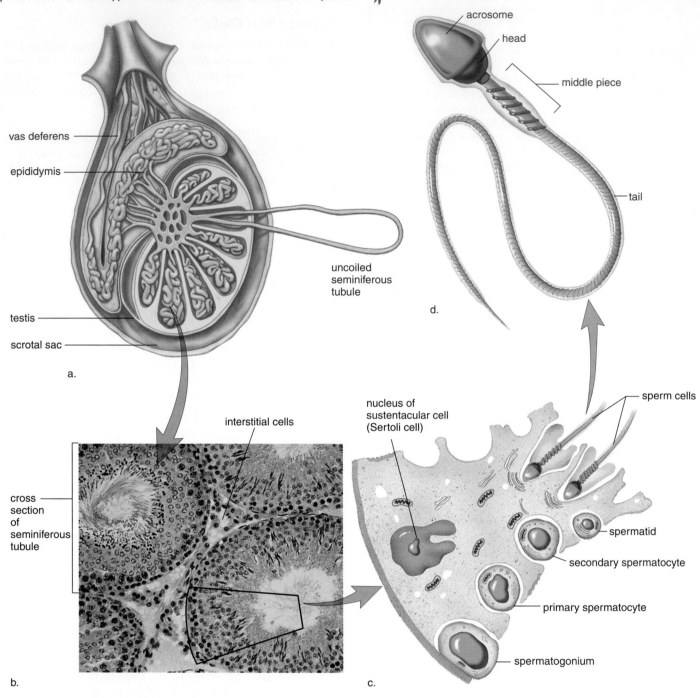

acrosome

head

middle piece

vas deferens

epididymis

tail

uncoiled seminiferous tubule

testis

scrotal sac

a.

d.

sperm cells

nucleus of sustentacular cell (Sertoli cell)

interstitial cells

cross section of seminiferous tubule

spermatid

secondary spermatocyte

primary spermatocyte

spermatogonium

b.

c.

figure 17.3 The penis. **a.** Structure of the penis, showing layers. **b.** Penis in cross section.

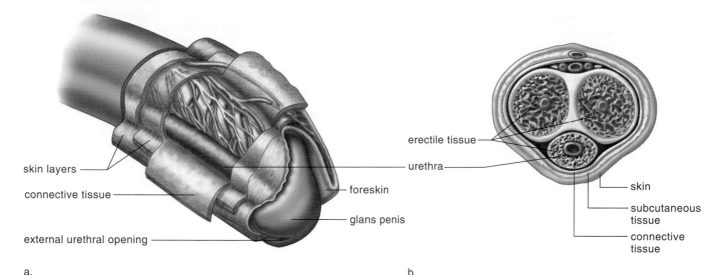

skin layers

connective tissue

external urethral opening

foreskin

glans penis

erectile tissue

urethra

skin

subcutaneous tissue

connective tissue

a.

b.

Each component of seminal fluid seems to have a particular function. Sperm are more viable in a basic solution, and seminal fluid, which is milky in appearance, has a slightly basic pH of about 7.5. Swimming sperm require energy, which is presumably provided by the sugar fructose contained in seminal fluid. Seminal fluid also contains prostaglandins, chemicals that cause the uterus to contract. Some investigators now believe that uterine contraction is necessary to help propel the sperm toward the egg.

> Sperm mature in the epididymis and are stored in the vas deferens before entering the urethra just prior to ejaculation. The accessory glands (seminal vesicles, prostate gland, and bulbourethral glands) produce seminal fluid. Semen, which contains sperm and seminal fluid, leaves the penis during ejaculation.

External Genitals

The penis and scrotum are the male external genitals. As discussed earlier, the scrotum contains the testes. The **penis** (fig. 17.3) is the organ of sexual intercourse in males. It has a long shaft and an enlarged tip called the glans penis. At birth, the glans penis is covered by a layer of skin called the **foreskin,** or prepuce. Sometime near puberty, small glands located in the foreskin and glans begin to produce an oily secretion. This secretion, along with dead skin cells, forms a cheesy substance known as smegma. **Circumcision** is the surgical removal of the foreskin, usually soon after birth.

When the male is sexually aroused, the penis becomes erect. **Erection** is achieved because blood sinuses within the erectile tissue of the penis fill with blood. Parasympa-thetic impulses dilate the arteries of the penis, while the veins are compressed passively so that blood flows into the erectile tissue under pressure. **Impotence** is a condition in which erection cannot be achieved. Medical and surgical remedies are available to treat impotence.

Orgasm in Males

As sexual stimulation intensifies, sperm enter the urethra from each vas deferens, and the glands add their fluids. Once seminal fluid is in the urethra, rhythmic muscle contractions expel it in spurts from the penis. During ejaculation, a sphincter closes off the bladder so that no urine enters the urethra.

The contractions that expel semen from the penis are a part of male **orgasm,** the physiological and psychological sensations that occur at the climax of sexual stimulation. The psychological sensation of pleasure is centered in the brain, but the physiological reactions involve the genital (reproductive) organs and associated muscles, as well as the entire body. Marked muscular tension is followed by contraction and relaxation.

Following ejaculation and/or loss of sexual arousal, the penis returns to its normal flaccid state. After ejaculation, a male typically experiences a refractory period, during which stimulation does not bring about an erection. The length of the refractory period increases with age.

In excess of 400 million sperm may be present in the 3.5 milliliters of semen expelled during ejaculation. The sperm count can be much lower than this, however, and fertilization (see fig. 17.7) still can take place.

Regulation of Male Hormone Levels

The hypothalamus has ultimate control of the testes' sexual functions because it secretes gonadotropic-releasing hormone (GnRH), which stimulates the anterior pituitary to produce the gonadotropic hormones. Two gonadotropic hormones, **FSH (follicle-stimulating hormone)** and **LH (luteinizing hormone)**, are named for their function in females but exist in both sexes, stimulating the appropriate gonads in each. FSH promotes spermatogenesis in the seminiferous tubules, and LH promotes testosterone production in the interstitial cells. LH in males is also called interstitial cell-stimulating hormone (ICSH).

The hormones mentioned are regulated by negative feedback (fig. 17.4), a mechanism that maintains testosterone production at a fairly constant level. For example, when the amount of testosterone in the blood rises to a certain level, it causes the anterior pituitary to decrease its secretion of LH. As the level of testosterone begins to fall, the anterior pituitary increases its secretion of LH, and stimulation of the interstitial cells recurs. Only minor fluctuations of the testosterone level occur in the male, and the feedback mechanism in this case acts to maintain testosterone at a normal level. The sustentacular cells in the wall of the seminiferous tubules produce a hormone called **inhibin** that blocks FSH secretion.

Testosterone

The male sex hormone, testosterone, has many functions. It is essential for normal development and function of the primary sex organs, those structures we have just discussed. It is also necessary for the sperm production. FSH causes spermatogenic cells to take up testosterone, which promotes spermatogenesis.

Greatly increased testosterone secretion at the time of puberty stimulates maturation of the penis and the testes. Testosterone also brings about and maintains the male secondary sexual characteristics, which develop at the time of puberty. Testosterone causes growth of a beard, axillary (underarm) hair, and pubic hair. It prompts the larynx and the vocal cords to enlarge, causing the voice to change. It is responsible for the greater muscular strength of males, which is why some athletes take a supplemental anabolic steroid, which is either testosterone or a related chemical. The disadvantages of anabolic steroid use are discussed in a Medical Focus reading in chapter 10. Testosterone also causes oil and sweat glands in the skin to secrete, thereby contributing to acne and body odor. A side effect of testosterone activity is baldness. Genes for baldness are probably

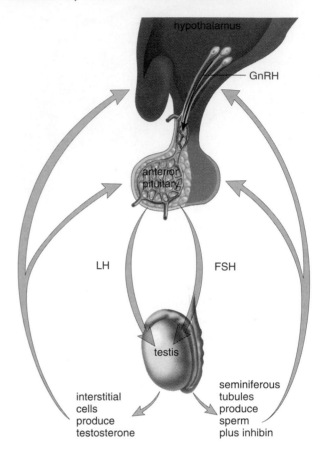

figure 17.4 Negative feedback. Regulation of testosterone secretion involves negative feedback (reverse arrows) by testosterone on GnRH and LH. Regulation of sperm production involves negative feedback by inhibin on GnRH and FSH.

hypothalamus

GnRH

anterior pituitary

LH

FSH

testis

interstitial cells produce testosterone

seminiferous tubules produce sperm plus inhibin

inherited by both sexes, but baldness is seen more often in males because of the presence of testosterone. This makes baldness a sex-influenced trait.

Testosterone is believed to be largely responsible for the sex drive. It may even contribute to the supposed aggressiveness of males.

> In males, FSH promotes spermatogenesis in the seminiferous tubules, and LH promotes testosterone production by the interstitial cells. Testosterone stimulates growth of the male genitals during puberty and is necessary for sperm maturation and development of the male secondary sexual characteristics.

figure 17.5 Female reproductive system in sagittal section.

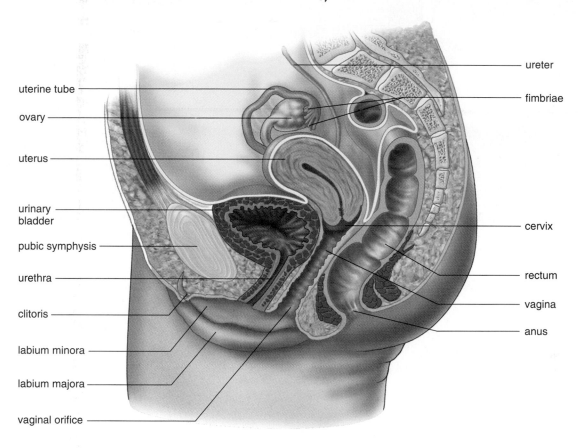

uterine tube

ovary

uterus

urinary bladder

pubic symphysis

urethra

clitoris

labium minora

labium majora

vaginal orifice

ureter

fimbriae

cervix

rectum

vagina

anus

Female Reproductive System

Figure 17.5 illustrates the female reproductive system, and table 17.2 lists the anatomical parts of this system.

Ovaries

The **ovaries** lie in shallow depressions, one on each side of the upper pelvic cavity. A longitudinal section through an ovary shows that it is made up of an outer cortex and an inner medulla. There are many saclike structures, called **follicles,** in the cortex, each of which contains an immature egg called an oocyte. A female is born with as many as 2 million follicles, but the number is reduced to 300,000 to 400,000 by puberty. Only a small number of follicles (about 400) reach maturation because a female usually produces only one egg per month during her reproductive years.

As the follicle undergoes maturation, it develops from a primary follicle into a secondary follicle and, finally, into a **vesicular** (Graafian) (graf´e-an) **follicle** (fig. 17.6).

table 17.2

Female Reproductive System	
Organ	**Function**
Ovary	Produces egg and sex hormones
Uterine tube (fallopian tube or oviduct)	Conducts egg toward uterus
Uterus (womb)	Houses developing fetus
Vagina	Receives penis during sexual intercourse and serves as birth canal

Oogenesis (o″o-jen´ĕ-sis) (production of eggs) takes place in the follicles. In a primary follicle, a primary **oocyte** (egg) has 46 chromosomes and divides meiotically into two cells, each having 23 chromosomes. One of these cells, termed the secondary oocyte (egg), receives almost all the cytoplasm. The other is a polar body that disintegrates. A secondary follicle contains the secondary oocyte, which is pushed to one side of a fluid-filled cavity. In a vesicular follicle, pressure within the fluid-filled cavity increases to the

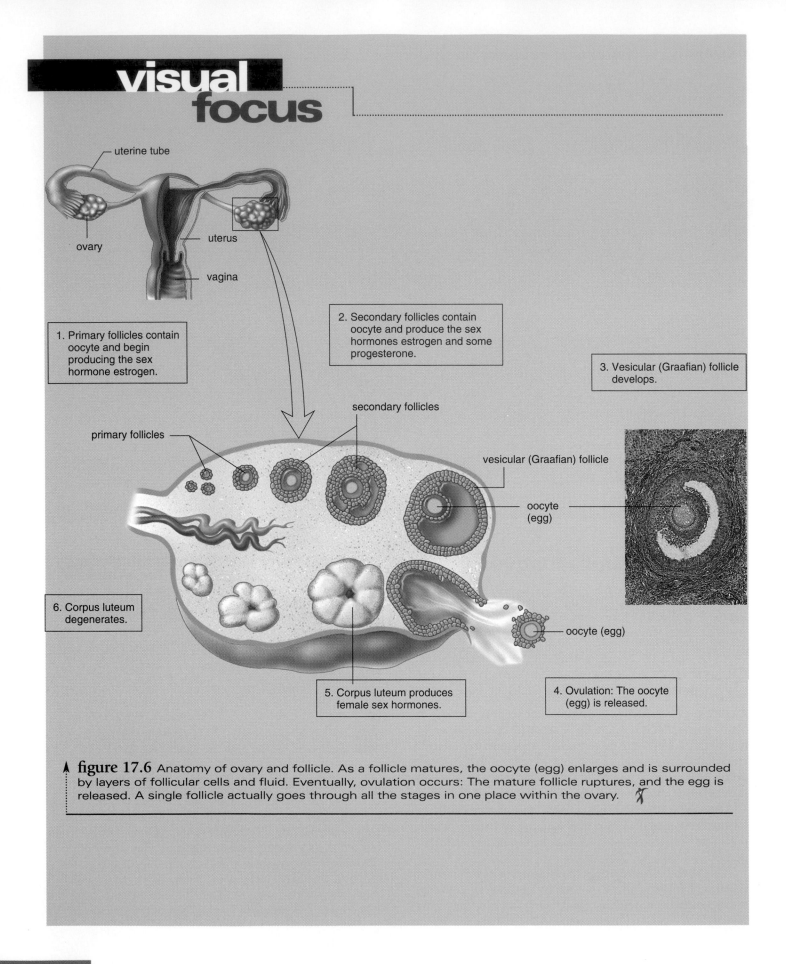

uterine tube

ovary

uterus

vagina

1. Primary follicles contain oocyte and begin producing the sex hormone estrogen.

2. Secondary follicles contain oocyte and produce the sex hormones estrogen and some progesterone.

3. Vesicular (Graafian) follicle develops.

secondary follicles

primary follicles

vesicular (Graafian) follicle

oocyte (egg)

6. Corpus luteum degenerates.

oocyte (egg)

5. Corpus luteum produces female sex hormones.

4. Ovulation: The oocyte (egg) is released.

figure 17.6 Anatomy of ovary and follicle. As a follicle matures, the oocyte (egg) enlarges and is surrounded by layers of follicular cells and fluid. Eventually, ovulation occurs: The mature follicle ruptures, and the egg is released. A single follicle actually goes through all the stages in one place within the ovary.

figure 17.7 Fertilization. **a.** Human sperm. **b.** Fertilization of egg by sperm.

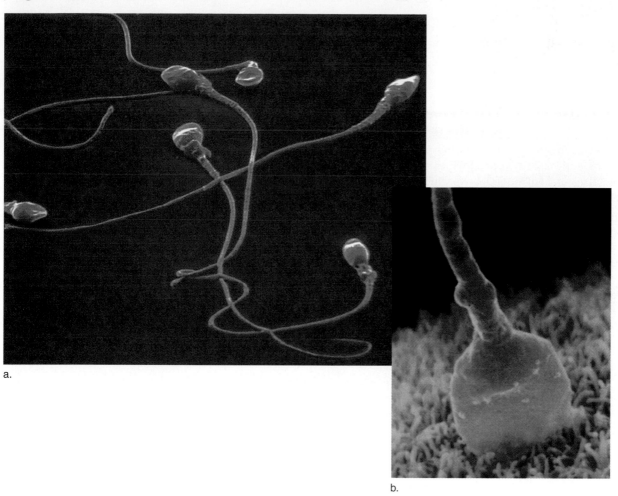

a.

b.

point that the follicular wall balloons out on the surface of the ovary. The wall then bursts, releasing the egg. The egg is surrounded by follicular cells, which collectively are called the corona radiata. This process is referred to as **ovulation.**

Once a follicle has lost its egg, it develops into a **corpus luteum** (kor′pus lu′te-um), a glandlike structure. If pregnancy does not occur, the corpus luteum begins to degenerate after about 10 days. If pregnancy does occur, the corpus luteum persists for three to six months. The follicle and the corpus luteum secrete the female sex hormones estrogen and progesterone, as discussed on page 347.

Cancer of an ovary, or ovarian cancer, which is discussed in the Medical Focus reading on page 346, causes more deaths than cervical and uterine cancer.

Genital Tract

The female genital tract includes the uterine tubes, uterus, and vagina.

Uterine Tubes

The uterine tubes, also called fallopian tubes or oviducts, extend from the uterus to the ovaries. The uterine tubes are not attached to the ovaries; instead, they have fingerlike projections called **fimbriae** (fim′bre-a) that sweep over the ovary at the time of ovulation. When the egg bursts from the ovary during ovulation (fig. 17.6), it is usually swept up into a uterine tube by the combined action of the fimbriae and the beating of cilia that line the uterine tubes.

Once in the uterine tube, the egg is propelled slowly toward the uterus by muscular contractions and the cilia of epithelial cells. Fertilization (fig. 17.7b), the completion of oogenesis, and zygote formation normally occur in a uterine tube. The developing embryo usually arrives at the uterus after several days and then embeds, or implants, itself in the uterine lining, which has been prepared to receive it.

Occasionally, the embryo becomes embedded in the wall of a uterine tube, where it begins to develop. Tubular

Ovarian Cancer

Ovarian cancer is often "silent," showing no obvious signs or symptoms until late in its development. The most common sign is enlargement of the abdomen, which is caused by the accumulation of fluid. Rarely is there abnormal vaginal bleeding. In women over 40, vague digestive disturbances (stomach discomfort, gas, distention) that persist and cannot be explained by any other cause may indicate the need for a thorough evaluation for ovarian cancer.

Risk for ovarian cancer increases with age. The highest rates are for women over age 60. Women who have never had children are twice as likely to develop ovarian cancer as those who have. Early age at first pregnancy, early menopause, and the use of oral contraceptives, which reduces ovulation frequency, appear to be protective against ovarian cancer. If a woman has had breast cancer, her chances of developing ovarian cancer double. Certain rare genetic disorders are associated with increased risk. With the exception of Japan, the highest incidence rates are reported in the more industrialized countries.

Early detection requires periodic, thorough pelvic examinations. The Pap smear, useful in detecting cervical cancer, does not reveal ovarian cancer. Women over age 40 should have a cancer-related checkup every year. Testing for the level of tumor marker CA-125, a protein antigen, is helpful.

Surgery, radiation therapy, and drug therapy are treatment options. Surgery usually includes the removal of one or both ovaries (**oophorectomy**) (o-ah-fah-rek'tah-me), the uterus (**hysterectomy**) (his"ter-ek'to-me), and the uterine tubes (**salpingectomy**) (sal-pin-jek'to-me). In some very early tumors, only the involved ovary is removed, especially in young women. In advanced disease, an attempt is made to remove all intraabdominal disease to enhance the effect of chemotherapy.

---

pregnancies cannot succeed because the tubes are not anatomically capable of allowing full development to occur. Such a pregnancy is called an **ectopic** (ek-tah'pik) **pregnancy** because it occurs outside the uterus.

Uterus

The **uterus** is a thick-walled, muscular organ about the size and shape of an inverted pear. Normally, it lies above and is tipped over the urinary bladder. The uterine tubes join the uterus anteriorly, while posteriorly, the **cervix,** the narrow end of the uterus, projects into the vagina at nearly a right angle. A small opening in the cervix leads to the lumen of the vagina.

Development of the embryo normally takes place in the uterus. This organ, sometimes called the womb, is approximately 5 centimeters (2 inches) wide in its usual state but is capable of stretching to over 30 centimeters (12 inches) to accommodate the growing baby. The lining of the uterus, called the **endometrium,** participates in the formation of the placenta (see chapter 18), which supplies nutrients needed for embryonic and fetal development. The endometrium has two layers: a basal layer and an inner functional layer. In the nonpregnant female, the functional layer of the endometrium varies in thickness according to a monthly reproductive cycle, called the uterine cycle, discussed later in the chapter.

Cancer of the cervix is a common form of cancer in women. Early detection is possible by means of a **Pap smear,** which entails the removal of a few cells from the region of the cervix for microscopic examination. If the cells are cancerous, a hysterectomy (the removal of the uterus) may be recommended. Removal of the ovaries in addition to the uterus is termed an **ovariohysterectomy** (o-var"e-o-his-ter-ek'to-me). Because the vagina remains intact, the woman still can engage in sexual intercourse.

Vagina

The **vagina** is a tube that makes a 45° angle with the small of the back. The mucosal lining of the vagina lies in folds that extend when the fibromuscular wall stretches. This capacity to extend is especially important when the vagina serves as the birth canal, and it can also facilitate intercourse, when the vagina receives the penis.

The egg enters the uterine tubes, which lead to the uterus, where implantation and development occur. The vagina is the organ of sexual intercourse in females.

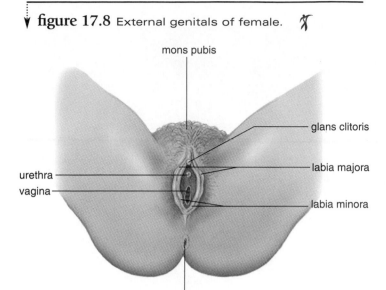

figure 17.8 External genitals of female.

mons pubis

glans clitoris

labia majora

urethra

labia minora

vagina

anus

External Genitals

The female external genitals (fig. 17.8) are known collectively as the **vulva.** The vulva includes two large, hair-covered folds of skin called the **labia majora.** They extend posteriorly from the *mons pubis,* a fatty prominence underlying the pubic hair. The **labia minora** are two small folds of skin lying just inside the labia majora. They extend forward from the vaginal opening to encircle and form a foreskin for the **clitoris** (klǐ′to-ris), an organ that is homologous to the penis. Although quite small, the clitoris has a shaft of erectile tissue and is capped by a pea-shaped glans. The clitoris also has sensory receptors that allow it to function as a sexually sensitive organ.

The *vestibule,* a cleft between the labia minora, contains the orifices of the urethra and the vagina. The vagina can be partially closed by a ring of tissue called the hymen. The hymen ordinarily is ruptured by initial sexual intercourse; however, it can also be disrupted by other types of physical activities. If the hymen persists after sexual intercourse, it can be surgically ruptured.

The urinary and reproductive systems in the female are entirely separate: The urethra carries only urine, and the vagina serves only as the birth canal and as the organ for sexual intercourse.

Orgasm in Females

Sexual response in the female is not as distinct as in the male, but there are certain similarities. The clitoris is an especially sensitive organ for initiating sexual sensations. It can become slightly erect as its erectile tissues become engorged with blood, but vasocongestion is more obvious in the labia minora, which expand and deepen in color. Erectile tissue within the vaginal wall also expands with blood, and the added pressure in these blood vessels causes small droplets of fluid to squeeze through the vessel walls and to lubricate the vagina.

During orgasm, females experience release from muscular tension, especially in the region of the vulva and vagina but also throughout the entire body. Increased uterine motility may assist the transport of sperm toward the uterine tubes. Since female orgasm is not signaled by ejaculation, there is a wide range in normalcy of sexual response.

Regulation of Female Hormone Levels

Hormone regulation in the female is quite complex, so this discussion has been simplified for easy understanding. The following glands and hormones are involved in hormonal regulation:

Hypothalamus secretes GnRH (gonadotropic-releasing hormone)

Anterior pituitary secretes FSH (follicle-stimulating hormone) and LH (luteinizing hormone), the gonadotropic hormones

Ovaries secrete **estrogen** and **progesterone,** the female sex hormones

The female sex hormones, estrogen and progesterone, have many effects on the body. In particular, estrogen secreted at the time of puberty stimulates the growth of the uterus and the vagina. Estrogen is necessary for egg maturation and is largely responsible for female secondary sexual characteristics. For example, it is responsible for the onset of the uterine cycle, as well as for female body hair and fat distribution. In general, females have a more rounded appearance than males because of a greater accumulation of fat beneath the skin. Also, the pelvic girdle enlarges in females so that the pelvic cavity has a larger relative size compared to males; this means that females have wider hips. Both estrogen and progesterone are also required for breast development.

1. The hypothalamus produces GnRH (gonadotropic-releasing hormone).

2. GnRH stimulates the anterior pituitary to produce FSH (follicle-stimulating hormone) and LH (luteinizing hormone).

3. FSH stimulates the follicle to produce estrogen, and LH stimulates the corpus luteum to produce progesterone.

4. Estrogen and progesterone affect the sex organs (e.g., uterus) and the secondary sex characteristics, and exert feedback control over the hypothalamus and the anterior pituitary.

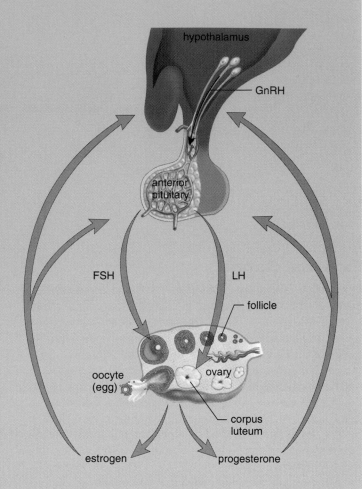

figure 17.9 Negative feedback. The secretion of estrogen and progesterone is regulated by negative feedback (reverse arrows). As the blood level of estrogen rises, the secretion of GnRH and FSH decrease; therefore, the secretion of estrogen lessens accordingly. As the blood level of progesterone rises, the secretion of GnRH and LH decrease; therefore the secretion of progesterone lessens accordingly.

table 17.3 Ovarian and Uterine Cycles (simplified)

Ovarian Cycle Phases	Events	Uterine Cycle Phases	Events
Follicular phase (days 1–13)	Follicle maturation and secretion of estrogen	Menstruation phase (days 1–5)	Endometrium breaks down.
		Proliferative phase (days 6–13)	Endometrium rebuilds.
Ovulation (day 14*)	Release of an egg from the ovary		
Luteal phase (days 15–28)	Corpus luteum formation and secretion of progesterone	Secretory phase (days 15–28)	Endometrium thickens, and glands are secretory.

*Assuming a twenty-eight-day cycle.

Ovarian Cycle

The **ovarian cycle** lasts an average of 28 days but may vary widely in individuals. For simplicity's sake, we will assume a 28-day cycle in the explanation that follows.

During the first half of a 28-day cycle (days 1–13, table 17.3), FSH secreted by the anterior pituitary promotes the development of a follicle in the ovary, and this follicle secretes estrogen. Rising blood estrogen levels exert negative feedback control over the anterior pituitary secretion of FSH so that this follicular phase ends (fig. 17.9). The end of the follicular phase is marked by ovulation on the fourteenth day of the 28-day cycle. During the last half of the ovarian cycle (days 15–28, table 17.3), anterior pituitary production of LH promotes the development of a corpus luteum, which secretes progesterone. A rising blood progesterone level exerts negative feedback control over anterior pituitary secretion of LH so that the corpus luteum degenerates. As the luteal phase ends, menstruation begins.

Uterine Cycle

The effect that the female sex hormones have on the endometrium of the uterus causes the uterus to undergo a cyclical series of events known as the **uterine cycle** (see table 17.3). Cycles that last 28 days are divided as follows:

During days 1–5, the level of female sex hormones in the body is low, causing the endometrium to disintegrate and its blood vessels to rupture. A flow of blood passes out of the vagina during **menstruation,** which is also known as **menses** or the menstrual period.

During days 6–13, increased estrogen production by an ovarian follicle causes the endometrium to thicken, becoming vascular and glandular. This series of events is called the *proliferative phase* of the uterine cycle.

Ovulation usually occurs on the 14th day of the 28-day cycle. During days 15–28, increased progesterone production by the corpus luteum causes the endometrium to double in thickness, and the uterine glands to mature, producing a thick, mucoid secretion. This is called the *secretory phase* of the uterine cycle. The endometrium now is prepared to receive the developing embryo. If pregnancy does not occur, the corpus luteum degenerates, and the low level of sex hormones in the female body causes the endometrium to break down. This is evidenced by the menstrual discharge that begins at this time. As discussed in the MedAlert reading on page 351, endometriosis is a cause of painful menstruation. Even while menstruation is occurring, the anterior pituitary begins to increase its production of FSH, and a new follicle begins to mature.

Table 17.3 lists the events in the ovarian and uterine cycles. Changes in the blood levels of the hormones controlling the ovarian and uterine cycles are shown in figure 17.10.

figure 17.10 Blood hormone levels associated with the ovarian and uterine cycles. During the follicular phase, FSH released by the anterior pituitary promotes the maturation of a follicle in the ovary. The follicle produces increasing estrogen levels, which causes the endometrium to thicken. After ovulation and during the luteal phase, LH promotes the development of the corpus luteum. This structure produces increasing progesterone levels, which causes the endometrial lining to become secretory. Menstruation begins when progesterone production declines to a low level.

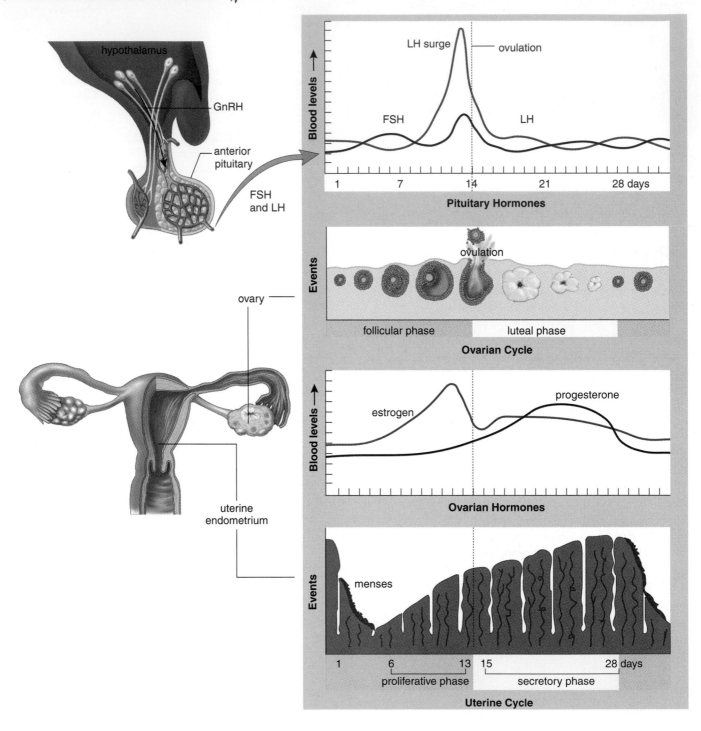

MedAlert

Endometriosis

*i*n the female reproductive tract, there is a small space between each uterine tube and ovary. The lining of the uterus, called the endometrium, loses its outer layer during menstruation. Sometimes, a portion of the menstrual discharge is carried backward up the uterine tube and into the abdominal cavity, instead of being discharged through the opening in the cervix. In the abdominal cavity, endometrial tissue can become attached to and implanted in various organs, such as the ovaries; the wall of the vagina, intestine, or bladder; or even on the nerves that serve the lower back or legs (fig. 17A). This painful condition, called **endometriosis** (en-do-me"tre-o′sis), affects 1–3% of women of reproductive age.

Women with uterine cycles of fewer than 27 days and with a menstrual flow lasting longer than one week have an increased chance of endometriosis. Women who have taken the birth control pill for a long time or who have had several pregnancies have a decreased risk of endometriosis.

When a woman has endometriosis, the displaced endometrial tissue reacts as if it were still in the uterus—thickening, becoming secretory, and then breaking down. The discomfort of menstruation is then felt in other organs of the abdominal cavity, resulting in pain. An area of endometriosis can degenerate and become a scar. Scars that hold two organs together are called **adhesions.** Adhesions can distort organs and lead to infertility.

Only direct observation of the abdominal organs can confirm endometriosis. First, a half-inch incision is made near the navel, and carbon dioxide gas is injected into the abdominal cavity to separate the organs. Then, a **laparoscope** (lă′pah-rah-skōp) (an optical instrument for viewing the peritoneal cavity) is inserted into the abdomen, allowing the physician to see the organs. Patches of endometriosis show up as purple, blue, or red spots, and there may be dark brown cysts filled with blood. Rupture of these cysts results in a great deal of pain. In the severest cases, there is scarring and the formation of adhesions and abnormal masses around the pelvic organs. A second incision, usually at the pubic hairline, allows the insertion of other instruments that can be used to remove endometrial implants.

Further treatment can take one of two courses. The drug nafarelin, administered as a nasal spray, acts through hormonal controls to stop estrogen production and, therefore, to stop the uterine cycle. Or, the ovaries can be removed, stopping the uterine cycle. In either case, the woman will suffer symptoms of menopause that can be relieved in the first instance by withdrawing the drug nafarelin or in the second instance by giving the woman estrogen in doses that do not reactivate the uterine cycle.

Questions

Why would you expect:

1. a person with short uterine cycles and increased days of menstrual flow to be at greater risk of endometriosis?
2. taking the birth control pill or having several pregnancies to decrease the risk of endometriosis?
3. removal of the ovaries to eliminate symptoms of endometriosis?

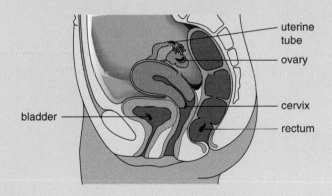

▲ **figure 17A** Endometriosis. Endometrial tissue (dark red) becomes attached to abdominal organs.

Pregnancy

Pregnancy occurs when the developing embryo embeds itself in the endometrial lining several days following fertilization. During implantation, an embryonic membrane surrounding the embryo begins to produce **human chorionic gonadotropic hormone (HCG)**, which prevents degeneration of the corpus luteum and instead causes it to secrete even larger quantities of progesterone. Progesterone (*pro* means for; *gestation* means pregnancy) also inhibits the motility of the uterus. Together with estrogen, it prepares the breasts for lactation (milk production). The corpus luteum may be maintained for as long as six months, even after the placenta is fully developed.

The **placenta** (see fig. 18.3) originates from both maternal and fetal tissue, and is the region of exchange of molecules between fetal and maternal blood, although the two blood types do not mix. After its formation, the placenta continues production of HCG and begins production of progesterone and estrogen. The latter hormones have two effects: They inhibit the anterior pituitary so that no new follicles mature, and they maintain the endometrium so that the corpus luteum is not needed. Menstruation does not occur during the nine months of pregnancy.

Hormonal regulation in the female results in an ovarian cycle. During the first half of the cycle, FSH causes maturation of the follicle, which secretes estrogen. After ovulation and during the second half of the cycle, LH converts the follicle into the corpus luteum, which produces progesterone. Estrogen and progesterone regulate the uterine cycle. Estrogen causes the endometrium to rebuild. Ovulation usually occurs on the fourteenth day of a 28-day cycle. As progesterone is produced by the corpus luteum, the endometrium thickens and becomes secretory. Then, a low level of hormones causes the endometrium to break down, as menstruation occurs. Menstruation does not occur during pregnancy because of placental hormones.

Control of Reproduction and Sexually Transmitted Diseases

Infertility

In some cases, couples do not need to prevent pregnancy; instead, conception, or fertilization, does not occur despite frequent intercourse. The American Medical Association estimates that 15% of all U.S. couples are unable to have children and therefore are termed **sterile**; another 10% have fewer children than they wish and therefore are termed **infertile**. The latter term assumes that the couple has been trying to become pregnant and has been unsuccessful for at least one year.

Causes of Infertility

The two major causes of female infertility are (1) blocked uterine tubes, possibly due to a previous infection, particularly gonorrhea or chlamydia; and (2) failure to ovulate due to low body weight. Endometriosis, which is discussed in the MedAlert reading on page 351, can also contribute to infertility.

In some cases, the causes of infertility can be corrected surgically and/or medically. If no obstruction is apparent and body weight is normal, females can be given a substance rich in FSH and LH extracted from the urine of postmenopausal women. This treatment causes multiple ovulations and, sometimes, multiple pregnancies.

The most frequent causes of male infertility are low sperm count and/or a large proportion of abnormal sperm. Disease, radiation, chemical mutagens, high testes temperature, and the use of psychoactive drugs can contribute to infertility.

When reproduction does not occur in the usual manner, many couples adopt a child. Other couples try one of the alternative reproductive methods discussed in the Medical Focus reading on page 353.

Corrective medical and surgical procedures can help people who are infertile but wish to have a child. Alternative methods of reproduction may also be considered.

Alternative Methods of Reproduction

Some couples are infertile due to various physical abnormalities. When corrective medical procedures fail, alternative methods of reproduction are available to couples who wish to become parents.

During **artificial insemination by a donor (AID)**, sperm are placed in the vagina by a physician. In some cases, a women is artificially inseminated by her husband's sperm. This is especially helpful if the husband has a low sperm count because the sperm can be collected over a period of time and concentrated so that the sperm count is sufficient for fertilization. Often, however, a woman is inseminated by sperm acquired from an anonymous donor. At times, a mixture of husband and donor sperm are used.

A variation of AID is **intrauterine insemination (IUI)**. IUI involves hormonal stimulation of the ovaries, followed by placement of the donor's sperm in the uterus, rather than in the vagina.

In the case of **in vitro fertilization (IVF)**, hormonal stimulation of the ovaries is followed by laparoscopy, in which an aspiratory tube is used to retrieve preovulatory eggs (see fig. 17.6). Alternately, a needle can be inserted through the vaginal wall and guided by the use of ultrasound to the ovaries, where it is used to retrieve eggs. This method is called transvaginal retrieval. Sperm from the male are then placed in a solution that approximates the conditions of the female genital tract. When eggs (sometimes treated to facilitate entrance of a sperm) are introduced, fertilization occurs. The resultant zygotes begin development, and after two to four days, the embryos are inserted into the uterus of the woman, who is now in the secretory phase of her menstrual cycle. If implantation is successful, normal fetal development continues to term.

Gamete intrafallopian transfer (GIFT) was devised as a means to overcome the low success rate (15–20%) of in vitro fertilization. The method is identical to in vitro fertilization except that the eggs and sperm are immediately placed in the uterine tubes after they have been brought together. This procedure is helpful to couples whose eggs and sperm never join in the uterine tubes; sometimes, the egg enters the abdominal cavity instead of the uterine tubes, and sometimes the sperm never reach the uterine tubes. GIFT has an advantage in that it is a one-step procedure for the woman: The eggs are removed and reintroduced all in the same time period. For this reason, it is less expensive—approximately $1,500 compared to $3,000 and higher for in vitro fertilization.

In some instances, women who are called surrogate mothers are paid to have babies by other individuals who contributed sperm (or eggs) to the fertilization process.

If all of these alternative methods of reproduction are considered, a baby potentially could have five parents: (1) a sperm donor, (2) an egg donor, (3) a surrogate mother, and (4) and (5) an adoptive mother and father.

table 17.4	Common Birth Control Methods				
Name	**Procedure**	**Methodology**	**Effectiveness***	**Risk**	
Abstinence	Refrain from sexual intercourse.	No sperm in vagina	100%	None	
Vasectomy	Vasa deferentia are cut and tied.	No sperm in semen	Almost 100%	Irreversible sterility	
Tubal ligation	Uterine tubes are cut and tied.	No eggs in uterine tube	Almost 100%	Irreversible sterility	
Oral contraception (the Pill)	Hormone medication is taken daily.	Anterior pituitary does not release FSH and LH.	Almost 100%	Thromboembolism, especially in smokers	
Depo-Provera	Four injections of progesterone-like steroid are given a year.	Anterior pituitary does not release FSH and LH.	About 99%	Breast cancer? Osteoporosis?	
Contraceptive implants	Tubes of progestin (form of progesterone) are implanted under the skin.	Anterior pituitary does not release FSH and LH.	More than 90%	Presently unknown	
Intrauterine device (IUD)	Plastic coil is inserted into uterus by physician.	Prevents implantation	More than 90%	Infection (pelvic inflammatory disease)	
Diaphragm	Latex cup is inserted into vagina to cover cervix before intercourse.	Blocks entrance of sperm to uterus	With jelly about 90%	Presently unknown	
Cervical cap	Latex cap is held by suction over cervix.	Delivers spermicide near cervix	Almost 85%	Cancer of cervix?	
Male condom	Latex sheath is fitted over erect penis at time of intercourse.	Traps sperm and prevents sexually transmitted diseases	About 85%	Presently unknown	
Female condom	Polyurethane tubing is fitted inside vagina.	Blocks entrance of sperm to uterus and prevents sexually transmitted diseases	About 85%	Presently unknown	
Coitus interruptus (withdrawal)	Male withdraws penis before ejaculation.	Prevents sperm from entering vagina	75%	Presently unknown	
Jellies, creams, foams	These spermicidal products are inserted before intercourse.	Kill a large number of sperm	About 75%	Presently unknown	
Natural family planning	Day of ovulation is determined by record keeping and various methods of testing.	Intercourse avoided on certain days of the month	About 70%	Presently unknown	
Douche	Vagina is cleansed after intercourse.	Washes out sperm	Less than 70%	Presently unknown	

*The percentage of women who are not expected to get pregnant within one year using this means of birth control.

Birth Control

The most reliable method of birth control is abstinence—that is, the absence of sexual intercourse. This form of birth control has the added advantage of preventing sexually transmitted diseases. Other, perhaps more common, means of birth control used in the United States are listed in table 17.4.

In the male, a **vasectomy** consists of cutting and sealing the vas deferens on each side so that the sperm are unable to reach the seminal fluid that is ejected at the time of orgasm. The sperm are then largely reabsorbed. Following this operation, which can be done in a doctor's office, the amount of ejaculate remains normal because sperm account for only about 1% of semen volume. Also, the secondary sexual characteristics are not affected, since the testes continue to produce testosterone.

In the female, **tubal ligation** consists of cutting and sealing the uterine tubes. Pregnancy rarely occurs because the passage of the egg through the uterine tubes has been blocked. Whereas major abdominal surgery was formerly required for a tubal ligation, today simpler procedures are available. Using a method called laparoscopy, which requires only two small incisions, the surgeon inserts a small, lit telescope to view the uterine tubes and a small surgical blade to sever them. An even newer method called **hysteroscopy** (his-tah-rŏ'sko-pe) uses a telescope within the uterus to seal the tubes by means of an electrical current.

There has been a revival of interest in barrier methods of birth control, including male and female condoms, because these methods offer some protection against sexually transmitted diseases. The female condom is essentially a large polyurethane tube with a closed end that fits over the cervix and an open end that covers the external genitals.

Investigators have long searched for a "male pill." Analogs of gonadotropic-releasing hormone have been used to prevent the hypothalamus from stimulating the anterior pituitary. Inhibin has also been used to prevent the anterior pituitary (see fig. 17.4) from producing FSH. Testosterone and/or related chemicals have been used to inhibit spermatogenesis in males, but feminizing side effects are common because the body changes an excess of testosterone to estrogen.

Contraceptive vaccines are now being developed. For example, a vaccine developed to immunize women against HCG, the hormone so necessary to implantation of the embryo, was successful in a limited clinical trial. Since HCG is not normally present in the body, no untoward autoimmunity reaction is expected, and the immunization

does wear off with time. Others believe that it would also be possible to develop a safe antisperm vaccine that would be used in women.

Morning-After Pills

Morning-after regimens are available that, depending on when the woman begins medication, either prevent fertilization altogether or stop the fertilized egg from ever implanting. These regimens involve taking pills containing synthetic progesterone and/or estrogen in a manner prescribed by a physician. Many women do not realize that this method of birth control is available, and yet use of these regimens could greatly reduce the number of unintended pregnancies. Effective treatment sometimes causes nausea and vomiting, which can be severe.

Mifepristone, better known as RU486, is a pill presently used in France and Great Britain that is now being considered for use in the United States. RU486 causes the loss of an implanted embryo by blocking the progesterone receptors of endometrial cells. Without functioning receptors for progesterone, the endometrium sloughs off, carrying the embryo with it. When taken in conjunction with a prostaglandin to induce uterine contractions, RU486 is 95% effective. Someday, the medication may be used by women who are experiencing delayed menstruation without knowing if they are actually pregnant.

> Numerous, well-known birth control methods and devices are available to people who want to prevent pregnancy, but their effectiveness varies. In addition, new methods are being developed.

Sexually Transmitted Diseases

Sexually transmitted diseases (STDs) are caused by organisms ranging from viruses to arthropods; however, we will discuss only certain STDs caused by viruses and bacteria. Unfortunately, for unknown reasons, humans cannot develop good immunity to any STDs. Therefore, prompt medical treatment is needed after exposure to an STD. Condoms help prevent the spread of STDs; the use of a spermicide containing nonoxynol 9 in conjunction with a condom gives added protection.

Curing the STDs caused by viruses (for example, AIDS, genital herpes, and genital warts) is difficult, but the symptoms can be treated. Those STDs caused by bacteria (for example, gonorrhea, chlamydia, and syphilis) are treatable with antibiotics.

AIDS

The virus that causes **acquired immunodeficiency syndrome (AIDS)** is called **human immunodeficiency virus (HIV)**. HIV attacks the type of lymphocyte known as helper T cells. Helper T cells, as you may recall from chapter 13, stimulate the activities of B lymphocytes, which produce antibodies. After an HIV infection sets in, helper T cells begin to decline in number, and the person becomes more susceptible to other types of infections.

AIDS has three stages of infection called category A, B, and C. During a category A stage, which may last a year, the individual is an asymptomatic carrier. There may be no symptoms, but the individual can pass on the infection. Immediately after infection, and before the blood test becomes positive, there is a large number of infectious viruses in the blood that could be passed on to another person. Even after the blood test becomes positive, the person remains well as long as the body produces sufficient helper T cells to keep the count higher than 500^3. During the category B stage, which may last six to eight years, the lymph nodes swell, and there may also be weight loss, night sweats, fatigue, fever, and diarrhea. Infections like thrush (white sores on the tongue and in the mouth) and herpes reoccur. Finally, the person may progress to category C which is AIDS, characterized by nervous disorders and by the development of an opportunistic disease, such as an unusual type of pneumonia or skin cancer. Opportunistic diseases occur only in individuals who have little or no capability of fighting an infection. Without intensive medical treatment, the AIDS patient dies about seven to nine years after infection. Now, with a combination therapy of several drugs, AIDS patients are beginning to live longer in the United States.

An HIV infection is transmitted by sexual contact including vaginal or rectal intercourse and oral/genital contact. Also, needle-sharing among intravenous drug users is high-risk behavior. HIV first spread through the homosexual community, and male-to-male sexual contact still accounts for the largest percentage of new AIDS cases in the United States. But the largest increases of new AIDS cases are occurring through heterosexual contact or by intravenous drug use. Now women account for 20% of all newly diagnosed cases of AIDS. The rise of the incidence among women of reproductive age is paralleled by a rise in the incidence of AIDS in children younger than 13. Babies can become infected before or during birth, or through breast feeding after birth.

Genital Herpes

Genital herpes is caused by herpes simplex virus (fig. 17.11). Type 1 usually causes cold sores and fever blisters, while type 2 more often causes genital herpes. Many times, a person infected with type 2 has no symptoms, but if symptoms are present, there are painful ulcers on the genitals that heal and then recur. The ulcers may be accompanied by fever, pain upon urination, and swollen lymph nodes. At this time, the individual has an increased risk of acquiring an AIDS infection. Exposure in the birth canal can cause an infection in the newborn, which leads to neurological disorders and even death. Birth by cesarean section prevents this possibility.

Genital Warts

Genital warts are caused by the human papillomaviruses (HPVs) (fig. 17.12). Many times, carriers do not have any

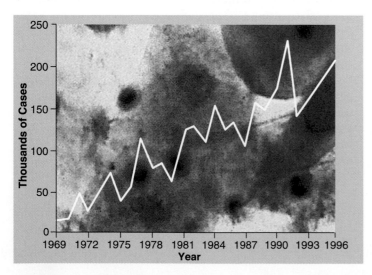

figure 17.11 Genital herpes. A graph depicting the incidence of new cases of genital herpes in the United States from 1969 to 1996 is superimposed on a photomicrograph of cells infected with the herpes virus.

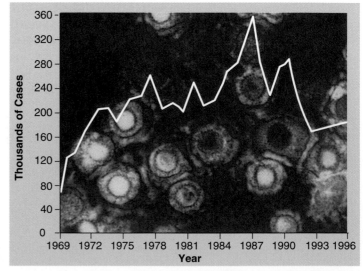

figure 17.12 Genital warts. A graph depicting the incidence of new cases of genital warts in the United States from 1969 to 1996 is superimposed on a photomicrograph of human papillomaviruses.

sign of warts, or merely flat lesions may be present. If visible warts are removed, they may recur. HPVs are now associated with cancer of the cervix, as well as tumors of the vulva, the vagina, the anus, and the penis. Some researchers believe that the viruses are involved in 90–95% of all cases of cancer of the cervix.

Gonorrhea

Gonorrhea (gah-nah-re'ah) is caused by the bacterium *Neisseria gonorrhoeae.* Diagnosis in the male is not difficult, since typical symptoms are pain upon urination and a thick, greenish yellow urethral discharge. In males and females, a latent infection leads to **pelvic inflammatory disease (PID),** in which the vasa deferentia or the uterine tubes are affected. As the inflamed tubes heal, they may become partially or completely blocked by scar tissue, resulting in sterility or infertility. If a baby is exposed during birth, an eye infection leading to blindness can result. All newborns are given eyedrops to prevent this possibility.

Chlamydia

Chlamydia (klah-mĭ'de-ah) is named for the tiny bacterium (*Chlamydia trachomatis*) that causes it. Chlamydia is the most common cause of **nongonococcal urethritis (NGU),** which is often difficult to distinguish from gonococcal urethritis. Since an infection can also cause PID, physicians routinely prescribe medicines for both gonorrhea and chlamydia at the same time. Chlamydia also causes cervical ulcerations, which increase the risk of acquiring AIDS. If a baby comes in contact with chlamydia during birth, inflammation of the eyes or pneumonia can result.

Syphilis

Syphilis (sĭ'ful-lis), which is caused by the bacterium *Treponema pallidum,* has three stages, which are typically separated by latent periods. In the primary stage, a hard chancre (ulcerated sore with hard edges) appears. In the secondary stage, a rash appears all over the body—even on the palms of the hands and the soles of the feet. During the tertiary stage, syphilis may affect the cardiovascular and/or nervous system. An infected person may become mentally retarded or blind, walk with a shuffle, or show signs of insanity. **Gummas,** which are destructive ulcers, may develop on the skin or within the internal organs. Syphilitic bacteria can cross the placenta, causing birth defects or a stillbirth. A blood test is available to diagnose syphilis.

Effects of Aging

Sex hormone levels decline with age in both men and women. **Menopause,** the period in a woman's life during which the ovarian and uterine cycles cease, is likely to occur between the ages of 45 and 55. The ovaries are no longer responsive to gonadotropic hormones, and they stop producing eggs and the female sex hormones. At the onset of menopause, the uterine cycle becomes irregular and then finally ceases. Hormonal imbalance often produces physical symptoms, such as "hot flashes" that are caused by circulatory irregularities, dizziness, headaches, insomnia, sleepiness, and depression. Menopausal symptoms vary greatly among women, and some symptoms may be absent altogether.

Following menopause, atrophy of the uterus, vagina, breasts, and external genitals is likely. The lack of estrogen also promotes the changes noted in previous chapters on the skin (for example, wrinkling), skeleton (for example, osteoporosis), and cardiovascular system (for example, increased incidence of heart attack).

In men, testosterone production diminishes steadily after age 50, which may be responsible for the enlargement of the prostate gland. Sperm production declines with age, yet men can remain fertile well into old age. The chance of impotency increases with age, which may be due to degenerative vascular changes in the penis.

Sexual desire and activity need not decline with age, and many older men and women enjoy sexual relationships. Men are likely to experience reduced erection until close to ejaculation, and women may experience a drier vagina.

Working Together

The accompanying illustration shows how the reproductive system works with other organ systems of the body to maintain homeostasis. All the organ systems of the body are interrelated.

BodyWorks CD-ROM
The module accompanying chapter 17 is Genitourinary System.

working together   Reproductive System

Integumentary System

Androgens activate oil glands; sex hormones stimulate fat deposition, affect hair distribution in males and females.

Skin receptors respond to touch; modified sweat glands produce milk; skin stretches to accommodate growing fetus.

How the Reproductive System works with other body systems

Circulatory System

Sex hormones influence cardiovascular health; sexual activities stimulate **cardiovascular** system.

Blood vessels transport sex hormones; vasodilation causes genitals to become erect; blood services the reproductive organs.

Skeletal System

Sex hormones influence bone growth and density in males and females.

Bones provide support and protection of reproductive organs.

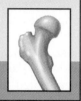

Lymphatic System/Immunity

Sex hormones influence immune functioning; acidity of vagina helps prevent pathogen invasion of body; milk passes antibodies to newborn.

Immune system does not attack sperm or fetus, even though they are foreign to the body.

Muscular System

Androgens promote growth of skeletal muscle.

Muscle contraction occurs during orgasm and moves gametes; abdominal and uterine muscle contractions occur during childbirth.

Respiratory System

Sexual activity increases breathing; pregnancy causes breathing rate and vital capacity to increase.

Gas exchange increases during sexual activity.

Nervous System

Sex hormones masculinize or feminize the brain, exert feedback control over the hypothalamus, and influence sexual behavior.

Brain controls onset of puberty; nerves are involved in erection of penis and clitoris, movement of gametes along ducts, and contraction of uterus.

Digestive System

Pregnancy crowds digestive organs and promotes heartburn and constipation.

Digestive tract provides nutrients for growth and repair of organs and for development of fetus.

Endocrine System

Gonads produce the sex hormones.

Hypothalamic, pituitary, and sex hormones control sex characteristics and regulate reproductive processes.

Urinary System

Penis in males contains the urethra and performs urination; prostate enlargement hinders urination.

Semen is discharged through the urethra in males; kidneys excrete wastes and maintain electrolyte levels for mother and child.

Basic Key Terms

clitoris (klĭ′to-ris), p. 347

corpus luteum (kor′pus lu′te-um), p. 345

ejaculation (e-jak″u-la′shun), p. 341

endometrium (en-do-me′tre-um), p. 346

epididymis (ep″ĭ-did′ĭ-mis), p. 339

erection (ĕ-rek′shun), p. 341

external genitals (eks-ter′nal jen′ĭ-talz), p. 338

fimbriae (fim′bre-a), p. 345

gamete (gam′ēt), p. 338

gonad (go′nad), p. 338

meiosis (mi-o′sis), p. 339

menopause (men′o-pawz), p. 357

menses (menstruation) (men′sēz) (men″stroo-a′shun), p. 349

oocyte (o′o-sīt), p. 343

oogenesis (o″o-jen′ĕ-sis), p. 343

ovarian cycle (o-va′re-an si′kl), p. 349

ovary (o′vah-re), p. 343

ovulation (o″vu-la′shun), p. 345

placenta (plah-sen′tah), p. 352

puberty (pu′ber-te), p. 338

scrotum (skro′tum), p. 338

seminal fluid (semen) (sem′ĭ-nal floo′id) (se′men), p. 339

seminiferous tubule (se″mĭ-nif′er-us tu′būl), p. 339

spermatogenesis (sper″mah-to-jen′ĕ-sis), p. 339

spermatozoa (sper″mah-to-zo′ah), p. 339

testis (tes′tis), p. 338

uterine cycle (u′ter-in si′kl), p. 349

vesicular (Graafian) follicle (graf′e-an fol′lĭ-kl), p. 343

vulva (vul′vah), p. 347

Clinical Key Terms

acquired immunodeficiency syndrome (AIDS) (ah-kwīrd ĭ-myu″no-dĭ-fĭ′shun-se sin′drōm), p. 356

adhesion (ad-he′zhun), p. 352

artificial insemination by a donor (AID) (ar″tĭ-fĭ′shul in-se-mah-na′shun), p. 353

chlamydia (klah-mĭ′de-ah), p. 357

circumcision (ser″kum-sizh′un), p. 341

cryptorchidism (krip-tor′kah-dizm), p. 338

ectopic pregnancy (ek-tah′pik preg′nun-se), p. 346

endometriosis (en-do-me″tre-o′sis), p. 351

gamete intrafallopian transfer (GIFT) (gam′ēt in″trah-fah-lo′pe-un tranz′fer), p. 353

genital herpes (jen′ĭ-tal her′pēz), p. 356

genital warts (jen′ĭ-tal worts), p. 356

gonorrhea (gah-nah-re′ah), p. 357

gumma (gŭ′mah), p. 357

hernia (her′ne-ah), p. 339

human immunodeficiency virus (HIV) (hu′mun ĭ-myu″no-dĭ-fĭ′shun-se vi′rus), p. 356

hysterectomy (his″ter-ek′to-me), p. 346

hysteroscopy (his-tah-rŏ′sko-pe), p. 355

impotence (im′pă-tents), p. 341

infertile (in-fer′tl), p. 352

intrauterine insemination (IUI) (in″trah-yu′tah-run in-sem-ĭ-na′shun), p. 353

in vitro fertilization (IVF) (in ve′tro fer″til-ĭ-za′shun), p. 353

laparoscope (lă′pah-rah-skōp), p. 351

nongonococcal urethritis (NGU) (non″gah-nah-kah′kul yur-ĭ-thri′tus), p. 357

oophorectomy (o-ah-fah-rek′tah-me), p. 346

ovarian cancer (o-va′re-an can′ser), p. 346

ovariohysterectomy (o-var″e-o-his-ter-ek′to-me), p. 346

Pap smear (pap smēr), p. 348

pelvic inflammatory disease (PID) (pel′vik in-flă′mah-tōr-e dĭ-zēz′), p. 357

prostate cancer (pros′tāt can′ser), p. 339

salpingectomy (sal-pin-jek′to-me), p. 346

sexually transmitted disease (STD) (sek′shu-ah-le tranz-mit′ted dĭ-zēz′), p. 355

sterile (ster′ul), p. 352

syphilis (sĭ′ful-lis), p. 357

tubal ligation (too′bul li-ga′shun), p. 355

vasectomy (vas-ek′tah-me), p. 355

Summary

I. Male Reproductive System

A. Testes. The testes contain the interstitial cells and seminiferous tubules, in which spermatogenesis occurs. Sperm, which are the reproductive cells of males, have an acrosome-capped head in which 23 chromosomes reside in a nucleus.

B. Genital tract. Sperm mature in the epididymis and are stored in the vas deferens before entering the urethra just prior to ejaculation. The accessory glands (seminal vesicles, prostate gland, and bulbourethral glands) produce seminal fluid. Semen, which contains sperm and seminal fluid, leaves the penis during ejaculation.

C. External genitals. The penis is the organ of sexual intercourse in males. The scrotum contains the testes.

D. Orgasm in males. In males, orgasm is a physical and emotional climax during sexual intercourse that results in ejaculation.

E. Regulation of male hormone levels. FSH promotes spermatogenesis in the seminiferous tubules, and LH promotes testosterone production by the interstitial cells. Testosterone stimulates growth of the male genitals during puberty and is necessary for sperm maturation and development of the male secondary sexual characteristics.

II. Female Reproductive System

A. Ovaries. In females, oogenesis occurs within the ovaries, where, typically, one follicle reaches maturity each month. This follicle balloons out of the ovary and bursts, releasing the egg. The ruptured follicle develops into a corpus luteum. The follicle and the corpus luteum produce the female sex hormones estrogen and progesterone.

B. Genital tract. The egg enters the uterine tubes, which lead to the uterus, where implantation and development occur. The vagina is the organ of sexual intercourse in females.

C. External genitals. The vagina opens into the vestibule, the location of female external genitals. The vestibule is bounded by the labia minora, which come together at the clitoris, a highly sensitive organ. Outside the labia minora are the labia majora.

D. Orgasm in females. In females, orgasm is not signaled by ejaculation, and normalcy of sexual response varies greatly.

E. Regulation of female hormone levels. Hormonal regulation in the female results in an ovarian cycle. During the first half of the cycle, FSH from the anterior pituitary causes maturation of the follicle, which secretes estrogen. After ovulation and during the second half of the cycle, LH from the anterior pituitary converts the follicle into the corpus luteum, which produces progesterone. Estrogen and progesterone regulate the uterine cycle. Estrogen causes the endometrium to rebuild. Ovulation usually occurs on the fourteenth day of a 28-day cycle. As progesterone is produced by the corpus luteum, the endometrium thickens and becomes secretory. Then, a low level of hormones causes the endometrium to break down, as menstruation occurs. Menstruation does not occur during pregnancy because of placental hormones.

III. Control of Reproduction and Sexually Transmitted Diseases

A. Infertility. Corrective medical and surgical procedures can help people who are infertile but wish to have a child. Alternative methods of reproduction may also be considered.

B. Birth control. Numerous, well-known birth control methods and devices are available to people who want to prevent pregnancy, but their effectiveness varies. In addition, new methods are being developed.

Study Questions

1. Discuss the anatomy and physiology of the testes. Describe the structure of sperm. (pp. 338–39)
2. Trace the path of sperm from their site of production to the site of fertilization. (p. 339)
3. What glands produce seminal fluid? (p. 339)
4. Discuss the anatomy and physiology of the penis. Describe ejaculation. (p. 341)
5. Discuss hormonal regulation in the male. Name three functions of testosterone. (p. 342)
6. Discuss the anatomy and physiology of the ovaries. Describe ovulation. (pp. 343–45)
7. Trace the path of the egg. Where do fertilization and implantation occur? Name two functions of the vagina. (pp. 345–46)
8. Describe the female external genitals. (p. 347)
9. Compare male and female orgasm. (pp. 341, 347)
10. Discuss hormonal regulation in the female, either simplified or detailed. Describe the events of the uterine cycle, and relate them to the ovarian cycle. In what way is menstruation prevented if pregnancy occurs? (pp. 347–50)
11. Name four functions of the female sex hormones. (p. 347)
12. Discuss the various means of birth control and their relative effectiveness. (pp. 354–55)

Fill in the blanks.

1. In tracing the path of sperm, the structure that follows the epididymis is the _____ .

2. The prostate gland, bulbourethral glands, and the _____ all contribute to seminal fluid.

3. An erection is caused by the entrance of _____ into sinuses within the penis.

4. The primary male sex hormone is _____ .

5. In the female reproductive system, the uterus lies between the uterine tubes and the _____ .

6. The female sex hormones are _____ and _____ .

7. In the ovarian cycle, once each month a _____ produces an egg. In the uterine cycle, the _____ is prepared to receive the zygote.

8. The most frequent causes of male infertility are _____ and _____ .

9. Spermatogenesis occurs within the _____ of the testes.

10. Androgens are secreted by the _____ cells that lie between seminiferous tubules.

11. Once a vesicular follicle has released an egg, it develops into a glandlike structure called a _____ .

12. The release of an egg from a vesicular follicle is called _____ .

13. Whereas AIDS and genital herpes are caused by _____ , gonorrhea and chlamydia are caused by _____ .

Medical Terminology Reinforcement Exercise

Consult Appendix B for help in pronouncing and analyzing the meaning of the terms that follow.

1. orchidopexy (or″kĭ-do-pek′se)
2. transurethral resection of prostate (TURP) (trans″u-re′thral re-sek′shun ov pros′tāt)
3. gonadotropic (gon″ah-do-trop′ik)
4. contraceptive (kon″trah-sep′tiv)
5. gynecomastia (jin″ĕ-ko-mas′te-ah)
6. hysterosalpingo-oophorectomy (his″ter-o-sal-pinj′go-o″of-o-rek′to-me)
7. endometriosis (en″do-me″tre-o′sis)
8. colporrhaphy (kol-por′ah-fe)
9. menometrorrhagia (men″o-met″ro-ra′je-ah)
10. multipara (mul-tip′ah-rah)
11. balanitis (bal″ah-ni′tis)

Website Link

For a listing of the most current Websites related to this chapter, please visit the Mader home page at:

www.mhhe.com/maderap

Human Development and Birth

Chapter Outline and Learning Objectives

After you have studied this chapter, you should be able to:

Fertilization (p. 363)

■ Explain the process of fertilization and the conversion of the egg into a zygote.

■ Explain the basis for a pregnancy test.

Development (p. 366)

■ Name the four extraembryonic membranes, and give a function for each.

■ Describe the structures and functions of the placenta and the umbilical cord.

■ Describe the events that occur during pre-embryonic and embryonic development.

■ Describe the events that occur during fetal development.

■ In general, describe the physical changes in the mother during pregnancy.

Birth (p. 373)

■ Describe the three stages of birth.

■ Describe the anatomy of the breast and the suckling reflex.

Medical Focus

Teratogens (p. 369)
Premature Babies (p. 374)

MedAlert

Detecting Birth Defects (p. 364)

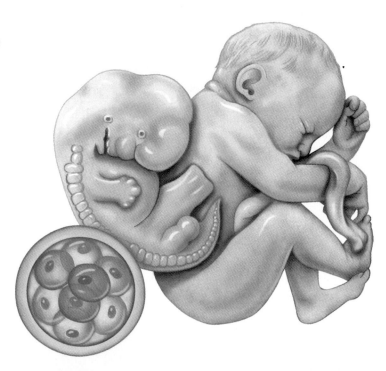

A human form appears as the pre-embryonic, embryonic, and fetal developmental stages occur.

figure 18.1 Stages in pre-embryonic development. Structures and events proceed clockwise. At ovulation (1), the oocyte (egg) leaves the ovary. A single sperm penetrates the egg, which then completes meiosis before fertilization (2) occurs in the uterine tube. As the zygote moves down the uterine tube, it undergoes cleavage (3) to produce a morula (4). The blastocyst (5) forms and implants itself in the endometrium (6).

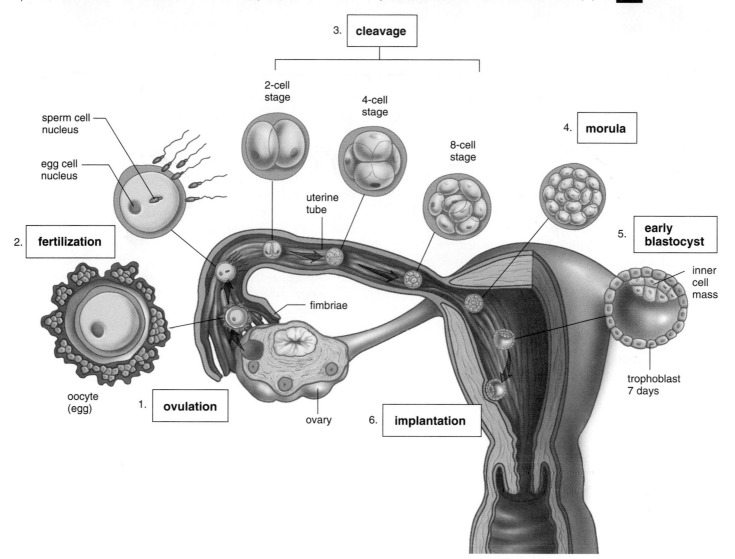

The male and female reproductive systems were discussed in chapter 17. The male continually produces sperm in the testes, and the female produces one secondary oocyte a month in the ovaries. For convenience sake, the secondary oocyte is often called an egg. Release of the oocyte, called **ovulation**, occurs on day 14 of an average ovarian cycle, which lasts 28 days. The hormones produced by the ovary control the uterine cycle, which consists of menstruation, during which the endometrium (uterine lining) breaks down; a proliferative phase, during which the endometrium rebuilds; and a secretory phase, during which the endometrium becomes secretory. Ovulation marks the end of the proliferative phase and the beginning of the secretory phase. Ovulation (fig. 18.1) precedes fertilization, development, and birth, the

topics discussed in this chapter. As discussed in the MedAlert reading on the next page, preovulatory eggs are sometimes tested for genetic defects and then used during in vitro fertilization.

Fertilization

Following ovulation, the egg lives for 12 to 24 hours. Most sperm usually live for 12 to 48 hours, although some have been known to live as long as 72 hours. Therefore, for **fertilization,** which is the fusion of egg and sperm, to occur, sexual intercourse must take place no earlier than 72 hours before ovulation and/or no later than 24 hours after ovulation.

Detecting Birth Defects

*t*hree methods for genetic defect testing before birth are amniocentesis, chorionic villi sampling, and obtaining eggs for screening (fig. 18A).

Amniocentesis (am"ne-o-sen-te'sis) is done during the fourteenth to sixteenth week of pregnancy because fetal cells (fibroblasts) are known to be present in amniotic fluid at this time. A long needle is passed through the abdominal wall in order to withdraw a small amount of amniotic fluid. Since only a few fetal cells are present in the fluid, testing must be delayed for four weeks until a cell culture produces enough cells for testing purposes. Chromosome analysis and biochemical tests for several different abnormalities are carried out on these cells.

Chorionic villi (ko"re-on'ik vil'i) **sampling** can be done as early as the fifth week of pregnancy. The doctor inserts a long, thin tube through the vagina into the uterus. With the help of **ultrasound,** which uses high-frequency sound waves to give a picture of the uterine contents, the tube is placed between the lining of the uterus and the chorion. Then, suction is used to remove a sampling of the chorionic villi cells. Chromosome analysis and biochemical tests to detect birth defects are carried out immediately on these cells.

Screening eggs for genetic defects is the newest testing technique. Preovulatory eggs are removed by aspiration after a laparoscope (a tiny tubular instrument with a lens and light source) is inserted into the abdominal cavity through a small incision in the region of the navel. Prior administration of the gonadotropic hormones FSH and LH ensures that several eggs will be available for screening. As a part of oogenesis, the eggs have already completed the first part of meiosis, the type of cell division that reduces the chromosome number. The unwanted chromosomes have been discarded in a polar body, and only the polar body chromosomes are tested for genetic defects. If a normal woman carries a hidden genetic defect and it is found in the polar body, then the egg must be normal. Only normal eggs then undergo in vitro fertilization and are placed in the prepared uterus. At present, only one in 10 attempts results in a birth, but an advantage is knowing ahead of time that the child will be normal.

Questions
1. A physician has withdrawn a small amount of blood from the chorionic villi. How would the physician determine if the fetus has sickle-cell disease?
2. Fluorescent antibodies are often used to test for the presence of a particular protein in fluids, such as amniotic fluid. With reference to figure 13.8, why would you expect an antibody-antigen reaction to be visible microscopically, even though both antibody and antigen are not individually visible?

After ovulation, the egg enters the uterine tube and begins to travel toward the uterus. Fertilization normally occurs in the upper third of the uterine tube. If intercourse is accompanied by ejaculation, some 200–600 million sperm are deposited at the rear of the vagina in the region of the cervix. Many of these sperm are killed by the acidic environment of the vagina, but perhaps a hundred thousand sperm manage to enter the uterus. This is especially likely on or about the day of ovulation because at that time, the cervical mucus is watery, and sperm can more easily pass through into the uterus.

Only several hundred sperm actually ever complete the journey through the female tract and reach the egg. Then, acrosomes (see fig. 17.2*d*) rupture and release enzymes that digest a pathway through the cells that surround the egg (the corona radiata). Only one sperm actually penetrates an elastic envelope (zona pellucida) and enters the egg. After this sperm has entered, the egg undergoes changes that prevent any more sperm from entering. When the sperm nucleus moves through the cytoplasm and fuses with the egg nucleus, fertilization is complete.

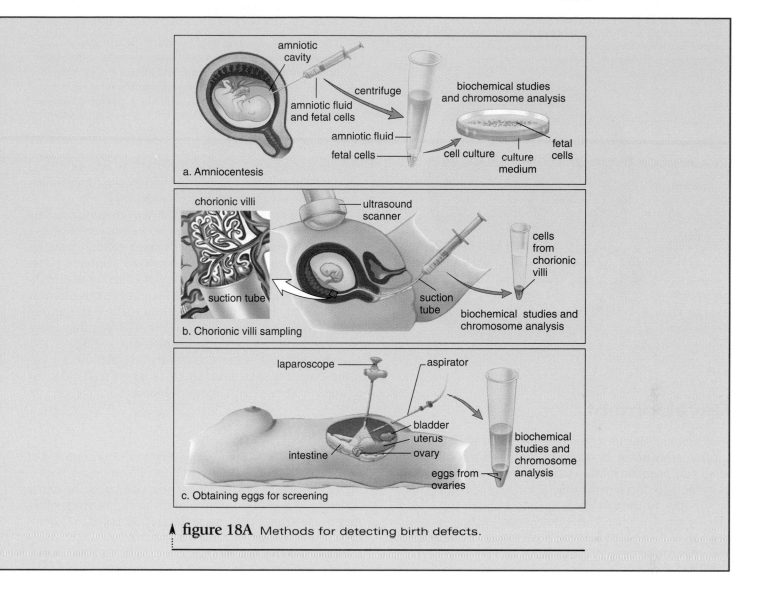

figure 18A Methods for detecting birth defects.

The fertilized egg, more properly called the **zygote,** begins to divide as it continues to travel very slowly down the uterine tube to the uterus, where it embeds itself in the prepared uterine lining or endometrium (see fig. 18.1). Once **implantation** of the developing mass of cells occurs, the female is pregnant. If implantation takes place, the endometrium is maintained because cells surrounding the embryo begin to produce a hormone called HCG (human chorionic gonadotropic hormone). This hormone prevents menstruation.

Physical signs that prompt a woman to have a pregnancy test include cessation of menstruation, increase in frequency of urination, morning sickness, and increase in the size and tenderness of the breasts. Another sign is the development of a dark coloration of the circular areas

surrounding the nipples of the breast, called the **areolae** (ah-re'o-le) (sing., *areola*).

Pregnancy tests are based on the presence of HCG in the blood and urine of a pregnant woman. Clinical urine tests are generally accurate 10 to 14 days after a missed menstruation. Blood tests are more expensive but can indicate a pregnancy earlier, often before the time of a missed menstruation. Home pregnancy tests measure HCG in the urine and are generally quite reliable.

During fertilization, a sperm nucleus fuses with the egg nucleus. The resulting zygote begins to develop into a mass of cells, which travels down the uterine tube and embeds itself in the endometrium. Cells surrounding the embryo produce HCG, the hormone whose presence indicates that the female is pregnant.

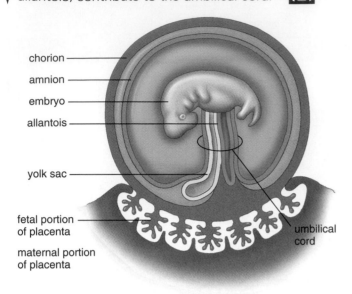

figure 18.2 The extraembryonic membranes. The chorion and amnion surround the embryo. The two other extraembryonic membranes, the yolk sac and allantois, contribute to the umbilical cord.

chorion

amnion

embryo

allantois

yolk sac

fetal portion of placenta

maternal portion of placenta

umbilical cord

Development

Development encompasses the time from fertilization to birth (parturition). In humans, the **gestation** period, or length of pregnancy, is approximately nine months. It is customary to calculate the time of birth by adding 280 days to the start of the last menstruation because this date is usually known, whereas the day of fertilization is usually unknown. Because the time of birth is influenced by many variables, only about 5% of babies actually arrive on the predicted date.

Human development can be divided into **embryonic development** (the second week through eighth week) and **fetal development** (the third through ninth month). During embryonic development, all the major organs form, and during fetal development, these structures are refined. The fetus, not the embryo, is recognizable as a human being. One of the major events in early development is the establishment of the extraembryonic membranes (fig. 18.2).

Extraembryonic Membranes

The term **extraembryonic membranes** is apt, because these membranes extend out beyond the embryo. One of the membranes, the **amnion,** provides a fluid environment for the developing embryo and fetus. It is a remarkable fact that all animals, even land-dwelling humans, develop in water. A. F. Guttmacher, in his book *Pregnancy, Birth and*

Family Planning (New York: New American Library, 1974), describes the functions of amniotic fluid in this way:

> The colorless amniotic fluid by which the fetus is surrounded serves many purposes. It prevents the walls of the uterus from cramping the fetus and allows it unhampered growth and movement. It encompasses the fetus with a fluid of constant temperature, which is a marvelous insulator against cold and heat. Above all, it acts as an excellent shock absorber. A blow on the mother's abdomen merely jolts the fetus, and it floats away.

Amniocentesis is a process by which amniotic fluid and the fetal cells floating in it are withdrawn for examination (see the MedAlert reading on page 364).

The **yolk sac** is another extraembryonic membrane. Yolk is a nutrient material utilized by other animal embryos—for example, the yellow of a chick's egg is yolk. In humans, the yolk sac is the first site of red blood cell formation. Part of this membrane becomes incorporated into the umbilical cord. Another extraembryonic membrane, the **allantois** (ah-lan'to-is), contributes to the circulatory system: Its blood vessels become umbilical blood vessels that transport fetal blood to and from the placenta (see fig. 12.15). The **chorion,** the outer extraembryonic membrane, becomes part of the **placenta** (fig. 18.3), where fetal blood exchanges molecules with maternal blood.

Placenta

The placenta has two portions, a fetal portion composed of chorionic tissue and a maternal portion composed of uterine tissue (see fig. 18.3). **Chorionic villi** cover the entire surface of the chorion until about the eighth week when they begin to disappear, except in the area beneath the embryo. These villi are surrounded by maternal blood, and it is here that exchanges of materials take place across the placental membrane. The *placental membrane* consists of the epithelial wall of an embryonic capillary and the epithelial wall of a chorionic villus. Maternal blood never mingles with fetal blood. Instead, oxygen and nutrient molecules, such as glucose and amino acids, diffuse from maternal blood across the placental membrane into fetal blood, and carbon dioxide and other wastes, such as urea, diffuse out of fetal blood into maternal blood.

Note that the digestive system, lungs, and kidneys do not function in the fetus. The functions of these organs are not needed because the placenta supplies the fetus with its nutritional and excretory needs.

The **umbilical cord** transports fetal blood to and from the placenta (see fig. 18.3). The umbilical cord is the fetal lifeline because it contains the umbilical arteries and vein, which transport waste molecules (carbon dioxide and urea) to the placenta for disposal and oxygen and nutrient molecules from the placenta to the rest of the fetal circulatory system.

figure 18.3 The placenta. Blood vessels within the umbilical cord lead to the placenta, which has a fetal portion and a maternal portion.

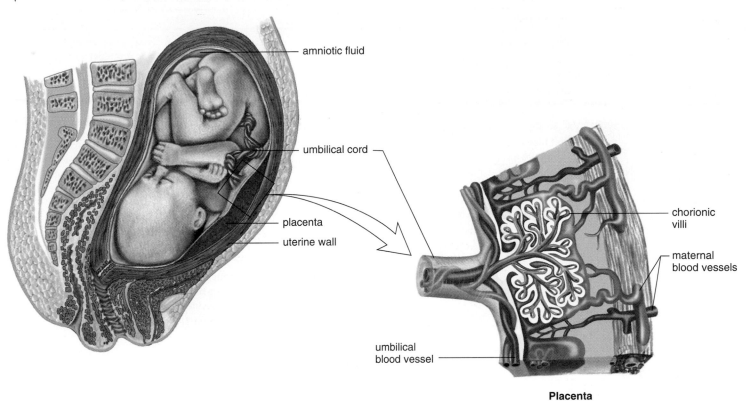

Placenta

The chorion and then the placenta produce HCG, the hormone detected by a pregnancy test. HCG prevents the normal degeneration of the corpus luteum and, instead, stimulates it to secrete even larger quantities of progesterone. Later, the placenta begins to produce progesterone and estrogen, and the corpus luteum degenerates—it is no longer needed. Placental estrogen and progesterone maintain the endometrium and have a negative feedback effect on the anterior pituitary so that it ceases to produce gonadotropic hormones during pregnancy. The ovarian cycle (see table 17.3) and menstruation do not occur during the length of pregnancy.

The extraembryonic membranes, placenta, and umbilical cord allow humans to develop internally within the uterus. These structures protect the embryo and allow it to exchange waste for nutrients with the mother's blood.

Pre-Embryonic and Embryonic Development

Development includes the following processes:

Cleavage Immediately after fertilization, the zygote begins to divide so that at first there are 2, then 4, 8, 16, and 32 cells, and so forth. Increase in size does not accompany these divisions (see fig. 18.1). Cell division during cleavage is mitotic, and each cell receives a full complement of chromosomes and genes.

Growth During embryonic development, cell division is accompanied by an increase in size of the daughter cells.

Morphogenesis Morphogenesis refers to the shaping of the embryo and is first evident when certain cells are seen to move, or migrate, in relation to other cells. By these movements, the embryo begins to assume various shapes.

Differentiation When cells take on a specific structure and function, differentiation occurs. The first system to become visibly differentiated is the nervous system.

Table 18.1 outlines the changes that occur during development. It also includes the changes that occur in the mother.

Pre-Embryonic Development

Immediately after fertilization, the zygote divides repeatedly as it passes down the uterine tube to the uterus. A **morula**

table 18.1

Human Development

Time	Events for Mother	Events for Baby
Pre-embryonic Development		
First week	Ovulation occurs.	Fertilization occurs. Cell division begins and continues. Morula becomes the blastocyst, and the chorion appears.
Embryonic Development		
Second week	Symptoms of early pregnancy (nausea, breast swelling and tenderness, tiredness) are present.	Implantation occurs. Amnion and yolk sac appear. Embryo has tissues. Placenta begins to form.
Third week	First menstruation is missed. Blood pregnancy test is positive.	Nervous system begins. Allantois and blood vessels are present. Placenta is well formed.
Fourth week	Urine pregnancy test is positive.	Limb buds begin. Heart is noticeable and beating. Nervous system is prominent. Embryo has tail. Other systems begin.
Fifth week	Uterus is the size of a hen's egg. Frequent need to urinate is due to pressure of growing uterus on bladder.	Embryo is curved. Head is large. Limb buds show divisions. Nose, eyes, and ears are noticeable.
Sixth week	Uterus is growing to the size of an orange.	Fingers and toes are present. Skeleton is cartilaginous.
Second month	Uterus can be felt above the pubic bone.	All systems are developing. Bone is replacing cartilage. Facial features are becoming refined. Embryo is about 1 1/2 inches (38 mm) long.
Fetal Development		
Third month	Uterus is the size of a grapefruit.	Sex of fetus is possible to distinguish. Fingernails develop.
Fourth month	Fetal movement is felt by women who have been pregnant before.	Skeleton is visible. Hair begins to appear.
Fifth month	Fetal movement is felt by women who have not been pregnant before. Uterus reaches up to level of umbilicus, and pregnancy is obvious.	Protective cheesy coating begins to be deposited. Heartbeat can be heard. Fetus is about 6 in. (150 mm) long and weighs about 6 oz. (200 g).
Sixth month	Doctor can tell where baby's head, back, and limbs are. Breasts have enlarged, nipples and areolae are darkly pigmented, and colostrum is produced.	Body is covered with fine hair. Skin is wrinkled and red.
Seventh month	Uterus reaches halfway between umbilicus and rib cage.	Testes descend into scrotum. Eyes are open. Fetus is about 12 in. (300 mm) long and weighs about 3 lbs. (1,350 g).
Eighth month	Weight gain is averaging about a pound a week. Standing and walking are difficult because center of gravity is thrown forward.	Body hair begins to disappear. Subcutaneous fat begins to be deposited.
Ninth month	Uterus is up to rib cage, causing shortness of breath and heartburn. Sleeping becomes difficult.	Fetus is ready for birth. It is about 21 in. (530 mm) long and weighs about 7 1/2 lbs. (3,885 g).

(mor'u-lah) is a compact ball of embryonic cells that becomes a **blastocyst.** The many cells of the blastocyst arrange themselves so that there is an inner cell mass surrounded by a layer of cells, the trophoblast (see fig. 18.1), which becomes the chorion. The early appearance of the chorion emphasizes the complete dependence of the developing embryo on this extraembryonic membrane. The inner cell mass is the **embryo.**

Each cell within the morula and blastocyst has the genetic capability of becoming a complete individual. Sometimes, these cells separate, or the inner cell mass splits, and two embryos start developing rather than one. These two embryos will be *identical twins* because they have inherited exactly the same chromosomes. *Fraternal twins,* who arise when two different eggs are fertilized by two different sperm, do not have identical chromosomes. It has even been known to happen that these "twins" have different fathers.

Embryonic Development

Second Week   Once it has arrived in the uterus on or about the seventh day, the blastocyst begins to implant

Teratogens

Some **congenital** (birth) **defects** are genetic in origin. Other congenital defects or illnesses are environmental in nature and are caused by substances that have crossed the placenta from the mother to the fetus. Agents that produce abnormalities during development are called **teratogens** (ter-ah'to-jenz).

Medications may pass into fetal blood at the placenta. The harmful developmental effects that can result from medications were evident in the thalidomide babies born in the 1960s (fig. 18B). An estimated 10,000 children were born with deformed arms and, sometimes, deformed legs because their mothers took the tranquilizer thalidomide during the first three months of pregnancy.

Babies of drug addicts and alcoholics commonly display withdrawal symptoms and have various abnormalities. Babies born to women who have about 45 drinks a month and as few as three drinks on one occasion are apt to have **fetal alcohol syndrome (FAS).** These babies have decreased weight, height, and head size, with malformation of the head and face. "Cocaine babies" now make up 60% of drug-affected babies. Cocaine babies have visual problems, lack coordination, and are mentally retarded.

Industrial chemicals in the environment can be harmful to the fetus. Some that have been implicated are chloro-biphenyls, vinyl chloride, formaldehyde, asbestos, and ben-zene. Cigarette smoke contains teratogens, and smokers typically have underweight babies. Also, pregnant women should avoid X-ray diagnosis because dividing cells in the embryo and fetus are particularly susceptible to damage by X ray.

In addition to drugs and other chemicals, infectious agents can pass from the expectant mother's blood into her baby's blood. For this reason, babies can be born with AIDS, toxoplasmosis, Lyme disease, and syphilis. A rubella virus (German measles) infection during the first three months of pregnancy can cause congenital malformations, particularly of the heart, eyes, and ears.

The embryo is more susceptible than the fetus to damage by drugs and disease. Unfortunately, the embryonic period is also the time when women most likely do not realize that they are pregnant. Therefore, especially when birth control is not being practiced, prospective mothers should carefully monitor their health and intake of drugs.

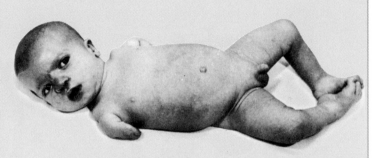

figure 18B Birth defects from thalidomide. If a pregnant woman took thalidomide during the embryo's period of limb formation, the infant was sometimes born with short, flipperlike arms and/or legs.

itself in the endometrium (see fig. 18.1). Implantation is completed, and embryonic development begins during the second week. With implantation, pregnancy has now taken place, and the placenta begins to form. The ever-growing number of cells are now arranged in tissues. The amniotic cavity is seen above the embryo, and the yolk sac is below (fig. 18.4a).

Third Week Another extraembryonic membrane, the allantois, makes a brief appearance, but later, it and the yolk sac become part of the developing umbilical cord (fig. 18.4c and d). Organs, including the spinal cord and heart, are already developed. The nervous system and circulatory system have begun to develop.

Fourth Week By the end of the first month, the chorionic villi project into the uterine wall, and the placenta is producing enough HCG to maintain the corpus luteum and the endometrium. Although the chorionic villi enlarge at one location only, they may eventually take up as much as 50% of the uterus (see fig. 18.3). Substances that sometimes cross the placenta can cause congenital defects, as discussed in the Medical Focus reading on this page.

The embryo has a nonhuman appearance, largely due to the presence of a tail, but also because the arms and legs, which begin as limb buds, resemble paddles. The head is much larger than the rest of the embryo, and the whole embryo bends under its weight (fig. 18.5). The eyes, ears,

figure 18.4 Formation of the extraembryonic membranes and the umbilical cord. **a.** At 14 days, the amniotic cavity appears. **b.** At 18 days, the chorion and the yolk sac are apparent. **c.** At 21 days, the body stalk and the allantois form. **d.** At 25 days, the embryo begins to take shape as the umbilical cord forms. **e.** Eventually, the umbilical cord is fully formed.

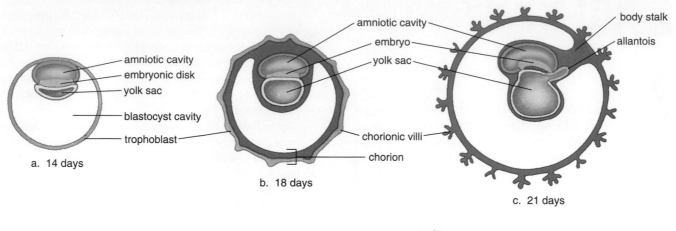

a. 14 days

amniotic cavity
embryonic disk
yolk sac
blastocyst cavity
trophoblast

b. 18 days

amniotic cavity
embryo
yolk sac
chorionic villi
chorion

c. 21 days

body stalk
allantois

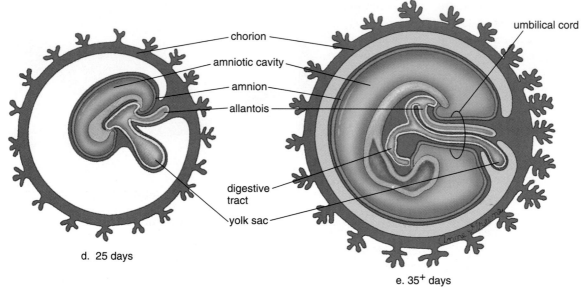

d. 25 days

chorion
amniotic cavity
amnion
allantois
digestive tract
yolk sac

e. 35⁺ days

umbilical cord

and nose are just appearing. The enlarged heart beats, and the bulging liver takes over the production of blood cells for blood that will carry nutrients and wastes to and from the developing organs.

Second Month At the end of two months, the embryo's tail has disappeared, and the arms and legs are more developed, with fingers and toes apparent (fig. 18.6). The head is very large, the nose is flat, the eyes are far apart, and the ears are distinctively present. Internally, all major organs have appeared. Embryonic development is now complete.

At the end of the embryonic period, all organ systems are established, and there is a mature and functioning placenta. The embryo is only about 1 1/2 inches (38 millimeters) long.

Fetal Development

Development of the **fetus** extends from the third through the ninth months.

Third and Fourth Months

At the beginning of the third month (fig. 18.7), head growth begins to slow down as the rest of the body increases in length. Epidermal refinements, such as eyelashes, eyebrows, hair on the head, fingernails, and nipples, appear.

Cartilage is replaced by bone as ossification centers appear in most of the bones. Cartilage remains at the ends of the long bones, and ossification is not complete until the age of 18 or 20 years. The skull has six large *fontanels* (membranous areas) that permit a certain amount of flexi-

figure 18.5 Human embryo at beginning of fifth week. **a.** Scanning electron micrograph. **b.** The embryo is curled so that the head touches the heart, the two organs whose development is further along than the rest of the body. The organs of the gastrointestinal tract are forming, and the arms and the legs develop from the bulges that are called limb buds.

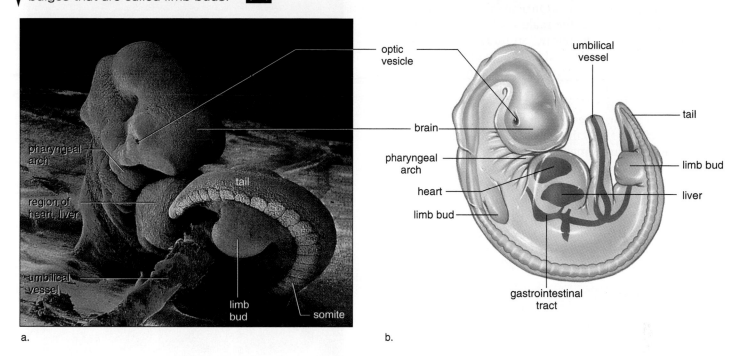

a.

b.

figure 18.6 Eight-week-old fetus. The embryonic period ends, and the fetus begins to develop a more human appearance.

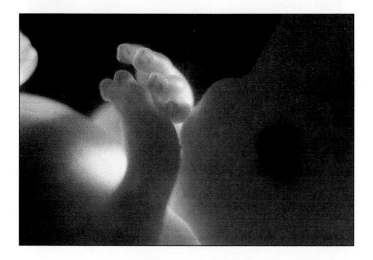

figure 18.7 Three- to four-month-old fetus looks human. Face, hands, and fingers are well defined.

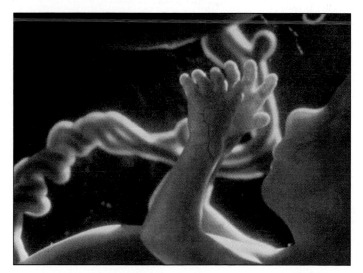

bility as the head passes through the birth canal and allow rapid growth of the brain during infancy. The fontanels disappear by two years of age.

Sometime during the third month, it is possible to distinguish males from females. Once the testes differentiate, they produce androgens, the male sex hormones. The androgens, especially testosterone, stimulate the growth of the male external genitals. In the absence of androgens, female genitals form. The ovaries do not produce estrogen because there is plenty of it circulating in the mother's bloodstream.

At this time, the testes or ovaries are located within the abdominal cavity. Later, the testes descend into the scrotal sacs of the scrotum. Sometimes, the testes fail to descend, in which case an operation can be performed to place them in their proper location.

During the fourth month, the fetal heartbeat is loud enough to be heard when a physician applies a stethoscope to the mother's abdomen. By the end of this month, the fetus is less than 6 inches (150 millimeters) in length and weighs a little more than 6 ounces (200 grams).

> Fetal development extends from the third through the ninth months. During the third and fourth months, the skeleton is becoming ossified. The sex of the fetus becomes distinguishable.

Fifth through Seventh Months

During the fifth through seventh months (fig. 18.8), the mother begins to feel movement. At first, there is only a fluttering sensation, but as the fetal legs grow and develop, kicks and jabs are felt. The fetus, though, is in the fetal position, with the head bent down and in contact with the flexed knees.

The wrinkled, translucent, pink-colored skin is covered by a fine down called **lanugo** (lah-nu′go). The lanugo is coated with a white, greasy, cheeselike substance called **vernix caseosa** (ver′niks ka″se-os′ah), which probably protects the delicate skin from the amniotic fluid. During these months, the eyelids open fully.

At the end of this period, the fetus is almost 12 inches (300 millimeters) long, and weight has increased to almost 3 pounds (1,350 grams). If born now, the baby would be considered premature and might not survive for the reasons listed in the Medical Focus reading on page 374.

Eighth and Ninth Months

As the time of birth approaches (fig. 18.9), the fetus usually rotates so that the head is pointed toward the cervix. However, if the fetus does not turn, then the likelihood of a **breech birth** (rump first) may necessitate a **cesarean sec-**

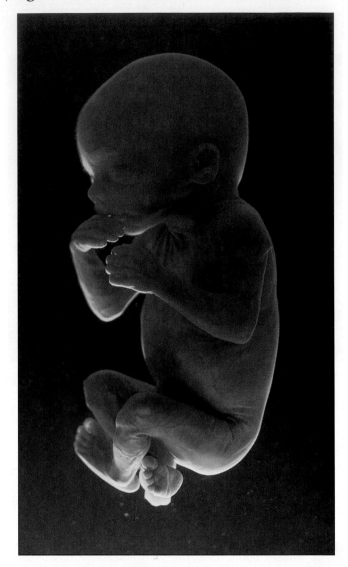

▼ **figure 18.8** Five- to seven-month-old fetus.

tion. It is very difficult for the cervix to expand enough to accommodate a breech birth, and asphyxiation of the baby is more likely.

At the end of nine months, the fetus is about 21 inches (530 millimeters) long and weighs about 7½ pounds (3,885 grams). Weight gain is due largely to an accumulation of fat beneath the skin. Full-term babies have a better chance of survival.

> During the fifth through ninth months, the fetus continues to grow and to gain weight. Babies born after six or seven months may survive, but full-term babies have a better chance of survival.

figure 18.9 Eight- to nine-month-old fetus.

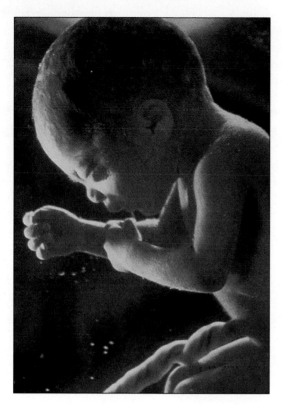

table 18.2	Approximate Weight Gain during Pregnancy	
	Full-term baby	7.7 lb
	Placenta	1.4 lb
	Amniotic fluid	1.8 lb
	Enlarged uterus	2.0 lb
	Enlarged breasts	0.9 lb
	Blood volume increase	4.0 lb
	Increased fluid retention	2.7 lb
	Maternal fat storage	3.5 lb
		24.0 lb

Source: H. S. Mitchel et al., *Nutrition in Health and Disease,* 16th edition. Copyright © 1976 J. B. Lippincott Company, Philadelphia, PA.

Effects of Pregnancy on the Mother

Table 18.1 also outlines the major changes in the mother during pregnancy. When first pregnant, the mother may experience nausea and vomiting, loss of appetite, and fatigue. Other changes that indicate pregnancy are swelling and tenderness of the breasts, increased urination, and irregular bowel movements. Some women, however, report increased energy levels and a general sense of well-being during this time.

The uterus enlarges greatly during pregnancy. In the nonpregnant female, the uterus is only about 2 inches (50 millimeters) by 3 inches (75 millimeters). Just prior to birth, the uterus almost fills the abdominal cavity and reaches the rib cage. Breasts increase as much as 25% in size. By the end of the sixth month, the nipples and areolae are darkly pigmented, and *colostrum,* a thin fluid secreted by the breasts, is produced.

Toward the end of pregnancy, the baby's enlarged size and extra weight cause various difficulties. (Table 18.2 gives the approximate weight gain to be expected during pregnancy.) The mother may have trouble breathing and may have an increased need to urinate. Her center of gravity is thrown forward; therefore, standing and walking are difficult. Sleeping may be disturbed not only due to the kicking of the baby but also because of an inability to find a comfortable position.

Birth

The uterus characteristically contracts throughout pregnancy. At first, light contractions, which last 20–30 seconds and occur every 15–20 minutes, are often unnoticed. Near the end of pregnancy, the contractions become stronger and more frequent so that the woman may think she is in labor. These "false-labor" contractions are called **Brackston-Hicks contractions.** However, the onset of true labor is marked by uterine contractions that occur regularly every 15–20 minutes and last for 40 seconds or more.

Parturition, which includes labor and expulsion of the fetus, is usually considered to have three stages. The events that cause parturition are still not known entirely, but evidence now suggests the involvement of prostaglandins. It may be, too, that the prostaglandins cause the release of oxytocin from the maternal posterior pituitary. Both prostaglandins and oxytocin cause the uterus to contract, and either hormone can be given to induce parturition.

Stages of Birth

During the first stage of parturition, the cervix dilates; during the second stage, the baby is born; and during the third stage, the afterbirth is expelled.

Premature Babies

*a*bout 7% of all newborns in the United States weigh less than 5½ pounds. Most of these are **premature babies** who face the following difficulties:

Respiratory distress syndrome (hyaline membrane disease) The lungs do not produce enough of a chemical surfactant that helps the alveoli stay open. Therefore, the lungs tend to collapse, instead of expanding to be filled with air.

Retinopathy of prematurity The high level of oxygen needed to ensure adequate gas exchange by the immature lungs can lead to proliferation of blood vessels within the eyes, with ensuing blindness.

Intracranial hemorrhage The delicate blood vessels in the brain are apt to break, causing swelling and inflammation of the brain. If not fatal, this can lead to brain damage.

Jaundice The immature liver fails to excrete the waste product bilirubin. Instead, it builds up in the blood, possibly causing brain damage.

Infections The level of antibodies in the body is low, and the various medical procedures performed could possibly introduce germs. Also, a bowel infection is common, along with perforation, bleeding, and shock.

Circulatory disorders Fetal circulation, discussed in chapter 12, has two features: the oval opening between the atria and the arterial duct that allows blood to bypass the lungs. If these features persist in the newborn, oxygenated blood will mix with deoxygenated blood, and blood circulation will be impaired, perhaps leading to the delivery of a "blue baby"—that is, a baby with cyanosis, a bluish cast to the skin. Heart failure can also result from these conditions.

The reasons for premature birth have been investigated, and researchers have concluded that prenatal care, including good nutrition and the willingness to refrain from excessive drinking of alcohol and smoking cigarettes, could reduce the incidence of premature birth and/or low birth weight.

Stage 1

Prior to or at the beginning of the first stage of parturition, there can be a "bloody show" caused by the expulsion of a mucous plug from the cervical canal. This plug prevents bacteria and sperm from entering the uterus during pregnancy.

Uterine contractions during the first stage of parturition occur in such a way that the cervical canal slowly disappears (fig. 18.10b) as the lower part of the uterus is pulled upward toward the baby's head. This process is called *effacement*, or "taking up the cervix." With further contractions, the baby's head acts as a wedge to assist cervical dilation. The baby's head usually has a diameter of about 4 inches (100 millimeters); therefore, the cervix has to dilate to this diameter in order to allow the head to pass through. If it has not occurred already, the amniotic sac is apt to rupture during this stage, releasing the amniotic fluid, which leaks out the vagina. The first stage of parturition ends once the cervix is dilated completely.

Stage 2

During the second stage of parturition, the uterine contractions occur every 1–2 minutes and last about 1 minute each. They are accompanied by a desire to push, or bear down. As the baby's head gradually descends into the vagina, the desire to push becomes greater. The baby's head turns so that the back of the head is uppermost (fig. 18.10c). Since the vaginal orifice may not expand enough to allow passage of the head without tearing, an **episiotomy** often is performed. This incision, which enlarges the opening, is sewn together later and heals more perfectly than a tear. As soon as the head is delivered, the baby's shoulders rotate so that the baby faces either to the right or the left. At this time, the physician may hold the head and guide it downward, while one shoulder and then the other emerges. The rest of the baby follows easily.

Once the baby is breathing normally, the umbilical cord is cut and tied, severing the child from the placenta. The stump of the cord shrivels and leaves a scar, which is the **navel.**

figure 18.10 Three stages of parturition. **a.** Position of fetus just before birth begins. **b.** Dilation of cervix. **c.** Birth of baby. **d.** Expulsion of afterbirth.

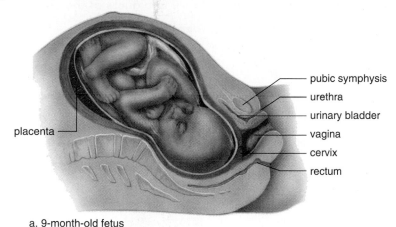

pubic symphysis
urethra
urinary bladder
vagina
cervix
rectum

placenta

a. 9-month-old fetus

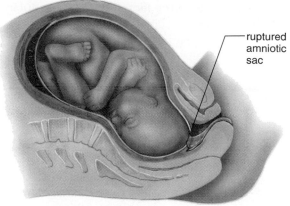

ruptured amniotic sac

b. First stage of birth: cervix dilates

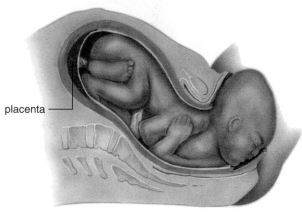

placenta

c. Second stage of birth: baby emerges

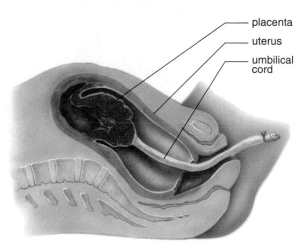

placenta
uterus
umbilical cord

d. Third stage of birth: expelling afterbirth

Stage 3

The placenta, or *afterbirth*, is delivered during the third stage of parturition (fig. 18.10*d*). About 15 minutes after delivery of the baby, uterine muscular contractions shrink the uterus and dislodge the placenta. The placenta then is expelled into the vagina. As soon as the placenta and its membranes are delivered, the third stage of parturition is complete.

During stage 1 of parturition, the cervix dilates. During stage 2, the child is born. During stage 3, the afterbirth is expelled.

Female Breast and Lactation

A female breast contains 15 to 25 lobules, each with its own milk duct, which begins at the nipple and divides into numerous other ducts that end in blind sacs called *alveoli* (fig. 18.11).

During pregnancy, the breasts enlarge as the ducts and alveoli increase in number and size. The same hormones that affect the mother's breasts can also affect those of the child. Some newborns, including males, even secrete a small amount of milk for a few days.

figure 18.11 Structure of the breast, anterior view. The female breast contains lobules consisting of ducts and alveoli. The alveoli are lined by milk-producing cells in the lactating (milk-producing) breast.

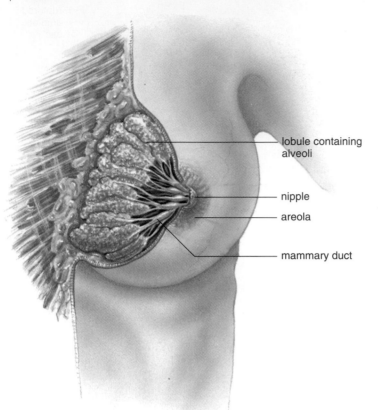

lobule containing alveoli

nipple

areola

mammary duct

figure 18.12 Suckling reflex. Suckling sets in motion the sequence of events that lead to milk letdown, the flow of milk into ducts of the breast.

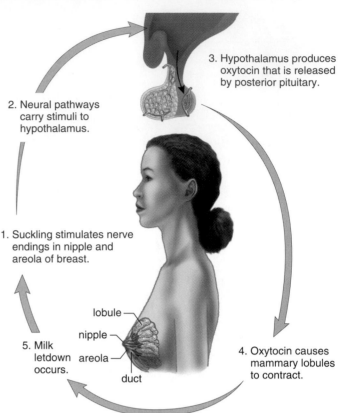

3. Hypothalamus produces oxytocin that is released by posterior pituitary.

2. Neural pathways carry stimuli to hypothalamus.

1. Suckling stimulates nerve endings in nipple and areola of breast.

lobule

nipple

5. Milk letdown occurs.

areola

duct

4. Oxytocin causes mammary lobules to contract.

Usually, there is no milk production (**lactation**) during pregnancy. The hormone **prolactin** is needed for lactation to begin, and production of this hormone is suppressed because of the feedback control that the increased amounts of estrogen and progesterone during pregnancy have on the pituitary. Once the baby is delivered, however, the pituitary begins secreting prolactin. It takes a couple of days for milk production to begin; in the meantime, the breasts produce *colostrum*, which is a thick secretion rich in protein and antibodies.

Continued lactation requires a suckling child. When a breast is suckled, the nerve endings in the areola are stimulated, and a nerve impulse travels along neural pathways from the nipple to the hypothalamus, which directs the pituitary gland to release the hormone oxytocin. When this hormone arrives at the breast, it causes contraction of the lobules so that milk flows into the ducts (called milk letdown), where it may be drawn out of the nipple by the suckling child. The more suckling, the more oxytocin released, and the more milk there is for the child (fig. 18.12).

In addition to initiating milk letdown, oxytocin also causes the uterus to contract, which helps it return to its nonpregnant state more quickly.

Lactation requires the hormone prolactin, and the suckling reflex involves the hormone oxytocin.

Basic Key Terms

allantois (ah-lan'to-is), p. 366
amnion (am'ne-on), p. 366
areola (ah-re'o-lah), p. 365
blastocyst (blas'to-sist), p. 368
chorion (ko're-on), p. 366
embryo (em'bre-o), p. 368
embryonic development (em"bre-on'ik de-vel'op-ment), p. 366
extraembryonic membrane (eks"trah-em"bre-on'ik mem'brān), p. 366
fertilization (fer'tĭ-lĭ-za'shun), p. 363
fetal development (fe'tal de-vel'op-ment), p. 366
fetus (fe'tus), p. 370
gestation (jes-ta'shun), p. 366
implantation (im"plan-ta'shun), p. 365
lactation (lak-ta'shun), p. 376
lanugo (lah-nu'go), p. 372
morula (mor'u-lah), p. 367
navel (na'vel), p. 374
ovulation (o"vu-la'shun), p. 363
parturition (par"tu-rish'un), p. 373

placenta (plah-sen'tah), p. 366
prolactin (pro-lak'tin), p. 376
umbilical cord (um-bil'ĭ-kal kord), p. 366
vernix caseosa (ver'niks ka"se-os'ah), p. 372
yolk sac (yōk sak), p. 366
zygote (zi'gōt), p. 365

Clinical Key Terms

amniocentesis (am"ne-o-sen-te'sis), p. 364
Brackston-Hicks contraction (brax'ton-hiks'con-trak'shun), p. 373
breech birth (brēch berth), p. 372
cesarean section (sĭ-zar'e-an sek'shun), p. 372
chorionic villi sampling (ko"re-on'ik vil'i sam'pling), p. 364
congenital defects (kon-jen'ĭ-tal de'feks), p. 369
episiotomy (e-piz"e-ot'o-me), p. 374
fetal alcohol syndrome (FAS) (fe'tal al'cah-hol sin'drōm), p. 369
premature baby (pre-mah-tyur' ba'be), p. 374
teratogen (ter-ah'to-jen), p. 369
ultrasound (ul'trah-sound), p. 364

Summary

I. Fertilization

During fertilization, a sperm nucleus fuses with the egg nucleus. The resulting zygote begins to develop into a mass of cells, which travels down the uterine tube and embeds itself in the endometrium. Cells surrounding the embryo produce HCG, the hormone whose presence indicates that the female is pregnant.

II. Development

A. The extraembryonic membranes, placenta, and umbilical cord allow humans to develop internally within the uterus. These structures protect the embryo and allow it to exchange waste for nutrients with the mother's blood.

B. At the end of the embryonic period, all organ systems are established, and there is a mature and functioning placenta. The embryo is only about 1½ inches (38 millimeters) long.

C. Fetal development extends from the third through the ninth months. During the third and fourth months, the skeleton is becoming ossified. The sex of the fetus becomes distinguishable.

D. During the fifth through ninth months, the fetus continues to grow and to gain weight. Babies born after six or seven months may survive, but full-term babies have a better chance of survival.

E. During pregnancy, the mother's uterus enlarges greatly, resulting in weight gain, standing and walking difficulties, and general discomfort.

III. Birth

A. During stage 1 of parturition, the cervix dilates. During stage 2, the child is born. During stage 3, the afterbirth is expelled.

B. Lactation requires the hormone prolactin, and the suckling reflex involves the hormone oxytocin.

Study Questions

1. Describe the process of fertilization and the events immediately following it. (pp. 363–65)
2. What is the basis of the pregnancy test? (p. 365)
3. Name the four extraembryonic membranes, and give a function for each. (p. 366)
4. Describe the structure and function of the umbilical cord. (p. 366)
5. Describe the structure and function of the placenta. (pp. 366–67)
6. Specifically, what events normally occur during embryonic development? What events normally occur during fetal development? (pp. 368–372)
7. In general, describe the physical changes in the mother during pregnancy. (pp. 372–73)
8. What are the three stages of birth? Describe the events of each stage. (pp. 373–75)
9. Describe the suckling reflex. (p. 376)

Objective Questions

Fill in the blanks.

1. Fertilization occurs when the _____ nucleus fuses with the _____ nucleus.
2. The _____ membranes include the chorion, the _____ , the yolk sac, and the allantois.
3. During development, the nutrient needs of the developing embryo (fetus) are served by the _____ .
4. The zygote divides as it passes down a uterine tube. This process is called _____ .
5. When cells take on a specific structure and function, _____ occurs.
6. Once the blastocyst arrives at the uterus, it begins to _____ itself in the endometrium.
7. During embryonic development, all major _____ form.
8. Fetal development begins at the end of the _____ month.
9. In most deliveries, the _____ appear(s) before the rest of the body.
10. The hormone _____ is required for milk letdown during the suckling reflex.

Medical Terminology Reinforcement Exercise

Consult Appendix B for help in pronouncing and analyzing the meaning of the terms that follow.

1. morphogenesis (mor″fo-jen′ĕ-sis)
2. neonatologist (ne″o-na-tol′o-jist)
3. prenatal (pre-na′tal)
4. hyperemesis gravidarum (hi″per-em′ĕ-sis grav-id-ar′um)
5. dysmenorrhea (dis″men-o-re′ah)
6. pseudocyesis (soo″do-si′ĕ-sis)
7. primigravida (pri″mĭ-grav′I-dah)
8. cryptorchidism (krip-tor′ki-dizm)
9. oligospermia (ol″i-go-sper′me-ah)
10. perineorrhaphy (per″i-ne-or′ah-fe)
11. abruptio placentae (ab-rup′she-o plah-sen′te)
12. dystocia (dis-to′se-ah)
13. galactostasis (gal″ak-tos′tah-sis)
14. polyhydramnios (pol″e-hi-dram′ ne-os)

Website Link

For a listing of the most current Websites related to this chapter, please visit the Mader home page at:

www.mhhe.com/maderap

chapter 19

Human Genetics

Chapter Outline and Learning Objectives

After you have studied this chapter, you should be able to:

Inheritance of Autosomes (p. 381)
■ Explain the normal chromosome inheritance of humans.
■ Describe the characteristics and causes of three autosomal abnormalities in humans.

Inheritance of Sex Chromosomes (p. 383)
■ Explain how sex is inherited in humans.
■ Describe the characteristics of the most common human conditions resulting from inheritance of abnormal combinations of sex chromosomes.

Gene Inheritance (p. 385)
■ Explain dominant, recessive, and X-linked gene inheritance in humans.
■ Give examples of dominant, recessive, and X-linked genetic disorders in humans.

Biochemical Genetics (p. 387)
■ Explain how DNA, with the help of mRNA and tRNA, controls protein synthesis and, therefore, the characteristics of the cell and organism.
■ Describe the contributions of biotechnology to the field of medicine.
■ Explain how gene therapy is being used to treat genetic disorders.

Visual Focus
Ex vivo Gene Therapy in Humans (p. 390)

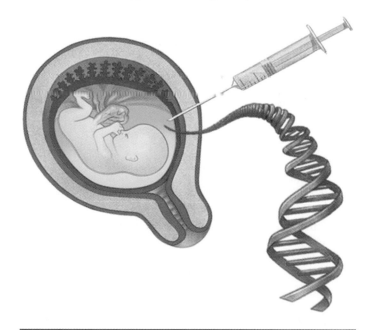

During amniocentesis, a needle is used to withdraw fetal cells floating in the fluid that surrounds the fetus. These cells can be tested for certain genetic defects.

figure 19.1 Human karyotype preparation. As illustrated here, the stain used can result in chromosomes with a banded appearance. The bands help researchers identify and analyze the chromosomes.

1. Blood is centrifuged to separate out blood cells.

2. Only white blood cells are transferred and treated to stop cell division.

3. Sample is fixed, stained, and spread on a microscope slide.

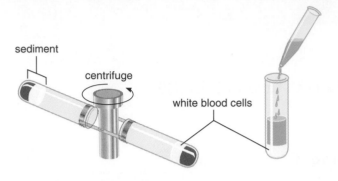

4. Slide is examined microscopically, and the chromosomes are photographed. Computer arranges the chromosomes into pairs.

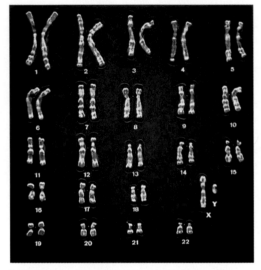

5. Karyotype: Chromosomes are paired by size, centromere location, and banding patterns.

Genetics is the study of how biological information is transferred from one generation to the next. This information is stored in the **genes,** which are composed of DNA, a molecule that contains encoded information for directing protein synthesis. Proteins perform structural and metabolic functions in cells. (Enzymes are proteins, for example.) Genes, which are sections of chromosomes, control body traits, such as height, eye color, or length of fingers.

A zygote receives 23 pairs of chromosomes when the gametes unite during fertilization. One of each pair is inherited from the male parent and the other from the female parent.

Thereafter, due to the process of cell division, each body cell contains copies of these chromosomes in the nucleus. It is possible to photograph the nucleus of a cell that is about to divide (the chromosomes are more visible then), so that a picture of the chromosomes is obtained. The picture may be entered into a computer and the chromosomes electronically arranged by pairs. The resulting display of chromosomes is called a **karyotype** (fig. 19.1). Twenty-two of the chromosome pairs in the karyotype are called the *autosomes,* and one pair is called the **sex chromosomes** because they determine the individual's sex. In males, one of the sex

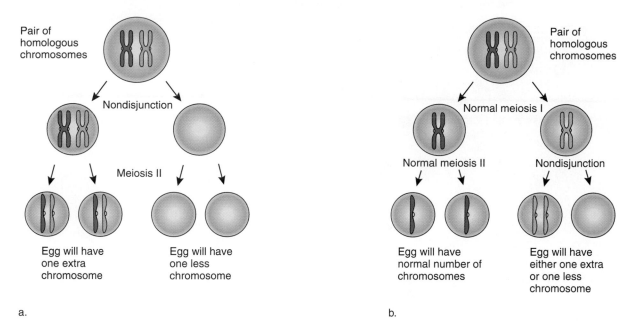

a.

b.

chromosomes is larger than the other. The larger of this pair is called the X chromosome, and the smaller is called the Y chromosome. Females have two X chromosomes in their karyotype.

> Normally, humans inherit 22 pairs of autosomal chromosomes and one pair of sex chromosomes.

Inheritance of Autosomes

Normally, an individual inherits 22 autosomes from the father and 22 autosomes from the mother. In a karyotype, the pairs of autosomes are numbered 1–22 according to size and by position of the centromere. Recall from chapter 3 that a chromosome contains two identical halves called chromatids. The centromere is the place where the chromatids are held together.

Sometimes, an individual inherits either too many or too few autosomes, most likely due to nondisjunction. **Nondisjunction** is the failure of chromosomes (or chromatids) to separate during meiosis. Nondisjunction causes the formation of abnormal eggs or sperm (fig. 19.2).

If a woman is concerned about the possibility of having an abnormal child, she can either have chorionic villi sampling or amniocentesis performed, as described in the chapter 18 MedAlert reading. Following this procedure, a karyotype will reveal whether or not the child has an abnormal chromosome number.

Down Syndrome

A *syndrome* is a group or pattern of symptoms that occur together in the same individual due to the presence of an abnormal condition. The most common autosomal abnormality is seen in individuals with **Down syndrome,** which results from **trisomy** 21—having three copies of chromosome 21 instead of the normal two (fig. 19.3). People with Down syndrome are very often mentally retarded and have an Asian-like fold above the eyes, flattened facial features, an unusual palm crease, muscular flaccidity, and short stature. With care, a person with Down syndrome can grow to adulthood and become a productive member of society. Women who are nearing the end of their reproductive years are more likely to have a child with Down syndrome than are younger women (table 19.1).

figure 19.3 Down syndrome. **a.** Child with Down syndrome. **b.** Down syndrome is due to the inheritance of three copies of chromosome 21. Investigators have determined that a gene on chromosome 21 called the Gart gene is responsible for the mental retardation that accompanies Down syndrome.

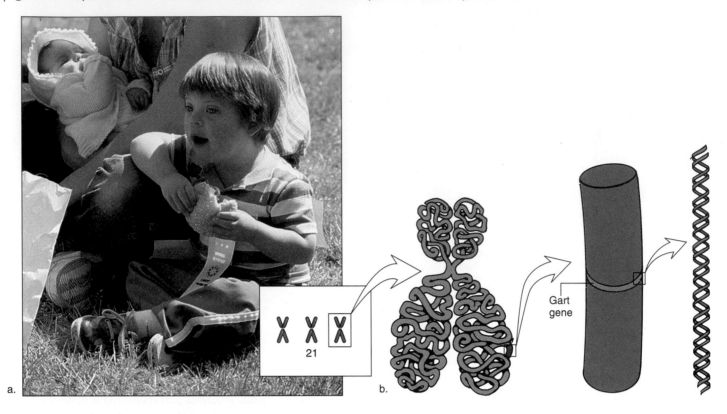

a.

21

b.

Gart gene

table 19.1	Incidence of Selected Chromosomal Abnormalities				
	Name	**Sex**	**Chromosomes**	**Frequency**	
	Autosomes			*Abortuses*	*Births*
	Down syndrome	M or F	Trisomy 21	1/40	1/800
	Edward syndrome	M or F	Trisomy 13	1/33	1/15,000
	Patau syndrome	M or F	Trisomy 18	1/200	1/6,000
	Sex Chromosomes				
	Turner syndrome	F	XO	1/18	1/6,000
	Triplo-X syndrome	F	XXX (or XXXX)	0	1/1,500
	Klinefelter syndrome	M	XXY (or XXXY)	0	1/1,500
	Jacobs syndrome	M	XYY	?	1/1,000

Other Autosomal Abnormalities

Two other trisomies are known among human beings, although they are far less common than Down syndrome. In **Edward syndrome**, or trisomy 18, the baby, usually female, has abnormal hands and feet, grows very slowly, and is mentally retarded. Most babies with Edward syndrome survive only two to four months. In **Patau syndrome**, or trisomy 13, the baby has a cleft lip and palate, eye defects, too many fingers and toes, and abnormal feet. The brain and heart are also malformed. Average survival time is six months.

> The most common autosomal abnormality is Down syndrome, due to the inheritance of an extra chromosome 21. Edward syndrome, or trisomy 18, and Patau syndrome, or trisomy 13, also occur.

Inheritance of Sex Chromosomes

Due to meiosis during spermatogenesis, a sperm carries either an X or Y chromosome. In contrast, an egg always carries an X chromosome. Therefore, the sex of the newborn depends on whether an X- or Y-bearing sperm fertilizes the egg. Theoretically, this also implies a fifty-fifty chance of having a boy or girl (fig. 19.4). However, for reasons that are not clear, more males than females are conceived. But from conception on, the death rate among males is higher. More males than females are spontaneously aborted (miscarried), and this trend continues after birth until the ratio of males to females is dramatically reversed (table 19.2).

The inheritance of an abnormal number of sex chromosomes (see table 19.1) can be due to nondisjunction. Nondisjunction of the sex chromosomes during oogenesis can lead to an egg with either two X chromosomes or no X chromosomes. Nondisjunction of the sex chromosomes during spermatogenesis can result in a sperm with no sex chromosome, both an X and a Y chromosome, two X chromosomes, or two Y chromosomes. Assuming that the other gamete is normal, the zygote could develop into an individual with one of the conditions shown in figure 19.5.

> In some cases, a person inherits an abnormal combination of sex chromosomes, usually due to nondisjunction during meiosis.

A female with **Turner syndrome** (XO) has only one sex chromosome, an X chromosome; the O signifies the absence of the second sex chromosome. Because the ovaries never become functional, these females do not undergo puberty or menstruation, and the breasts do not develop (fig. 19.6a). Generally, females with Turner syndrome have

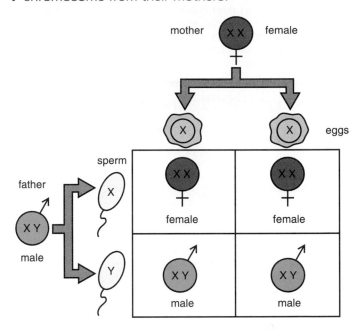

figure 19.4 Inheritance of sex. Males inherit a Y chromosome and females an X chromosome from their fathers. Both males and females inherit an X chromosome from their mothers.

table 19.2	Ratio of Males to Females in the United States	
	Age	**Sex Ratio (Males to Females)**
	Birth	106 : 100
	18 years	100 : 100
	50 years	85 : 100
	85 years	50 : 100
	100 years	20 : 100

a stocky build, a webbed neck, and difficulty recognizing various spatial patterns.

When an egg having two X chromosomes is fertilized by an X-bearing sperm, the **triplo-X syndrome** results. It might be supposed that the XXX female with 47 chromosomes would be especially feminine, but this is not the case. Although there is a tendency toward learning disabilities, most metafemales have no apparent physical abnormalities, and many are fertile and have children with a normal chromosome count.

When an egg having two X chromosomes is fertilized by a Y-bearing sperm, a male with **Klinefelter syndrome**

figure 19.5 Nondisjunction of sex chromosomes. **a.** Nondisjunction during oogenesis produces eggs that have no X chromosome or two X chromosomes. Fertilization with normal sperm results in the conditions noted. **b.** Nondisjunction during meiosis I of spermatogenesis *(left)* produces sperm with no sex chromosomes or sperm with both an X and a Y chromosome. Nondisjunction during meiosis II *(right)* of spermatogenesis produces sperm that have two X chromosomes when a secondary spermatocyte from meiosis I received a single X chromosome, and both chromatids go into the same sperm. Nondisjunction during meiosis II produces sperm that have two Y chromosomes when a secondary spermatocyte from meiosis I received a single Y chromosome, and both chromatids go into the same sperm. Fertilization of a normal egg results in the conditions noted.

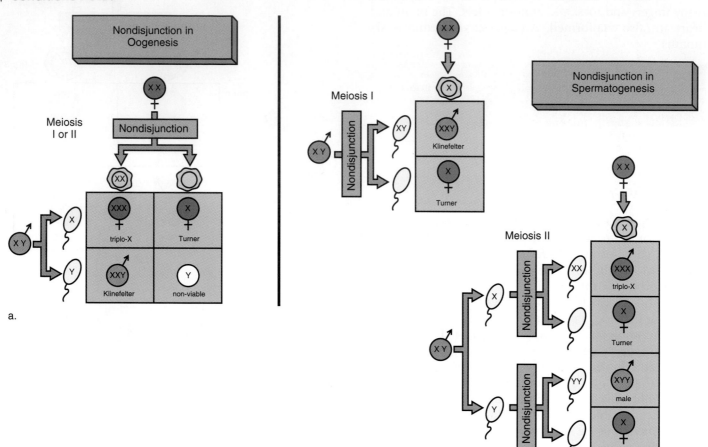

figure 19.6 Abnormal sex chromosome inheritance. **a.** Female with Turner (XO) syndrome. **b.** Male with Klinefelter (XXY) syndrome.

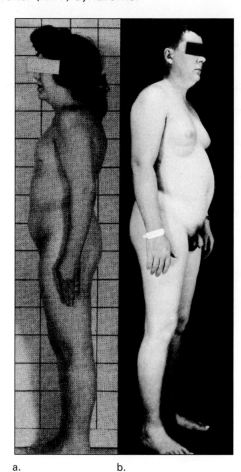

a.   b.

table 19.3	Inheritance of Some Human Traits
Characteristic	**d = dominant** **r = recessive**

Hairline
  Widow's peak (d)
  Continuous hairline (r)
Eye Color
  Brown (d)
  Blue (r)
Hair Color
  Black (d)
  Brown, blond (r)
Earlobes
  Unattached (d)
  Attached (r)
Skin Pigmentation
  Freckles (d)
  No freckles (r)
Hair on Back of Hand
  Present (d)
  Absent (r)
Thumb Hyperextension
  Last segment cannot be bent backward (d)
  Last segment can be bent back to 60° (r)
Bent Little Finger
  Little finger bends toward ring finger (d)
  Straight little finger (r)
Interlace Fingers
  Left thumb over right (d)
  Right thumb over left (r)

results. This individual is male in general appearance, but the testes are underdeveloped, and the breasts may be enlarged (fig. 19.6b). The limbs of XXY males tend to be longer than average, muscular development is poor, body hair is sparse, and many XXY males have learning disabilities.

XYY males with Jacobs syndrome also can result from nondisjunction during spermatogenesis. These males are usually taller than average, are muscular, suffer from persistent acne, and tend to have slightly below normal intelligence. At one time, it was suggested that XYY males were likely to be criminally aggressive, but the incidence of such behavior has been shown to be no greater than that among normal XY males.

Gene Inheritance

Just as individuals receive pairs of chromosomes (one from the egg and one from the sperm), they receive pairs of genes. One gene can be **dominant** to the other, which is called **recessive**. Table 19.3 lists some common human traits and indicates which characteristics are dominant and which are recessive. Figure 19.7 shows how dominant versus recessive genes control the inheritance of type of earlobe.

The sex chromosomes carry genes, just as the autosomal chromosomes do. One gene on the Y chromosome determines the sex of the individual (that is, whether the individual has testes or ovaries), the rest of the genes on

figure 19.7 Genetic inheritance. **a.** Female has inherited two dominant genes (*EE*) and has unattached earlobes. **b.** Female has inherited two recessive genes (*ee*) and has attached earlobes. **c.** Female has inherited one dominant gene (*E*) and one recessive gene (*e*) and has unattached earlobes.

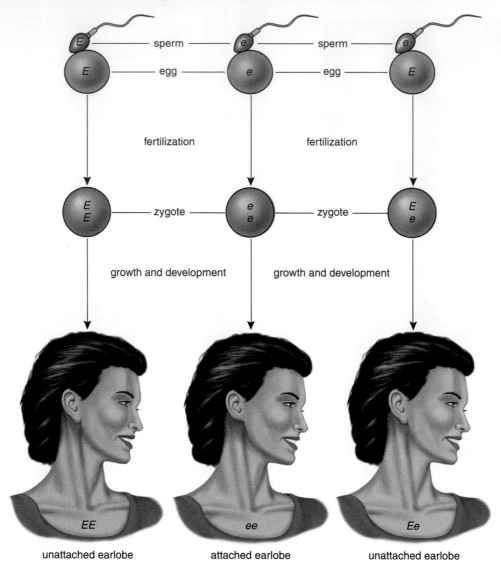

| unattached earlobe | attached earlobe | unattached earlobe |

the sex chromosomes control traits unrelated to sexual characteristics. They are called sex-linked genes because they are on the sex chromosomes. A few sex-linked genes are on the Y chromosome, but the most important ones discovered so far are only on the much larger X chromosome. Since, in a male, the Y chromosome is blank for **X-linked genes,** a recessive gene present on the X chromosome will be expressed.

> Genes ultimately control human traits. Dominant genes are expressed when a single copy is present; recessive genes require two copies in order to be expressed.

Genetic Disorders

Commonly inherited dominant and recessive genetic disorders are listed in table 19.4. When an abnormal gene is inherited for one of the dominant disorders listed in table 19.4, the individual has the disorder, even if the other gene is normal. (In these cases, one of the parents will have the disorder.) To have one of the recessive disorders, the individual must receive two abnormal genes. (In most cases, neither parent will have the disorder, since each will have a dominant normal gene.)

table 19.4 Inheritance of Some Genetic Disorders

Dominant	Recessive	X-Linked	Multiple Genes
Currently, some 1,489 dominantly inherited disorders have been cataloged. Examples include: • Neurofibromatosis—benign tumors in skin or deeper • Achondroplasia—a form of dwarfism • Chronic simple glaucoma (some forms)—a major cause of blindness if untreated • Huntington disease—progressive nervous system degeneration • Familial hypercholesterolemia—high blood cholesterol levels, propensity to heart disease • Polydactyly—extra fingers or toes	Among the 1,117 recessively inherited disorders cataloged are: • Cystic fibrosis—disorder affecting function of mucous and sweat glands • Galactosemia—inability to metabolize milk sugar • Phenylketonuria—essential liver enzyme deficiency • Sickle-cell disease—blood disorder primarily affecting blacks • Thalassemia—blood disorder primarily affecting persons of Mediterranean ancestry • Tay-Sachs disease—lysosomal storage disease leading to nervous system destruction	Among the 205 cataloged disorders transmitted by a gene or genes on the X chromosome are: • Agammaglobulinemia—lack of immunity to infections • Color blindness—inability to distinguish certain colors • Hemophilia—defect in blood-clotting mechanisms • Muscular dystrophy (some forms)—progressive wasting of muscles • Spinal ataxia (some forms)—spinal cord degeneration	The number of defects due to multifactorial inheritance is unknown. Some that are thought to be multifactorial are: • Cleft lip and/or palate • Clubfoot • Congenital dislocation of the hip • Spina bifida—open spine • Hydrocephalus (with spina bifida)—water on the brain • Pyloric stenosis—narrowed or obstructed opening from stomach into small intestine

Source: Data from the National Foundation/March of Dimes.

Males are more likely than females to have the X-linked disorders listed in table 19.4. When a mother carries the abnormal gene on one of her X chromosomes, males have a 50% chance of inheriting either the normal or the abnormal gene and of having the disorder. Males only receive a blank Y from their fathers. Females also have a 50% chance of receiving the abnormal gene from their mothers, but most likely, the X they receive from their fathers will have a normal dominant gene. Several disorders do not appear unless multiple abnormal genes are inherited. The inheritance of these conditions, listed in the "Multiple Genes" column in table 19.4, is complex, and they are mentioned here only.

Dominant genetic disorders are due to the inheritance of at least one dominant gene that is abnormal. Recessive genetic disorders are due to the inheritance of two recessive genes that are abnormal. X-linked disorders are due to the inheritance of only one abnormal recessive gene in males (two in females) on the X chromosome(s).

Biochemical Genetics

The chromosomes located in the nuclei of the sperm and egg carry parental genes to the child. As mentioned in chapter 2, genes are made up of the nucleic acid DNA, which contains a code that directs protein synthesis. The code is found in the sequence of nitrogen bases in DNA.

The genes and therefore the DNA code are located in the cell nucleus. Protein synthesis takes place in the cytoplasm at the ribosomes of the rough endoplasmic reticulum (fig. 19.8a). The DNA code reaches the ribosomes via the nucleic acid messenger RNA (mRNA), which carries the code from the nucleus to the ribosomes in the cytoplasm. Messenger RNA is formed in the nucleus, and the sequence of its bases is patterned after the sequence of bases in DNA. The mRNA then travels through the pores of the nuclear envelope to the ribosomes. The sequence of nitrogen bases in mRNA specifies the order of the amino acids in a protein. The RNA called transfer RNA (tRNA)

figure 19.8 Protein synthesis. **a.** Protein synthesis takes place at the ribosomes often located on the rough endoplasmic reticulum (ER). **b.** Messenger RNA (mRNA), which is made in the nucleus, carries a sequence of bases patterned after the sequence of bases in DNA. Every three bases in mRNA is a codon. The order of the codons determines the order in which amino acids are linked together. Transfer RNAs (tRNAs) carry amino acids to the ribosomes; each type contains an anticodon that pairs with a codon. The arrows indicate that after an amino acid has joined the growing protein, the tRNA leaves the ribosome.

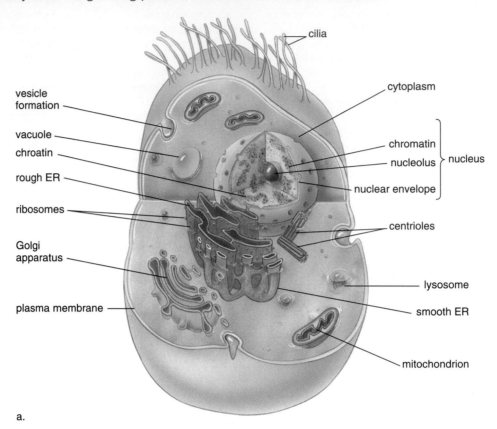

cilia

cytoplasm

vesicle formation

chromatin

nucleolus

nucleus

vacuole

chroatin

rough ER

nuclear envelope

ribosomes

centrioles

Golgi apparatus

lysosome

plasma membrane

smooth ER

mitochondrion

a.

Protein Synthesis

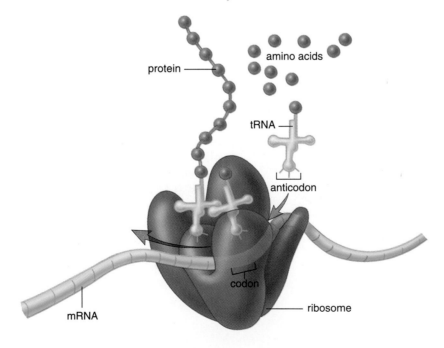

protein

amino acids

tRNA

anticodon

codon

mRNA

ribosome

b.

table 19.5 Representative Biotechnology Products

Hormones and Similar Types of Proteins		Vaccines
Treatment of Humans	**For**	**Use in Humans to Prevent**
Insulin	Diabetes	AIDS
Growth hormone	Pituitary dwarfism	Herpes (oral and genital)
tPA (tissue plasminogen activator)	Heart attack	Hepatitis A, B, and C
Interferons	Cancer	Lyme disease
Erythropoietin	Anemia	Whooping cough
Interleukin–2	Cancer	Chlamydia
Clotting factor VIII	Hemophilia	Malaria
Human lung surfactant	Respiratory distress syndrome	
Atrial natriuretic factor	High blood pressure	
Tumor necrosis factor	Cancer	
Ceredase	Gaucher disease	

brings amino acids to the ribosome, where they are joined in the correct sequence (fig. 19.8b).

Many proteins are enzymes, and since no reaction occurs in a cell unless its enzyme is present, the enzymes determine what the cell and the organism will be like. Therefore, by controlling enzyme synthesis, the genes control the characteristics of human beings.

When an abnormal gene is inherited, the enzyme and then the metabolic activity of the cell are abnormal. If geneticists know which gene is abnormal, the condition sometimes can be cured by gene therapy, as discussed later in the chapter.

Not all birth defects are due to the inheritance of abnormal genes. Some are due to teratogens, agents that cross the placenta and affect DNA or cellular metabolism after the zygote receives normal chromosomes and genes from each parent. Development does not proceed normally, and the child is born with a birth defect. Teratogens are discussed in a chapter 18 Medical Focus, page 369.

Biotechnology

Biotechnology is the use of a natural biological system to produce a product or to achieve an end desired by humans. Today, bacteria are being genetically engineered to produce the hormones and proteins listed in table 19.5. The human genes for these substances are inserted into bacteria, which then mass produce them.

One impressive advantage of biotechnology is that it allows mass production of proteins that are very difficult to obtain otherwise. For example, the human growth hormone that is used to treat people who are growing more slowly than normal was previously extracted from the pituitary glands of cadavers, and it took 50 glands to obtain enough for one dose. Now human growth hormone is produced in quantity by biotechnology. Insulin, to treat diabetics, was previously extracted from the pancreatic glands of slaughtered cattle and pigs. It was expensive and sometimes caused allergic reactions in recipients. Today, however, biotechnology is responsible for mass production of insulin.

Vaccines produced through biotechnology do not cause illness. Bacteria and viruses have surface proteins, and a gene for just one of these can be used to genetically engineer bacteria. The copies of the surface protein that result can be used as a vaccine. A vaccine for hepatitis B is now available, and potential vaccines for chlamydia, malaria, and AIDS are in experimental stages.

The last few entries of table 19.5 list other troublesome and serious afflictions in humans that may soon be treatable by biotechnology products: Clotting factor VIII treats hemophilia; human lung surfactant treats respiratory distress syndrome in premature infants; atrial natriuretic factor helps control hypertension. And the list will grow because bacteria (or other cells) can be engineered to produce virtually any protein.

> Biotechnology products include hormones and similar types of proteins and vaccines. These products are enormously important in the treatment and prevention of a variety of human afflictions.

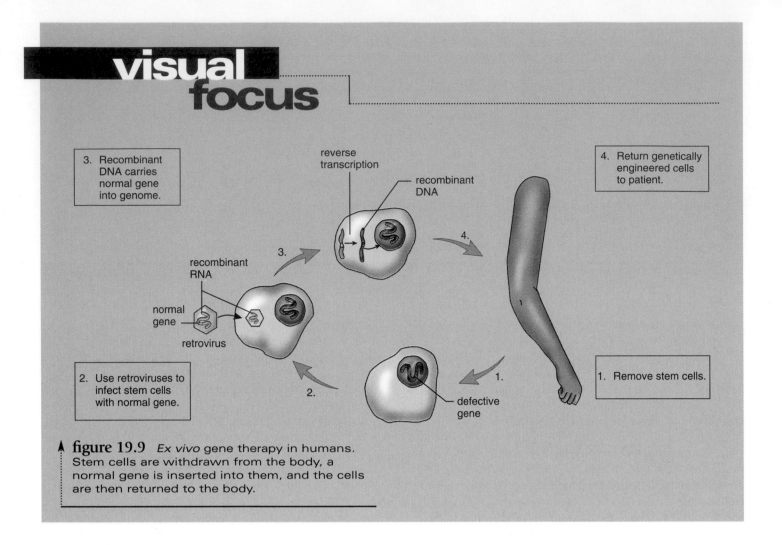

3. Recombinant DNA carries normal gene into genome.

reverse transcription

recombinant DNA

4. Return genetically engineered cells to patient.

recombinant RNA

normal gene

retrovirus

2. Use retroviruses to infect stem cells with normal gene.

defective gene

1. Remove stem cells.

▲ **figure 19.9** *Ex vivo* gene therapy in humans. Stem cells are withdrawn from the body, a normal gene is inserted into them, and the cells are then returned to the body.

Gene Therapy

Gene therapy involves replacing defective genes with healthy genes. It also includes the use of genes to treat genetic disorders and various other human illnesses.

During *ex vivo* (outside the living organism) therapy, cells are removed from a patient, treated, and returned to the patient. Viruses called retroviruses are often used as vectors to carry normal genes into the cells of the patient because they normally insert their genes into host chromosomes (fig. 19.9).

Two young girls with severe combined immunodeficiency syndrome (SCID) underwent *ex vivo* gene therapy several years ago. (SCID is often called the "bubble-baby" disease after David, a young person who lived under a plastic dome to prevent infection.) These girls lacked an enzyme that is involved in the maturation of T and B cells, and therefore, they were subject to life-threatening infections. White blood cells were removed from their blood and infected with a retrovirus that carried a normal gene for the enzyme. Then the cells were returned to the girls. Recently, researchers have reported that many of the girls' T cells carry the healthy gene and that the girls are in good health. However, for ethical reasons, the girls have also been receiving the traditional treatment for SCID (injections of the missing enzyme), and their good health may be in response to this treatment, gene therapy, or both.

Also, one of these girls and two newborn boys have recently received bone marrow stem cells genetically engineered in the same way. Bone marrow stem cells are capable of producing all types of blood cells, including the white blood cells of the immune system. Genetically engineered stem cells are preferred because their use may result in a permanent cure.

Gene therapy is also being used for treatment of familial hypercholesterolemia, a condition that develops when liver cells lack a receptor for removing cholesterol from the blood. The high levels of blood cholesterol make the patient subject to fatal heart attacks at a young age. In a newly developed procedure, a small portion of the liver is surgically excised and infected with a retrovirus containing a normal gene for the receptor. In one gene therapy patient, the cholesterol level dropped from 448 to 366 (milligrams/deciliter). Along the same lines, it may be possible to give patients with failing hearts genetically engineered cardiac muscle cells that carry extra receptor proteins for adrenaline, a heart stimulant.

> Gene therapy, which involves replacing defective genes with healthy genes, is now a reality. Researchers are envisioning various applications aimed at curing human genetic disorders, as well as many other types of illnesses.

Selected New Terms

Basic Key Terms

dominant gene (dom'ĭ-nant jen), p. 385
gene (jēn), p. 380
genetics (jĕ-net'iks), p. 380
karyotype (kar'e-o-tīp), p. 380
nondisjunction (non"dis-junk'shun), p. 381
recessive gene (re-ses'iv jēn), p. 385
sex chromosome (seks kro'mo-sōm), p. 380
trisomy (trī-so'-me), p. 381
X-linked gene (x-linkt jēn), p. 386

Clinical Key Terms

biotechnology (bi"o-tek-nol'o-ge), p. 389
Down syndrome (down sin'drom), p. 381
Edward syndrome (ed'werd sin'drōm), p. 383
gene therapy (jēn ther'ah-pe), p. 390
Klinefelter syndrome (klīn'fel-ter sin'drōm), p. 383
Patau syndrome (pa'tow sin'drōm), p. 383
triplo-X syndrome (trip"lo-x sin'drōm), p. 383
Turner syndrome (tur'ner sin'drōm), p. 383
XYY male (māl), p. 385

Summary

Normally, humans inherit 22 pairs of autosomal chromosomes and one pair of sex chromosomes. Males are XY, and females are XX.

I. **Inheritance of Autosomes**
   The most common autosomal abnormality is Down syndrome, due to the inheritance of an extra chromosome number 21. Edward syndrome, or trisomy 18, and Patau syndrome, or trisomy 13, also occur.

II. **Inheritance of Sex Chromosomes**
   A. In some cases, a person inherits an abnormal combination of sex chromosomes, usually due to nondisjunction during meiosis.
   B. Abnormal combinations of sex chromosomes include XO (Turner syndrome), XXX (triplo-X), XXY (Klinefelter syndrome), and XYY (Jacob's syndrome). Individuals with a Y chromosome are always male, no matter how many X chromosomes there are. However, at least one X chromosome is needed for survival.

III. **Gene Inheritance**
   Genes ultimately control human traits. Dominant genes are expressed when a single copy is present; recessive genes require two copies in order to be expressed.
   A. Dominant genetic disorders (for example, neurofibromatosis, Huntington disease) are due to the inheritance of at least one dominant gene that is abnormal.
   B. Recessive genetic disorders (for example, cystic fibrosis, Tay-Sachs disease) are due to the inheritance of two recessive genes that are abnormal.
   C. X-linked disorders (for example, color blindness, hemophilia) are due to the inheritance of only one abnormal recessive gene in males (two in females) on the X chromosome(s).

IV. **Biochemical Genetics**
   A. Biotechnology products include hormones and similar types of proteins and vaccines. These products are enormously important in the treatment and prevention of a variety of human afflictions.
   B. Gene therapy, which involves replacing defective genes with healthy genes, is now a reality. Researchers are envisioning various applications aimed at curing human genetic disorders, as well as many other types of illnesses.

Study Questions

1. What is the normal chromosome inheritance of humans? (p. 381)
2. What are the characteristics and cause of Down syndrome? (pp. 381–82)
3. How is sex inherited in humans? (p. 383)
4. What are the characteristics of the most common human conditions resulting from inheritance of abnormal combinations of sex chromosomes? (pp. 383–85)
5. Explain dominant, recessive, and X-linked gene inheritance in humans. (pp. 385–86)
6. Give examples of dominant, recessive, and X-linked genetic disorders in humans. (pp. 386–87)
7. Describe the function of DNA, the genetic material. (p. 387)
8. Explain how genes control the structure and function of cells. (p. 388–89)
9. How is biotechnology helping to treat and cure human disorders? (p. 389)
10. What is gene therapy? Explain how defective genes are replaced. (pp. 390–91)

Objective Questions

Fill in the blanks.

1. The genes are on the _____ .
2. A karyotype shows the individual's _____ .
3. The sex chromosomes of a male are labeled _____ .
4. A person with Down syndrome has inherited _____ copies of chromosome number 21.
5. A person with Klinefelter syndrome has the chromosomes _____ .
6. A dominant genetic disorder only requires the inheritance of _____ (one or two) abnormal gene(s).
7. If a person inherits a genetic disease, and both parents are unaffected by the disorder, the disorder is _____ .
8. DNA is found in the _____ , but protein synthesis takes place at the ribosomes in the cytoplasm.
9. The use of bacteria to produce human insulin is an example of _____ .
10. Replacing defective genes with healthy genes is the goal of _____ .

Medical Terminology Reinforcement Exercise

Consult Appendix B for help in pronouncing and analyzing the meaning of the terms that follow.

1. neogenesis (ne″o-jen′ĕ-sis)
2. regeneration (re-jen″er-a′shun)
3. amniocentesis (am″ne-o-sen-te′sis)
4. fetoscope (fe′to-skōp)
5. chromosome (kro′mo-sōm)
6. polydysplasia (pol″e-dis-pla′ze-ah)
7. congenital (kon-jen′ĭ-tal)
8. hyperplasia (hi″per-pla′ze-ah)
9. atrophy (at′ro-fe)
10. agammaglobulinemia (a-gam″ah-glo″bu-li-ne′me-ah)

Website Link

For a listing of the most current Websites related to this chapter, please visit the Mader home page at:

www.mhhe.com/maderap

Reference Figures
The Human Organism

▼ **Plate 1** Anterior view of the human torso with the superficial muscles exposed. (m. = muscles; v. = vein.)

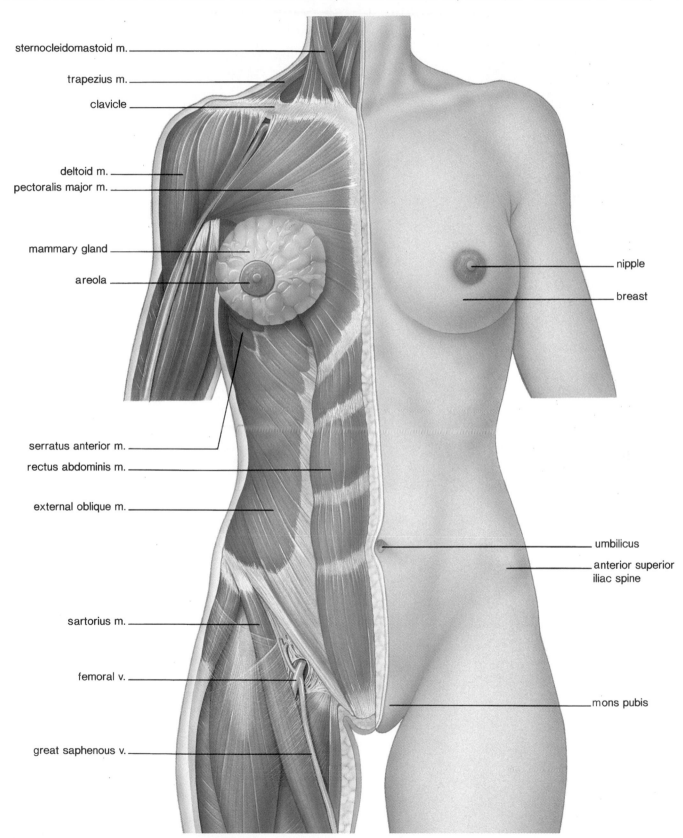

sternocleidomastoid m.

trapezius m.

clavicle

deltoid m.

pectoralis major m.

mammary gland

areola

nipple

breast

serratus anterior m.

rectus abdominis m.

external oblique m.

umbilicus

anterior superior iliac spine

sartorius m.

femoral v.

mons pubis

great saphenous v.

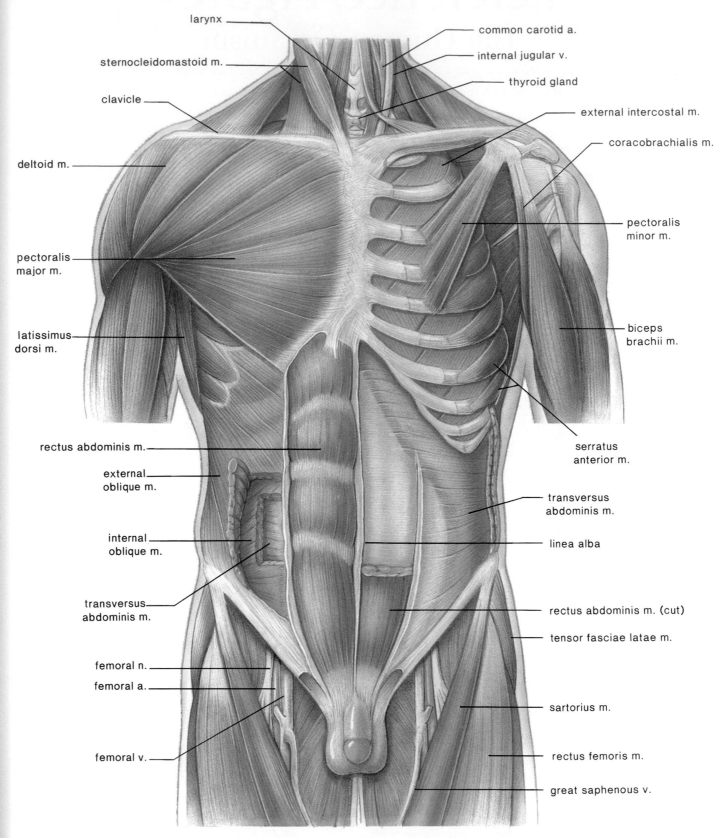

Plate 2 The torso with the deep muscles exposed. (m. = muscle; n. = nerve; a. = artery; v. = vein.)

larynx

common carotid a.

sternocleidomastoid m.

internal jugular v.

thyroid gland

clavicle

external intercostal m.

coracobrachialis m.

deltoid m.

pectoralis minor m.

pectoralis major m.

latissimus dorsi m.

biceps brachii m.

rectus abdominis m.

serratus anterior m.

external oblique m.

transversus abdominis m.

internal oblique m.

linea alba

transversus abdominis m.

rectus abdominis m. (cut)

tensor fasciae latae m.

femoral n.

femoral a.

sartorius m.

rectus femoris m.

femoral v.

great saphenous v.

Plate 3 The torso with the anterior abdominal wall removed to expose the abdominal viscera. (a. = artery; v. = vein; m. = muscle; n. = nerve.)

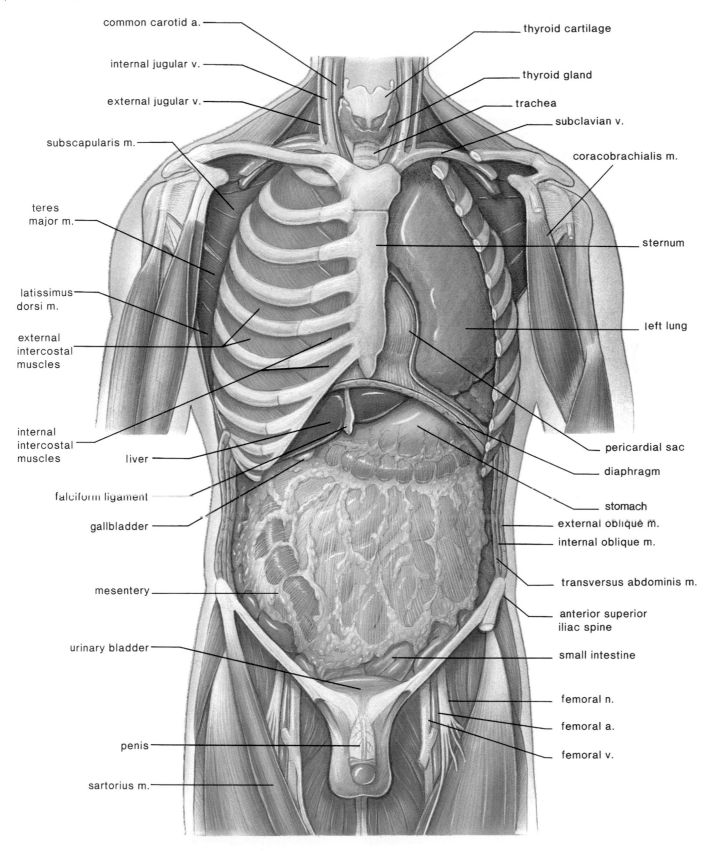

common carotid a.

internal jugular v.

external jugular v.

subscapularis m.

teres major m.

latissimus dorsi m.

external intercostal muscles

internal intercostal muscles

liver

falciform ligament

gallbladder

mesentery

urinary bladder

penis

sartorius m.

thyroid cartilage

thyroid gland

trachea

subclavian v.

coracobrachialis m.

sternum

left lung

pericardial sac

diaphragm

stomach

external oblique m.

internal oblique m.

transversus abdominis m.

anterior superior iliac spine

small intestine

femoral n.

femoral a.

femoral v.

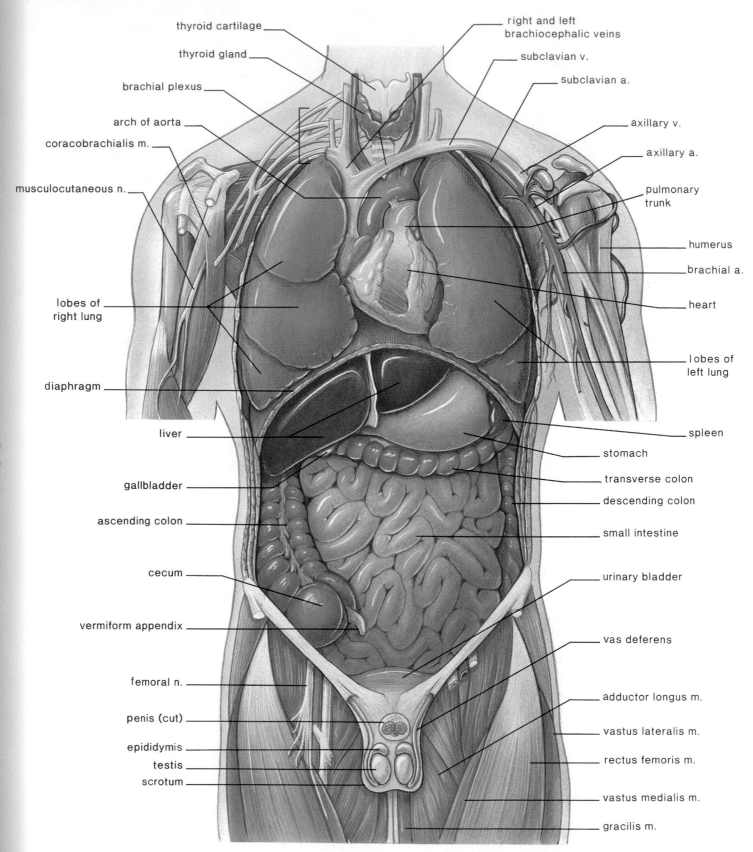

Plate 4 The torso with the anterior thoracic wall removed to expose the thoracic viscera. (a. = artery; m. = muscle; n. = nerve; v. = vein.)

thyroid cartilage

thyroid gland

brachial plexus

arch of aorta

coracobrachialis m.

musculocutaneous n.

lobes of right lung

diaphragm

liver

gallbladder

ascending colon

cecum

vermiform appendix

femoral n.

penis (cut)

epididymis

testis

scrotum

right and left brachiocephalic veins

subclavian v.

subclavian a.

axillary v.

axillary a.

pulmonary trunk

humerus

brachial a.

heart

lobes of left lung

spleen

stomach

transverse colon

descending colon

small intestine

urinary bladder

vas deferens

adductor longus m.

vastus lateralis m.

rectus femoris m.

vastus medialis m.

gracilis m.

Plate 5 The torso as viewed with the thoracic viscera sectioned in a coronal plane, and the abdominal viscera as viewed with most of the small intestine removed. (a. = artery; m. = muscle; v. = vein.)

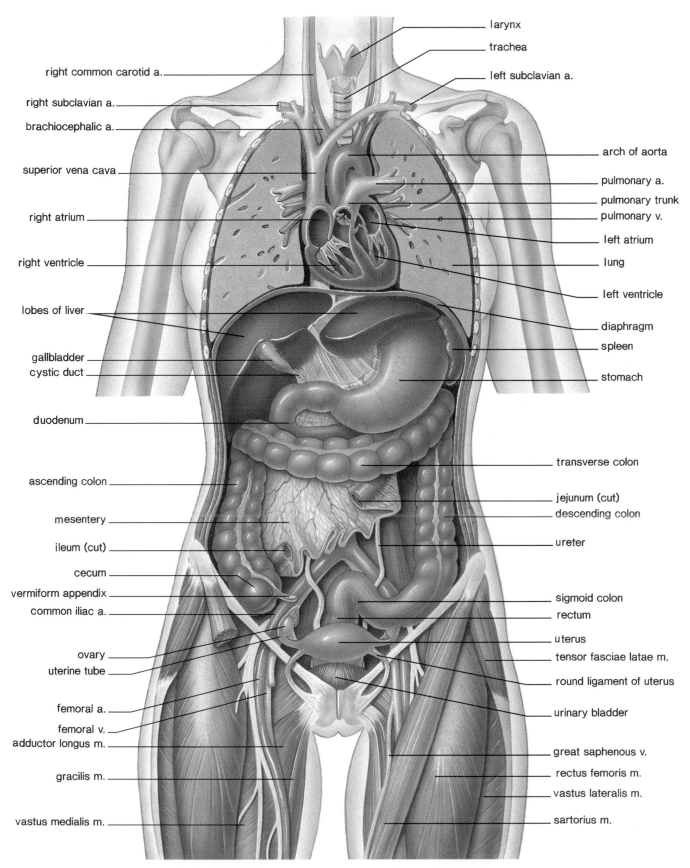

larynx

trachea

left subclavian a.

right common carotid a.

right subclavian a.

brachiocephalic a.

arch of aorta

superior vena cava

pulmonary a.

pulmonary trunk

pulmonary v.

right atrium

left atrium

lung

right ventricle

left ventricle

lobes of liver

diaphragm

spleen

gallbladder

cystic duct

stomach

duodenum

transverse colon

ascending colon

jejunum (cut)

descending colon

mesentery

ileum (cut)

ureter

cecum

vermiform appendix

sigmoid colon

common iliac a.

rectum

uterus

ovary

tensor fasciae latae m.

uterine tube

round ligament of uterus

femoral a.

urinary bladder

femoral v.

adductor longus m.

great saphenous v.

gracilis m.

rectus femoris m.

vastus lateralis m.

sartorius m.

vastus medialis m.

Plate 6 The torso as viewed with the heart, liver, stomach, and portions of the small and large intestines removed. (a. = artery; m. = muscle; v. = vein.)

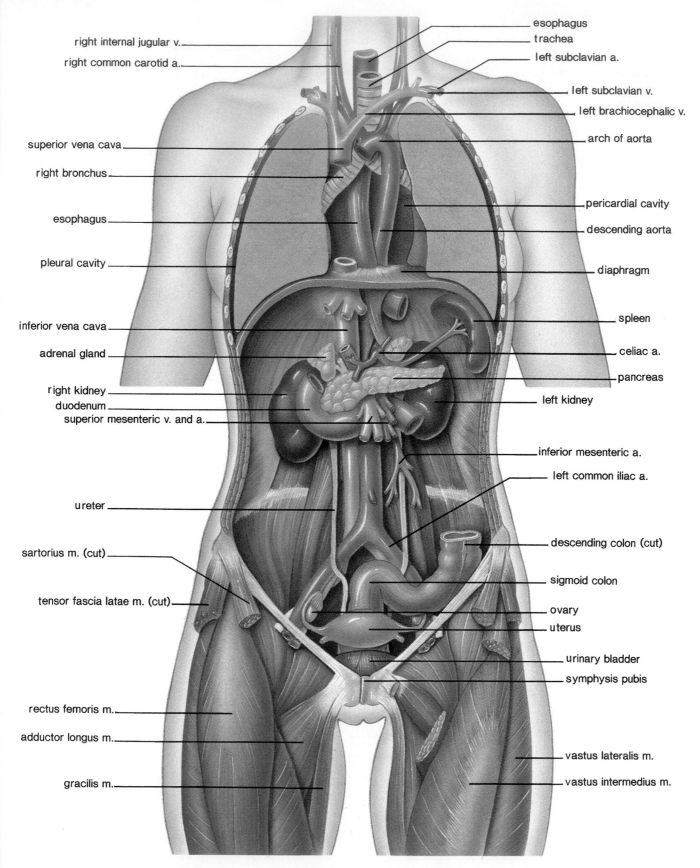

right internal jugular v.

right common carotid a.

superior vena cava

right bronchus

esophagus

pleural cavity

inferior vena cava

adrenal gland

right kidney

duodenum

superior mesenteric v. and a.

ureter

sartorius m. (cut)

tensor fascia latae m. (cut)

rectus femoris m.

adductor longus m.

gracilis m.

esophagus

trachea

left subclavian a.

left subclavian v.

left brachiocephalic v.

arch of aorta

pericardial cavity

descending aorta

diaphragm

spleen

celiac a.

pancreas

left kidney

inferior mesenteric a.

left common iliac a.

descending colon (cut)

sigmoid colon

ovary

uterus

urinary bladder

symphysis pubis

vastus lateralis m.

vastus intermedius m.

Plate 7 The torso with the anterior thoracic and abdominal walls removed, along with the viscera, to expose the posterior walls and body cavities. (a. = artery; m. = muscle.)

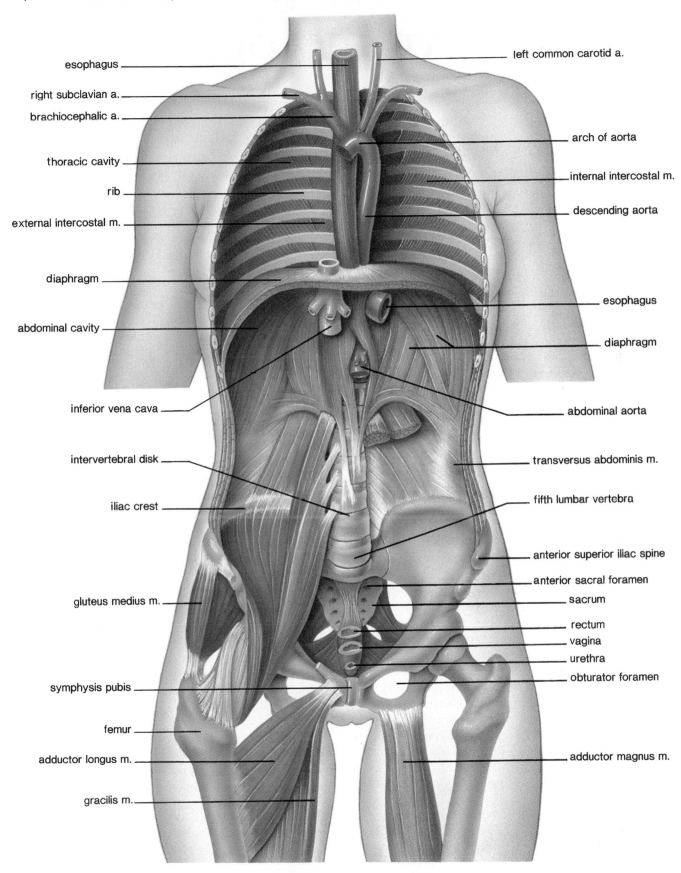

esophagus

right subclavian a.

brachiocephalic a.

thoracic cavity

rib

external intercostal m.

diaphragm

abdominal cavity

inferior vena cava

intervertebral disk

iliac crest

gluteus medius m.

symphysis pubis

femur

adductor longus m.

gracilis m.

left common carotid a.

arch of aorta

internal intercostal m.

descending aorta

esophagus

diaphragm

abdominal aorta

transversus abdominis m.

fifth lumbar vertebra

anterior superior iliac spine

anterior sacral foramen

sacrum

rectum

vagina

urethra

obturator foramen

adductor magnus m.

Understanding Medical Terminology

Learning Objectives

Upon completion of this section, you should be able to:

1. Discuss the importance of medical terminology and how it can be incorporated into the study of the human body.
2. Differentiate between a prefix, suffix, root word, and compound word.
3. Link word parts to form medical terms.
4. Differentiate between singular and plural endings of medical terms.
5. Practice pronunciation of medical words.
6. Dissect (cut apart) compound medical words into parts to analyze the meaning.
7. Recognize the more commonly used prefixes, suffixes, and root words used in medical terminology.

Introduction to Medical Terminology

As students of medical science, we are the inheritors of a vast fortune of knowledge. This fortune, amassed by giants of eighteenth- and nineteenth-century scholarship, was nurtured largely in the atmospheres of universities in which Latin and Greek were the languages of lecture and writing. Scientists then strove to define a universal language in which to communicate their findings. Latin and Greek, studied throughout Europe, became the languages of choice for scholars whose native tongue was English, German, French, Spanish, and so on, because they all read Latin and Greek. So, many seminal works in medicine were first penned in Latin, and their vocabularies remain to this day.

Anatomy and physiology were born in the eighteenth century in the midst of a glut of quacks, frauds, charlatans, myths, and superstitions. Honest scholars sought proofs to banish practices that should have been questioned by reason and proved wrong by experience. These scholars were among the first to connect disease with the failure of function or structure of body tissue; thus, the race to name and define all anatomical structures began.

Problems arose, inevitably, with the discovery of heretofore unknown tissue. Names were virtually created from parts or existing words by combining parts until they approximated an acceptable description. Medical terminology is simply a catalog of parts that allows us to take apart and reassemble the special language of medicine. The study of medical terminology is easier than it first seems.

Medical words have three basic parts: prefix, root word, and suffix. A prefix comes before a root word and alters the meaning. For example, the prefix *hyper-* means over or above. Hyper/kinetic means overactive, hyper/esthesia is overly sensitive, hyper/tension is high blood pressure, and hyper/trophy is overdevelopment.

A suffix is attached to the end of a root word and changes the meaning of the word. For example, the suffix *-itis* means inflammation. Inflammation can occur at almost any part of the body, so *-itis* can be added to root words to make hundreds of words. Dermat/itis is inflammation of the skin, rhin/itis is inflammation of the nose, gastr/itis is inflammation of the stomach, and so on.

A root word is the main part of the word. Once the root word is known for each part of the anatomy, the prefixes and suffixes can be used to analyze and/or build many medical words. The root word for heart is *cardi*. A few terms in which *cardi* appears are: cardi/algia means pain in the heart, cardio/omegaly means enlarged heart, brady/cardia means slow heart, and peri/cardio/centesis means puncture to aspirate fluid from around the heart.

Many medical words have, in addition to a prefix and/or a suffix, more than one word part. These are called compound words and can be analyzed by breaking them into parts. For example, hysterosalpingo-oophorectomy is made up of three root words and a suffix. *Hyster* is the root word for uterus, *salping* is the root word for tube, *oophor* is the root word for ovary, and *-ectomy* is the suffix for cut out. Now we know that hysterosalpingo-oophorectomy means the surgical excision of the uterus, tube, and ovary.

To facilitate pronunciation, word parts need to be linked together. The linkage for word parts is *o* and may be referred to as a combining form. For example, linking the root *cardi* with the suffix *-pathy* would produce a word that would be difficult to pronounce; therefore, an *o* is used to link the root word with the suffix. The complete word is written cardiopathy and pronounced kar″dēop′ah-the, and the combining form is cardi/o.

When a word is only a root or ends with a root, the word ending depends on whether the word is a noun or an adjective. For example, duodenum (noun) is a part of the small intestine. Duodenal (adjective) is related to the duodenum (for example, duodenal ulcer).

Accurate spelling of each word part is essential:

1. Changing one letter may change the word part. For example, *ileum* is a part of the small intestine, whereas *ilium* is a pelvic bone.
2. Finding a word in the dictionary requires a knowledge of spelling—at least of the beginning of the word. For example, *pneumonia* and *psychology* have a silent *p*, *rhinitis* (inflammation of the nose) has a silent *h*, and *eupnea* (easy breathing) has an initial silent *e*.

Plural Endings

In many English words, the plurals are formed by adding *s* or *es*, but in Greek and Latin, the plural may be designated by changing the ending:

Singular Ending	Plural Ending	Examples
a	ae	aorta—aortae
ax	aces	thorax—thoraces
en	ena	lumen—lumena
ex, ix	ices	cortex—cortices appendix—appendices
is	es	testis—testes
on, um	a	phenomenon—phenomena medium—media
ur	ora	femur—femora
us	i	bronchus—bronchi
x	ces	calyx—calyces
y	ies	anomaly—anomalies
ma	mata	adenoma—adenomata

If a word ends in *s* and the vowel in the last syllable is short, the word is singular. If the word ends in *s* and the vowel in the last syllable is long, the word is plural.

Any word ending in a consonant is singular (for example, *um, us, at*).

In Latin, *a* is a singular ending (for example, aorta). *Al, as, ir* and *i* are plural endings.

Pronunciation

1. Words are made up of syllables.
2. Syllables are made up of letters—consonants and vowels (*a,e,i,o,u*).
3. Vowels have a long sound (pronounced by saying its name) when:
   a. a syllable ends with an unmarked vowel. For example, glomeruli (glo-mer′u-li).
   b. it has a macron (¯) over it. For example, gastroscopy (gas-tros′ko-pē).
4. A short sound (ah, eh, ih, oo, uh) is used when:
   a. there is an unmarked vowel in a syllable ending with a consonant. For example, anginal (an′je-nal).
   b. the vowel constitutes a syllable or ends a syllable, and is indicated by a breve (˘). For example, effect (ĕ-fekt′).
5. The syllable *ah* is used for the sound *a* in open, unaccented syllables: For example, abortion (ah-bor′shun) or amenorrhea (ah-men′o-re′ah).
6. The primary accent in a word is indicated by a single accent.
7. The secondary accent is indicated by a double accent. For example, duodenostomy (du″od-e-nos′to-me).
8. The accent on medical terms is generally on the third from the last syllable.

Practice pronouncing the following words:

1. hematemesis (hem″ah-tem′ĕ-sis) —vomiting blood
2. hysterosalpingo-oophorectomy (his″ter-o-sal-ping″go: o″of-o-rek′to-me) —surgical excision of the uterus, tube, and ovary
3. phrenohepatic (fren″o-hĕ-pat′ih) —pertaining to the diaphragm and liver
4. gastropathy (gas″trop′ah-the) —disease of the stomach
5. metatarsus (met″ah-tar′sus) —part of the foot between the tarsus (ankle) and toes

Commonly Used Prefixes

Prefix	Meaning	Example
a-, an-, in-	without, negative	a/men/orrhea—without a monthly flow
ab-	from, away from	ab/normal—away from normal
ad-, ac-, as-, at-	to, toward	ad/duct—carry toward
aniso-	unequal	an/iso/cyt/osis—abnormal condition of unequal cells
ante-, pre-	before	anterior—front; pre/natal—before birth
anti-, ant-, ob-	against	anti/pyre/tic—agent used against fever
bi-	two	bi/lateral—two sides
bio-	life	bio/logy—study of life
brachy-	short	brachy/dactyl/ism—short fingers and toes
brady-	slow	brady/cardia—slow heart rate
cent-	hundred	centi/meter—1/100 of a meter
circum-	around	circum/cis/ion—to cut around
co-, com-, con-	with, together	con/genital—born with
contra-	against	contra/indicated—against indication
de-	away from	de/hydrate—loss of water
dextr-	right	dextr/o/cardia—heart displaced to right
dia-	through	dia/rrhea—flow through
dis-	apart	dis/sect—to cut apart
dys-	bad, difficult	dys/pnea—difficult breathing
e-, ex-	out, out from	ex/cise—to cut out
ect-, exo-, extra-	outside	extra/corporeal—outside the body
en-	in, on	en/capsulated—in a capsule
end-	within	endo/scopy—visualization within
epi-	upon	epi/dermis—upon the skin
eu-	good	eu/phonic—good sound
hemi-, semi-	half	hemi/gastr/ectomy—surgical removal of half of the stomach
hyper-	over, above	hyper/kinetic—overactive
hypo-	under, below	hypo/glossal—under the tongue
immun-	free, exempt	immun/ity—exempt from the effects of specific disease-causing agents
infra-	beneath	infra/mammary—beneath the breast
inter-	between	inter/cellular—between the cells
intra-	within	intra/cranial—within the cranium
kil-	thousand	kilo/gram—1,000 grams
lyso-	dissolution, disintegration	lyso/some—organelle that degrades worn cell parts
macr-	large	macro/cyte—large cell
mal-	bad	mal/nutrition—bad nourishment
mes-	middle	me/entery—middle of intestine
meta-	after, beyond	meta/carpals—beyond the carpals (wrist)
micr-	small	micro/cephal/ic—having a small head
milli-	one-thousandth	milli/liter—1/1,000 of a liter
multi-	many	multi/para—one who has many children
neo-	new	neo/plasm—new growth
olig-	scanty, few	olig/uria—scanty amount of urine
onc-	tumor	onc/ology—study of tumors
per-	through	per/cutaneous—through the skin
peri-	around	peri/tonsillar—around the tonsil
pleur-	rib, side	pleur/al membranes—serous membranes that enclose the lungs
poly-	much, many	poly/cystic—many cysts
post-	after	post/mortem—after death
pre-	before	pre/natal—before birth
presby-	old	presby/opia—old vision
primi-	first	primi/gravida—first pregnancy

Appendix B

Commonly Used Prefixes cont'd

Prefix	Meaning	Example
pro-	before	pro/gnosis—foreknowledge, predict outcome
re-	back, again	re/generate—produce, develop again
retr-	behind	retro/sternal—behind the sternum
sub-	under	sub/lingual—below the tongue
super-, supra-	above	superior—above
syn-, sym-	with, together	syn/ergism—working together
tachy-	fast	tachy/phasia—fast speech

Commonly Used Suffixes

Suffix	Meaning	Example
-algia	pain	dent/algia—pain in the tooth
-atresia	without an opening	proct/atresia—rectum without an opening
-cele	hernia	omphalo/cele—umbilical hernia
-centesis	puncture to aspirate fluid	arthro/centesis—puncture to aspirate fluid from a joint
-cept	take, receive	re/cept/or—something that receives again
-cide	kill	bacteri/cidal—able to kill bacteria
-cis	cut	circum/cis/ion—cutting around
-cyte	cell	erythro/cyte—red cell
-denia	pain	cephalo/denia—pain in the head
-desis	fusion	arthro/desis—fusion of a joint
-ectasia	expansion	cor/ectasis—expanding/dilating pupil
-ectomy	cut out, excise	nephr/ectomy—surgically remove kidney
-edema	swelling	cephal/edema—swelling of head
-emesis	vomiting	hyper/emesis—excessive vomiting
-emia	blood	hyper/glyc/emia—elevated blood sugar
-gnosis	knowledge	dia/gnosis—knowledge through examination (determining cause of disease)
-gram	record	myelo/gram—X ray of the spinal cord
-graphy	making a record	angio/graphy—making a record of vessels
-iasis	condition	chole/lith/iasis—condition of gallstones
-ist	one who	opto/metr/ist—one who measures vision
-itis	inflammation	aden/itis—inflammation of a gland
-lepsy	seizures	narco/lepsy—seizures of numbness
-logist	one who specializes	ophthalmo/log/ist—one specializing in eyes
-logy	study of	bio/logy—study of life
-lysis, -lytic, -lyze	break down, dissolve	teno/lysis—destruction of tendons
-lyt	dissolvable	electro/lyte—substance that ionizes in water solution
-malacia	abnormal softening	osteo/malacia—abnormal softening of bone
-mania	madness	pyro/mania—irresistible urge to set fires
-megaly	enlargement	spleno/megaly—enlargement of spleen
-meter	measure	thermo/meter—instrument to measure temperature
-oid	resembling	muc/oid—resembling mucus
-oma	tumor	neur/oma—nerve tumor
-opia	vision	ambly/opia—dim vision
-osis	abnormal condition	nephr/osis—abnormal condition of kidney
-osme	smell	an/osmia—inability to smell
-ostomy	create an opening	col/ostomy—create an opening in the colon
-otia	ear	macr/otia—large ear
-pathy	disease	encephalo/pathy—disease of the brain
-penia	deficiency, poor	leuko/cyto/penia—deficiency of white cells
-pepsia	digestion	dys/pepsia—bad digestion

Commonly Used Suffixes cont'd

Suffix	Meaning	Example
-pexy	surgical fixation	nephro/pexy—surgical fixation of kidney
-phasia	speak, say	a/phasia—without ability to speak
-philia	love, attraction	chromo/philic—attracted to color
-phobia	abnormal fear	agora/phobia—abnormal fear of crowds
-plasia	formation	hyper/plasia—excessive formation
-plasm	substance	proto/plasm—original substance
-plasty	make, shape	rhino/plasty—to shape the nose
-plegia	paralysis	hemi/plegia—paralysis of one half of body
-pnea	breath	tachy/pnea—fast breathing
-ptosis	prolapse, dropping	hystero/ptosis—prolapse of uterus
-rrhagia	burst forth	metro/rrhagia—hemorrhage from uterus
-rrhaphy	suture, sew	hernio/rrhaphy—suture a hernia
-rrhea	flow, discharge	oto/rrhea—discharge from ear
-rrhexis	rupture	spleno/rrhexis—rupture of the spleen
-scope	instrument for viewing	oto/scope—instrument to look in ears
-scopy	visualization	laryngo/scopy—visualization of larynx
-some, -soma	body	lyso/some—body that lyses/dissolves
-spasm	twitching	blepharo/spasm—twitching of eyelid
-stasis	stop, control	hemo/stasis—control bleeding
-therapy	treatment	hydro/therapy—treatment with water
-tome	instrument to cut	osteo/tome—instrument to cut bone
-tomy	to cut	laparo/tomy—to cut into the abdomen
-tripsy	crushing	nephro/litho/tripsy—crushing stone in kidney
-trophy, -trophic, -trophin	development	hyper/trophy—overdevelopment
-uria	urine	hemat/uria—blood in the urine

Commonly Used Root Words*

Roots	Meaning	Example
acr-	extremity, peak	acro/megaly—enlarged extremities
		acro/phobia—abnormal fear of heights
aden-	gland	adeno/pathy—disease of a gland
aer-	air	aero/phagia—swallowing air
angi-	vessel	angi/oma—tumor of a vessel
arthr-	joint	arthr/algia—pain in the joint
blast-	bud, growing thing	neuro/blast—growing nerve cell
blephar-	eyelid	blepharo/ptosis—drooping of eyelid
brachi-	arm	brachial—pertaining to the arm
bronch-	windpipe	bronch/us—a branch of the trachea (windpipe)
carcin-	cancer	adeno/carcin/oma—cancerous tumor of a gland
cardi-	heart	myo/cardi/tis—inflammation of heart muscle
carp-	wrist	flexor carpi—muscle to bend wrist
caud-	tail	caudal—pertaining to tail
celio-	abdomen	celio/tomy—incision of the abdomen
cephal-	head	cephalo/dynia—pain in the head
cervic-	neck	cervic/itis—inflammation of the neck of uterus
cheil-	lip	cheilo/plasty—shaping the lip
cheir-, chir-	hand	chiro/megaly—large hands
chol-	bile, gall	chole/cyst/ectomy—surgical removal of the gallbladder
chondr-	cartilage	chondro/malacia—softening of cartilage
chrom-	color	poly/chromatic—having many colors
chron-	time	syn/chron/ous—occurring at the same time
col-	colon	mega/colon—enlarged colon

Commonly Used Root Words* cont'd

Roots	Meaning	Example
colp-	vagina	colp/orrhaphy—suture of vagina
cost-	rib	inter/costal—between the ribs
crani-	skull	crani/otomoy—incision into the skull
cry-	cold	cryo/philic—cold loving
crypt-	hidden	crypt/orchid/ism—hidden (undescended) testicle
cutan-, cut-	skin	sub/cutaneous—below the skin
cyan-	blue	acro/cyan/osis—abnormal condition of blueness of extremities
cyst-	bladder	cysto/cele—bladder hernia
cyt-	cell	thrombo/cyte—clotting cell (platelet)
dacry-	tear	dacryo/rrhea—flow of tears
dactyl-	fingers, toes	poly/dactyl/ism—too many fingers and toes
dent-, odont-	tooth	peri/odontal—around the teeth
		dent/algia—toothache
derm-, dermat-	skin	intra/dermal—within the skin
dextr-	right	dextro/cardia—heart displaced to the right
dips-	thirst	poly/dipsia—excessive thirst
dors-	back	dorsal—pertaining to the back
duct-	carry	ovi/duct—tube to carry ova (eggs)
encephal-	brain	encephalo/cele—hernia of the brain
enter-	intestine	gastro/enter/itis—inflammation of stomach and intestine
erg-	work	en/ergy—working with
erthyr-	red	erthyro/ctyo/penia—deficiency of red cells
esthe-	sensation	an/esthe/tic—agent to eliminate sensation
esthen-	weakness	my/esthenia—muscle weakness
febr-	fever	a/febrile—without a fever
flex-	bend	dorsi/flex—bend backward
gastr-	stomach	gastro/scopy—visualization of the stomach
gen-	produce	patho/genic—agent that produces disease
gingiv-	gums	gingiv/ectomy—removal of gums
gloss-	tongue	hypo/glossal—under the tongue
glyc-, glu-	sugar	hypo/glyc/emia—low blood sugar
gnath-	jaw	micro/gnath/ism—small jaw
grav-	heavy	secundi/gravida—second pregnancy
gynec-	female	geneco/logy—study of female conditions
hem-, hemat-	blood	hemat/emesis—vomiting blood
hepat-	liver	hepato/megaly—enlarged liver
heter-	different	hetero/genous—different origins
hidr-	perspiration	hidro/rrhea—flow of perspiration
hist-	tissue	histo/logy—study of tissue
home-, hom-	same	homeo/stasis—stay same, equilibrium
hydr-, hydra-	water	de/hydra/tion—process of losing water
hyster-	uterus	hyster/ectomy—removal of the uterus
iatr-	physician	iatro/genic—produced by the physician
irid-	iris	irid/ectomy—surgical removal of iris
is-	equal	iso/tonic—equal in pressure
kary-	nut, nucleus	mega/kayro/cyte—cell with large nucleus
kerat-	cornea	kerato/plasty—repair of cornea
kin-	move	kinesio/logy—study of movement
lacrim-	tear	lacrima/tion—crying
lact-, galact-	milk	lacto/genic—milk producing
lapar-	abdomen	laparo/rrhaphy—suture of the abdomen
laryng-	larynx	laryngo/scopy—visualization of larynx
later-	side	bi/lateral—two sides
leuk-, leuc-	white	leuko/rrhea—white discharge
lingu-	tongue	sub/lingual—under the tongue

Commonly Used Root Words* cont'd

Roots	Meaning	Example
lip-	fat	lip/oma—tumor of fat
lith-	stone	litho/tripsy—crushing a stone
mast-, mamm-	breast	mast/itis—inflammation of the breast
		mammo/gram—X ray of breast
melan-	black	melan/oma—black tumor
men-	monthly, menses	dys/meno/rrhea—difficult monthly flow
mening-	membrane	mening/es—membranes that cover the brain and spinal cord
metr-	uterus	endo/metr/ium—lining of uterus
morph-	shape, form	poly/morphic—pertaining to many shapes
my-	muscle	myo/metr/itis—inflammation of muscle of uterus
myc-	fungus	onycho/myc/osis—fungus condition of the nails
myel-	marrow, spinal cord	myelo/gram—X-ray record of spinal cord
myring-	eardrum	myringo/tomy—opening into eardrum
nas-	nose	naso/pharyng/eal—pertaining to nose and throat
nat-	to be born	pre/nat/al—before birth
necr-	dead	necr/opsy—examining dead bodies; autopsy
nephr-, ren-	kidney	hydro/nephr/osis—abnormal condition of water in the kidney
neur-	nerve	neur/algia—nerve pain
noct-, nyct-	night	noct/uria—voiding at night
nucle-	kernel	nucle/us—dense core (kernel) of an atom
null-	none	nulli/gravida—woman who has had no pregnancies
ocul-	eye	mon/ocular—pertaining to one eye
omphal-	umbilicus	omphalo/rrhea—discharge from the navel
onych-	nail	onycho/crypt/osis—condition of hidden nail (ingrown)
oo-	ova, egg	oo/genesis—producing eggs
oophor-	ovary	oophoro/cyst/ectomy—removal of cyst from ovary
ophthalm-	eye	ex/ophthalmos—condition of protruding eyes
or-	mouth	oro/pharyngeal—pertaining to mouth and throat
orchid-	testis	orchid/ectomy—removal of testis
orexis-	appetite	an/orexis—absence of appetite
orth-	straight	orth/odont/ist—one who straightens teeth
oste-, oss-	bone	osteo/chondr/oma—tumor of bone and cartilage
ot-, aur-	ear	ot/itis—inflammation of the ear
		post/auricular—behind the ear
para-	to bear	primi/para—to bear first child
path-	disease	patho/physio/logy—study of effects of disease on body functioning
pect-	chest	pectoralis—chest muscle
ped-	child	ped/iatrician—doctor who specializes in children
peps-	digest	dys/pepsia—bad digestion
phag-	swallow, eat	a/phagia—inability to swallow
pharmac-	drug	pharmaco/logy—study of drugs
pharyng-	throat	pharyng/itis—inflammation of the throat
phas-	speak, say	tachy/phasia—speaking fast
phleb-	vein	phlebo/thromb/osis—abnormal condition of clot in vein
phon-	voice	a/phonic—absence of voice
phren-	diaphragm	phreno/hepatic—pertaining to the diaphragm and liver
pil-, trich-	hair	tricho/glossia—hairy tongue
pneum-	air, breath	pneumo/thorax—air in the chest
pneumon-	lung	pneumon/ectomy—surgical removal of the lung
pod-	foot	pod/iatrist—one who specializes in foot problems
proct-	rectum	procto/scopy—visualization of the rectum
pseud-	false	pseudo/cyesis—false pregnancy
psych-	mind	psycho/somatic—pertaining to the mind and the body
pulmo(n)-	lung	cardio/pulmonary—pertaining to heart and lungs
py-	pus	pyo/rrhea—flow of pus
pyel-	kidney pelvis	pyelo/nephr/itis—inflammation of the kidney pelvis

Commonly Used Root Words* cont'd

Roots	Meaning	Example
pyl-	door, orifice	pyl/oric sphincter—ring of muscle that controls food entry into duodenum
pyr-	fire, fever	anti/pyretic—agent used against fever
quadri-	four	quadri/plegia—paralysis of all four extremities
rhin-	nose	rhino/plasty—revision of the nose
salping-	tube	salping/itis—inflammation of the uterine tube
sanguin-	blood	ex/sanguina/tion—process of bleeding out (bleed to death)
sarc-, sarco-	flesh, striated muscle	sarco/lemma—cell membrane of a muscle fiber
scler-	hard	arterio/scler/osis—condition of hardening of arteries
sect-	cut	dis/section—cutting apart
sept-	contamination	anti/septic—agent used against contamination
sial-	saliva	poly/sialia—excessive salivation
sten-	narrow, constricted	pyloric stenosis—narrowing of pylorus
stomat-	mouth	stoma/itis—inflammation of the mouth
strict-	draw tight	vaso/con/strict/or—agent that compresses vessels
tax-	order, arrange	a/taxic—uncoordinated
ten-	tendon	teno/rrhaphy—suture a tendon
therm-	heat	hyper/thermia—raising body heat
thorac-	chest	thoraco/centesis—puncture to aspirate fluid from chest
thromb-	clot	thrombo/cyte—clotting cell
tox-	poison	tox/emia—poison in the blood
trache-	windpipe	tracheo/malacia—softening of tracheal cartilages
trachel-	neck	trachel/orrhaphy—suture of cervix (neck of uterus)
traumat-	wound	traumat/ology—study of trauma
tri-	three	tri/geminal—having three beginnings
troph-	turn	ec/tropion—turned out
ur-	urine	ur/emia—urine constituents in the blood
vas-	vessel	vaso/constriction—narrowing of a vessel
vert-	turn	retro/vert/ed—turned backward
vesic-	bladder	vesico/cele—hernia of the bladder
viscer-	internal organs	e/viscera/tion—process of viscera protruding from abdominal wall
vita-	life	vital—necessary for life

*Words that are the same as anatomical terms used in English (for example, pancreas, tonsil, and so on) have been omitted.

Now you are ready to apply your knowledge! Think of the word parts when you encounter new words in the text, and try to analyze the meaning. At the end of each chapter, in the Medical Terminology Reinforcement Exercise, you are given an opportunity to reinforce your knowledge by pronouncing the words and dissecting them into parts to arrive at a meaning. You can also begin to build medical words and use them in your everyday conversation. You will be amazed at how rapidly your vocabulary will grow and how your study of the human body will become easier and more enjoyable.

Answers to MedAlert Questions

Chapter 3

Cell Structure and Function

1. The longer we live, the more time there is to acquire "promoters" of cancer.
2. Smokers take carcinogens into the respiratory tract.
3. Certain foods are known to inhibit cancer, while others are known to promote cancer.

Chapter 6

The Skeletal System

1. It takes time for a joint to be "overworked."
2. Artificial hips do not have the flexibility of natural ones.

Chapter 7

The Muscular System

1. Exercise promotes regular bowel movements because it encourages movement of intestinal contents.
2. Exercise requires energy; therefore, it uses up body fat.
3. It would improve longevity because the heart does not work as hard.

Chapter 8

The Nervous System

1. Neurons that release this neurotransmitter are damaged in AD.
2. AD runs in families.
3. The AD neuron has neurofibrillary tangles and amyloid plaques.

Chapter 9

The Senses

1. and 2. If the macula lutea degenerates, a person cannot see detail or color.
3. Vision in dim light is dependent on the rods, which are found outside the macula.

Chapter 12

The Circulatory System

1. Such a diet reduces blood cholesterol levels. High blood cholesterol levels cause plaque, particularly in coronary arteries.
2. During a myocardial infarction, a thromboembolism, or clot, lodges in a coronary artery that has already been narrowed by plaque. The portion of the heart deprived of blood dies, and surrounding tissue may be damaged.

Chapter 13

The Lymphatic System and Immunity

1. An HIV blood test is used to detect the presence of antibodies in the blood that are directed against HIV. A positive HIV test indicates prior exposure to the virus. A T4 cell count examines the number of T4 cells in the blood. AIDS is characterized by a T4 cell count below 200 per cubic millimeter.
2. Individuals with AIDS have a drastically weakened immune system and are unable to fend off infections that are normally nonfatal.

Chapter 14

The Respiratory System

1. Cigarette smoke first weakens and finally destroys cilia that protect the lungs from bacteria and viruses.
2. When arterioles constrict, less nutrient- and oxygen-rich blood is delivered to the heart or placenta.
3. The impurities bring about lung cancer and cause the alveoli to break down so that emphysema results.

Chapter 15

The Digestive System

1. A reflex action signals the urge to defecate, but relaxation of the sphincters is under voluntary control.
2. The defecation reflex is sometimes damaged by overuse of laxatives.

Chapter 16

The Urinary System

1. The drug nafarelin stops testosterone production, and finasteride stops the action of an enzyme that acts on testosterone. Terazosin relaxes the muscular wall of the urethra, leading to better urine flow.
2. Drug therapy interferes with testosterone metabolism.

Chapter 17

The Reproductive System

1. The more days of menstrual flow, the greater the chance of abnormal flow occurring.
2. Menstruation ceases during pregnancy and may decrease in duration when using birth control pills.
3. Removal of the ovaries causes the ovarian and uterine cycles to cease.

Chapter 18

Human Development and Birth

1. The physician would examine the sample microscopically to see if the red blood cells have a sickle shape.
2. Antibodies and antigens form a complex that can be seen microscopically.

Appendix C

Further Readings

Achord, J. L. March/April 1995. Alcohol and the liver. *Scientific American Science & Medicine* 2(2):16.

Alcamo, I. E. 1997. *AIDS: The biological basis.* 2d ed. Dubuque, Iowa: The McGraw-Hill Companies, Inc.

American Chemical Society. 2000. *Chemistry in context: Applying chemistry to society.* 3d ed. Dubuque, Iowa: The McGraw-Hill Companies, Inc.

Applegate, E. J. 1995. *The anatomy and physiology learning system.* Philadelphia: W. B. Saunders.

Bartecchi, C. E. et al. May 1995. The global tobacco epidemic. *Scientific American* 272(5):44.

Bayley, H. September 1997. Building doors into cells. *Scientific American* 277(3):62. Protein engineers are designing artificial pores for drug delivery.

Beardsley, T. January 1994. A war [on cancer] not won. *Scientific American* 270(1):130.

Beardsley, T. March 1996. Vital data. *Scientific American* 274(3):100. DNA tests for a wide array of conditions are becoming available.

Beardsley, T. August 1997. The machinery of thought. *Scientific American* 277(2):78. Researchers have identified the area of the brain responsible for memory.

Becker, W. M., and D. W. Deamer. 1995. *The world of the cell.* 3d ed. Redwood City, Calif.: Benjamin/Cummings Publishing.

Berns, M. W. April 1998. Laser scissors and tweezers. *Scientific American* 278(62):4. New laser techniques allow manipulation of chromosomes and other structures inside cells.

Bikle, D. D. March/April 1995. A bright future for the sunshine hormone. *Scientific American Science & Medicine.* 2(2):58.

Blumenthal, M. et al. March 1999. Discoveries in allergy and asthma. *Discover* 20(3):S-1. This supplement examines advances in understanding and treating allergy and asthma.

Boon, T. March 1993. Teaching the immune system to fight cancer. *Scientific American* 268(3):82.

Borek, C. November/December 1997. Antioxidants and cancer. *Scientific American Science & Medicine* 4(6):52. The importance of supplemental antioxidant vitamins depends on factors such as diet and lifestyle.

Borén, T., and P. Falk. September/October 1994. *Helicobacter pylori* binds to blood group antigens. *Scientific American Science & Medicine* 1(4):28.

Brown, J. L., and E. Pollitt. February 1996. Malnutrition, poverty and intellectual development. *Scientific American* 274(2):38. Article discusses the complex role of essential nutrients in a child's mental development.

Caret, R. L. et al. 1997. *Principles and applications of organic and biological chemistry.* 2d ed. Dubuque, Iowa: The McGraw-Hill Companies, Inc. This text emphasizes material unique to health-related studies.

Cavanee, W. K., and R. L. White. March 1995. The genetic basis of cancer. *Scientific American* 272(3):72.

Chiu, R. C. J. November/December 1994. Using skeletal muscle for cardiac assistance. *Scientific American Science & Medicine* 1(5):68.

Clemente, C. D. 1996. *Anatomy: A regional atlas of the human body.* 4th ed. Philadelphia: Lea and Febiger.

Cooper, G. 1993. *The cancer book.* Boston: Jones and Bartlett Publishers.

Cooper, G. M. 1992. *Elements of human cancer.* Boston: Jones and Bartlett Publishers.

Crowley, L. 1997. *Introduction to human disease.* Boston: Jones & Bartlett Publishers. This well-illustrated text for study in the allied health fields describes diseases and their symptoms, diagnoses, and treatments.

Curiel, T. September/October 1997. Gene therapy: AIDS-related malignancies. *Scientific American Science & Medicine* 4(5):4. The field of AIDS-related gene therapies is advancing.

Dickman, S. July 1997. Mysteries of the heart. *Discover* 18(7):117. Article discusses why coronary arteries may still become blocked after treatment for atherosclerosis.

Duan, L., and R. J. Pomerantz. May/June 1996. Intracellular antibodies for HIV-1 gene therapy. *Scientific American Science & Medicine* 3(3):24. Article discusses the cloning of synthetic antibody fragments that can inhibit the function of viral proteins.

Eisenbarth, G. S., and D. Bellgrau. May/June 1994. Autoimmunity. *Scientific American Science & Medicine* 1(2):38.

Emini, E. A. May/June 1995. Hurdles in the path to an HIV-1 vaccine. *Scientific American Science & Medicine* 2(3):38.

Fox, S. I. 1999. *Human physiology.* 6th ed. Dubuque, Iowa: The McGraw-Hill Companies, Inc.

Frank, J. September/October 1998. How the ribosome works. *American Scientist* 86(5):428. New imaging techniques using cryo-electron microscopy allows researchers to study a three-dimensional map of the ribosome.

Frenay, A. C. F., and R. M. Mahoney. 1998. *Understanding medical terminology.* 10th ed. Dubuque, Iowa: The McGraw-Hill Companies, Inc.

Garnick, M. B. April 1994. The dilemmas of prostate cancer. *Scientific American* 270(4):72.

Garnick, M. B., and W. R. Fair. December 1998. Combating prostate cancer. *Scientific American* 279(6):74. Article details the recent developments in diagnosis and treatment of prostate cancer.

Gibbs, W. W. August 1996. Gaining on fat. *Scientific American* 275(2):88. Some weight problems are genetic or physiological in origin. New treatments might help.

Glausiusz, J. October 1997. The good bugs on our tongues. *Discover* 18(10):32. Without the friendly bacteria that live on our tongues, we would be vulnerable to bacteria such as *Salmonella*.

Glausiusz, J. September 1998. Infected hearts. *Discover* 19(9):30. Infectious bacteria may play a role in heart disease; antibiotics could prevent the need for heart surgery.

Glausiusz, J. January 1999. The genes of 1998. *Discover* 20(1):33. Nine important human genes that were identified in 1998 through the Human Genome Project are examined.

Glover, D. M. et al. June 1993. The centrosome. *Scientific American* 268(6):62.

Goldberg, J. April 1998. A head full of hope. *Discover* 19(4):70. A new gene therapy for killing brain cancer cells is presented.

Golde, D. W. December 1991. The stem cell. *Scientific American* 265(6):86.

Golub, E. S., and D. R. Green. 1992. *Immunology: A synthesis.* 2d ed. Sunderland, Mass.: Sinauer Associates.

Green, H. November 1991. Cultured cells for the treatment of disease. *Scientific American* 265(5):96.

Grillner, S. January 1996. Neural networks for vertebrate locomotion. *Scientific American* 274(1):64. Discoveries about how the brain coordinates muscle movement raise hopes for restoration of mobility for some accident victims.

Gunstream, S. E. 2000. *Anatomy and physiology.* 2d ed. Dubuque, Iowa: Wm. C. Brown Publishers.

Guyton, A. C. 1992. *Human physiology and mechanisms of disease.* 5th ed. Philadelphia: Saunders College Publishing.

Hales, C. N. July/August 1994. Fetal nutrition and adult diabetes. *Scientific American Science & Medicine* 1(3):54.

Hales, D. 1994. *An invitation to health.* 6th ed. Redwood City, Calif.: Benjamin/Cummings Publishing.

Halstead, L. S. April 1998. Post-polio syndrome. *Scientific American* 278(4):42. Recovered polio victims are experiencing fatigue, pain, and weakness, resulting from degeneration of motor neurons.

Hanson, L. A. November/December 1997. Breast feeding stimulates the infant immune system. *Scientific American Science & Medicine* 4(6):12. Long-lasting protection against some infectious diseases has been reported in breast-fed infants.

Harken, A. H. July 1993. Surgical treatment of cardiac arrhythmias. *Scientific American* 269(1):68.

Harvard Health Letter. April 1998. A special report: Parkinson's disease. This overview presents the symptoms and diagnosis of Parkinson disease, and discusses medications and surgical methods of treatment.

Hirshhorn, N., and W. B. Greenbough, III. May 1991. Progress in oral rehydration therapy. *Scientific American* 264(5):50.

Hole, J. W., Jr. 1995. *Essentials of human anatomy and physiology.* 5th ed. Dubuque, Iowa: Wm. C. Brown Publishers.

Holloway, M. March 1991. Rx for addiction. *Scientific American* 264(3):94.

Jensen, M. M., and D. N. Wright. 1992. *Introduction to microbiology for the health sciences.* 3d ed. Englewood Cliffs, N.J.: Prentice-Hall.

Johnson, H. M. et al. April 1992. Superantigens in human disease. *Scientific American* 266(4):92.

Johnson, H. M. et al. May 1994. How interferons fight disease. *Scientific American* 270(5):68.

Jordan, V. C. October 1998. Designer estrogens. *Scientific American* 279(4):60. Selective estrogen receptor modulators may protect against breast and endometrial cancers, osteoporosis, and heart disease.

Julien, R. M. 1992. *A primer of drug action.* 6th ed. New York: W. H. Freeman.

Kempermann, G., and F. Gage. May 1999. New nerve cells for the adult brain. *Scientific American* 280(5):48. The knowledge that the human brain can produce new nerve cells in adulthood could lead to better treatments for neurological diseases.

Kher, U. January 1998. A manmade chromosome. *Discover* 18(1):40. Researchers announce a promising new gene carrier, a human artificial chromosome, for use in gene therapy.

Klatsky, A. L. March/April 1995. Cardiovascular effects of alcohol. *Scientific American Science & Medicine* 2(2):28.

Koff, R. S. March/April 1994. Solving the mysteries of viral hepatitis. *Scientific American Science & Medicine* 1(1):24.

Koprowski, J. May/June 1995. Visit to an ancient curse (rabies). *Scientific American Science & Medicine* 2(3):48.

Krauskopf, S. January 1999. Doing the meiosis shuffle. *American Biology Teacher* 61(1):60. A playing card demonstration walks students through the stages of meiosis.

Kunzig, R. February 1999. What's a pinna for? *Discover* 20(2):24. Article examines the function of the folds of the outer ear.

Lacy, P. E. July 1995. Treating diabetes with transplanted cells. *Scientific American* 273(1):50.

Lang, F., and Waldegger, S. September/October 1997. Regulating cell volume. *American Scientist* 85(5):456. Changes in cell volume may threaten organ or tissue function.

Lasic, D. D. May/June 1996. Liposomes. *Science & Medicine* 3(3):34. Liposomes can be used to deliver drugs or genes for gene therapy.

Leffell, D. J., and D. E. Brash. July 1996. Sunlight and skin cancer. *Scientific American* 275(1):52. Discusses the sequence of changes that may occur in skin cells after exposure to UV rays.

Lienhard, G. E. et al. January 1992. How cells absorb glucose. *Scientific American* 266(1):86.

Life, death, and the immune system (special issue). 1993. *Scientific American* 269(3).

Liotta, L. A. February 1992. Cancer cell invasion and metastasis. *Scientific American* 266(2):54.

Little, R. C., and W. C. Little. 1989. *Physiology of the heart and circulation.* 4th ed. Chicago: Year Book Medical Publishers, Inc. A good reference resource that gives an in-depth look at cardiovascular physiology.

MacDonald, P. C., and M. L. Casey. March/April 1996. Preterm birth. *Scientific American Science & Medicine* 3(2):42. Article discusses the role of oxytocin, prostaglandins, and infections in the initiation of human labor.

Mader, S. S. 1992. *Human reproductive biology.* 2d ed. Dubuque, Iowa: The McGraw-Hill Companies, Inc. An introductory text covering human reproduction in a clear, easily understood manner.

Mader, S. S. 2000. *Human biology.* 7th ed. Dubuque, Iowa: The McGraw-Hill Companies, Inc. A student-friendly text that covers the principles of biology with emphasis on human anatomy and physiology.

Marcus, D. M., and M. W. Camp. May/June 1998. Age-related macular degeneration. *Scientific American Science & Medicine* 5(3):10. New therapies are needed for this common cause of vision loss in the elderly.

Mattson, M. P. March/April 1998. Experimental models of Alzheimer's disease. *Scientific American Science & Medicine* 5(2):16. In Alzheimer disease, mutations accelerate changes that occur during normal aging.

Mayor, M. B., and J. Collier. May/June 1994. The technology of hip replacement. *Scientific American Science & Medicine* 1(2):58.

Melzack, R. April 1992. Phantom limbs. *Scientific American* 266(4):120.

Moore, K., and T. Persaud. 1998. *Before we are born: Essentials of embryology and birth defects.* 5th ed. Philadelphia: W. B. Saunders and Co. For medical and associated health students, this text presents the essentials of normal and abnormal human embryological development.

Moore, P. S., and C. V. Broome. November 1994. Cerebrospinal meningitis epidemics. *Scientific American* 271(5):38.

Nature Medicine Vaccine Supplement, May 1998, vol. 4, no. 5. Entire issue is devoted to the topic of vaccines, including history, recent developments and research in malaria, cancer, and HIV vaccines.

Nestler, J. E. September/October 1994. Assessment of insulin resistance. *Scientific American Science & Medicine* 1(4):58.

Nicholls, J. G., A. R. Martin, and B. G. Wallace. 1992. *From neuron to brain.* Sunderland, Mass.: Sinauer Associates.

Nilsson, L. 1990. *A child is born.* 2d ed. New York: Delacorte Press.

Nolte, J. 1998. *The human brain.* 4th ed. St. Louis: Mosby-Year Book, Inc. Beginners are guided through the basic aspects of brain structure and function.

Nowak, M. A., and A. J. McMichael. August 1995. How HIV defeats the immune system. *Scientific American* 273(2):58.

Nucci, M. L., and A. Abuchowski. February 1998. The search for blood substitutes. *Scientific American* 278(2):72. Artificial blood substitutes based on hemoglobin are being developed from synthetic chemicals.

Opie, L. H. July/August 1994. ACE inhibitors: Almost too good to be true. *Scientific American Science & Medicine* 1(3):14.

Packer, C. July/August 1998. Why menopause? *Natural History* 107(6):24. Article addresses possible reasons why menopause occurs so early in life, compared to other aging processes.

Packer, L. March/April 1994. Vitamin E is nature's master antioxidant. *Scientific American Science & Medicine* 1(1):54.

Perls, T. T. January 1995. The oldest old. *Scientific American* 272(1):70.

Plomerin, R., and J. C. DeFries. May 1998. The genetics of cognitive abilities and disabilities. *Scientific American* 278(5):62. The search is underway for the genes involved in cognitive abilities and disabilities, including dyslexia.

Pool, R. May 1998. Saviors. *Discover* 19(5):52. Genetic engineering may make animal organs compatible for human transplants.

Powledge, T. M. July 1999. Addiction and the brain. *BioScience* 49(7):513. Drug use changes the biochemistry and anatomy of neurons and alters the way they work.

Ray, O., and C. Ksir. 1993. *Drugs, society, & human behavior.* 6th ed. St. Louis: Mosby-Year Book.

Rennie, J. December 1990. The body against itself. *Scientific American* 263(6):106.

Rodgers, G. P. et al. October 1994. Sickle cell anemia. *Scientific American Science & Medicine* 1(4):48.

Ross, F. C. 1997. *Foundation of allied health sciences: An introduction to chemistry and cell biology.* 4th ed. Dubuque, Iowa: The McGraw-Hill Companies, Inc. This introductory text provides the background necessary for students in allied health sciences.

Sack, R. L. September/October 1998. Melatonin. *Science & Medicine* 5(8):8. Certain mood and sleep disorders can be managed with melatonin treatments.

Sarrel, P. M. et al. July/August 1994. Estrogen actions in arteries, bone, and brain. *Scientific American Science & Medicine* 1(3):44.

Sataloff, R. T. December 1992. The human voice. *Scientific American* 267(6):108.

Schiller, L. R. November/December 1994. Peristalsis. *Scientific American Science & Medicine* 1(5):16.

Schultz, J. S. August 1991. Biosensors. *Scientific American* 265(2):64.

Schwartz, A. T. et al. 1997. *Chemistry in context: Applying chemistry to society.* 2d ed. Dubuque, Iowa: Wm. C. Brown Publishers. This introductory text is designed for students in the allied health fields.

Schwartz, R. H. August 1993. T cell energy. *Scientific American* 269(2):62.

Schwartz, W. J. May/June 1996. Internal timekeeping. *Scientific American Science & Medicine* 3(3):44. Circadian rhythm mechanisms are discussed.

Scientific American September 1996. What you need to know about cancer (special issue). *Scientific American* 275(3). The entire issue is devoted to the causes, prevention, and early detection of cancer, and cancer therapies—conventional and future.

Scientific American editors. June 1997. Special report: Making gene therapy work. *Scientific American* 276(6):95. Obstacles must be overcome before gene therapy is ready for widespread use.

Scientific American July 1998. Defeating AIDS: What will it take? *Scientific American* 279(1):81. Nine separate articles address AIDS problems and issues.

Scientific American April 1999. The promise of tissue engineering. *Scientific American* 280(4):59–89. Much of the issue examines the hopes and challenges of tissue engineering for use in gene therapy and for the growth of new organs.

Scrimshaw, N. S. October 1991. Iron deficiency. *Scientific American* 265(4):46.

Selkoe, D. J. November 1991. Amyloid protein and Alzheimer's disease. *Scientific American* 265(5):68.

Shier, D., J. Butler, and R. Lewis. 2000. *Essentials of human physiology.* 7th ed. Dubuque, Iowa: The McGraw-Hill Companies, Inc.

Sloane, E. 1994. *Anatomy and physiology: An easy learner.* Boston: Jones and Bartlett Publishers.

Smith, K. A. March 1990. Interleukin-2. *Scientific American* 262(3):50.

Smith, R. March 1999. The timing of birth. *Scientific American* 280(3):68. Research shows that a hormone found in the human placenta influences the timing of delivery. This knowledge could yield ways to prevent premature labor.

Snider, D. E. et al. May/June 1994. Multi-drug-resistant tuberculosis. *Scientific American Science & Medicine* 1(2):16.

Stamler, J., and J. D. Neaton. May/June 1994. Benefits of lower cholesterol. *Scientific American Science & Medicine* 1(2):28.

Stix, G. October 1997. Growing a new field. *Scientific American* 277(4):15. Tissue engineers try to grow organs in the laboratory.

Sussman, N. L., and J. H. Kelly. May/June 1995. The artificial liver. *Scientific American Science & Medicine* 2(3):68.

Swerdlow, J. L. June 1995. The brain. *National Geographic* 187(6):2.

Tate, P., R. A. Seeley, and T. D. Stepens. 1994. *Understanding the human body.* St. Louis: Mosby.

Taubes, G. February 1999. The cold warriors. *Discover* 20(2):40. Article examines research in the cure for the common cold.

Thibodeau, G. A., and K. T. Patton. 1997. *The human body in health & disease.* 2nd ed. St. Louis: Mosby-Year Book.

Tortora, G. J., and S. R. Grabowski. 2000. *Principles of anatomy and physiology.* 9th ed. New York: John Wiley & Sons.

Valtin, H. 1994. *Renal function.* 3d ed. Boston: Little, Brown and Company. A good reference resource that discusses renal mechanisms for preserving fluid and solute balance.

Valtin, H., and J. Schafer. 1995. *Renal function.* 3d ed. Boston: Little, Brown.

Van De Graaff, K. M. 1995. *Human anatomy.* 4th ed. Dubuque, Iowa: Wm. C. Brown Publishers.

Van De Graaff, K. M., and S. I. Fox. 1999. *Concepts of human anatomy & physiology.* 5th ed. Dubuque, Iowa: The McGraw-Hill Companies, Inc.

Vander, A. J., J. H. Sherman, and D. S. Luciano. 2001. *Human physiology: The mechanism of body function.* 8th ed. New York: The McGraw-Hill Companies, Inc.

Van Noorden, C. J. F. et al. March/April 1998. Metastasis. *American Scientist* 86(2):130. The mechanisms by which cancer cells metastasize are discussed.

vonBoehmer, H., and P. Kisielow. October 1991. How the immune system learns about self. *Scientific American* 265(4):74.

Wardlaw, G. et al. 1996. *Contemporary nutrition.* 3d ed. St. Louis: Mosby-Year Book, Inc. This text gives a clear understanding of nutritional information found on product labels.

Weindruch, R. January 1996. Caloric restriction and aging. *Scientific American* 274(1):64. Consuming fewer calories may increase longevity. Digestion/Development.

Weiss, R. November 1997. Aging—new answers to old questions. *National Geographic* 192(5):2. The mechanics of human aging are studied.

Weissman, G. January 1991. Aspirin. *Scientific American* 264(1):84.

West, J. B. 1999. *Respiratory physiology—The essentials.* 6th ed. Baltimore: Williams & Wilkins. A good reference resource that discusses all aspects of respiratory physiology including breathing, and external and internal respiration.

White, R. J. September 1998. Weightlessness and the human body. *Scientific American* 279(3):58. Space medicine is providing new ideas about treatment of anemia and osteoporosis.

Winkonkal, N. M., and D. E. Brash. September/October 1998. Squamous cell carcinoma. *Scientific American Science & Medicine* 5(5):18. Mutations of tumor-suppressor gene p53 are commonly found in squamous cell carcinomas.

Woolf, N., and M. J. Davies. September/October 1994. Arterial plaque and thrombus formation. *Scientific American Science & Medicine* 1(4):38.

Youdim, M. B., and P. Riederer. January 1997. Understanding Parkinson's disease. *Scientific American* 276(1):52. The tremors and immobility of Parkinson disease can be traced to damage in a part of the brain that regulates movement.

Zivin, J. A., and D. W. Choi. July 1991. Stroke therapy. *Scientific American* 265(1):56.

Glossary

A

abdomen Portion of the body between the diaphragm and the pelvis. 6

abdominal Pertaining to the abdomen. 6

abdominopelvic Pertaining to the abdominal and pelvic regions. 6

abduction Movement of a body part away from the midline. 105

ABO system System of typing blood according to the presence (or absence) of two antigens on red blood cell membranes. 216

accommodation Adjustment of the lens for close vision. 177

acetabulum Socket in the lateral surface of the hipbone into which the head of the femur articulates. 96

acetylcholine (ACh) Neurotransmitter substance secreted at the ends of many neurons; responsible for the transmission of a nerve impulse across a synaptic cleft. 140

acetylcholinesterase (AChE) Enzyme in the membrane of postsynaptic cells that breaks down acetylcholine; this enzymatic reaction inactivates the neurotransmitter. 140

ACh See *acetylcholine.* 140

AChE See *acetylcholinesterase.* 140

acid Solution in which pH is less than 7; substance that contributes or liberates hydrogen ions in a solution; opposite of *base.* 21

acidosis Excessive accumulation of acids in body fluids. 22, 198

acne vulgaris Inflammation of sebaceous glands; the common form of acne. 67

acquired immunodeficiency syndrome See *AIDS* 264, 356

acromegaly Condition resulting from an increase in growth hormone production after adult height has been achieved. 192

acrosome Covering on the tip of a sperm cell's nucleus that is believed to contain enzymes necessary for fertilization. 339

ACTH See *adrenocorticotropin.* 192

actin One of the two major proteins of muscle; makes up thin myofilaments in myofibrils of muscle cells. See *myosin.* 112

action potential Change in potential propagated along the membrane of a neuron; the nerve impulse. 140

active immunity Resistance to disease due to the immune system responding to a microorganism or a vaccine. 262

active transport Transfer of a substance into or out of a cell against a concentration gradient by a process that requires a carrier protein and an expenditure of energy. 41

acute Sudden in onset and severe. 13

acute lymphoblastic leukemia (ALL) Cancer of the blood in which immature lymphocytes proliferate in bone marrow, the thymus, and lymph nodes. 210

acute pancreatitis Blockage preventing pancreatic juice from entering the duodenum, resulting in digestion of pancreatic tissue. 304

Addison disease Condition resulting from a deficiency of adrenal cortex hormones. 196

adduction Movement of a body part toward the midline. 105

adenosine triphosphate (ATP) Molecule used by cells when energy is needed. 29

ADH See *antidiuretic hormone.* 330

adhesion Scar that holds two organs together. 352

adrenalin See *epinephrine.* 195

adrenocorticotropin (ACTH) Hormone secreted by the anterior lobe of the pituitary gland that stimulates the adrenal cortex to produce cortisol. 192

aerobic cellular respiration Metabolic process that uses nutrients and oxygen within mitochondria to produce ATP, the type of chemical energy needed by cells. 38

agglutination Clumping of cells, particularly in reference to red blood cells involved in an antigen-antibody reaction. 217

agranular leukocyte White blood cell with poorly visible cytoplasmic granules. 209

AID See *artificial insemination by a donor.* 353

AIDS (acquired immunodeficiency syndrome) Disease caused by a retrovirus and transmitted via body fluids; characterized by failure of the immune system. 264

albinism Gene disorder characterized by a defect in pigment production. 64

aldosterone Hormone secreted by the adrenal cortex that functions in regulating sodium and potassium excretion by the kidneys. 196, 330

alimentary canal Tubular portion of the digestive tract. 297

alkalosis Excessive accumulation of bases in body fluids. 22

ALL See *acute lymphoblastic leukemia.* 210

allantois Extraembryonic membrane that serves as a source of blood vessels for the umbilical cord. 366

allergy Immune response to substances that usually are not recognized as foreign. 266

all-or-none law Law that states that muscle fibers either contract maximally or not at all, and that neurons either conduct a nerve impulse completely or not at all. 114

alopecia Loss of hair. 65

alveolus Air sac of a lung (pl., *alveoli*). 276

Alzheimer disease Brain disorder characterized by a general loss of mental abilities. 142, 160

amino acid Unit of protein that takes its name from the fact that it contains an amino group ($-NH_2$) and an acid group ($-COOH$). 27

amniocentesis Method of retrieving fetal cells for genetic testing in which a long needle is used to withdraw a sample of amniotic fluid. 364

amnion One of the extraembryonic membranes; a fluid-filled sac around the embryo. 366

amphiarthrosis Slightly movable joint. 100

ampulla Expansion at the end of each semicircular canal that contains receptors for dynamic equilibrium. 180

amylase Starch-digesting enzyme secreted by the salivary glands (salivary amylase) and the pancreas (pancreatic amylase).   298, 304

anatomy Branch of science dealing with the form and structure of body parts.   2

androgen Male sex hormone.   199

anemia Condition characterized by a deficiency of red blood cells or hemoglobin. See also *iron deficiency anemia, pernicious anemia.*   210

angina pectoris Condition characterized by thoracic pain resulting from occluded coronary arteries; precedes a heart attack.   232

ankle-jerk reflex Automatic, involuntary response initiated by tapping the Achilles tendon just above its attachment to the calcaneus (heel bone).   143

ANS See *autonomic nervous system.*   152, 155

antagonist Muscle that acts in opposition to a prime mover, or an agonist.   115

anterior Pertaining to the front; the opposite of *posterior.*   3, 4

anterior pituitary Front lobe of the pituitary gland.   190

antibody Protein produced in response to the presence of some foreign substance in the blood or tissues.   217, 258

antibody-mediated immunity Resistance to disease-causing agents resulting from the production of specific antibodies by B lymphocytes; humoral immunity.   259

antibody titer Amount of antibody present in a sample of blood serum.   262

antidiuretic hormone (ADH) Hormone released from the posterior lobe of the pituitary gland that enhances water conservation by the kidneys; sometimes called vasopressin.   190, 330

antigen Foreign substance, usually a protein, that stimulates the immune system to produce antibodies.   258

anus Outlet of the digestive tube.   307

aorta Major systemic artery that receives blood from the left ventricle.   227

aortic body Receptor in the aortic arch sensitive to oxygen content, carbon dioxide content, and blood pH.   278

aplastic anemia An insufficient number of red blood cells brought on by damage to the red bone marrow due to radiation or chemicals.   210

apnea Temporary cessation of breathing.   278

appendicitis An infected, swelling of the appendix.   309

appendicular Pertaining to the upper limbs (arm) and lower limbs (legs).   5

appendicular skeleton Part of the skeleton forming the upper limbs, pectoral girdle, lower limbs, and pelvic girdle.   91

aqueous humor Watery fluid that fills the anterior cavity of the eye.   172

arachnoid membrane Weblike middle covering (one of the three meninges) of the central nervous system.   150

areola Dark, circular area surrounding the nipple of the breast.   365

arrector pili Smooth muscle in the skin associated with a hair folicle.   66

arrhythmia Abnormal heart rhythm.   230

arterial duct See *ductus arteriosus.*   242

arteriole Branch from an artery that leads into a capillary.   234

arteriosclerosis Thickening and hardening of arterial walls.   235

artery Vessel that takes blood away from the heart; characteristically possesses thick elastic walls.   234

articular cartilage Hyaline cartilaginous covering over the articulating surface of the bones of synovial joints.   79

articulation Joining together of bones at a joint.   79

artificial insemination by a donor (AID) Placement of donated sperm in the vagina so that fertilization followed by pregnancy might occur.   353

ascending colon Portion of the large intestine that travels superiorly as it extends from the entry of the small intestine to the transverse colon.   307

asthma Condition in which bronchioles constrict and cause difficulty in breathing.   277, 287

atherosclerosis Condition in which fatty substances accumulate abnormally beneath the inner linings of the arteries.   232

athlete's foot Skin disease caused by fungal infection, usually of the toes and soles of the foot.   68

atlas First cervical vertebra; it supports and balances the head.   89

atom Smallest unit of matter.   2

ATP See *adenosine triphosphate.*   29

atrial natriuretic hormone (ANH) Substance secreted by the atria of the heart that accelerates sodium excretion so that blood volume decreases.   196, 331

atrioventricular (AV) bundle Part of the cardiac conduction system that extends from the AV node to the bundle branches.   229

atrioventricular (AV) node Small region of neuromuscular tissue located near the septum of the heart that transmits impulses from the SA node to the ventricular walls.   229

atrioventricular (AV) valve Valve located between the atrium and the ventricle.   224

atrium Chamber; particularly an upper chamber of the heart that lies above the ventricles (pl., *atria*).   223

atrophy Wasting away or decrease in size of an organ or tissue.   116

auditory canal Tube in the outer ear that leads to the tympanic membrane.   180

auditory (eustachian) tube Air tube that connects the pharynx to the middle ear.   180

autonomic nervous system (ANS) Sympathetic and parasympathetic portions of the nervous system that function to control the actions of the visceral organs and skin.   152, 155

AV bundle See *atrioventricular bundle.*   229

AV node See *atrioventricular node.*   229

AV valve See *atrioventricular valve.*   224

axial Pertaining to the body's axis.   5

axial skeleton Portion of the skeleton that supports and protects the organs of the head, neck, and trunk.   82

axis Second cervical vertebra upon which the atlas rotates, allowing the head to turn.   5

axon Process of a neuron that conducts nerve impulses away from the cell body.   137

B

ball-and-socket joint The most freely movable type of joint (for example, the shoulder or hip joint).   102

balloon angioplasty Procedure for treating a blocked coronary artery: A flexible guide wire is pushed into coronary artery, and a miniature balloon catheter is pushed down the wire to the blockage; repeated inflations of the balloon decrease or relieve the blockage.   233

basal cell carcinoma Form of skin cancer that begins in the epidermis and rarely metastasizes but has the capacity to invade local tissues.   70

base Solution in which pH is more than 7; a substance that contributes or liberates hydroxide ions in a solution; alkaline; opposite of *acid.*   21

basophil Leukocyte with a granular cytoplasm and that is able to be stained with a basic dye.   212

benign prostatic hyperplasia (BPH) Enlargement of the prostate gland.   332

benign tumor Mass of cells derived from a single mutated cell that has repeatedly undergone cell division but remained at the site of origin.   44

bicarbonate ion The form in which carbon dioxide is carried in the blood; HCO_3^-.   284

bicuspid valve Atrioventricular valve between the left atrium and the left ventricle; also known as the mitral valve.   225

bifocals Corrective lenses with an upper part for distant vision and a lower part for close vision. 178

bile Secretion of the liver that is temporarily stored in the gallbladder before being released into the small intestine, where it emulsifies fat. 304

biopsy Removal of sample tissue by plungerlike devices to diagnose a disease. 57

biotechnology Use of a natural biological system to produce a commercial product. 389

blastocyst Early stage of embryonic development that consists of a hollow ball of cells. 368

blind spot Area where the optic nerve passes through the retina and where vision is not possible due to the lack of rods and cones. 173

blood Connective tissue composed of cells separated by plasma. 55

blood pressure Force of blood against a blood vessel wall. 244

B lymphocyte Type of lymphocyte that is responsible for antibody-mediated immunity. 258

bolus Small lump of food that has been chewed and swallowed. 298

bone Connective tissue having a hard matrix of calcium salts deposited around protein fibers. 53

brachiocephalic Pertaining to the arm and head, as in the brachiocephalic artery. 238

Brackston-Hicks contractions Strong, late-term uterine contractions prior to cervical dilation; also called false labor. 373

bradycardia Slow heart rate, characterized by fewer than 60 heartbeats per minute. 230

breech birth Birth in which the baby is positioned rump first. 372

bronchiole Smaller air passages in the lungs. 276

bronchitis Acute or chronic inflammation of the bronchi. 287

bronchus One of the two major divisions of the trachea; leads to the lungs (pl., *bronchi*). 276

buffer Substance or compound that prevents large changes in the pH of a solution. 22, 331

bulbourethral (Cowper's) gland Gland located in the pelvic cavity that adds secretions to seminal fluid within the urethra. 339

bursa Saclike, fluid-filled structure, lined with synovial membrane, that occurs near a joint (pl., *bursae*). 101

bursitis Inflammation of any of the friction-easing sacs called bursae within the knee joint. 101

C

calcaneus Heelbone. 100

calcitonin Hormone secreted by the thyroid gland that helps regulate the level of blood calcium. 194

capillary Microscopic vessel located in the tissues connecting arterioles to venules; molecules either exit or enter the blood through the thin walls of capillaries. 235

carbaminohemoglobin Hemoglobin carrying carbon dioxide. 284

carbohydrate Organic compounds with the general formula $(CH_2O)_n$, including sugars and glycogen. 23

carcinogen Any agent that causes cancer. 44

carcinoma Cancer arising in epithelial tissue. 57

cardiac Of or pertaining to the heart. 229

cardiac conduction system Neuromuscular tissue and fibers that control the cardiac cycle; includes the SA node, the AV node, the AV bundle and its branches, and the Purkinje fibers. 229

cardiac control center Portion of the medulla oblongata that regulates the heartbeat rate. 230

cardiac muscle Heart muscle (myocardium) consisting of striated muscle cells that interlock. 57, 111

caries Destruction of tooth enamel by oral bacteria. 311

carotid Either of two arteries branching off the aortic arch and supplying the head and neck. 238

carotid body Structure located at the branching of the carotid arteries that contains chemoreceptors. 278

carpals Bones of the wrist. 278

carrier Molecule that combines with a substance and actively transports it through the plasma membrane. 42

cartilage Connective tissue, usually part of the skeleton, that is composed of cells in a flexible matrix. 53

cataract Opaqueness of the lens of the eye, making the lens incapable of transmitting light. 177

CCK See *cholecystokinin*. 301

cecum Blind pouch, such as the one below where the small intestine enters the large intestine. 307

cell Structural and functional unit of an organism; smallest structure capable of performing all the functions necessary for life. 2

cell body Portion of a nerve cell that includes a cytoplasmic mass and a nucleus, and from which the nerve fibers extend. 137

cell-mediated immunity Immunological defense provided by killer T cells, which destroy virus-infected cells, foreign cells, and cancer cells. 260

central Situated at the center of the body or an organ. 4

central canal Tube within the spinal cord that is continuous with the ventricle of the brain and contains cerebrospinal fluid. 151

central nervous system (CNS) Brain and spinal cord. 136, 144

centriole Short, cylindrical organelle that contains microtubules in a 9 + 0 pattern and is associated with the formation of the spindle during cell division. 39

cerebellum Part of the brain that controls muscular coordination. 146

cerebral hemisphere One of the large, paired structures that together constitute the cerebrum of the brain. 146

cerebral palsy Spastic weakness of the arms and legs due to damage to the motor areas of the cerebral cortex. 148

cerebrospinal fluid (CSF) Fluid found within ventricles of the brain and surrounding the CNS in association with the meninges. 151

cerebrovascular accident (CVA) Condition resulting when an arteriole in the brain bursts or becomes blocked by an embolism; stroke. 235

cerebrum Main portion of the vertebrate brain that is responsible for consciousness. 146

cervix Narrow end of the uterus that projects into the vagina. 346

cesarean section Birth by surgical incision of the abdomen and uterus. 372

chemotherapy Use of drugs to kill cancer cells. 45

Cheyne-Stokes respiration Type of respiration characterized by alternate periods of deep, labored breathing and no breathing at all. 278

chlamydia Sexually transmitted disease caused by the bacterium *Chlamydia trachomatis*; often causes painful urination and swelling of the testes in men; is usually symptomless in women but can cause inflammation of the cervix or uterine tubes. 357

cholecystokinin (CCK) Hormone secreted by the small intestine that stimulates the release of pancreatic juice from the pancreas and bile from the gallbladder. 301

chordae tendineae Tough bands of connective tissue that attach the papillary muscles to the atrioventricular valves within the heart. 224

chorion Extraembryonic membrane that forms an outer covering around the embryo and contributes to the formation of the placenta.   366

chorionic villi Projections from the chorion that appear during implantation and that in one area contribute to the development of the placenta.   366

chorionic villi sampling Method of retrieving fetal cells for genetic testing in which a long, thin tube is passed through the vagina into the uterus, and suction is used to obtain a sample of chorionic villi cells.   364

choroid Vascular, pigmented middle layer of the wall of the eye.   172

chromatids Two identical parts of a chromosome following replication of DNA.   43

chromatin Threadlike network in the nucleus that condenses to become the chromosomes just before cell division.   36

chromosome Rod-shaped body in the nucleus, particularly during cell division, that contains the hereditary units, or genes.   36

chronic Long and continued but not acute.   13

chronic obstructive pulmonary disease (COPD) Continued interference with airflow in the lungs due to chronic bronchitis or emphysema.   287

chyme Semifluid food mass leaving the stomach.   300

cilia Membrane-bound microtubular structures that project from a cell and in multicellular animals facilitate the flow of materials over the cell surface.   39

ciliary muscle Muscle that controls the curvature of the lens of the eye.   172

circle of Willis Arterial ring located on the ventral surface of the brain.   241

circumcision Removal of the foreskin of the penis.   341

circumduction Conelike movement of a body part, such that the distal end moves in a circle, while the proximal portion remains relatively stable.   105

cirrhosis Chronic, irreversible injury to liver tissue; commonly caused by frequent alcohol consumption.   306

clavicle Bone extending from the sternum to the scapula.   91

cleavage Cell division of the fertilized egg that is unaccompanied by growth so that numerous small cells result.   367

clitoris Small, erectile, female organ located in the vulva and homologous to the penis.   347

CNS See *central nervous system.*   136

coccyx Caudal end of the vertebral column formed by the fusion of four vertebrae; tailbone.   89

cochlea Portion of the inner ear that contains the receptors of hearing.   180

cochlear canal Canal within the cochlea that bears small hair cells that function as hearing receptors.   180

cochlear implant Prosthetic device used to help persons with severe hearing impairment; the device converts sound to an electrical impulse that directly stimulates the auditory nerve.   182

collecting duct Tube that receives urine from several distal convoluted tubules.   324

colon Large intestine.   324

color blindness Deficiency in one or more of the three kinds of cones responsible for color vision.   176

colostomy Attachment of a shortened colon to a surgical opening in the abdominal wall.   309

columnar epithelium Pillar-shaped cells usually having the nuclei near the bottom of each cell and found lining the digestive tract.   49

compact bone Hard bone consisting of Haversian systems cemented together.   55, 81

complement system Group of proteins in plasma that aid the general defense of the body by destroying bacteria; often called complement.   256

compound Chemical substance having two or more different elements in fixed ratio.   18

concha Shell-shaped structure, such as that seen in the bones of the nasal cavity.   274

condensation Chemical change resulting in the covalent bonding of two monomers with the accompanying loss of a water molecule.   23

conduction deafness Hearing impairment due to fusion of the ossicles or other damage to the middle ear, thereby restricting the ability to transmit and magnify sound.   182

condyle Large, rounded surface at the end of a bone.   98

cone Color receptor located in the retina of the eye.   173

congenital defect Bodily abnormality arising from birth and due to hereditary factors.   369

congestive heart failure Inability of the heart to maintain adequate circulation, especially of the venous blood returned to it.   227

connective tissue Type of tissue, characterized by cells separated by a matrix, that often contains fibers.   51

constipation Infrequent, difficult defecation caused by insufficient water in the feces.   308

COPD See *chronic obstructive pulmonary disease.*   287

cornea Transparent, anterior portion of the outer layer of the eyeball.   172

coronary artery Artery that supplies blood to the wall of the heart (myocardium).   241

coronary bypass operation Therapy for blocked coronary arteries in which part of a blood vessel from another part of the body is grafted around the obstructed artery.   233

corpus callosum Mass of white matter within the brain, composed of nerve fibers connecting the right and left cerebral hemispheres.   146

corpus luteum Structure that forms from the tissues of a ruptured ovarian follicle and functions to secrete female hormones.   345

cortex Outer layer of an organ, such as the convoluted cerebrum, adrenal gland, or kidney.   322

cortisol Glucocorticoid secreted by the adrenal cortex.   195

covalent bond Chemical bond created by the sharing of electrons between atoms.   18

coxal bone Bone of the pelvic girdle.   95

cranial Pertaining to the cranium.   6

cranial nerve Nerve that arises from the brain.   152

crenation Shrinking of red blood cells often caused by osmotic conditions.   41

cretinism Condition resulting from a lack of thyroid hormone in an infant.   193

cryptorchidism Failure of the testes to descend into the scrotum.   338

CSF See *cerebrospinal fluid.*   151

cuboidal epithelium Cube-shaped cells found lining the kidney tubules.   49

Cushing syndrome Condition characterized by thin arms and legs and a "moon face," and accompanied by high blood glucose and sodium levels due to hypersecretion of cortical hormones.   196

cutaneous Pertaining to the skin.   60, 63

cyanosis Bluish cast to the skin due to an increased amount of deoxhemoglobin in the blood; sometimes due to a defective atrial septum, which incompletely closes the foramen ovale after birth.   243

cytoplasm Ground substance of cells that is located between the nucleus and the plasma membrane.   36

D

dandruff Skin disorder characterized by flaking, itchy scalp; caused by accelerated keratinization of the scalp. 68

decubitus ulcer Skin sore due to restricted blood flow to the area in bedridden patients; also called bedsores. 64

deep Located away from the surface of the body or an organ. 4

dendrite Process of a neuron, typically branched, that conducts nerve impulses toward the cell body. 137

deoxyhemoglobin Hemoglobin not carrying oxygen. 208

deoxyribonucleic acid (DNA) Nucleic acid; the genetic material found in the nucleus of a cell. 29

depolarization Loss in polarization, as when a nerve impulse occurs. 140

depression A lowering movement. 105

dermatome Segment of skin supplied by a particular spinal nerve. 152

dermis Thick skin layer that lies beneath the epidermis. 64

descending colon That portion of the large intestine that travels inferiorly as it extends from the transverse colon to the sigmoid colon. 307

diabetes insipidus Condition characterized by an abnormally large production of urine, due to a deficiency of antidiuretic hormone. 23, 190

diabetes mellitus Condition characterized by a high blood glucose level and the appearance of glucose in the urine, due to a deficiency of insulin. 23, 197

diagnosis Decision based on an examination to determine the nature of a diseased condition. 57

diaphragm Sheet of muscle that separates the thoracic cavity from the abdominopelvic cavity; also, a birth control device inserted in front of the cervix in females. 278

diaphysis Shaft of a long bone. 79

diarrhea Frequent, water defecation, often caused by digestive infection or stress. 308

diarthrosis Freely movable joint. 101

diastole Relaxation of heart chambers. 228

diastolic pressure Arterial blood pressure during the diastolic phase of the cardiac cycle. 246

diencephalon Portion of the brain in the region of the third ventricle that includes the thalamus and hypothalamus. 146

differential white blood cell count Microscopic examination of a blood sample in which each type of white blood cell is counted. 210

differentiation Process by which a cell becomes specialized for a particular function. 367

diffusion Passive movement of molecules from an area of greater concentration to an area of lesser concentration. 39

disaccharide Sugar that contains two units of a monosaccharide; for example, maltose. 23

disease Any abnormal condition considered harmful to the body; an illness or disorder. 12

distal Further from the midline or origin; opposite of *proximal*. 4

distal convoluted tubule Highly coiled region of a nephron that is distant from the glomerular capsule. 324

diuretic Drug used to counteract hypertension by inhibiting Na$^+$ reabsorption so that less water is reabsorbed in the nephron. 330

diverticulosis Presence of diverticula, or saclike pouches, of the colon. 309

DNA See *deoxyribonucleic acid.* 29

DNA replication Duplication of DNA; occurs when the cell is not dividing. 42

dominant gene Hereditary factor that expresses itself even when there is only one copy in the genotype. 385

dorsal Pertaining to the back or posterior portion of a body part; opposite of *ventral*. 3

dorsal-root ganglion Mass of sensory neuron cell bodies located in the dorsal root of a spinal nerve. 142

Down syndrome Human congenital disorder associated with an extra chromosome 21. 381

ductus arteriosus Fetal connection between the pulmonary artery and the aorta; venous artery. 242

ductus venosus Fetal connection between the umbilical vein and the inferior vena cava; also called venous duct. 242

duodenum First portion of the small intestine into which ducts from the gallbladder and pancreas enter. 302

dura mater Tough outer layer of the meninges; membranes that protect the brain and spinal cord. 150

E

ECG See *electrocardiogram.* 230

ectopic pregnancy Implantation of the embryo in a location other than the uterus, most often in a uterine tube. 346

eczema Form of noncontagious dermatitis that begins with itchy red patches that thicken and crust over. 68

edema Swelling due to tissue fluid accumulation in the intercellular spaces. 254

Edward symdrome Genetic defect marked by abnormal hands and feet, slow growth, and mental retardation, due to three copies of chromosome 18. 383

effector Structure, such as a muscle or gland, that allows a response to environmental stimuli. 143

ejaculation Ejection of seminal fluid. 341

EKG See *electrocardiogram.* 230

elastic cartilage Cartilage composed of elastic fibers, allowing greater flexibility. 53

electrocardiogram (ECG) Recording of the electrical activity that accompanies the cardiac cycle. 230

electroencephalogram (EEG) Graphic recording of the brain's electrical activity. 149

electrolyte Any substance that ionizes and conducts electricity; electrolytes are present in the body fluids and tissues. 21, 331

element The simplest of substances, consisting of only one type of atom (for example, carbon, hydrogen, oxygen). 17

elephantiasis Swelling of the arms, legs, or external genitalia due to failure of the lymphatic system to remove excess fluid. 254

elevation A raising movement. 105

embolus Moving blood clot that is carried through the bloodstream. 216

embryo Organism in its early stages of development; in humans, the organism in its second week to two months of development. 368

embryonic development Period of development from the second through eighth weeks. 366

emphysema Lung impairment caused by deterioration of the bronchioles, which traps air in alveoli. 287

endocardium Inner layer of the heart wall. 223

endochondral Formed or situated within cartilage. 81

endocrine gland Gland that secretes hormones directly into the bloodstream or body fluids. 187

endocytosis Process in which a vesicle is formed at the plasma membrane to bring a substance into the cell. 42

endometriosis Implantation of uterine endometrial tissue in the abdominal cavity, possibly as a result of irregular menstrual flow. 352

endometrium Lining of the uterus that becomes thickened and vascular during the uterine cycle. 346

endoplasmic reticulum (ER) Complex system of tubules, vesicles, and sacs in cells; sometimes has attached ribosomes. 37

enzyme Protein catalyst that speeds up a specific reaction or a specific type of reaction. 25

eosinophil Granular leukocyte capable of being stained with the dye eosin. 211

epidermis Organism's outer layer of cells. 63

epididymis Coiled tubules next to the testes where sperm mature and may be stored for a short time. 339

epidural hematoma Bleeding between the dura mater and the bone, as a result of a head injury. 150

epigastric Pertaining to the upper middle portion of the abdomen. 6

epiglottis Stucture that covers the glottis during the process of swallowing. 275

epinephrine Hormone produced by the adrenal medulla that stimulates "fight-or-flight" reactions; also called adrenalin. 195

epiphyseal disk Cartilaginous layer within the epiphysis of a long bone that functions as a growing region. 81

epiphysis End segment of a long bone, separated from the diaphysis early in life by an epiphyseal plate, but later becoming part of the larger bone. 79

episiotomy Surgical procedure performed during childbirth in which the opening of the vagina is enlarged to avoid tearing. 374

epithelial tissue Type of tissue that lines the body's internal cavities and covers the body's external surface. 49

erection State in which the penis is erect and prepared for copulation. 341

ERV See *expiration reserve volume.* 281

erythrocyte Non-nucleated, hemoglobin-containing blood cell capable of carrying oxygen; the red blood cell. 55, 208

esophagus Tube that transports food from the mouth to the stomach. 300

essential amino acid Amino acid that is necessary in the diet because the body is unable to manufacture it. 312

essential fatty acid Fatty acid that is necessary in the diet because the body is unable to manufacture it. 312

estrogen Female sex hormone secreted by the ovaries that, along with progesterone, promotes the development and maintenance of the primary and secondary female sexual characteristics. 199, 347

eversion Movement of the foot in which the sole is turned outward. 105

exocrine gland Particular glands with ducts, such as salivary glands, whose secretions are deposited into cavities. 51

exocytosis Process in which an intracellular vesicle fuses with the plasma membrane so that the vesicle's contents are released outside the cell. 42

exophthalmic goiter Enlargement of the thyroid gland, accompanied by an abnormal protrusion of the eyes. 193

expiration Process of expelling air from the lungs; exhalation. 280

expiratory reserve volume (ERV) Volume of air that can be forcibly exhaled after normal exhalation. 281

extension Movement that increases the angle between parts at a joint. 105

external auditory meatus Opening through the temporal bone that connects with the tympanum and the middle ear chamber and through which sound vibrations pass. 83

external ear Portion of the ear consisting of the pinna and the auditory canal. 180

external genitals Sex organs that occur outside the body in the groin. 338

external respiration Exchange of oxygen and carbon dioxide between alveoli and blood. 282

extraembryonic membranes Membranes that are not a part of the embryo but that are necessary to the embryo's continued existence and health. 366

extrinsic muscle Muscle that anchors and moves the eye. 170

F

familial hypercholesterolemia Inability to remove cholesterol from the bloodstream; predisposes individual to heart attack. 25

fascia Tough sheet of fibrous tissue that binds the skin to underlying muscles or that supports and separates muscles. 112

fascicle Small bundle of muscle fibers. 112

fat Organic molecule that the body uses for long-term energy storage. 24, 64

fatigue Failure of a muscle fiber to continue to contract, due to exhaustion of ATP. 114

fatty acid Molecule that contains a hydrocarbon chain and ends with an acid group. 24

feces Indigestible wastes expelled from the digestive tract; excrement. 307

femur Thighbone found in the upper leg. 98

fertilization Union of a sperm nucleus and an egg nucleus, which creates a zygote with the diploid number of chromosomes. 363

fetal alcohol syndrome (FAS) Babies born with decreased weight, height, and head size and with malformation of the head and face due to the mothers consumption of alcohol during pregnancy. 369

fetal development Period of human development from the ninth week through birth. 366

fetus Human in its later developmental stages (from three months to term), following the embryonic stage. 370

fiber Dendrites and axons of neurons. 139

fibrillation Rapid but uncoordinated heartbeat. 230

fibrin Insoluble protein threads formed from fibrinogen during blood clotting. 215

fibrinogen Plasma protein that is converted into fibrin threads during blood coagulation. 212

fibrocartilage Cartilage with a matrix of strong collagenous fibers. 53

fibrous connective tissue Tissue composed mainly of closely packed collagenous fibers and found in tendons and ligaments. 53

fibula Long, slender bone located on the lateral side of the tibia. 99

filament Protein molecule that makes up part of a myofibril. 112

filtration Passage of fluid through a membrane because of mechanical pressure, such as when blood pressure forces water out of a capillary. 41

fimbria Fingerlike extension from the uterine tube near the ovary (pl., *fimbriae*). 345

flagellum Slender, long process used for locomotion—for example, by sperm (pl., *flagella*). 39

flexion Bending at a joint so that the angle between bones is decreased. 104

floating kidney Kidney that has been dislodged from its normal position. 322

follicle Structure in the ovary that produces the egg and particularly the female sex hormone estrogen. 343

follicle-stimulating hormone (FSH) Hormone secreted by the anterior pituitary gland that stimulates the development of an ovarian follicle in a female or the production of sperm cells in a male. 342

fontanel Membranous region located between certain cranial bones in the skull of a fetus or infant. 82

foramen Opening, usually in a bone or membrane (pl., *foramina*). 85

foramen ovale Oval-shaped opening between the atria in the fetal heart. 242

foreskin Skin covering the glans penis in uncircumcised males.   341

formed element Cellular constituent of blood.   208

fovea centralis Region of the retina that consists of densely packed cones and that is responsible for the greatest visual acuity.   173

fracture A break in a bone.   82

free radicals Atoms or molecules with an unpaired electron in their outermost shell; linked to several diseases and play a role in the aging process.   313

frontal plane Plane or section that divides a structure lengthwise into anterior and posterior portions; pertaining to the region of the forehead.   5

frontal lobe Area of the cerebrum responsible for voluntary movements and higher intellectual processes.   146

FSH See *follicle-stimulating hormone.*   342

G

gallbladder Saclike organ associated with the liver that stores and concentrates bile.   306

gamete Sex cell (egg or sperm) that joins in fertilization to form a zygote.   338

gamete intrafallopian transfer (GIFT) Method of achieving pregnancy in which eggs retrieved from the ovary are mixed with sperm and immediately placed into a uterine tube.   353

gamma globulin Type of globulin, a class of proteins in the blood; includes all antibodies.   212

ganglion (sing., ganglia) Collection of neuron cell bodies outside the central nervous system.   142

gastric gland Gland within the stomach wall that secretes gastric juice.   300

gene Unit of heredity located on a chromosome.   380

gene therapy Method of replacing a defective gene with a healthy gene.   390

genetic disease Illness that results from the inheritance of faulty DNA.   36

genetics Science of heredity, including the ways by which traits are passed from one generation to the next.   380

genital herpes Sexually transmitted disease caused by herpes simplex virus and sometimes accompanied by painful ulcers on the genitals.   356

genital wart Raised growth on the genitals due to a sexually transmitted disease caused by human papilloma virus.   356

gestation Period of development, from the start of the last menstrual cycle until birth; in humans, typically 280 days.   366

gestational diabetes mellitus Temporary presence of sugar in the urine during pregnancy, probably due to the presence of placental hormones.   199

GH See *growth hormone.*   192

gingivitis Inflammation of the gums.   311

glaucoma Increasing loss of field of vision, caused by blockage of the ducts that drain the aqueous humor, creating pressure buildup and nerve damage.   172

globin Protein portion of a hemoglobin molecule.   208

glomerular capsule Double-walled cup that surrounds the glomerulus at the beginning of the kidney tubule; also known as Bowman's capsule.   324

glomerulus Cluster—for example, the cluster of capillaries surrounded by the glomerular capsule in a kidney nephron.   324

glottis Slitlike opening between the vocal cords.   275, 299

glucagon Hormone secreted by the pancreatic islets of Langerhans that causes the release of glucose from glycogen.   197

glucose Blood sugar that is broken down in cells to acquire energy for ATP production.   23

glycerol Three-carbon molecule that joins with fatty acids to form fat.   24

glycogen Polysaccharide that is the principal storage compound for sugar in animals.   23

Golgi apparatus Organelle that consists of concentrically folded membranes and that functions in the packaging and secretion of cellular products.   38

gonad Organ that produces sex cells: the ovary, which produces eggs, and the testis, which produces sperm.   338

gonadotropin Type of hormone that regulates the activity of the ovaries and testes; principally follicle-stimulating hormone and luteinizing hormone.   192

gonorrhea Sexually transmitted disease caused by the bacterium *Neisseria gonorrhoeae* that causes painful urination and swollen testes in men and is usually symptomless in women, but can cause inflammation of the cervix and uterine tubes.   357

gout Joint inflammation caused by accumulation of uric acid.   327

Graafian follicle Mature follicle within the ovaries that houses a developing egg.   343

granular leukocyte White blood cell with prominent granules in the cytoplasm.   209

Graves' disease Autoimmune disease with swollen throat due to an enlarged, hyperactive thyroid gland; patients often have protruding eyes and are underweight, hyperactive, and irritable.   193

gray matter Nonmyelinated nerve fibers in the central nervous system.   142

greater sciatic notch Indentation in the posterior coxal bone through which pass the blood vessels and the large sciatic nerve to the lower leg.   96

growth Increase in the number of cells and/or the size of these cells.   367

growth hormone (GH) Hormone released by the anterior lobe of the pituitary gland that promotes the growth of the organism; also known as somatotropin.   192

gumma Destructive ulcer accompanying the third stage of syphilis that may appear on any organ of the body.   357

gyrus Convoluted elevation or ridge (pl., *gyri*).   146

H

hair follicle Tubelike depression in the skin in which a hair develops.   65

hard palate Anterior portion of the roof of the mouth that contains several bones.   298

hCG See *human chorionic gonadotropic hormone.*   351

head Enlargement on the end of a bone.   98

heart Muscular organ located in thoracic cavity that is responsible for maintenance of blood circulation.   223

heart attack See *myocardial infarction.*   233

heartburn Burning pain in the chest occurring when part of the stomach contents escapes into the esophagus.   300

heart transplant Replacement of a diseased heart with a healthy heart from another individual.   227

heart valve Valve found between the chambers of the heart or between a chamber and a vessel leaving the heart.   224

Heimlich maneuver Strategy for dislodging an object stuck in the larynx.   282

hemarthrosis Bleeding into joints caused by insufficient clotting factors (pl., *hemarthroses*).   216

hematocrit Volume percentage of red blood cells within a sample of whole blood.   208

hematopoiesis Production of blood cells.   81

heme Iron-containing portion of a hemoglobin molecule.   208

hemodialysis Mechanical way to remove nitrogenous wastes and to regulate blood pH when the kidneys are unable to perform these functions.   40

hemoglobin Pigment of red blood cells responsible for oxygen transport.   208

hemolysis Bursting of red blood cells with the release of hemoglobin; can be caused by osmotic conditions. 41

hemolytic anemia Insufficient number of red blood cells caused by an increased rate of red blood cell destruction. 210

hemolytic disease of the newborn Destruction of a fetus's red blood cells by the mother's immune system, caused by differing Rh factors between mother and fetus. 210

hemophilia A Most common of the severe clotting disorders caused by the absence of a blood clotting factor. 216

hemorrhage An escape of blood from blood vessels. 214

hemorrhoids Abnormally dilated blood vessels of the rectum. 236

hepatic Referring to the liver. 238

hepatic portal system Portal system that begins at the villi of the small intestine and ends at the liver. 242

hepatic portal vein Vein leading to the liver and formed by the merging blood vessels of the small intestine. 242

hepatitis Inflammation of the liver; often due to a serious infection by any of a number of viruses. 306

hernia Protrusion of an organ through an abnormal opening, such as the intestine through the abdominal wall near the scrotum (inguinal hernia) or the stomach through the diaphragm (hiatal hernia). 339

herniated disk Fibrous ring of cartilage between two vertebrae that has ruptured. 90

hinge joint Type of joint characterized by a convex surface of one bone fitting into a concave surface of another so that movement is confined to one place, such as in the knee or interphalangeal joint. 104

hirsutism Excessive body and facial hair in women. 65

histamine Substance produced by basophil-derived mast cells in connective tissue that causes capillaries to dilate; causes many of the symptoms of allergy. 256

Hodgkin's disease Cancer of the lymph glands that is normally localized in the neck region. 254

homeostasis Constancy of conditions, particularly the environment of body cells: constant temperature, blood pressure, pH, and other body conditions. 12

hormone Substance secreted by an endocrine gland that is transmitted in the blood or body fluids. 187

human chorionic gonadotropic hormone (HCG) Hormone produced by the placenta that helps maintain pregnancy and is the basis for the pregnancy test. 351

human immunodeficiency virus (HIV) Virus responsible for AIDS. 356

humerus Heavy bone that extends from the scapula to the elbow. 92

Huntington disease Genetic disease marked by progressive deterioration of the nervous system due to deficiency of a neurotransmitter. 141

hyaline cartilage Cartilage composed of very fine collagenous fibers and a matrix of a glassy, white, opaque appearance. 53

hydrocephalus Enlargement of the brain due to abnormal accumulation of cerebrospinal fluid. 151

hydrogen bond Weak attraction between a partially positive hydrogen and a partially negative oxygen or nitrogen some distance away; found in proteins and nucleic acids. 21

hydrolysis Splitting of a bond by the addition of water. 23

hyperglycemia Excessive glucose in the blood. 198

hyperopia Inability to see nearby objects. 178

hyperpnea Deep and labored breathing. 278

hypertension Elevated blood pressure, particularly the diastolic pressure. 18, 246

hyperthermia Abnormally high body temperature. 68

hypertonic solution Solution that has a higher concentration of solute and a lower concentration of water than the cell. 41

hypertrophy Increase in the size of an organ, usually by an increase in the size of its cells. 116

hypochondriac Pertaining to the abdominal region beneath the ribs. 6

hypodermic needle Slender, hollow instrument for introducing material into or removing material from or below the skin. 64

hypogastric Pertaining to the abdominal region beneath the stomach. 6

hypoglycemia Insufficient amount of glucose in the blood. 198

hypothalamus Region of the brain; the floor of the third ventricle that helps maintain homeostasis. 146, 190

hypothermia Abnormally low body temperature. 68

hypotonic solution Solution that has a lower concentration of solute and a higher concentration of water than the cell. 41

hysterectomy Surgical removal of the uterus. 346

hysteroscopy Method of sealing the uterine tubes with an electrical current. 355

I

ileum Lower portion of the small intestine. 302

iliac Pertaining to the ilium. 6

ilium One of the bones of a coxal bone or hipbone. 302

immunity Resistance to disease-causing organisms. 256

immunization Strategy for achieving artificial immunity to the effects of specific disease-causing agents. 262

immunosuppression Inactivation of the immune system to prevent organ rejection. 267

impetigo Contagious skin disease caused by bacteria in which vesicles erupt and crust over. 68

implantation Attachment and penetration of the embryo to the lining (endometrium) of the uterus. 365

impotence Failure of the penis to achieve erection. 341

incontinence Involuntary loss of urine. 333

incus Middle of three ossicles of the ear; serves with the malleus and the stapes to conduct vibrations from the tympanic membrane to the oval window of the inner ear. 180

inferior Situated below something else; pertaining to the lower surface of a part. 4

infertile Inability to have as many children as desired. 351

inflammatory reaction Tissue response to injury that is characterized by dilation of blood vessels and an accumulation of fluid in the affected region. 256

influenza Acute viral infection of the respiratory system that is accompanied by fever and aches and pains in the joints. 285

inhibin Hormone secreted by seminiferous tubules that inhibits the release of follicle-stimulating hormone from the anterior pituitary. 342

inner ear Portion of the ear, consisting of a vestibule, semicircular canals, and the cochlea, where balance is maintained and sound is transmitted. 180

insertion End of a muscle that is attached to a movable part. 115

inspiration The act of breathing in; inhalation. 279

inspiratory reserve volume (IRV) Volume of air that can be forcibly inhaled after normal inhalation. 281

insulin Hormone produced by the pancreas that regulates glucose storage in the liver and glucose uptake by cells. 197

integument Pertaining to the skin. 63

integumentary system Pertaining to the skin and accessory organs. 9, 63

interferon Protein formed by a cell infected with a virus that can increase the resistance of other cells to the virus. 258

internal respiration Exchange of oxygen and carbon dioxide between blood and tissue fluid. 282

interneuron Neuron found within the central nervous system that takes nerve impulses from one portion of the system to another. 137

interstitial cell Hormone-secreting cell located between the seminiferous tubules of the testes. 339

intervertebral disk Layer of cartilage located between adjacent vertebrae. 88

intrauterine insemination (IUI) Process of achieving pregnancy in which donated sperm are deposited in the uterus. 353

inversion Movement of the foot in which the sole is turned inward. 105

in vitro fertilization (IVF) Process of achieving pregnancy in which eggs retrieved from an ovary are fertilized in a laboratory; viable embryos are then placed into the woman's uterus. 353

ion A charged atom. 18

ionic bond Chemical attraction between a positive ion and a negative ion. 18

ionize Breaking of a chemical bond such that ions are released. 21

iris Muscular ring that surrounds the pupil and regulates the passage of light through this opening. 172

iron deficiency anemia Abnormally low amount of red blood cells or hemoglobin, due to a lack of iron in the diet. 18, 210

IRV See *inspiratory reserve volume.* 281

ischemic heart disease Insufficient oxygen delivery to the heart, usually caused by partially blocked coronary arteries. 232

ischial spine Projection of the coxal bone into the pelvic cavity. 339

isometric contraction Muscular contraction in which the muscle fails to shorten and there is no movement. 115

isotonic contraction Muscle contraction producing movement of an object, thereby maintaining constant tension. 115

isotonic solution Solution that contains the same concentration of solutes and water as does the cell. 41

isotope One of two or more atoms with the same atomic number that differs in the number of neutrons and, therefore, in weight. 18

IVF See *in vitro fertilization.* 353

IVI See *intrauterine insemination.* 353

J

jaundice Yellowish tint to the skin caused by an abnormal amount of birubin in the blood, indicating liver malfunction. 306

jejunum Middle portion of the small intestine. 302

joint Union of two or more bones; an articulation. 100

jugular Any of four veins that drain blood from the head and neck. 238

K

karyotype Arrangement of all the chromosomes from a nucleus by pairs in a fixed order. 380

keratin Insoluble protein present in the epidermis and in epidermal derivatives, such as hair and nails. 64

ketone Acidic molecule. 197

kidney Organ in the urinary system that forms, concentrates, and excretes urine. 322

Klinefelter syndrome Condition caused by the inheritance of XXY chromosomes. 383

knee-jerk reflex Automatic, involuntary response initiated by tapping the ligaments just below the patella (kneecap). 143

kyphosis Increased roundness in the thoracic curvature of the spine; also called "hunchback." 88

L

labia majora Two large, hairy folds of skin of the female external genitalia. 347

labia minora Two small folds of skin inside the labia majora and encircling the clitoris. 347

lacrimal apparatus Structures that provide tears to wash the eye, consisting of the lacrimal gland and the lacrimal sac with its ducts. 171

lactation Production and secretion of milk by the mammary glands. 376

lacteal Lymph vessel in a villus of the wall of the small intestine. 303

lacuna Small pit or hollow cavity, as in bone or cartilage, where a cell or cells are located. 53

Langerhans' cell Specialized epidermal cells that assist the immune system. 63

lanugo Short, fine hair that is present during the later portion of fetal development. 372

laparoscope Type of endoscope for viewing the periotoneal cavity. See also *endoscope.* 352

large intestine Portion of the digestive tract that extends from the small intestine to the anus. 307

laryngitis Inflammation of the larynx. 286

larynx Structure that contains the vocal cords; also known as the voice box. 275

lateral Pertaining to the side. 4

lens Clear, membranelike structure that is found in the eye behind the iris and that brings objects into focus. 172

leukemia Form of cancer characterized by uncontrolled production of leukocytes in red bone marrow. 57, 210

leukocyte Several types of colorless, nucleated blood cells that, among other functions, resist infection; white blood cells. 55, 208

leukocytosis Abnormally large increase in the number of white blood cells. 210

leukopenia Abnormally low number of leukocytes in the blood. 210

LH See *luteinizing hormone.* 342

ligament Strong connective tissue that joins bone to bone. 53, 101

limbic system System that involves many different centers of the brain and that is concerned with visceral functioning and emotional responses. 149

lipase Enzyme secreted by the pancreas that digests or breaks down fats. 304

lipid Group of organic compounds that are insoluble in water—notably, fats, oils, and steroids. 24

liver Largest organ in the body, located in the abdominal cavity below the diaphragm; performs many vital functions that maintain homeostasis of blood. 304

loose connective tissue Tissue that is composed mainly of fibroblasts separated by collagenous and elastin fibers and that is found beneath epithelium. 51

lordosis Exaggerated lumbar curvature of the spine; also called "swayback." 88

lumbar Pertaining to the loin region. 6

lung cancer 277

lunula Pale, half-moon–shaped area at the base of nails. 66

luteinizing hormone (LH) Hormone produced by the anterior pituitary that stimulates the development of the corpus luteum in females and the production of testosterone in males. 342

lymph Fluid having the same composition as tissue fluid; carried in lymph vessels. 252

lymphadenitis Infection of the lymph nodes. 254

lymphangitis Infection of the lymphatic vessels. 254

lymphatic system Vascular system that takes up excess tissue fluid and transports it to the bloodstream. 9, 252

lymph node Mass of lymphoid tissue located along the course of a lymphatic vessel. 254

lymphocyte Type of white blood cell characterized by agranular cytoplasm; lymphocytes usually constitute 20–25% of the white cell count.   212

lymphokine Chemical secreted by T lymphocytes that has the ability to affect the characteristics of monocytes.   265

lymphoma Cancer of lymphoid tissue (reticular connective tissue).   57, 254

lysosome Organelle involved in intracellular digestion; contains powerful digestive enzymes.   38

M

macromolecule Large molecule composed of smaller molecules.   2

macrophage Enlarged monocyte that ingests foreign material and cellular debris.   256

macular degeneration Disruption of the macula lutea, a central part of the retina, causing blurred vision.   174

major histocompatibility complex (MHC) protein Protein on the plasma membrane of a macrophage that identifies the cell as belonging to the individual and that aids in the stimulation of T cells.   260

malignant The power to threaten life; cancerous.   44

malleolus Rounded projection from a bone.   96

malleus First of three ossicles of the ear; serves with the incus and stapes to conduct vibrations from the tympanic membrane to the oval window of the inner ear.   180

marrow Connective tissue that occupies the spaces within bones.   81

mastication Chewing, usually of food.   298

mastoiditis Inflammation of the mastoid sinuses of the skull.   82

matrix Secreted basic material or medium of biological structures, such as the matrix of cartilage or bone.   51

medial Toward or near the midline.   4

mediastinum Tissue mass located between the lungs.   6

medulla oblongata Lowest portion of the brain; concerned with the control of internal organs.   144

medullary cavity Within the diaphysis of a long bone, cavity occupied by yellow marrow.   79

meiosis Type of cell division in which the daughter cells have 23 chromosomes; occurs during spermatogenesis and oogenesis.   44, 339

melanin Pigment found in the skin and hair of humans that is responsible for their coloration.   64

melanocyte Melanin-producing cell.   64

melanoma Deadly form of skin cancer that begins in the melanocytes, pigment cells present in the epidermis.   68

melatonin Hormone, secreted by the pineal gland, that is involved in biorhythms.   199

memory B cell Cells derived from B lymphocytes that remain within the body for some time and account for the presence of active immunity.   259

meninges Protective membranous coverings around the brain and spinal cord (sing., *meninx*).   60, 150

meniscus Piece of fibrocartilage that separates the surfaces of bones in the knee (pl., *menisci*).   101

menopause Termination of the ovarian and uterine cycles in older women.   357

menses (menstruation) Loss of blood and tissue from the uterus at the beginning of a female uterine cycle.   349

mesentery Fold of peritoneal membrane that attaches an abdominal organ to the abdominal wall.   59

messenger RNA (mRNA) Nucleic acid (ribonucleic acid) complementary to genetic DNA; has codons, which direct cell protein synthesis at the ribosomes.   36

metabolism All of the chemical changes that occur within cells, considered together.   36

metacarpal Bone of the hand between the wrist and finger bones.   95

metafemale Female who has three X chromosomes.   383

metastasis Mechanism of cancer spread in which cancer cells break off from the initial tumor, enter the blood vessels or lymphatic vessels, and start new tumors elsewhere in the body.   44

metatarsals Bones found in the foot between the ankle and the toes.   100

MHC protein See *major histocompatibility complex*.   260

micturition Emptying of the bladder; urination.   324

midbrain Small region of the brain stem located between the forebrain and the hindbrain; contains tracts that conduct impulses to and from the higher parts of the brain.   146

middle ear Portion of the ear consisting of the tympanic membrane, the oval and round windows, and the ossicles, where sound is amplified.   180

mineral Inorganic substance; certain minerals must be in the diet for normal metabolic functioning of cells.   313

mitochondrion Organelle in which cellular respiration produces the energy molecule ATP.   38

mitosis Type of cell division in which two daughter cells receive 46 chromosomes; occurs during growth and repair.   42

mitral stenosis Narrowing of the opening of the bicuspid (mitral) valve.   229

mixed nerve Nerve that contains both the long dendrites of sensory neurons and the long axons of motor neurons.   142

mole Raised growth on the skin due to an overgrowth of melanocytes.   68

molecule Smallest quantity of a substance that retains its chemical properties.   2, 18

monoclonal antibody Antibody of one type that is produced by cells derived from a lymphocyte that has fused with a cancer cell.   264

monocyte Type of white blood cell that functions as a phagocyte.   212

mononucleosis Viral disease characterized by the presence of an increase in atypical lymphocytes in the blood.   210

monosaccharide Simple sugar; a carbohydrate that cannot be decomposed by hydrolysis.   23

morphogenesis Establishment of shape and structure in an organism.   367

morula Early stage in development in which the embryo consists of a mass of cells, often spherical.   367

motor nerve Nerve containing only the long axons of motor neurons.   142

motor neuron Neuron that takes nerve impulses from the central nervous system to an effector; also known as an efferent neuron.   137

motor unit Motor neuron and all the muscle fibers it innervates.   112

mouth Opening through which food enters the body.   298

MS See *multiple sclerosis*.   111

mucous membrane Membrane that lines a cavity or tube that opens to the outside of the body; also called mucosa.   59

multiple sclerosis (MS) Disease in which the outer, myelin layer of nerve fiber insulation becomes scarred, interfering with normal conduction of nerve impulses.   139, 267

muscle fiber Muscle cell.   111

muscle spindle Modified skeletal muscle fiber that can respond to changes in muscle length.   115

muscular tissue Major type of tissue that is adapted to contract; the three kinds of muscle are cardiac, smooth, and skeletal.   55

mutation Permanent change in DNA that is passed to future generations of cells or offspring.   36

myasthenia gravis Muscle weakness due to an inability to respond to the neurotransmitter acetylcholine.   171, 267

myelin sheath Fatty plasma membranes of Schwann cells that cover long neuron fibers and give them a white, glistening appearance. 58

myocardial infarction Damage to the myocardium due to blocked circulation in the coronary arteries; a heart attack. 233

myocardium Heart (cardiac) muscle consisting of striated muscle cells that interlock. 223

myofibril Contractile portion of muscle fibers. 112

myopia Inability to see distant objects clearly. 178

myosin Thick myofilament in myofibrils that is made of protein and is capable of breaking down ATP; see also *actin*. 112

myringotomy Incision of the tympanic membrane and insertion of a small tube; performed on small children who have frequent ear infections. 180

myxedema Condition resulting from a deficiency of thyroid hormone in an adult. 193

N

navel Abdominal scar that marks the site of attachment of the umbilical cord. 374

negative feedback Mechanism that is activated by a surplus imbalance and acts to correct it by stopping the process that brought about the surplus. 12

nephron Anatomical and functional unit of the kidney; kidney tubule. 322

nerve Bundle of long nerve fibers that run to and/or from the central nervous system. 58, 142

nerve deafness Hearing impairment that usually occurs when the cilia on the sense receptors within the cochlea have worn away. 182

nerve impulse Change in polarity that flows along the membrane of a nerve fiber. 140

nervous tissue Tissue of the nervous system, consisting significantly of neurons and neuroglial cells. 58

neurilemmal sheath Sheath on the outside of some nerve fibers, due to the presence of Schwann cells. 139

neuroglial cell Support cell of the nervous system. 58

neuromuscular junction Junction between a neuron and a muscle fiber. 112

neuron Nerve cell that characteristically has three parts: dendrite, cell body, and axon. 58, 137

neurotransmitter Chemical made at the ends of axons that is responsible for transmission across a synapse. 140

neutrophil Phagocytic white blood cell that normally constitutes 60–70% of the white blood cell count. 211

NGU See *nongonococcal urethritis*. 357

node of Ranvier Gap in the myelin sheath of a nerve fiber. 58

nondisjunction Failure of the chromosomes (or chromatids) to separate during meiosis. 381

nongonococcal urethritis (NGU) Infection of the urethra, commonly caused by chlamydia. 357

norepinephrine Hormone secreted by the adrenal medulla to help initiate the "fight-or-flight" reaction. 195

nose Specialized structure on the face that serves as the sense organ of smell and as part of the respiratory system. 274

nucleic acid Large organic molecule found in the nucleus (DNA and RNA) and cytoplasm (RNA). 29

nucleolus Organelle found inside the nucleus and composed largely of RNA for ribosome formation (pl., *nucleoli*). 36

nucleus Large organelle that contains the chromosomes and acts as a cell control center. 35

O

occipital Pertaining to the inferior, dorsal portion of the head. 6

occipital lobe Area of the cerebrum responsible for vision, visual images, and other sensory experiences. 146

occluded coronary arteries Blockage of the blood vessels that serve the needs of the heart. 233

olfactory cell Cell located high in the nasal cavity that bears receptor sites on cilia for various chemicals and whose stimulation results in smell. 169

oncogene Gene that contributes to the transformation of a normal cell into a cancer cell. 45

oocyte Developing female gamete. 343

oogenesis Production of eggs in females by the process of meiosis and maturation. 44, 343

oophorectomy Surgical removal of one or both ovaries. 346

opportunistic infection Disease that arises in the presence of a severely impaired immune system. 265

optic nerve Nerve composed of the ganglion cell fibers that form the innermost layer of the retina. 173

organ Structure consisting of a group of tissues that perform a specialized function; a component of an organ system. 3

organelle Part of a cell that performs a specialized function. 36

organism Individual living thing. 3

organ of Corti Organ that contains the hearing receptors in the inner ear. 180

organ system Group of related organs working together. 3

organ transplantation Replacement of a diseased or defective organ with a healthy one from another. 9

orgasm Physical and emotional climax during sexual intercourse; results in ejaculation in the male. 341

origin End of a muscle that is attached to a relatively immovable part. 115

osmosis Movement of water from an area of greater concentration to an area of lesser concentration across the plasma membrane. 41

ossicles Tiny bones found in the middle ear; malleus (hammer), incus (anvil), and stapes (stirrup). 180

ossification Formation of bone. 81, 370

osteoarthritis Disintegration of the cartilage between bones at a synovial joint. 104

osteoblast Bone-forming cell. 81

osteoclast Cell that causes the erosion of bone. 81

osteocyte Mature bone cell. 81

osteoporosis Weakening of bones due to decreased bone mass. 105

otitis media Inflammation of the middle ear. 286

otolith Granule that lies above and whose movement stimulates ciliated cells in the utricle and saccule. 180

otosclerosis Overgrowth of bone that causes the stapes to adhere to the oval window, resulting in conductive deafness. 182

ototoxic Damaging to any of the elements of hearing or balance. 182

oval window Membrane-covered opening between the stapes and the inner ear. 180

ovarian cancer Cancer of an ovary. 346

ovarian cycle Monthly changes in the ovary that affect the level of sex hormones in the blood of females. 349

ovariohysterectomy Surgical removal of the ovaries and uterus. 346

ovary Female gonad; the organ that produces eggs, estrogen, and progesterone. 343

ovulation Discharge of a mature egg from the follicle within the ovary. 345, 363

oxygen debt Amount of oxygen needed to metabolize the lactic acid that accumulates during vigorous exercise. 114

oxyhemoglobin Hemoglobin bound to oxygen in a loose, reversible way. 208, 283

oxytocin Hormone released by posterior pituitary that causes contraction of uterus and milk letdown. 190

P

pacemaker Small region of neuromuscular tissue that initiates the heartbeat; also called the *SA node*.   229

pancreas Endocrine organ located near the stomach that secretes digestive enzymes into the duodenum and produces hormones, notably insulin.   304

pancreatic islets (of Langerhans) Distinctive groups of cells within the pancreas that secrete insulin and glucagon.   197

Pap smear Sample of cells removed from the tip of the cervix and then stained and examined microscopically.   57, 346

paranasal sinus One of several air-filled cavities in the maxillary, frontal, sphenoid, and ethmoid bones that is lined with mucous membrane and drains into the nasal cavity.   274

paraplegia Paralysis of the lower body and legs, due to injury to the spinal cord between vertebrae T1 and L2.   150

parasympathetic division Portion of the autonomic nervous system that usually promotes those activities associated with a normal state.   158

parathyroid hormone (PTH) Hormone secreted by the parathyroid glands that raises the blood calcium primarily by stimulating reabsorption of bone.   194

parietal Pertaining to the wall of a cavity.   59

parietal lobe Area of the cerebrum responsible for sensations involving temperature, touch, pressure, pain, and speech.   146

Parkinson disease Progressive deterioration of the central nervous system due to a deficiency in the neurotransmitter dopamine; also called paralysis agitans.   141

parturition Processes that lead to and include the birth of a human and the expulsion of the extraembryonic membranes through the terminal portion of the female reproductive tract.   373

passive immunity Protection against infection acquired by transfer of antibodies to a susceptible individual.   262

Patau syndrome Genetic defect marked by mental retardation, cleft palate, too many fingers and toes, and eye defects, due to three copies of chromosome 13.   383

patella Bone of the kneecap.   98

pathogen Disease-causing agents like bacteria and viruses.   256

pathologist Person trained in knowledge of diseases and their symptoms, allowing for the diagnosis of disease.   57

pattern baldness Progressive symmetrical loss of scalp hair; genetically caused but also androgen-dependent; also called male pattern alopecia.   199

pectoral girdle Portion of the skeleton that provides support and attachment for the upper limbs.   91

pelvic Pertaining to the pelvis.   6

pelvic girdle Portion of the skeleton to which the lower limbs are attached.   95

pelvic inflammatory disease (PID) Latent infection of gonorrhea or chlamydia in the vasa deferentia or uterine tubes.   357

pelvis Bony ring formed by the sacrum and coxal bones; also, a hollow chamber in the kidney that lies inside the medulla and receives freshly prepared urine from the collecting ducts.   95

penis Male excretory and copulatory organ.   341

pepsin Protein digesting enzyme produced by the stomach.   300

peptide bond Bond that joins two amino acids.   27

pericardial Around the heart.   6, 223

pericardium Protective serous membrane that surrounds the heart.   223

periodontitis Inflammation of the periodontal membrane that lines tooth sockets, causing loss of bone and loosening of teeth.   311

periosteum Fibrous connective tissue covering the surface of bone.   79

peripheral Situated away from the center of the body or an organ.   4

peripheral nervous system (PNS) Nerves and ganglia of the nervous system that lie outside the brain and spinal cord.   136

peristalsis Rhythmical contraction that serves to move the contents along in tubular organs, such as the digestive tract.   300

peritoneum A serous membrane that lines the abdominopelvic cavity and encloses the abdominal viscera.   59

peritubular capillary network Capillary network that surrounds a nephron and functions in reabsorption during urine formation.   324

pernicious anemia Insufficiency of mature red blood cells, due to poor absorption of vitamin B_{12}.   210

peroxisome Membranous vesicle containing enzymes that catalyze reactions producing and decomposing hydrogen peroxide.   37

pH Measure of the hydrogen ion concentration; any pH below 7 is acidic, and any pH above 7 is basic.   22

phagocytosis Taking in of bacteria and/or debris by engulfing; also called cell eating.   42, 211

phalanges Bones of the fingers and thumb in the hand and of the toes in the foot (sing., *phalanx*).   95

pharynx Common passageway (throat) for both food intake and air movement.   274, 299

phlebitis Inflammation of a vein.   236

physiology Branch of science dealing with the study of body functions.   2

pia mater Innermost meningeal layer that is in direct contact with the brain and spinal cord.   150

pineal gland Small endocrine gland, located in the third ventricle of the brain, that secretes melatonin and is involved in biorhythms.   199

pinna Outer, funnel-like structure of the ear that picks up sound waves.   180

pinocytosis Formation of a vesicle that brings molecules into a cell; also called cell drinking.   42

pituitary dwarf Person of normal proportions but small stature, caused by inadequate growth hormone.   192

placenta Structure formed from the chorion and uterine tissue, through which nutrient and waste exchange occurs for the embryo and later the fetus.   351, 366

plaque Accumulation of soft masses of fatty material, particularly cholesterol, beneath the inner linings of arteries.   232

plasma Liquid portion of blood.   55, 208

plasma cell Cell derived from a B lymphocyte that is specialized to mass-produce antibodies.   259

plasma membrane Membrane that surrounds the cytoplasm of cells and regulates the passage of molecules into and out of the cell.   34

platelet Cell-like disks formed from fragmentation of megakaryocytes that initiate blood clotting.   212

platelet plug Platelets that stick and cling to each other in order to seal a break in a blood vessel wall.   215

pleural Pertaining to the lungs.   6

pleural membrane Serous membrane that encloses the lungs.   59, 278

pneumonectomy Surgical removal of all or part of a lung.   289

pneumonia Infection of the lungs that causes alveoli to fill with mucus and pus.   287

PNS See *peripheral nervous system.*   136

polycythemia Abnormally high number of red blood cells in the blood.   210

polyp Small, abnormal growth on any mucous membrane, such as the digestive tract.   309

polyphagia Excessive eating.   197

polysaccharide Carbohydrate composed of many bonded glucose units—for example, glycogen.   23

pons Portion of the brain stem above the medulla oblongata and below the midbrain; assists the medulla oblongata in regulating the breathing rate.   146

posterior Toward the back; opposite of *anterior*.   4

posterior pituitary Portion of the pituitary gland connected by a stalk to the hypothalamus.   190

postganglionic fiber In the autonomic nervous system, the axon that leaves, rather than goes to, a ganglion.   155

preganglionic fiber In the autonomic nervous system, the axon that goes to, rather than leaves, a ganglion.   155

premature baby Child born before full term and weighing 5 pounds, 8 ounces, or less.   374

presbycusis Loss of hearing that accompanies old age.   183

prime mover Muscle most directly responsible for a particular movement; an agonist.   115

PRL See *prolactin*.   192, 376

progesterone Female sex hormone secreted by the ovaries that, along with estrogen, promotes the development and maintenance of the primary and secondary female sexual characteristics.   199, 347

prolactin (PRL) Hormone secreted by the anterior pituitary that stimulates milk production in the mammary glands; also known as lactogenic hormone.   192, 376

pronation Rotation of the forearm so that the palm faces backward.   105

proprioceptor Sensory receptor that assists the brain in knowing the position of the limbs.   167

prostaglandins Hormones that have various and powerful effects, often within the cells that produce them.   202

prostate cancer Cancer of the prostate gland.   339

prostate gland Gland in males that is located about the urethra at the base of the bladder; contributes to the seminal fluid.   339

prosthesis Artificial substitute for a missing part of the body.   103

protein Macromolecule composed of amino acids.   25

prothrombin Plasma protein made by the liver that must be present in blood before clotting can occur.   215

prothrombin activator Enzyme that catalyzes the transformation of the precursor prothrombin to the active enzyme thrombin.   215

proximal Closer to the midline or origin; opposite of *distal*.   4

proximal convoluted tubule Highly coiled region of a nephron near glomerular capsule.   324

pseudostratified Appearance of layering in some epithelial cells when, actually, each cell touches a baseline and true layers do not exist.   49

psoriasis Common chronic, inherited skin disease in which red patches are covered with scales, occurs most often on the elbows, knees, scalp, and trunk.   68

PTH See *parathyroid hormone*.   194

puberty Stage of development in which the reproductive organs become functional.   338

pulmonary artery Blood vessel that takes blood away from the heart to the lungs.   227

pulmonary circulation Path of blood through vessels that take deoxygenated blood to and oxygenated blood away from the lungs.   236

pulmonary edema Excessive fluid in the lungs caused by congestive heart failure.   254

pulmonary embolism Blockage of a pulmonary artery by a blood clot that commonly originates in a vein of the lower legs.   236

pulmonary fibrosis Accumulation of fibrous connective tissue in the lungs; caused by inhaling irritating particles, such as silica, coal dust, or asbestos.   287

pulmonary vein Blood vessel that takes blood away from the lungs to the heart.   227

pulse Vibration felt in arterial walls due to expansion of the aorta following ventricle contraction.   244

pupil Opening in the center of the iris that controls the amount of light entering the eye.   172

Purkinje fiber Specialized muscle fiber that conducts the cardiac impulse from the AV bundle into the ventricular walls.   229

pus Thick, yellowish fluid composed of dead phagocytes, dead tissue, and bacteria.   256

Q

quadriplegia Paralysis of the entire body and all four limbs, due to injury to the spinal cord between vertebrae C4 and T1.   150

R

radioactive isotope Atom whose nucleus undergoes degeneration and in the process gives off radiation.   18

radius Elongated bone located on the thumb side of the lower arm.   93

recessive gene Hereditary factor that expresses itself only when two copies are present in the genotype.   385

rectum Terminal portion of the intestine.   307

red blood cell See *erythrocyte*.   55, 208

red blood cell count Number of red blood cells per cubic millimeter of blood.   208

red bone marrow Blood cell–forming tissue located in spaces within certain bones.   81, 255

referred pain Pain perceived as having come from a site other than that of its actual origin.   167

reflex Inborn autonomic response to a stimulus that is dependent on the existence of fixed neural pathways.   142

REM sleep Rapid eye movement sleep; a stage in sleep that is characterized by eye movements and dreaming.   149

renal calculi Stones that form in the kidney or bladder; also called kidney stones.   324

renal cortex Outer, primarily vascular portion of the kidney.   322

renal medulla Inner portion of the kidney, including the renal pyramids.   322

renal pelvis Inner cavity of the kidney formed by the expanded ureter and into which the collecting ducts open.   322

renin Secretion from the kidney that activates angiotensinogen to angiotensin I.   196

repolarization Recovery of a neuron's polarity to the resting potential after the neuron ceases transmitting impulses.   140

residual volume (RV) Volume of air that remains in the lungs after normal exhalation.   281

respiration Transport and exchange of gases between the atmosphere and the cells via the lungs and blood vessels.   273

respiratory center Group of neurons in the medulla oblongata that regulates respiration.   278

respiratory distress syndrome Insufficiency of lung function due to a lack of surfactant, which leads to lung collapse.   277

respiratory membrane Alveolar wall plus the capillary wall, across which gas exchange occurs.   277

retina Innermost layer of the eyeball that contains the rods and cones.   173

retinopathy of prematurity Disease of premature babies in which the blood vessels of the retina have dilated and multiplied, potentially causing blindness.   374

rheumatoid arthritis Persistent inflamation of synovial joints, often causing cartilage destruction, bone erosion, and joint deformities.   104, 267

Rh factor Type of antigen on red blood cells. 218

rib Bone hinged to the vertebral column and sternum that, along with muscle, defines the top and sides of the chest cavity. 90

ribonucleic acid (RNA) Nucleic acid that helps DNA in protein synthesis. 29

ribosomal RNA (rRNA) RNA (ribonucleic acid) occurring in ribosomes, structures involved in protein synthesis. 36

ribosome Minute particle, found attached to the endoplasmic reticulum or loose in the cytoplasm, that is the site of protein synthesis. 36

rickets Defective mineralization of the skeleton, usually due to inadequate vitamin D in the body. 67

RNA See *ribonucleic acid.* 29

rod Dim-light receptor in the retina of the eye that detects motion but no color. 173

rotation Movement of a bone around its own longitudinal axis. 105

round window Membrane-covered opening between the inner ear and the middle ear. 180

rugae Deep folds, as in the wall of the stomach. 300

RV See *residual volume.* 281

S

saccule Saclike cavity of the inner ear that contains receptors for static equilibrium. 180

sacroiliac joint Connection between the coxal bone and the sacrum. 95

sacrum Bone consisting of five fused vertebrae that form the posterior wall of the pelvic girdle. 89

sagittal plane Plane or section that divides a structure into right and left portions. 5

salivary gland Gland associated with the mouth that secretes saliva. 298

salpingectomy Surgical removal of the uterine tubes. 346

SA node See *sinoatrial node.* 229

sarcoma Cancer that arises in striated muscle, cartilage, or bone. 57

scapula Large bone in the posterior shoulder area. 91

Schwann cell Cell that surrounds a fiber of a peripheral nerve and forms the neurilemmal sheath and myelin. 58, 139

SCID See *severe combined immunodeficiency disease.* 267

sclera White, fibrous outer layer of the eyeball. 172

scoliosis Abnormal lateral (side-to-side) curvature of the vertebral column. 88

scrotum Pouch of skin that encloses the testes. 6, 338

sebaceous gland Gland of the skin that secretes sebum. 66

sebum Oily secretion of the sebaceous glands. 66

sella turcica Saddle-shaped area of the sphenoid bone; houses the pituitary gland. 84

semen Sperm-containing secretion of males; seminal fluid plus sperm. 339

semicircular canal Tubular structure within the inner ear with ampullae that contain the receptors responsible for the sense of dynamic equilibrium. 180

semilunar valve Valve resembling a half-moon located between the ventricles and their attached vessels. 224

seminal fluid Sperm-containing secretion of males; also called semen. 339

seminal vesicle Convoluted, saclike structure attached to vas deferens near the base of the bladder in males; contributes to seminal fluid. 339

seminiferous tubule Highly coiled duct within the male testes that produces and transports sperm. 339

sensory nerve Nerve containing only sensory neuron dendrites. 142

sensory neuron Neuron that takes the nerve impulse to the central nervous system; also known as the afferent neuron. 137

sensory receptor Sensory structure specialized to receive information from the environment and to generate nerve impulses; also a protein located in the plasma membrane or within the cell that binds to a substance that alters some aspect of the cell. 142

septum Partition or wall, such as the septum in the heart, that divides the right half from the left half. 223

serous membrane Membrane that covers internal organs and lines cavities without an opening to the outside of the body; also called serosa. 59

serum Light-yellow liquid left after clotting of the blood. 216

severe combined immunodeficiency disease (SCID) Congenital illness in which both antibody and cell-mediated immunity are lacking or inadequate. 267

sex chromosome Chromosome responsible for the development of characteristics associated with maleness or femaleness; an X or Y chromosome. 380

sexually transmitted disease Illnesses communicated primarily or exclusively through sexual intercourse. 355

sickle-cell disease Hereditary disease in which red blood cells are narrow and curved so that they are unable to pass through capillaries and are destroyed, which causes chronic anemia. 210

sigmoid colon Portion of the large intestine that is S-shaped and extends from the descending colon to the rectum. 307

simple goiter Condition in which an enlarged thyroid produces low levels of thyroxin. 193

sinoatrial (SA) node Small region of neuromuscular tissue that initiates the heartbeat; also called the pacemaker. 229

sinus A cavity, as the sinuses in the human skull. 82

sinusitis Inflammation of the mucous membrane lining a paranasal sinus. 286

skeletal muscle Contractile tissue that comprises the muscles attached to the skeleton; also called striated muscle. 55, 111

SLE See *systemic lupus erythematosus* 267

small intestine Portion of the digestive tract that extends from the lower opening of the stomach to the large intestine. 302

smooth muscle Contractile tissue that comprises the muscles found in the walls of internal organs. 55, 111

SNS See *somatic nervous system.* 152

soft palate Entirely muscular posterior portion of the roof of the mouth. 298

somatic nervous system (SNS) Portion of the peripheral nervous system containing motor neurons that control skeletal muscles. 152

somatotropin See *growth hormone.* 192

spermatogenesis Sperm production in males by the process of meiosis and maturation. 44, 339

spermatozoa Developing male gametes. 339

sphincter Muscle that surrounds a tube and closes or opens the tube by contracting and relaxing. 300

spinal Pertaining to the spinal cord or to the vertebral canal. 6

spinal nerve Nerve that arises from the spinal cord. 152

spindle Apparatus composed of microtubules to which the chromosomes are attached during cell division. 43

spleen Large, glandular organ located in the upper left region of the abdomen that stores and purifies blood. 253

spongy bone Bone found at the ends of long bones and consists of bars and plates separated by irregular spaces. 55, 81

squamous cell carcinoma Tumor that begins in squamous epithelium, such as that of the skin, and that grows slowly and can be removed surgically.   70

squamous epithelium Flat cells found lining the lungs and blood vessels.   49

stapes Last of three ossicles of the ear; serves with the malleus and incus to conduct vibrations from the tympanic membrane to the oval window of the inner ear.   180

stereoscopic vision Product of two eyes and both cerebral hemispheres functioning together so that depth perception results.   177

sterile Inability to have children.   351

sternum Breastbone to which the ribs are ventrally attached.   90

steroid Lipid-solution, biologically active molecules having four interlocking rings; examples are cholesterol, progesterone, and testosterone.   25

stomach Saclike, expandable digestive organ located between the esophagus and the small intestine.   300

stratified Layered, as in stratified epithelium, which contains several layers of cells.   49

stratum basale Deepest layer of the epidermis, where cell division occurs.   63

stratum corneum Uppermost keratinized layer of the epidermis.   64

striations Characteristic alternating light and dark bands of cardiac and skeletal muscle.   111

stroke See *cerebrovascular accident.*   149

sty Inflammation of a sebaceous gland.   170

subclavian Either of two arteries branching off the aortic arch and supplying the arms.   238

subcutaneous Tissue beneath the dermis that tends to contain fat cells.   64

subcutaneous injection Introduction of a substance beneath the skin, using a syringe.   64

subdural hematoma Accumulation of blood between the dura mater and the brain.   150

superficial Near the surface.   4

superior Toward the upper part of a structure or toward the head.   4

supination Rotation of the forearm so that the palm faces forward, as in the anatomical position.   105

surfactant Agent that reduces the surface tension of water; in the lungs, a surfactant prevents the alveoli from collapsing.   277

suture Type of immovable joint articulation found between bones of the skull.   82

sweat gland Skin gland that secretes a fluid substance for evaporative cooling; also called sudoriferous glands.   66

sympathetic division Part of the autonomic nervous system that generally causes effects associated with emergency situations.   155, 158

symphysis Slightly movable joint between bones separated by a pad of fibrocartilage.   100

synapse Region between two nerve cells where the nerve impulse is transmitted from one to the other, usually from axon to dendrite.   140

synaptic cleft Small gap between the synaptic knob on one neuron and the dendrite on another neuron.   140

synarthrosis Immovable joint.   100

synergist Muscle that assists the action of the prime mover.   115

synovial fluid Fluid secreted by the synovial membrane.   101

synovial joint Freely movable joint.   101

synovial membrane Membrane that forms the inner lining of a capsule of a freely movable joint.   60, 101

synthesis To build up, such as the combining together of two small molecules to form a larger molecule.   27

syphilis Sexually transmitted disease caused by the bacterium *Treponema pallidum* that causes a painless chancre on the penis or cervix; if untreated, can lead to cardiac and central nervous system disorders.   357

systemic Affecting the entire body or involving several organ systems.   13

systemic circulation Part of the circulatory system that serves body parts other than the gas-exchanging surfaces in the lungs.   236

systemic disease Illness that involves the entire body or several body systems.   13

systemic lupus erythematosus (SLE) Syndrome involving the connective tissues and various organs including kidney failure.   267

systole Contraction of the heart chambers, particularly the left ventricle.   228

systolic pressure Arterial blood pressure during the systolic phase of the cardiac cycle.   246

T

tachycardia Abnormally rapid hearbeat.   230

talus Ankle bone.   230

tarsal Bone of the ankle in humans.   100

taste bud Organ containing the receptors associated with the sense of taste.   168

Tay Sachs Lethal genetic disease in which the newborn has a faulty lysosomal digestive enzyme.   38

tectorial membrane Membrane in the organ of Corti that lies above and makes contact with the receptor cells for hearing.   180

temporal lobe Area of the cerebrum responsible for hearing and smelling and for the interpretation of sensory experience and memory.   146

tendon Tissue that connects muscle to bone.   101, 111

teratogen Any substance that produces abnormalities during human development.   369

testis Male gonad; the organ that produces sperm and testosterone (pl. *testes*).   338

testosterone Most potent of the androgens, the male sex hormones.   199

tetanus Sustained muscle contraction without relaxation.   114

tetany Severe twitching caused by involuntary contraction of the skeletal muscles due to a lack of calcium.   194

thalamus Mass of gray matter located at the base of the cerebrum in the wall of the third ventricle; receives sensory information and selectively passes it to the cerebrum.   146

thoracic Pertaining to the chest.   6

thoracic cage Bones enclosing the chest cavity that protect the heart and lungs, play a role in breathing, and support the bones of the shoulders.   90

thrombin Enzyme derived from prothrombin that converts fibrinogen to fibrin threads during blood clotting.   215

thrombocytopenia Insufficient number of platelets in the blood.   212

thromboembolism Obstruction of a blood vessel by a thrombus that has dislodged from the site of its formation.   216, 232

thrombolytic therapy Any procedure to relieve an occluded coronary artery.   233

thrombus Blood clot that remains in the blood vessel where it formed.   216

thymosins Hormones secreted by the thymus.   202

thymus Lobular gland that lies in the neck and chest area and is necessary to the development of immunity.   202

thyroid-stimulating hormone (TSH) Hormone that causes the thyroid to produce thyroxin.   192

thyroxin Hormone produced by the thyroid that speeds up the metabolic rate.   193

tibia Shinbone found in the lower leg.   99

tidal volume (TV) Amount of air that enters the lungs during a normal, quiet inspiration.   281

tissue Group of similar cells that performs a specialized function.   3, 49

tissue fluid Fluid found about tissue cells containing molecules that enter from or exit to the capillaries.   235, 252

T lymphocyte One of two types of lymphocytes; a killer T cell that interacts directly with antigen-bearing cells and is responsible for cell-mediated immunity, or a helper T cell that stimulates other immune cells.   258

tone Continuous, partial contraction of muscle; also, the quality of a sound.   115

tonsil Partly encapsulated lymph nodule located in the pharynx.   254

tonsilectomy Surgical removal of the tonsils.   286

tonsilitis Inflammation of the tonsils.   286

trachea Windpipe that serves as a passageway for air.   276

tracheostomy Creation of an artificial airway by incision of the trachea and insertion of a tube.   276

tract Bundle of neurons forming a transmission pathway through the brain and spinal cord.   142

transfer RNA (tRNA) Molecule of RNA (ribonucleic acid) that carries an amino acid to a ribosome engaged in the process of protein synthesis.   36

transfusion Introduction of whole blood or other blood components into the bloodstream of a patient.   216

transverse colon Portion of the large intestine that travels transversely as it extends from the ascending colon to the descending colon.   307

transverse plane Plane or section that divides a structure horizontally to give a cross section.   5

tricuspid valve Atrioventricular valve between the right atrium and the right ventricle.   224

triplo-x syndrome   383

trisomy State of having an extra chromosome—three instead of the normal two.   381

trypsin Protein digesting enzyme produced by the pancreas.   304

TSH See *thyroid-stimulating hormone.*   192

tubal ligation Method for preventing pregnancy in which the uterine tubes are cut and sealed.   355

tuberculosis Infection by the tubercle bacillus *Mycobacterium tuberculosis.*   287

tumor Abnormal growth of tissue that serves no useful purpose.   44

tumor-suppressor gene Gene that suppresses the development of a tumor; the mutated form contributes to the development of cancer.   45

Turner syndrome Condition caused by the inheritance of a single X chromosome.   383

TV See *tidal volume.*   281

twitch Brief muscular contraction followed by relaxation.   114

tympanic membrane Membrane located between the external and middle ear; the eardrum.   180

U

ulcer Open sore in the lining of the stomach; frequently caused by bacterial infection.   300

ulna Elongated bone found within the lower arm.   93

ultrasound Method of viewing internal organs (or a developing fetus) by using high-frequency sound waves.   364

umbilical Pertaining to the umbilicus.   6

umbilical cord Cord through which blood vessels that connect the fetus to the placenta pass.   366

urea Primary nitrogenous waste of mammals.   304

uremia High level of urea nitrogen in the blood.   327

ureters Tubes that take urine from the kidneys to the bladder.   324

urethra Tube that takes urine from the bladder to the outside of the body.   324

urinalysis Examination of a urine sample to determine its chemical, physical, and microscopic aspects.   327

urinary bladder Organ where urine is stored before being discharged by way of the urethra.   324

urticaria Skin eruption characterized by the development of welts as a result of capillary dilation.   68

uterine cycle Female reproductive cycle characterized by regularly occurring changes in the uterine lining.   349

uterus Female organ in which the fetus develops.   346

utricle Saclike cavity of the inner ear that contains receptors for static equilibrium.   180

V

vaccine Treated antigens that can promote active immunity when administered.   262

vagina Female copulatory organ and birth canal.   344

valve Structure that opens and closes, ensuring one-way flow; common to vessels, such as systemic veins, lymphatic veins, and veins to the heart.   236

varicose vein Irregular dilation of a superficial vein, seen particularly in the lower legs, due to weakened valves within the veins.   236

vascularization Process of a tumor becoming supplied with blood vessels.   44

vas (ductus) deferens Tube connecting epididymis to the urethra; sperm duct (pl., *vasa deferentia*).   339

vasectomy Method for preventing pregnancy in which the vasa deferentia are cut and sealed.   355

VC See *vital capacity.*   281

vein Blood vessel that takes blood to the heart.   236

vena cava One of two large veins that convey deoxygenated blood to the right atrium of the heart (pl., *venae cavae*).   227

venous duct See *ductus venosus.*   242

ventilation Breathing; the process of moving air into and out of the lungs.   278

ventral Toward the front or belly surface; the opposite of *dorsal.*   6

ventricle Cavity in an organ, such as the ventricles of the brain or the ventricles of the heart.   144, 223

venule Type of blood vessel that takes blood from capillaries to veins.   236

vermiform appendix Small, tubular appendage that extends outward from the cecum of the large intestine.   307

vernix caseosa Cheeselike substance covering the skin of the fetus.   372

vertebra Bone of the vertebral column.   88

vertebral column Backbone of vertebrates, composed of individual bones called vertebrae.   88

vesicle Small, membranous sac that stores substances within a cell.   37

vestibule Space or cavity at the entrance of a canal, such as the cavity that lies between the semicircular canals and the cochlea.   180

villus Fingerlike projection that lines the small intestine and functions in absorption (pl., *villi*).   302

visceral Pertaining to the contents of a body cavity.   6, 59

vital capacity (VC) Maximum amount of air a person can exhale after taking the deepest breath possible.   281

vitamin Organic molecules (usually coenzymes) that must be in the diet and are necessary in trace amounts for normal metabolic functioning of cells.   312

vitreous humor Substance that occupies the posterior cavity of the eye.   172

vomiting Forcible expulsion of the stomach contents back through the mouth.   300

vulva External genitalia of the female that lie near the opening of the vagina.   347

W

wart Raised growth on the skin due to a viral infection.   71

white blood cell See *leukocyte*.   55, 208

white blood cell count Number of white blood cells per cubic millimeter of blood.   208

white matter Myelinated nerve fibers in the central nervous system.   142

X

X-linked gene Gene found on the X chromosome that controls traits other than sexual traits.   386

XYY male Male who has an X chromosome and two Y chromosomes in each nucleus.   385

Y

yolk sac Extraembryonic membrane that serves as the first site of red blood cell formation.   366

Z

zygote Cell formed by the union of the sperm and egg; the product of fertilization.   365

Credits

Text

Text Pronunciation Guide: Pronunciation guides for chapter terms are from *Dorland's Illustrated Medical Dictionary*, 27th ed., 1988. Philadelphia, W. B. Saunders Co. Reprinted with permission.

Photos

Chapter 1
1.6a, 1.7a: © Dr. Sheril D. Burton; **1A:** Courtesy Childrens Hospital of Pittsburgh, photo by Ed Eckman.

Chapter 2
2A: © Catherine/CNRI/SPL/Photo Researchers, Inc.; **2B:** © Scott Camazine/Photo Researchers, Inc.; **2.8b:** © Don W. Fawcett/Photo Researchers, Inc.

Chapter 3
3.6: © John Walsh/SPL/Photo Researchers, Inc.

Chapter 4
4.1c, 4.2c: © Ed Reschke; **4.3c:** © Manfred Kage/Peter Arnold, Inc.; **4.4c, 4.5b, 4.6c, 4.7c, 4.8c, 4.9c, 4.11c, 4.12c, 4.13c, 4.14c:** © Ed Reschke.

Chapter 5
5.2: Courtesy Kent Van De Graaff, photo by James M. Clayton; **5.3b:** © CNRI/SPL/Photo Researchers, Inc.; **5.6:** Courtesy Dr. George Bogumill; **5.8a:** © James Stevenson/SPL/Photo Researchers, Inc.; **5.8b:** © Ken Green/Visuals Unlimited; **5.8c:** © Dr. P. Marazzi/SPL/Photo Researchers, Inc.; **5A:** © Ira Wyman/Sygma.

Chapter 6
6.2d: © Biophoto Associates/Photo Researchers, Inc.; **6.23b-e:** © Ed Reschke; **6.23f:** © Ed Reschke/Peter Arnold, Inc.

Chapter 7
7.1b: Courtesy Dr. Hugh Huxley; **7.2b:** © Ed Reschke/Peter Arnold, Inc.

Chapter 9
9.4: © Omikron/SPL/Photo Researchers, Inc.; **9.10:** © Lennart Nilsson/The Incredible Machine.

Chapter 10
10.4: © Ewing Galloway, Inc.; **10.5:** © Lester Bergman/The Bergman Collection; **10.7, 10.8:** © Ken Green/Visuals Unlimited.

Chapter 11
11.1: © CNRI/Photo Researchers, Inc.; **11A:** © Bill Longcore/Photo Researchers, Inc.; **11.3a,b:** © SPL/Photo Researchers, Inc.; **11.9a,b:** Courtesy Stuart I. Fox.

Chapter 12
12.4a: © The McGraw-Hill Companies, Inc., Karl Rubin photographer; **12b:** © Biophoto Associates/Photo Researchers, Inc.; **12.8d:** © Ed Reschke.

Chapter 13
13.1b: © John D. Cunningham/Visuals Unlimited; **13.1c:** © Astrid & Hans Frieder-Michler/SPL/Photo Researchers, Inc.; **13.4a-e:** © Ed Reschke/Peter Arnold, Inc.; **13.8b:** © R. Feldman/D.McCoy/Rainbow; **13.10:** © Lennart Nilsson/Boehringer Ingelheim International; **13A:** © Michael Newman/PhotoEdit.

Chapter 14
14.3c: © CNRI/Phototake, NYC; **14.5:** © John Watney Photo Library; **14.9a:** © SIU/Visuals Unlimited; **14b(both):** © Martin Rotker.

Chapter 15
15.4b: © Ed Reschke/Peter Arnold, Inc.; **15.4c:** © St. Bartholomew's Hospital/SPL/ Photo Researchers, Inc.; **15.6b:** © Manfred Kage/Peter Arnold, Inc.; **15.6c:** Photo by Susumu Ito, from Charles Flickinger, *Medical Cellular Biology*, W.B. Saunders, 1979.

Chapter 16
16.5b: © David M. Phillips/Visuals Unlimited; **16.5c:** © 1966 Academic Press, from A.B. Maunsbach, J. Ultrastruct. Res. 15:242–282.

Chapter 17
17.2b: © Biophoto Associates/Photo Researchers, Inc.; **17.6b:** © Ed Reschke/ Peter Arnold, Inc.; **17.7a,b:** © David M. Phillips/Visuals Unlimited; **17.11:** © Charles Lightdale/Photo Researchers, Inc.; **17.12:** © CDC/Peter Arnold, Inc.

Chapter 18
18B: Courtesy Almuth Lenz; **18.5a:** © Lennart Nilsson/A Child Is Born; **18.6:** © Petit Format/Science Source/Photo Researchers, Inc.; **18.7:** © Claude Edelmann, Petit Format et Guigorz from "First Days of Life, Black Star"; **18.8:** © James Stevenson/SPL/Photo Researchers, Inc.; **18.9:** © Petit Format/Photo Researchers, Inc.

Chapter 19
19.1b,c: © CNRI/SPL/Photo Researchers, Inc.; **19.3a:** © Jill Cannefax/EKM-Nepenthe; **19.6a,b:** From R. Kampheier, "Physical Examination of Health Diseases," © 1958 F.A. Davis Company.

Index